Methods of
Experimental Physics

VOLUME 20

BIOPHYSICS

METHODS OF EXPERIMENTAL PHYSICS:

C. Marton, *Editor-in-Chief*

Volume 20

Biophysics

Edited by

GERALD EHRENSTEIN
and
HAROLD LECAR

Laboratory of Biophysics
National Institutes of Health
Bethesda, Maryland

1982

ACADEMIC PRESS
A Subsidiary of Harcourt Brace Jovanovich, Publishers

New York   London
Paris   San Diego   San Francisco   São Paulo   Sydney   Tokyo   Toronto

ACADEMIC PRESS, INC.
111 Fifth Avenue, New York, New York 10003

United Kingdom Edition published by
ACADEMIC PRESS, INC. (LONDON) LTD.
24/28 Oval Road, London NW1 7DX

Library of Congress Cataloging in Publication Data
Main entry under title:

Biophysics.

(Methods of experimental physics ; v. 20)
Includes bibliographical references and
index.
1. Biophysics. I. Lecar, H. (Harold)
II. Ehrenstein, G. (Gerald) III. Series.
QH505.B472   574.19'1   82-6642
ISBN 0-12-475962-9   AACR2

PRINTED IN THE UNITED STATES OF AMERICA

82 83 84 85   9 8 7 6 5 4 3 2 1

CONTENTS

CONTRIBUTORS   xvii

PUBLISHER'S FOREWORD   xix

FOREWORD   xxi

PREFACE   xxiii

LIST OF VOLUMES IN TREATISE   xxvii

1. Nuclear Magnetic Resonance
by JOSEPH R. SCHUH AND SUNNEY I. CHAN

1.1. Introduction   1

1.2. Phenomenon of Magnetic Resonance   2
    1.2.1. Quantum Description   2
    1.2.2. Classical Description   4
    1.2.3. Detection of Magnetic Resonance   7

1.3. Spin Relaxation   12
    1.3.1. Spin–Lattice Relaxation   13
    1.3.2. Spin–Spin Relaxation   14
    1.3.3. Correlation Time   14
    1.3.4. Relaxation Mechanisms   16
    1.3.5. Effects of Anisotropic and Restricted
            Motions   24

1.4. Experimental Methods   25
    1.4.1. Pulse Techniques   25
    1.4.2. Magic-Angle Spinning   30
    1.4.3. Double Resonance   32
    1.4.4. Two-Dimensional FT-NMR   39

1.5. Selected Studies on Biological Systems   41

1.5.1. Proteins: Probing the Active Site of an
Enzyme 41
1.5.2. Nucleic Acids: The Dynamic Structure of
tRNA 44
1.5.3. Membranes: The Restricted Motions of Lipid
Chains 47
1.5.4. Intact Cells: Metabolism Studied *in Vivo* 49

2. Nitroxide Spin Labels
by GILLIAN M. K. HUMPHRIES AND HARDEN M. McCONNELL

2.1. Introduction 53

2.2. Spin-Label Theory: A Descriptive Treatment 54
2.2.1. Fundamentals Including the Resonance
Condition.................................... 54
2.2.2. Effect of Orientation......................... 59
2.2.3. Effect of Motion 62
2.2.4. Effect of the Electrostatic Environment.......... 63
2.2.5. Effect of Distribution or Concentration of
Spin Labels: Spin-Exchange Broadening 65
2.2.6. Effect of Distribution or Concentration of
Spin Labels: Dipole–Dipole Interactions......... 65

2.3. Spin-Label Theory: A Mathematical Treatment 65
2.3.1. The Resonance Condition 65
2.3.2. Absorption Probabilities and Paramagnetic
Relaxation.................................. 68
2.3.3. The Bloch Equations 72
2.3.4. Nuclear Hyperfine Structure................. 75
2.3.5. The Spectroscopic Splitting Factor g 83
2.3.6. The Spin Hamiltonian....................... 84
2.3.7. Effect of Orientation: Single-Crystal
Spectra..................................... 85
2.3.8. Isotropic Distributions of Strongly
Immobilized Labels: Powder Spectra 90
2.3.9. Effects of Isotropic Motion on Spectra.......... 93
2.3.10. Line Shapes with Fast Anisotropic Motion
and Partial Orientation—Frequency
Amplitude Order Parameters.................. 98
2.3.11. Effect of Electron–Electron Spin–Spin
Interaction: Dipolar Interactions in

Biradicals with Fixed Distances between
Spins.................................... 102
2.3.12. Effect of Electron–Electron Spin–Spin
Interactions: Spin–Spin Interaction
Determined by Translational Motion and
Resulting from Colliding Spins 104
2.3.13. Enhancement of Nuclear Relaxation by
Spin Labels................................ 109

2.4. Use of Spin Labels as Antigenic Determinants
Capable of Reporting Their Physical State.............. 110
2.4.1. Background................................ 110
2.4.2. Spin-Label Lipid Haptens 111
2.4.3. Determination of the Rate of Inside–Outside
Transitions of Phospholipids in Vesicle
Membranes 112
2.4.4. Determination of Lateral Diffusion
of Spin-Label Lipid Haptens in Membranes 112
2.4.5. Determination of the Surface Area of Lipid
Membranes 113
2.4.6. Antibodies Specific for Spin-Label
Determinants 114
2.4.7. Physical Factors Affecting the Binding of
Spin-Label Specific Antibodies to
Hapten-Bearing Bilayers 116
2.4.8. Efferent Immune Responses Controlled by
Antibodies Bound to Antigen-Bearing
Membranes 121

3. Raman Spectroscopy
   by MARGARET R. BUNOW

3.1. Introduction 123

3.2. Origin of the Raman Spectrum....................... 124

3.3. Analysis of the Raman Spectrum..................... 125

3.4. Resonance Raman Scattering 128
     3.4.1. A-Term Scattering............................ 129
     3.4.2. B-Term Scattering............................ 130
     3.4.3. The Polarizability Tensor..................... 130

3.5. Instrumentation 131
    3.5.1. Sample Handling 132
    3.5.2. Optimization and Standardization of Signal 132
    3.5.3. Management of Fluorescence 133

3.6. Strategy of Raman Spectroscopic Applications in
    Biology.. 134

3.7. Conformational Studies 137
    3.7.1. Conformation of Lipids in Biological
          Membranes 137
    3.7.2. Conformation of Nucleic Acids 143
    3.7.3. Raman Spectral Analysis of
          Polysaccharides 145
    3.7.4. Raman Vibrational Analysis of Protein
          Conformation 146
    3.7.5. Vibrational Markers of Protein
          Side-Groups 149
    3.7.6. Resonance Raman Studies of Porphyrins
          and Hemoproteins 150
    3.7.7. Resonance Raman Parameters for Nonheme
          Metalloproteins 154
    3.7.8. Extrinsic Chromophores for Resonance
          Raman Studies of Proteins.................... 155
    3.7.9. Resonance-Enhanced Vibrational Behavior
          of Biological Polyenes 155

3.8. Kinetic Studies.................................. 156

3.9. Nonlinear Phenomena............................ 159

3.10. The Raman Microscope 161

3.11. Conclusions and Prognostications 161

4. Picosecond Laser Spectroscopy
   by TAKAYOSHI KOBAYASHI

4.1. Introduction 163

4.2. Nanosecond and Picosecond Spectroscopy 166
    4.2.1. Comparison of Nanosecond and Picosecond
          Spectroscopy 166
    4.2.2. Picosecond Light Pulses..................... 170

4.2.3. Various Methods for Picosecond Absorption
         and Emission Spectroscopy.................... 172

4.3. Applications to Photosynthesis and Vision 181
     4.3.1. Photosynthesis 181
     4.3.2. Vision...................................... 185

Appendix. Nonlinear Optical Phenomena, Optical
         Elements, and Detectors Related to Techniques of
         Picosecond Spectroscopy........................... 190
     A.1. Nonlinear Optical Phenomena................... 190
     A.2. Optical Elements and Detectors................. 194

5. Fluorescence Methods for Studying Membrane
   Dynamics
   by JOSEPH SCHLESSINGER AND ELLIOT L. ELSON

5.1. Introduction 197

5.2. Molecular Rotation in Membranes.................... 199
     5.2.1. Steady-State Fluorescence Polarization 201
     5.2.2. Nanosecond Time-Dependent Fluorescence
            Polarization................................. 203
     5.2.3. Decay of Transient Dichroism 204
     5.2.4. Special Features and Limitations............... 205

5.3. Macroscopic Membrane Motions..................... 208
     5.3.1. Basic Concepts.............................. 208
     5.3.2. Fluorescence Photobleaching Recovery 209
     5.3.3. Fluorescence Correlation Spectroscopy 212

5.4. Applications 216
     5.4.1. Translational and Rotational Diffusion of
            Molecules in Membranes and Membrane
            Viscosity 216
     5.4.2. Possible Significance of Mobility and
            Immobility of Membrane Proteins 222

5.5. Recent Developments............................... 224
     5.5.1. Technical Developments 224
     5.5.2. Origin of Constraints on Membrane Protein
            Mobility.................................... 225
     5.5.3. Receptor Mobility............................ 226

6. Structure Determination of Biological Macromolecules
   Using X-Ray Diffraction Analysis
   by Eaton E. Lattman and L. Mario Amzel

   6.1. Introduction 229

   6.2. Diffraction by a General Object 230

   6.3. Crystallography 237
        6.3.1. Diffraction by Crystals........................ 237
        6.3.2. The Patterson Function 240

   6.4. Protein Crystallography 244
        6.4.1. Introduction 244
        6.4.2. Crystallization 246
        6.4.3. Collection of X-Ray Diffraction Data 247
        6.4.4. Phase Determination Using Isomorphous
               Replacement................................... 252
        6.4.5. Interpretation of Electron Density Maps 256
        6.4.6. Difference Fourier Syntheses 259
        6.4.7. Molecular Replacement 260
        6.4.8. Refinement of Protein Structures............... 263
        6.4.9. A Case History: Rubredoxin.................... 268

   6.5. Fibers... 275
        6.5.1. Definition and Description..................... 275
        6.5.2. Fiber Diffraction Theory 276
        6.5.3. Solution of Simple Fiber Structures 287
        6.5.4. Tobacco Mosaic Virus Structure 288
        6.5.5. Structure of Microtubules 293
        6.5.6. Conclusion 296

7. Laser Light Scattering
   by Ralph Nossal

   7.1. Introduction 299

   7.2. Theory... 300
        7.2.1. Classical Light Scattering 304
        7.2.2. Intensity Fluctuation Spectroscopy............. 306

7.3. Instrumentation and General Techniques............... 310
      7.3.1. Relationship between Photon Autocorrelation
             and Special Analysis........................... 310
      7.3.2. Characteristics of Measured Autocorrelation
             Functions..................................... 312
      7.3.3. Equipment..................................... 313

7.4. Diffusion Coefficients 316

7.5. Large Particles..................................... 319

7.6. Determination of Electrophoretic Mobilities 324

7.7. Motility Measurements.............................. 326

7.8. Applications in Cell Biology 330

7.9. Blood Flow... 331

7.10. Gels and Solutions of Fibrillous Proteins.............. 333

8. Small-Angle Scattering Techniques for the Study
   of Biological Macromolecules and Macromolecular
   Aggregates
   by PETER B. MOORE

8.1. Introduction 337
     8.1.1. The Nature of the Technique 337
     8.1.2. Historical Comments 338
     8.1.3. The Sources of Current Interest................ 338
     8.1.4. Purpose 339
     8.1.5. Earlier Reviews 340

8.2. The Experimental Problem 341
     8.2.1. The Scattering Signal 341
     8.2.2. Isolation of the Macromolecular Signal 342
     8.2.3. Small-Angle Apparatus: General Features....... 343
     8.2.4. The Small-Angle Compromise 344
     8.2.5. X Rays...................................... 345
     8.2.6. Monochromatization of X Rays 346
     8.2.7. Collimation 347
     8.2.8. Flight Path.................................. 349

8.2.9.   Detection.................................   349
8.2.10. Beam Monitors............................   350
8.2.11. Apparatus Designs.........................   351
8.2.12. Neutrons   352
8.2.13. Monochromatization of Neutrons   353
8.2.14. Pulsed Reactors   355
8.2.15. General Design: Collimation   356
8.2.16. Flight Path and Detection   357
8.2.17. Available Instruments........................   358
8.2.18. Absolute Intensity Measurements   358
8.2.19. Data Correction   360

8.3. Data Analysis....................................   362
8.3.1.   Molecular Weight and Partial Specific
         Volume   362
8.3.2.   Molecular Size and Shape: Radius of
         Gyration.................................   364
8.3.3.   Quantities Related to the Radius of
         Gyration.................................   365
8.3.4.   Length Distributions   367
8.3.5.   Characteristic Function Parameters............   368
8.3.6.   Molecular Shape   369
8.3.7.   Spherical Objects..........................   371
8.3.8.   Contrast..................................   372
8.3.9.   The Dependence of Radius of Gyration on
         Contrast.................................   372
8.3.10. Dependence of the Extended Scattering
         Profile on Contrast   374
8.3.11. Absolute Scale and Contrast Variation.........   374
8.3.12. Neutrons versus X Rays   375
8.3.13. The Hydrogen Exchange Problem..............   376
8.3.14. Contrast Variation in Practice: Radius of
         Gyration.................................   378
8.3.15. Contrast Variation in Practice: Extended
         Scattering   380
8.3.16. Macromolecular Aggregates: Synthesis
         Methods..................................   381
8.3.17. Macromolecular Aggregates: Distance
         Finding...................................   383
8.3.18. Solution Scattering Studies on Molecules of
         Known Structure...........................   385

8.4. Concluding Remarks.............................. 388
    8.4.1. Summary.................................. 388
    8.4.2. Critique 389

9. Electron Microscopy
   by ROBERT M. GLAESER

   9.1. Electron Microscopy as a Tool for Structure
      Determination...................................... 391

   9.2. Image Formation in the Electron Microscope........... 395
    9.2.1. A Simplified Picture: The Mass Thickness
       Approximation 395
    9.2.2. The "Weak Phase Object" Model.............. 397
    9.2.3. Complications............................... 398

   9.3. Contrast Transfer Function Theory................... 401
    9.3.1. Origin of Phase Contrast in Electron
       Microscopy.................................. 401
    9.3.2. The Envelope Function: Partial Coherence....... 403
    9.3.3. Image Restoration 404
    9.3.4. Complications............................... 406

   9.4. Three-Dimensional Reconstruction 409
    9.4.1. Fourier (Crystallographic) Methods............. 409
    9.4.2. Direct-Space Methods 413
    9.4.3. The Hollow Cone Problem 416

   9.5. Radiation Damage 418
    9.5.1. Radiation Physics and Radiation Chemistry 419
    9.5.2. Empirical Studies of the Radiation Damage
       Effect under Electron Microscope
       Conditions.................................. 422
    9.5.3. Shot-Noise Limitation for High-Resolution
       Electron Microscope Work.................... 427
    9.5.4. Solution of the Shot-Noise Problem by
       Image Superposition 427

   9.6. Specimen Hydration within the Vacuum of the
      Instrument.. 430
    9.6.1. The Need for Maintaining the Hydrated
       State with Complex Biological Structures 430
    9.6.2. Difficulty of the Problem 430

9.6.3. Technical Solutions to the Hydration
         Problem 431

9.7. Recent Innovations in Experimental Methods.......... 433
     9.7.1. Dark-Field Methods 433
     9.7.2. Single-Sideband Images and Holography 438
     9.7.3. Image Formation with Inelastically Scattered
             Electrons.................................... 439
     9.7.4. High-Voltage Electron Microscopy 441
     9.7.5. Aberration Correction........................ 444

10. Voltage Clamping of Excitable Membranes
by FRANCISCO BEZANILLA, JULIO VERGARA,
AND ROBERT E. TAYLOR

10.1. Introduction 445

10.2. General Principles of Voltage Clamp................. 447

10.3. Axial-Wire Voltage Clamp.......................... 451
      10.3.1. Cable Theory of an Axon with Axial Wire
               in Voltage Clamp.......................... 452
      10.3.2. Giant Axon Preparation 455
      10.3.3. Measurement of Membrane Current 459
      10.3.4. Electronics for the Voltage Clamp System 461
      10.3.5. Series-Resistance Compensation 463
      10.3.6. Pulse Generation and Data Acquisition 470

10.4. Voltage Clamp with Microelectrodes 473
      10.4.1. Two Microelectrodes 473
      10.4.2. Three Microelectrodes 481

10.5. Voltage Clamp of an Isolated Patch Using
      External Pipettes.................................. 482
      10.5.1. External Patch Isolation.................... 482
      10.5.2. Internal Access 484

10.6. Voltage Clamp with Gap Isolation Techniques........ 486
      10.6.1. Node of Ranvier 487
      10.6.2. Vaseline-Gap Techniques in Single Muscle
               Fibers.................................... 497
      10.6.3. Sucrose-Gap Methods....................... 500
      10.6.4. Errors Introduced by the Finite Length
               of the Gap................................ 501

10.7. Concluding Remarks 504

10.A. Appendix 504
    10.A.1. Circuit Equations of Vaseline-Gap
            Voltage Clamp 504
    10.A.2. Potential Distribution for a Fiber in a Gap
            Voltage Clamp 508

11. Lipid Model Membranes
by G. SZABO AND R. C. WALDBILLIG

11.1. Biological Membranes 513

11.2. Lipid Monolayers 515

11.3. Spherical Lipid Bilayers 516

11.4. Planar Lipid Bilayers 519
    11.4.1. Lipid/Solvent Systems 520
    11.4.2. Solvent-Filled Bilayers 521
    11.4.3. Solvent-Depleted Bilayers 523
    11.4.4. Pure Lipid Bilayers 524

11.5. Ion Transport Mechanisms 525

11.6. Techniques for the Measurement of Ion
Transport 528

11.7. Direct Transport of Hydrophobic Ions 529

11.8. Carrier-Mediated Ion Transport 530

11.9. Molecular Channels 535
    11.9.1. Single Channels in Bilayers 536
    11.9.2. Multichannel Bilayers 539

11.10. Relationship between Channel Structure
and Function 541

AUTHOR INDEX 545

SUBJECT INDEX 565

CONTRIBUTORS

Numbers in parentheses indicate the pages on which the authors' contributions begin.

L. MARIO AMZEL, *Department of Biophysics, Johns Hopkins University School of Medicine, Baltimore, Maryland 21205* (229)

FRANCISCO BEZANILLA, *Department of Physiology, School of Medicine, University of California, Los Angeles, California 90024* (445)

MARGARET R. BUNOW, *National Institutes of Health, Bethesda, Maryland 20205* (123)

SUNNEY I. CHAN, *A. A. Noyes Laboratory of Chemical Physics, California Institute of Technology, Pasadena, California 91125* (1)

ELLIOT L. ELSON,* *Department of Chemistry, Cornell University, Ithaca, New York 14853* (197)

ROBERT M. GLAESER, *Department of Biophysics and Medical Physics, University of California, Berkeley, California 94720* (391)

GILLIAN M. K. HUMPHRIES,† *Stauffer Laboratory for Physical Chemistry, Stanford University, Stanford, California 94305* (53)

TAKAYOSHI KOBAYASHI, *Department of Physics, University of Tokyo, Bunkyo-Ku, Tokyo 113 Japan* (163)

EATON E. LATTMAN, *Department of Biophysics, Johns Hopkins University School of Medicine, Baltimore, Maryland 21205* (229)

HARDEN M. McCONNELL, *Stauffer Laboratory for Physical Chemistry, Stanford University, Stanford, California 94305* (53)

PETER B. MOORE, *Departments of Chemistry and Molecular Biophysics and Biochemistry, Yale University, New Haven, Connecticut 06511* (337)

RALPH NOSSAL, *National Institutes of Health, Bethesda, Maryland 20205* (299)

---

* Present address: Department of Biological Chemistry, Washington University School of Medicine, St. Louis, Missouri 63110.

† Present address: Institute for Medical Research, San Jose, California 95128.

JOSEPH SCHLESSINGER, *Department of Chemical Immunology, The Weizmann Institute for Science, Rehovot, Israel* (197)

JOSEPH R. SCHUH, *A. A. Noyes Laboratory of Chemical Physics, California Institute of Technology, Padadena, California 91125* (1)

G. SZABO, *Department of Physiology and Biophysics, University of Texas Medical Branch, Galveston, Texas 77550* (513)

ROBERT E. TAYLOR, *Laboratory of Biophysics, National Institutes of Neurological and Communicative Disorders and Stroke, National Institutes of Health, Bethesda, Maryland 20205* (445)

JULIO VERGARA, *Department of Physiology, School of Medicine, University of California, Los Angeles, California 90024* (445)

R. C. WALDBILLIG, *Department of Physiology and Biophysics, University of Texas Medical Branch, Galveston, Texas 77550* (513)

PUBLISHER'S FOREWORD

It is with sadness that we inform readers of *Methods of Experimental Physics* of the passing of Dr. Claire Marton. Both Claire and her late husband, Bill, were associated with Academic Press almost from its founding. Their passing is both a personal and professional loss to us.

Future volumes will be edited by Dr. Robert Celotta of the Electron Physics Group, Radiation Physics Division, National Bureau of Standards, Washington, D.C., and by Dr. Judah Levine of the Joint Institute for Laboratory Astrophysics, University of Colorado, Boulder, Colorado. Dr. Celotta and Dr. Levine had already started to work with Claire. We are fortunate to have such qualified and experienced people join us.

FOREWORD

It would seem that for more than any of the other volumes of this treatise the choice of topics to be included in this volume was the most difficult to make. Given the subject matter, in retrospect, this may not appear surprising. Certainly, more volumes on this subject are called for.

The high quality of this book is testimony to the effort and good judgment of the editors and the fine cooperation of the authors. This volume is a proud addition to the treatise.

C. Marton

PREFACE

This volume describes three types of measurements that have been of intense interest to physicists working in biology—spectroscopy of macromolecules, structure determination by the scattering of particles or radiation, and electrical measurements on cell membranes.

Parts 1–4 discuss several types of spectroscopy, and are primarily concerned with the determination of molecular information. Parts 5–9 are concerned with methods used to visualize the structures and motions of large molecules and cells. Parts 10 and 11 discuss methods for studying membranes, either directly or by means of model systems.

Almost every type of spectroscopy has been used with some success to study biological macromolecules, and any sampling of methods is somewhat arbitrary. We have emphasized methods which can be used to study molecular motion and conformational changes of macromolecules. Magnetic resonance spectroscopies have found considerable use in studies of biological systems, and the first two parts are about these methods.

In nuclear magnetic resonance, the absorption frequency depends on the magnetic moment of a nucleus. Since protons have large magnetic moments and are prevalent in biological material, early work centered on the proton. With the improved sensitivity now available, other nuclei with magnetic moments are often studied. Part 1 includes an example of the use of natural abundance ^{13}C NMR, as well as other new and more sensitive methods. A major theme of Part 1 is the use of nuclear magnetic resonance to measure relaxation times.

Electron spin resonance (ESR) is similar to nuclear magnetic resonance in concept, but utilizes the interaction of a magnetic field with an electron spin rather than a nuclear spin. Thus ESR spectroscopy is restricted to molecules with unpaired electron spins. Unpaired spins are present in biological molecules containing transition metals and in free radicals created as intermediates in biochemical reactions or by radiation damage. For systems without unpaired electron spins, it is sometimes possible to add covalently a group that contains an unpaired spin. This approach, called spin labeling, is the subject of detailed discussion in Part 2. Successful spin labeling requires that the label be incorporated into the desired part of the molecule under study and also that it not interfere with the normal functioning of that molecule. When these conditions are met, the spectrum of the spin label can be used to determine a number of important parameters, such as molecular motions, distances between groups, and rates of reactions.

A number of IR, visible, and UV spectroscopies have been successfully used for biological studies. We have chosen to include a paper on Raman spectroscopy (Part 3) because it combines several important advantages, such as the possibility of recording from small samples, the lack of interference from liquid water, and sensitivity to homonuclear bonds. Raman spectroscopy has been utilized effectively in studies of heme groups, nucleic acids, and lipid membranes. Other types of spectroscopies which have been used extensively to study macromolecules, but are not described in this volume, include fluorescence, optical rotatory dispersion (ORD), and circular dichroism (CD). Fluorescence spectroscopy has been very successful because of the prevalence of chromophoric groups in biological molecules, the possibility of incorporating fluorescent markers, and the high sensitivity inherent in fluorescence measurements. ORD and CD signals, which are sensitive to helical content of macromolecules, are especially valuable spectral indicators of conformation.

Picosecond laser spectroscopy, which is discussed in Part 4, is a relatively new method that has considerable promise for several types of biological problems. Part 4 discusses the new technology needed to produce and record pulses of light in the picosecond range and provides examples of applications to photosynthesis and vision, where previous methods have been inadequate to resolve the rapid transitions which occur immediately after the absorption of a photon.

As previously mentioned, fluorescence can be a very valuable tool in the measurement of absorption spectra. The fluorescence methods presented in Part 5, however, are based not on the particular frequency of absorption, but on the bleaching of fluorescent molecules and the interdiffusion of bleached and unbleached molecules. These methods allow the determination of molecular motion within a cell membrane. Although similar information can be obtained from spin labeling (Part 2), the fluorescence methods are somewhat more flexible in that they can be used to determine motions over relatively long distances.

Parts 6–9 describe different approaches to the problem of obtaining images of small objects. Whether the objects are macromolecules, cell organelles, or whole cells, the scattering and diffraction methods all share a common conceptual basis—the distribution of scattered radiation or particles gives the Fourier transform of the density of the object under study. Thus these parts are representative of the current emphasis on Fourier optics.

Part 6 deals with x-ray diffraction, which has been the prime source of our knowledge of the detailed structure of biological macromolecules. For those macromolecules that can be crystallized, x-ray diffraction can give a complete description of the spatial molecular structure to a level of resolution better than 3 Å. Much of our pictorial knowledge of the relations between structure and function for proteins and for DNA comes from the x-ray crystallographic analyses of approximately 100 prototypical macromolecules. Even for structures (such as fibrous macromolecules, viruses, and ribosomes) that cannot be grown as crystals, x-ray diffraction provides a wealth of information. For example, it complements the method of neutron diffraction (Part 8) to provide a good description of the structure of ribosomes.

The laser, because of its ability to generate coherent light, has had a strong impact on all of optical technology. Part 7 surveys the many applications to cell biology of laser light scattering and emphasizes a particularly promising area—the interpretation of light scattering from motile bacteria. Different types of motion produce different Doppler-shift characteristics, and these can be distinguished by means of autocorrelation calculations for the different motions.

The concepts involved in small-angle x-ray and neutron scattering, discussed in Part 8, are similar to those involved in light scattering. However, since the wavelength of radiation is so small (of the order of 1 Å), scattering by macromolecules is characteristic of objects much larger than the wavelength of the incident radiation. Consequently, the scattering is confined to "small" angles near the forward direction. These short-wavelength methods provide information complementary to that obtained by crystallography. In particular, they can be used to analyze the shapes and interactions of macromolecules in solution.

Part 9 emphasizes those aspects of electron microscopy that have grown out of physical considerations of the process of image formation. High-energy, high-resolution microscopy, scanning electron microscopy, and electron diffraction—techniques that provide new and striking images of key supramolecular structures—are considered here. Two examples are cytoskeletal architecture and membrane–protein complexes such as bacterial rhodopsin.

The last two parts describe methods for studying membrane transport. Part 10 describes voltage clamping of excitable cell membranes, and Part 11 describes methods for making lipid model membranes and for determining their properties.

Present membrane-transport research focuses on characterizing those molecular structures in the membrane that are responsible for the various transport processes. Molecular structures of particular interest are the ion-selective channels responsible for excitability and ion "pumps," the membrane-bound enzymes that use metabolic energy to maintain concentration gradients of ions. In electrically excitable cells, such as nerve axons, the electrical behavior is extremely nonlinear. In fact, the nonlinearity is crucial to excitability. As discussed in Part 10, the voltage clamp has proven to be the major tool for studying such highly nonlinear systems, because it enables the experimenter to control the key variable— the transmembrane potential—and thus determine the time dependence and potential dependences of membrane conductance. Without this control, membrane potential and membrane conductance would vary in an extremely complex manner, making it virtually impossible to disentangle them. The concept of voltage clamping has also been useful in several recent important advances in the study of membrane ionic channels—the measurement of currents from individual channels, the use of noise analysis to determine the properties of the channels as they switch on and off at random, and the measurement of gating current (the displacement current caused by molecular conformational changes during channel opening or closing).

Part 11 discusses methods for using synthetic lipid bilayers as analogs of cell membranes. This approach has contributed greatly to the currently accepted paradigm for membrane structure—a lipid bilayer matrix resembling a two-dimensional fluid, in which various membrane proteins are embedded. In addition, the many studies of synthetic ionophores implanted in bilayers have yielded major insights into the transport mechanisms of membrane channels and carriers. For example, lipid bilayer studies provided the first measurements showing that membrane ionic current passes through discrete channels. More recently, significant progress has been made in reconstituting functional entities from biological membranes into lipid bilayers or lipid vesicles. Among the systems that have been reconstituted are Na-K ATPase pumps, postsynaptic acetylcholine-activated channels, and photosynthetic reaction centers.

We hope that the sample of methods presented in this volume shows some of the technological ingenuity that biophysical problems have generated. We should like to thank the editors of this series for their encouragement, support, and patience. Dr. Claire Marton and Dr. L. Marton, both recently deceased, have taken considerable interest in furthering interactions between physicists and biophysicists. We should also like to thank the staff of Academic Press for their support. Finally, we thank the authors for their effort and for their willingness to accommodate their individual papers to the needs of this volume.

GERALD EHRENSTEIN
HAROLD LECAR

METHODS OF EXPERIMENTAL PHYSICS

Editor-in-Chief

C. Marton

Volume 1. Classical Methods
Edited by Immanuel Estermann

Volume 2. Electronic Methods, Second Edition (in two parts)
Edited by E. Bleuler and R. O. Haxby

Volume 3. Molecular Physics, Second Edition (in two parts)
Edited by Dudley Williams

Volume 4. Atomic and Electron Physics—Part A: Atomic Sources
and Detectors, Part B: Free Atoms
Edited by Vernon W. Hughes and Howard L. Schultz

Volume 5. Nuclear Physics (in two parts)
Edited by Luke C. L. Yuan and Chien-Shiung Wu

Volume 6. Solid State Physics—Part A: Preparation, Structure, Mechani-
cal and Thermal Properties, Part B: Electrical, Magnetic, and Optical
Properties
Edited by K. Lark-Horovitz and Vivian A. Johnson

Volume 7. Atomic and Electron Physics—Atomic Interactions (in
two parts)
Edited by Benjamin Bederson and Wade L. Fite

Volume 8. Problems and Solutions for Students
Edited by L. Marton and W. F. Hornyak

Volume 9. Plasma Physics (in two parts)
Edited by Hans R. Griem and Ralph H. Lovberg

Volume 10. Physical Principles of Far-Infrared Radiation
By L. C. Robinson

Volume 11. Solid State Physics
Edited by R. V. Coleman

Volume 12. Astrophysics—Part A: Optical and Infrared
Edited by N. Carleton
Part B: Radio Telescopes, Part C: Radio Observations
Edited by M. L. Meeks

Volume 13. Spectroscopy (in two parts)
Edited by Dudley Williams

Volume 14. Vacuum Physics and Technology
Edited by G. L. Weissler and R. W. Carlson

Volume 15. Quantum Electronics (in two parts)
Edited by C. L. Tang

Volume 16. Polymers—Part A: Molecular Structure and Dynamics, Part
   B: Crystal Structure and Morphology, Part C: Physical Properties
Edited by R. A. Fava

Volume 17. Accelerators in Atomic Physics
Edited by P. Richard

Volume 18. Fluid Dynamics (in two parts)
Edited by R. J. Emrich

Volume 19. Ultrasonics
Edited by Peter D. Edmonds

Volume 20. Biophysics
Edited by Gerald Ehrenstein and Harold Lecar

1. NUCLEAR MAGNETIC RESONANCE

By Joseph R. Schuh and Sunney I. Chan

1.1. Introduction

Nuclear magnetic resonance (NMR) spectroscopy has proven to be a most powerful tool for unraveling the fine details of molecular structure and interactions in solution. Since 1946, when the first successful NMR experiments were reported,[1,2] this method has undergone continuous development toward the realization of its full potential. Initial studies were done, out of necessity, on relatively simple chemical systems, such as small organic and inorganic molecules. Low spectrometer sensitivity and the increased spectral widths associated with larger molecules rendered it difficult to extend the method to macromolecules of biological interest. However, recently developed techniques and improved instrumentation have altered this state of affairs considerably.

In this part we review some of the advances in NMR spectroscopy that have resulted in its successful application to problems in biology. Particular emphasis is placed on recently developed techniques and their usefulness in studies of biological macromolecules and systems. In the next two chapters we briefly outline those aspects of magnetic resonance theory necessary to facilitate the discussions of these newer NMR techniques. This review will be followed by a presentation of selected techniques that have found applications in biological NMR. We conclude Part 1 with a number of illustrative examples taken from the recent literature to highlight the achievements of the method to date.

In the discussions to follow, it is assumed that the reader is familiar with elementary magnetic resonance, including the concepts of magnetic shielding, chemical shifts, magnetic dipolar interactions, electron-coupled spin–spin interaction, etc., as well as the manner in which these phenomena manifest themselves in liquid-state spectra and have proven useful for structural analysis. Instead, we emphasize nuclear magnetic relaxation phenomena, particularly those techniques and features of NMR relaxation which permit

---

[1] F. Bloch, W. W. Hansen, and M. Packard, *Phys. Rev.* **69**, 127 (1946).
[2] E. M. Purcell, H. C. Torrey, and R. V. Pound, *Phys. Rev.* **69**, 37 (1946).

METHODS OF EXPERIMENTAL PHYSICS, VOL. 20

one to sample the details of motional processes in molecules. Biophysics deals in part with the structure and dynamics of biological molecules and assemblies, especially those aspects which pertain to biological function. Biomolecular structure has received considerable attention in biological NMR in the past. We concentrate on the dynamical aspects here, as a number of recently developed techniques are bringing these measurements within the realm of possibility.

Those readers interested in a more general and detailed treatment of magnetic resonance are referred to a previous chapter in this treatise[3] as well as to the classical textbook by Abragam[4] and the review series *Advances in Magnetic Resonance*, edited by J. S. Waugh. For a more comprehensive treatment of NMR in biology the books by James,[5] Dwek,[6] Wüthrich[7], and Knowles *et al.*[8] should be consulted.

1.2. Phenomenon of Magnetic Resonance

1.2.1. Quantum Description

A large number of atomic nuclei possess an intrinsic spin angular momentum. For a nucleus of spin I, the angular momentum is denoted by $\mathbf{J} = \hbar\mathbf{I}$ and its magnitude is $\hbar\sqrt{I(I + 1)}$, where $\hbar$ is Planck's constant divided by 2π. Only half integral and whole integral values of I are allowed. In quantum mechanical terms, the angular momentum vector of an isolated spin is characterized by $2I + 1$ distinct spatial orientations. These quantized states correspond to the $2I + 1$ projections of $\hbar\mathbf{I}$ along an arbitrary axis in space. Each of these projections will assume a value $m_I\hbar$ with $m_I = -I$, $-I + 1, \ldots, I - 1, I$.

The spin angular momentum of a nucleus imparts to it a magnetic moment $\boldsymbol{\mu}$ along an axis collinear with $\hbar\mathbf{I}$. These two quantities are related by

$$\boldsymbol{\mu} = \gamma\hbar\mathbf{I}, \tag{1.2.1}$$

[3] J. D. Memory and G. W. Parker, *Methods Exp. Physics* **3B**, 465 (1974).

[4] A. Abragam, "The Principles of Nuclear Magnetism." Oxford Univ. Press, London and New York, 1961.

[5] T. L. James, "Nuclear Magnetic Resonance in Biochemistry." Academic Press, New York, 1975.

[6] R. A. Dwek, "Nuclear Magnetic Resonance (Nmr) in Biochemistry." Oxford Univ. Press (Clarendon), London and New York, 1973.

[7] K. Wüthrich, "NMR in Biological Research: Peptides and Proteins." Elsevier, North-Holland, Amsterdam, 1976.

[8] P. F. Knowles, D. Marsh, and H. W. E. Rattle, "Magnetic Resonance of Biomolecules." Wiley, New York, 1976.

where γ is the magnetogyric ratio. In the absence of electrostatic and magnetic interactions, the $2I + 1$ spin states are degenerate. However, when an external magnetic field $\mathbf{H}_0$ is applied, this degeneracy is totally lifted. As a result of the interaction between the magnetic moment and the applied magnetic field, the projections of $\mathbf{I}$ will be quantized along $\mathbf{H}_0$, and each m_I state will assume a different energy given by

$$E = \gamma \hbar \mathbf{H}_0 m_I. \tag{1.2.2}$$

Nuclear magnetic resonance is a spectroscopic method which permits one to monitor the transitions between the various spin states in a magnetic field. For an isolated spin, quantum mechanics allows only transitions between adjacent levels. Since contiguous spin states under this circumstance are separated by an equivalent amount of energy, transitions will occur only at

$$\Delta E = \gamma \hbar \mathbf{H}_0. \tag{1.2.3}$$

Typically, $\Delta E/h$ is in the radiofrequency (rf) range. Table I summarizes the NMR frequencies for a number of biologically important nuclei.

In an NMR experiment one actually samples an assembly of spins, so

TABLE I. Properties of Some Biologically Relevant Nuclei

Spin	Isotope	NMR frequency[a] (MHz)	Natural abundance (%)	Relative sensitivity[b]
$I = \frac{1}{2}$	^{1}H	360.00	99.98	1.000
	^{13}C	90.52	1.11	0.0159
	^{15}N	36.49	0.365	0.00104
	^{19}F	338.68	100	0.833
	^{31}P	145.74	100	0.0663
$I = 1$	^{2}H	55.26	0.0156	0.00965
	^{14}N	26.01	99.64	0.00101
$I = \frac{3}{2}$	^{23}Na	95.23	100	0.0925
	^{35}Cl	35.28	75.4	0.00470
	^{39}K	16.80	93.08	0.000508
$I = \frac{5}{2}$	^{17}O	48.81	0.037	0.0291
	^{25}Mg	22.04	10.05	0.00268
	^{55}Mn	89.23	100	0.178
$I = \frac{7}{2}$	^{43}Ca	24.22	0.13	0.064

[a] In a field of 84.6 kG.

[b] For an equal number of nuclei at constant field.

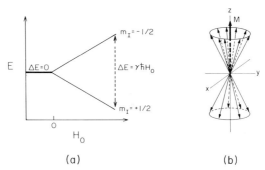

(a)   (b)

FIG. 1. (a) Zeeman splitting of spin states for nuclei having $I = \frac{1}{2}$. (b) Classical depiction of precessing nuclear magnetic moments with $I = \frac{1}{2}$, in the presence of exterior field $\mathbf{H}_0$. The net magnetization M, determined by the Boltzmann distribution, is along the z axis aligned with the field.

the measurement represents the net energy absorbed or emitted by a population of nuclei at the resonance condition. Inasmuch as the induced transitions between any two levels occur with equal probability, the signal one detects is proportional to the difference in the populations of nuclei occupying the different spin states. Thus for an $I = \frac{1}{2}$ system (see Fig. 1a) we may write

$$\text{signal} \propto (n_+ - n_-). \tag{1.2.4}$$

When the driving rf field used to observe the transition is sufficiently small so as not to perturb the populations of the spin states from their values at thermal equilibrium,

$$\text{signal} \propto (n_+^0 - n_-^0) = (1 - e^{-\gamma \hbar H_0/kT})/(1 + e^{-\gamma \hbar H_0/kT}). \tag{1.2.5}$$

For a proton NMR experiment at room temperature and 360 MHz, the Boltzmann factor deviates from unity by only 3 parts in 10^5, so that the energy change being measured is only about 10^{-7} kcal/mole of spins. Thus NMR is a spectroscopic method of low sensitivity, a limitation that has hampered the application of this technique to biological problems.

1.2.2. Classical Description

Many aspects of NMR theory can be appropriately described by either classical mechanics or quantum mechanics. Since the former affords a more graphic explanation, it will be used whenever possible. However, it should be noted that the classical approach is applicable only when discussing the net magnetic properties of a system; thus it will be necessary to turn to quantum mechanics when discussing other aspects of magnetic resonance, such as those relating to the properties of an individual nuclear spin.

1.2.2.1. The Concept of Magnetization. In order to illustrate the principles of NMR spectroscopy more fully, we now consider a population of N identical nuclei having $I = \frac{1}{2}$. In the presence of the external magnetic field $\mathbf{H}_0$, the nuclear magnetic moment vector $\mathbf{\mu}$ of each spin will assume one of two orientations with respect to the field, corresponding to the two quantum mechanically allowed energy states. In the low-energy state $\mathbf{\mu}$ will be aligned with the field, while for those nuclei in the high-energy state $\mathbf{\mu}$ will oppose $\mathbf{H}_0$.

For a given distribution of spins between these two states, the sample will exhibit a net magnetization $\mathbf{M}$, where $\mathbf{M} = \sum_i \mathbf{\mu}_i$. In the absence of other external forces an assembly of noninteracting spins exhibits no preference for the azimuthal angle. Since the distribution of the azimuthal angles would then be uniform,

$$M_x = \sum_i \mu_{xi} = 0, \qquad M_y = \sum_i \mu_{yi} = 0. \qquad (1.2.6)$$

Although experimental situations may still be found where these components are nonvanishing, the result given by Eq. (1.2.6) does hold for a system of spins at thermal equilibrium with the external degrees of freedom. Under these conditions, the population distributions among spin states are given by the Boltzmann factor, and the magnetization associated with the N, $I = \frac{1}{2}$, spins at thermal equilibrium is a vector of magnitude

$$\mathbf{M}_0 = \tfrac{1}{2}N\gamma\hbar \tanh(\gamma\hbar H_0/2kT) \qquad (1.2.7)$$

aligned in the direction of the external magnetic field (see Fig. 1b).

1.2.2.2. Motion of a Spin in a Magnetic Field. Classically, each of the individual spins that make up $\mathbf{M}$ experiences a torque in the presence of $\mathbf{H}_0$. The equation of motion is

$$d\mathbf{J}/dt = \langle\mathbf{\mu}\rangle \times \mathbf{H}_0, \qquad (1.2.8)$$

where $\langle\mathbf{\mu}\rangle$ denotes the classical magnetic moment. In view of the relationships $\mathbf{J} = \hbar\mathbf{I}$ and $\langle\mathbf{\mu}\rangle = \gamma\hbar\mathbf{I}$, the effect of the torque on the magnetic moment vector is

$$d\langle\mathbf{\mu}\rangle/dt = \gamma\langle\mathbf{\mu}\rangle \times \mathbf{H}_0. \qquad (1.2.9)$$

This equation describes the precession of $\langle\mathbf{\mu}\rangle$ about an axis parallel to the imposed field. The frequency of this motion is given by the Larmor equation

$$\mathbf{\omega}_0 = -\gamma\mathbf{H}_0. \qquad (1.2.10)$$

To demonstrate this result, we rewrite Eq. (1.2.9) in a frame of reference which rotates about the z axis at angular velocity $\mathbf{\omega}$. The result is

$$(\partial\langle\mathbf{\mu}\rangle/\partial t)_{\text{rot}} = \gamma\langle\mathbf{\mu}\rangle \times (\mathbf{H}_0 + \mathbf{\omega}/\gamma), \qquad (1.2.11)$$

where $(\partial \langle \boldsymbol{\mu} \rangle / \partial t)_{\mathrm{rot}}$ is the derivative computed in the rotating frame. From this equation it can be readily seen that $\langle \boldsymbol{\mu} \rangle$ will be stationary in the rotating frame when

$$\mathbf{H}_0 + \boldsymbol{\omega}/\gamma = 0 \qquad \text{or} \qquad \boldsymbol{\omega} = -\gamma \mathbf{H}_0 = \boldsymbol{\omega}_0. \qquad (1.2.12)$$

1.2.2.3. Magnetic Resonance. In magnetic resonance one typically applies a linear $\mathbf{H}_1$ field perpendicular to $\mathbf{H}_0$, oscillating sinusoidally with radio frequency $\boldsymbol{\omega}_{\mathrm{rf}} \approx \boldsymbol{\omega}_0$. Such a linearly polarized $\mathbf{H}_1$ field can be decomposed into two circularly polarized fields rotating in opposite directions. The component rotating in the same sense as the precessing spin will therefore appear stationary with respect to the magnetic moment when $\boldsymbol{\omega}_{\mathrm{rf}} = \boldsymbol{\omega}_0$, i.e., at resonance, while the counterclockwise component produces no net effect.

In the rotating frame, the torque exerted on $\langle \boldsymbol{\mu} \rangle$ by the $\mathbf{H}_1$ field will cause the magnetic moment to precess about the $\mathbf{H}_1$ axis at frequency $\boldsymbol{\omega}_1 = \gamma \mathbf{H}_1$. As a result of this precession the projection of $\langle \boldsymbol{\mu} \rangle$ along the Zeeman axis will vary and it will become opposed to the external field every half cycle, i.e., when

$$\omega_1 t = n\pi, \qquad (1.2.13)$$

where $n = 1, 3, 5, \ldots$. This is the classical analogy to the quantum-mechanical transition that occurs between two nuclear spin states when $I = \frac{1}{2}$. Since the energy of $\langle \boldsymbol{\mu} \rangle$ depends on its orientation *vis-à-vis* the magnetic field, there must be an exchange of energy between the rf field and the nuclear spin as it precesses about $\mathbf{H}_1$.

1.2.2.4. The Bloch Equations. The above results can be generalized to an assembly of noninteracting or weakly interacting spins. For such a system the torque equation is

$$d\mathbf{M}/dt = \gamma \mathbf{M} \times \mathbf{H}. \qquad (1.2.14)$$

When the effects of $\mathbf{H}_0$ and $\mathbf{H}_1$ on the components of $\mathbf{M}$ are explicitly spelled out, the equations become

$$dM_x/dt = \gamma(M_y H_0 + M_z H_1 \sin \omega_{\mathrm{rf}} t),$$

$$dM_y/dt = \gamma(M_z H_1 \cos \omega_{\mathrm{rf}} t - M_x H_0), \qquad (1.2.15)$$

$$dM_z/dt = -\gamma(M_x H_1 \sin \omega_{\mathrm{rf}} t + M_y H_1 \cos \omega_{\mathrm{rf}} t).$$

These are the equations of motion for the macroscopic magnetization vector in the presence of a static magnetic field $\mathbf{H}_0$ and an oscillating rf field $\mathbf{H}_1$. These equations, however, do not take into consideration relaxation processes which attempt to return the magnetization vector to its magnitude and direction at thermal equilibrium. Through interactions of the spins among themselves and with the surroundings, including the Zeeman field, the electronic cloud, vibrational/rotational/translational degrees of freedom, etc., both the precessional frequencies of the spins as well as their lifetimes in the spin states can be affected. In this manner, the spins can readjust their relative phases and redistribute themselves among the Zeeman levels of the system. Two types of relaxation processes may be distinguished corresponding to the decays of the M_z and M_x (M_y) components of the magnetization. Associated with each decay is a characteristic time constant. The spin–lattice or longitudinal relaxation time, denoted T_1, is the time constant for the return of the M_z component to its equilibrium value M_0. The spin–spin or transverse relaxation time, denoted T_2, is the time constant for the complete decay of the net magnetization in the xy plane. It is important to note that these two types of relaxation processes are intrinsically different; thus, for example, the relaxation of the M_z component involves an energy exchange, between the spins and the surroundings, while the decay of the M_x and M_y components does not.

Taking these relaxation processes into consideration one obtains the well-known Bloch equations,

$$dM_x/dt = \gamma(M_y H_0 + M_z H_1 \sin \omega_{rf} t) - M_x/T_2,$$
$$dM_y/dt = \gamma(M_z H_1 \cos \omega_{rf} t - M_x H_0) - M_y/T_2, \qquad (1.2.16)$$
$$dM_z/dt = -\gamma(M_x H_1 \sin \omega_{rf} t + M_y H_1 \cos \omega_{rf} t) - (M_z - M_0)/T_1.$$

These component equations of motion for the magnetization vector have proven to be extremely useful in the development of advanced experimental techniques; consequently, we shall have reason to return to them in subsequent sections. The Bloch equations are also useful for describing the effects of molecular motions on electron spin resonance signals, and explicit solutions are given in Part 2 of this treatise.

1.2.3. Detection of Magnetic Resonance

In the introduction to his classic paper on nuclear induction, Bloch noted that magnetic resonance could be observed in two ways.[9] In the first method, a frequency spectrum is obtained when the magnetic field $\mathbf{H}_0$ is slowly varied through the resonance condition while a radio frequency field $\mathbf{H}_1$ is imposed. This procedure is called the continuous wave (CW) method because of the requirement for an uninterrupted rf field. In the second method, all the nuclei of interest are excited simultaneously by a short, powerful pulse of rf energy at the resonance frequency. After the pulse is turned off, magnetic resonance is observed as the decay of the induced signal, called the free-induction decay (FID). Both the FID and the frequency spectrum contain the same

[9] F. Bloch, *Phys. Rev.*, **70**, 460 (1946).

information characterizing the system under study. However, since this information is more explicit when expressed in the frequency domain (i.e., the spectrum), it is more convenient. On the other hand, the FID (the time domain function) is obtained in a fraction of the time (10^{-2}–10^{-3}) necessary to acquire a frequency spectrum by the CW method.

In order to take advantage of the more efficient FID signal, Ernst and Anderson introduced Fourier-transform (FT) NMR spectroscopy in 1966.[10] This method is based on the finding that the Fourier transform of an FID is the frequency response spectrum acquired from a CW experiment.[11] Thus, using a mathematical operation, one can obtain the NMR frequency spectrum of a sample from its FID. This gives FT-NMR a distinct advantage over the CW method in that, for the same data acquisition time, the experiment can be repeated many more times. Subsequent averaging of the accumulated data produces a marked improvement in the signal-to-noise ratio. This enhancement is due to the fact that under appropriate experimental conditions, the signal response from a population of resonating nuclei is consistent for each FID, while the signal due to the background noise will fluctuate randomly. Thus by averaging the data from N free-induction decays, the signal-to-noise ratio in the transformed spectrum is increased by a factor of $N/N^{1/2}$.

It was previously noted that the low sensitivity of NMR spectroscopy is one of the major limitations of this method when it is applied to biological systems. As a result, the introduction of techniques utilizing FT-NMR to increase sensitivity has had a particularly significant impact on biological studies. Since this method is the basis for many of the techniques that will be discussed subsequently, its theory is included in this section.

1.2.3.1. Continuous-Wave NMR. Prior to the development of FT-NMR, to obtain a frequency spectrum experimentally it was necessary that a continuous $\mathbf{H}_1$ field be imposed. The spectrum is then obtained by varying either the $\mathbf{H}_0$ field strength (field sweep) or the rf frequency (frequency sweep). Under these conditions, the effective field in the rotating frame is

$$\mathbf{H}_{\text{eff}} = \mathbf{H}_0 + \mathbf{H}_1 + \omega/\gamma. \tag{1.2.17}$$

Initially the system is far from resonance, so that $\mathbf{H}_{\text{eff}} \cong \mathbf{H}_0$. However, as the resonance condition is approached, the fictitious field, ω/γ, and $\mathbf{H}_0$ cancel each other [cf. Eq. (1.2.10)] and $\mathbf{H}_{\text{eff}}$ tips toward the xy axis under the influence of the continuous $\mathbf{H}_1$ field. If the sweep rate is slow enough, the net magnetization will follow $\mathbf{H}_{\text{eff}}$, thus giving $\mathbf{M}$ a measurable component in the xy plane. This component is the resonance signal that is detected experimentally. Typically, this is done by measuring either the energy exchange

[10] R. R. Ernst and W. A. Anderson, *Rev. Sci. Instrum.* **37**, 93 (1966).
[11] I. J. Lowe and R. E. Norberg, *Phys. Rev.*, **107**, 46 (1957).

between the spin system and the rf coil[2] or the voltage induced in a separate receiver coil placed perpendicular to both $\mathbf{H}_0$ and $\mathbf{H}_1$.[1]

1.2.3.2. Pulse NMR. The pulse method differs from CW-NMR in that all the nuclei of interest are excited and observed simultaneously rather than individually. In order to accomplish this, the rf field $\mathbf{H}_1$ must have a magnitude such that

$$\gamma H_1 \gg 2\pi\Delta, \qquad (1.2.18)$$

where Δ is the frequency range spanned by the nuclei under examination. In general, Δ is determined by the range of chemical shifts for a given type of nucleus. As is well known, a nuclear moment is magnetically screened by the electrons in a molecule, and magnetic nuclei of the same kind exhibit different resonance frequencies depending upon the chemical environment. This dispersion of the NMR frequencies according to chemical environment has been widely exploited for structural analysis and determination.

The $\mathbf{H}_1$ field for a pulse experiment is usually several orders of magnitude larger than it is in the CW method. In addition, the pulse experiment is always done very close to resonance, i.e.,

$$\mathbf{H}_1 \gg \mathbf{H}_0 + \omega/\gamma. \qquad (1.2.19)$$

Accordingly, $\mathbf{H}_{\text{eff}} \cong \mathbf{H}_1$ for the period in which $\mathbf{H}_1$ is turned on. However, since the duration of the rf pulse is usually on the order of microseconds, the magnetization vector cannot follow $\mathbf{H}_{\text{eff}}$. Instead $\mathbf{M}$ precesses about $\mathbf{H}_1$ (see Fig. 2). Inasmuch as the detector only measures the components of $\mathbf{M}$ in the xy plane, the strongest signal will be observed when the duration of $\mathbf{H}_1$ is long enough to rotate $\mathbf{M}$ by $90°$ ($\pi/2$). The length of this $90°$ pulse, t_p, is

$$t_p = \pi/2\gamma H_1. \qquad (1.2.20)$$

In practice, H_1 is dictated by the range of nuclear frequencies to be observed, and the restriction in Eq. (1.2.18) requires that

$$t_p \ll 1/4\Delta. \qquad (1.2.21)$$

Immediately following a $90°$ pulse, the individual magnetic moments will begin to dephase freely, because of field inhomogeneity effects and other instrinsic relaxation processes. This results in the decay of the induced voltage in the receiver coil (the FID). Since the dephasing occurs in the xy plane, its duration is of the order of T_2^*, where the asterisk indicates that this T_2 value includes contributions from field inhomogeneities and may be dominated by them. In general, T_2^* is quite short and we have both T_1 and $T_2 \gg T_2^*$.

In order to obtain the intrinsic T_2 values a number of multiple-pulse techniques have been developed which eliminate the contribution from magnetic field inhomogeneity. Similar pulse methods have also been designed

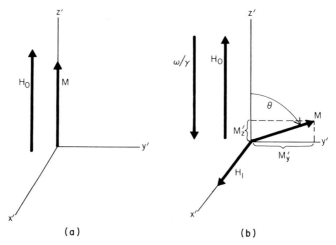

FIG. 2. (a) Net magnetization M aligned along the Zeeman axis in a rotating frame of reference. (b) Net magnetization at resonance with rf field $\mathbf{H}_1$ rotating at the Larmor frequency. The effect of $\mathbf{H}_1$ is to tip M toward the y axis, giving the magnetic vector a component along both the z and the y axes.

which allow one to accurately determine the longitudinal relaxation time, T_1. We discuss these techniques in more detail in Section 1.4.1.1.

1.2.3.3. Fourier-Transform NMR. The decay of an NMR signal in the xy plane, following a 90° pulse, arises from variations in local electrostatic and magnetic interactions. These interactions cause the transition frequencies to be different for different resonant nuclei. Classically, this variation in the transition frequencies corresponds to differences in the Larmor frequencies of the precessing magnetic moments under the influence of $\mathbf{H}_0$ (provided, of course, that the Zeeman interaction dominates). It is these various frequency components that interfere constructively and destructively to give the decay function in the time domain, the FID.

Mathematically, one can describe a time domain function as the Fourier transform of a frequency domain function $F(\omega)$. The latter is a measure of the frequency components that comprise the response function $f(t)$. In keeping with the properties of Fourier integrals these two functions are merely Fourier transforms of each other; thus

$$F(\omega) = \frac{1}{\sqrt{2\pi}} \int_{-\infty}^{\infty} f(t)e^{-i\omega t}\, dt, \qquad (1.2.22)$$

and

$$f(t) = \frac{1}{\sqrt{2\pi}} \int_{-\infty}^{\infty} F(\omega)e^{i\omega t}\, d\omega. \qquad (1.2.23)$$

Naturally, these two functions contain the same information about the nuclear spin system.

To illustrate this, we consider a simple spin system which exhibits an NMR transition at frequency $\omega = \omega_0$. We shall assume that the resonance line is Lorentzian with a full width at half height $\Delta\omega$, equal to $2/T_2$, so that

$$F(\omega) = AT_2/[1 + T_2^2(\omega - \omega_0)^2]. \qquad (1.2.24)$$

Taking the Fourier transform of this, we obtain the response function

$$f(t) = \frac{1}{\sqrt{2\pi}} \int_{-\infty}^{\infty} \frac{AT_2}{1 + T_2^2(\omega - \omega_0)^2} e^{i\omega t} \, d\omega$$

$$= \sqrt{\frac{\pi}{2}} \, A \, e^{-t/T_2} \, e^{i\omega_0 t}. \qquad (1.2.25)$$

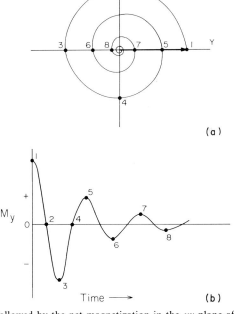

(a)

(b)

FIG. 3. (a) Path followed by the net magnetization in the xy plane after a 90° pulse. The trace is circular because the resonant nuclei precess at a frequency different from that of the rotating frame. Inward spiraling is caused by transverse relaxation of the moments in the xy plane. (b) FID of the system depicted in (a). The recorded signal is induced in a phase-sensitive receiver coil positioned to sample the net magnetization along y. The numbered points on the FID correspond to the points in (a) marking positions of the net moment in the xy plane.

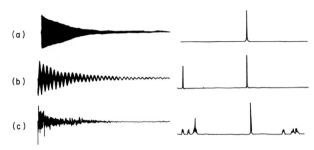

FIG. 4. Free-induction decay signals (left-hand side) and their Fourier transforms (right-hand side) for (a) single-line, (b) two-line, and (c) multiline spectra. From James.[5]

This result indicates that the FID signal, following a 90° pulse, is merely a sinusoidal function with frequency ω_0, except that the amplitude decays with a time constant of T_2.

This important result can be graphically illustrated in two ways. In Fig. 3a we plot the magnetization of our simple spin system as it appears in a frame of reference rotating at ω, where $\omega \neq \omega_0$. In this frame of reference, the magnetization will be precessing with frequency $|\omega - \omega_0|$, but since its magnitude is concurrently decreasing exponentially because of the T_2 relaxation processes, the precesssing magnetization will describe a spiraling arc in the xy plane. Alternatively, one can plot either the in-phase or out-of-phase component of the induced voltage across a receiver tuned to frequency ω as a function of time (see Fig. 3b). The resulting interference or beat pattern one observes here measures the decay of the transverse component of the precessing magnetization as it is being detected at the receiver. Note that from the periodicity of this decay and a knowledge of ω we can calculate the frequency of the NMR transition ω_0.

In practice, of course, one is usually sampling more than one nuclear species exhibiting various transition frequencies. Each of these will contribute to the induced signal, and the FID will appear to be quite complex. This is illustrated in Fig. 4, where the FIDs for three different samples are shown, along with their respective Fourier transforms.

1.3. Spin Relaxation

Following an rf pulse one typically creates a nonequilibrium magnetization along the Zeeman axis and in the xy plane. The decay of each of these two components is described by a distinct characteristic time. The return of the magnetization to its equilibrium value along the Zeeman axis is described by the spin–lattice or T_1 relaxation time. Since this component of the magnetization is a measure of the distribution of spins among the energy-dependent Zeeman states there must necessarily be an energy exchange between the spin system and other degrees of freedom. The application of an H_1 field in the xy plane can also perturb the phases of the wave functions for the nuclear spins and lead to a nonzero component of the magnetization in the xy plane. This nonequilibrium state decays to zero with a characteristic time T_2, which is distinct from the spin–lattice relaxation time since only the phases are involved. Although in T_2 relaxation there is no energy exchange between the total spin system and the surroundings, the phases of the spins can be modified via energy exchange among the individual spins themselves.

1.3.1. Spin–Lattice Relaxation

The nonradiative decay mechanisms responsible for T_1 relaxation have their origin in the interactions between resonant nuclei and fluctuating local electric and magnetic fields arising from the atoms and molecules present in the sample. For molecules in solution the fluctuating nature of these fields is caused by the rotational and translational motion of the molecules as they undergo thermally induced Brownian motions. These fields act on the nucleus in a manner analogous to the applied H_1 field, except that only those Fourier components which match the frequencies of the quantum mechanically allowed transitions will be effective. Thus the efficiency of any relaxation mechanism depends upon the nature of the interaction and the frequency with which it fluctuates.

In the simplest case, where we have a spin manifold of weakly interacting nuclei of $I = \frac{1}{2}$, a nonequilibrium distribution of the spin population can only return to equilibrium through transitions between the two Zeeman levels. The fluctuating local magnetic field acting at the site of the resonant nucleus can be decomposed into its Fourier components and written

$$H(t) = \frac{1}{\sqrt{2\pi}} \int_{-\infty}^{\infty} g(\omega)e^{i\omega t}\, d\omega, \tag{1.3.1}$$

where $g(\omega)$ describes the distribution of frequencies in the field. Since energy exchange between spin and lattice systems can only occur in quantized units of the spacing $\hbar\omega_0$ between the two spin levels, it follows that

$$1/T_1 \propto |g(\omega_0)|^2. \tag{1.3.2}$$

Therefore spin–lattice relaxation will be most efficient when the value of the distribution function at the Larmor frequency is at a maximum. Such a condition obtains only when the stochastic modulation time of the fluctuating field is approximately $(\omega_0)^{-1}$. This characteristic time constant is the well-known correlation time τ_c, which we discuss in more detail in Section 1.3.3.

1.3.2. Spin–Spin Relaxation

For an assembly of noninteracting spins the equilibrium value of the net magnetization in the xy plane is exactly equal to zero, since there can be no phase coherence of the individual magnetic moments under these conditions. The introduction of phase coherence by an appropriate rf pulse results in the creation of a nonzero xy magnetization in the rotating frame, M_{xy}, which will decay to zero in a characteristic time T_2.

In discussing the relaxation mechanisms responsible for T_2, it is useful to distinguish two major contributions to this process. For the so-called inhomogeneous contribution to T_2, the decay of M_{xy} occurs as the individual magnetic moments lose phase coherence by precessing at different frequencies. This is caused by static field inhomogeneity across the volume of the sample and by first-order interactions such as those responsible for chemical shifts and spin–spin splitting. Dephasing of an otherwise homogeneous set of spin vectors can also occur due to random fluctuations in local electric and magnetic fields. In this case the spin vectors precess about the field at the same average frequency but exhibit a statistical oscillation in their phase coherence, caused by transient variations in the frequency of precession for individual spins. The magnitude of this homogeneous contribution to T_2 depends upon both the time scale and nature of these fluctuations, analogous to the way local fields affect T_1. However, only local field components collinear with $\mathbf{H}_0$ along the z-axis contribute to homogeneous T_2 processes. Of course, if the spin–lattice relaxation rate for the system is sufficiently fast, then the nonzero M_{xy} must decay within the limit of T_1. It follows, under this condition, that the energy levels will be broadened by a width of the order of

$$\Delta v \cong 1/2T_1 \qquad (1.3.3)$$

and the M_{xy} component will decay to zero according to this time constant. In many cases, such as in liquids and gases, this contribution to the dephasing will be included as part of the homogeneous T_2.

1.3.3. Correlation Time

In the preceding discussion of spin–lattice relaxation, we introduce the concept of a motional correlation time τ_c. The length of this period may be considered the "memory time" of a molecule. This means that for the interval τ_c, motional changes in the position of the nucleus bear some relation to each other. At times longer than τ_c, the positional correspondence is lost and the motion appears to be random. The value of τ_c can be approximated as the time required for the molecule containing the resonant nucleus to either rotate 1 radian (rotational correlation time) or diffuse a distance equivalent to its own dimensions (translational correlation time). Thus the magnitude of τ_c will be a function of the size of the molecule, the viscosity of the solvent, and the temperature of the sample.

In order to illustrate this more quantitatively, we now consider the energy of a local fluctuating field, described by the Hamiltonian operator $\mathcal{H}_1$. If this field fluctuates in a random manner, its value at time t, $\mathcal{H}_1(t)$, will differ from that at $\mathcal{H}_1(t + \tau)$. For very small values of τ, we expect $\mathcal{H}_1(t) \approx \mathcal{H}_1(t + \tau)$, but for $t + \tau \gg t$, the similarity will be lost. A measure of the correlation between $\mathcal{H}_1$ values as they vary with time is the autocorrelation function $G(\tau)$. For a nuclear magnetic moment which can occupy two energy states, say E_m and E_k, the probability W_{km} of a transition induced by $\mathcal{H}_1$ between these states can be shown to be related to $G(\tau)$ by

$$W_{km} = \frac{1}{\hbar^2} \int_{-\infty}^{\infty} G_{mk}(\tau) e^{-i(E_m - E_k)\tau/\hbar} \, d\tau, \tag{1.3.4}$$

where $G_{mk}(\tau)$ is the autocorrelation function, described by

$$G_{mk}(\tau) = \overline{\langle m | \mathcal{H}_1(t + \tau) | k \rangle \langle k | \mathcal{H}_1(t) | m \rangle}. \tag{1.3.5}$$

While the function $G(\tau)$ tells us something about the rate at which a randomly fluctuating field loses its positional correspondence, its Fourier transform gives the spectral density function

$$J(\omega) = \int_{-\infty}^{\infty} G(\tau) e^{i\omega\tau} \, d\tau, \tag{1.3.6}$$

which describes the probability that a fluctuating field will occur at a particular frequency. Since relxation will be most efficient when the frequencies of the fluctuating fields lie close to the Larmor frequency of the resonant nuclei, T_1 will be at a minimum when $J(\omega_0)$ is maximal. This condition obtains when $(\tau_c)^{-1} = \omega_0$, as shown graphically in Fig. 5a where $J(\omega)$ is plotted for

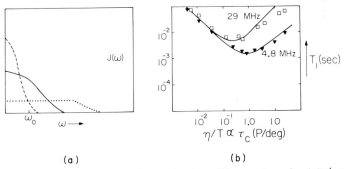

(a)   (b)

FIG. 5. (a) The spectral density function $J_{(\omega)}$ for three different values of τ_c: $(\tau_c)^{-1} \ll \omega_0$ (- - -); $(\tau_c)^{-1} = \omega_0$ (—); $(\tau_c)^{-1} \gg \omega_0$ ($\cdots$). (b) Proton spin–lattice relaxation times for glycerine at different temperatures and two different frequencies. As the temperature decreases, the viscosity η increases; thus τ_c becomes longer. From Farrar and Becker.[12]

three different correlation times.[12] Thus when τ_c is short, the transition probability will be equally low for a wide spectrum of frequencies, including ω_0. However, when τ_c is long, $J(\omega)$ increases for low frequencies, but the transition probability decreases at higher frequencies.

The dependence of T_1 on τ_c is well demonstrated by an experiment from the classic work of Bloembergen et al.[13] This is shown in Fig. 5b, where the proton spin–lattice relaxation time for glycerine is plotted at two different frequencies as a function of temperature. At high temperatures, where thermal motion is rapid, τ_c is short and the spectral densities of the random fields at either 29 or 4.8 MHz are equally low; accordingly, T_1 is relatively long and frequency independent. As τ_c increases so that $(\tau_c)^{-1}$ approaches the resonant frequencies, the spin–lattice relaxation rate becomes frequency dependent, reaching a maximum at ω_0. The T_1 minimum is shorter at 4.8 MHz because the maximum value for $J(\omega_0)$ increases as the resonant frequency decreases (note that the area under the spectral density function in Fig. 5a remains constant!). At lower temperatures, τ_c increases further, and $J(\omega_0)$ begins to decrease, first at the higher frequencies then at the lower ones. For this reason, the increasing T_1 values depicted on the right side of the curves in Fig. 5b are greater at 29 than at 4.8 MHz.

Thus far we have assumed that molecules undergo isotropic Brownian rotational and translational diffusion in the liquid state. However, in the case where molecular reorientation or translational diffusion is anisotropic, a plot of T_1 versus τ_c may yield a surface exhibiting multiple minima, corresponding to temporally distinct correlation times. For example, with a rigid molecule undergoing axially symmetric anisotropic motion (where $\tau_\parallel \ll \tau_\perp$) two minima may occur. Similarly, for motionally restricted molecules, such as those found in membrane systems, multiple correlation times can contribute to the spin–lattice relaxation, although T_1 may exhibit an additional dependence on the average orientation of the molecule relative to the magnetic field.

1.3.4. Relaxation Mechanisms

In this section we discuss a number of mechanisms that have been found to contribute significantly to nuclear magnetic relaxation in biological systems. For convenience we assume that all motions are sufficiently rapid to average the various interactions to first order. However, it should be noted that for large macromolecular complexes exhibiting long correlation times

[12] T. C. Farrar and E. D. Becker, "Pulse and Fourier Transform NMR." Academic Press, New York, 1971.

[13] N. Bloembergen, E. M. Purcell, and R. V. Pound, Phys. Rev. 73, 679 (1948).

the expressions used here to describe relaxation will not necessarily be valid.

1.3.4.1. Magnetic Dipole–Dipole Interactions. The predominant relaxation mechanism for nuclei having spin number $I = \frac{1}{2}$ (e.g., ^{1}H, ^{13}C, ^{31}P, ^{19}F) is through magnetic dipole–dipole interactions with other nuclear magnetic moments. For a two-spin system consisting of $\mu_1 = \gamma_I \hbar \mathbf{I}$ and $\mu_2 = \gamma_s \hbar \mathbf{S}$ separated by a distance r, the energy of interaction between the spins is[14]

$$\mathcal{H}_D = (\mu_1 \cdot \mu_2)/r^3 - 3(\mu_1 \cdot \mathbf{r})(\mu_2 \cdot \mathbf{r})/r^5. \tag{1.3.7}$$

This dipolar Hamiltonian can be conveniently rewritten in a polar coordinate system involving the interspin vector $\mathbf{r}$ and the applied magnetic field. If $\mathbf{r}$ makes an angle θ with respect to the field and ϕ is the azimuthal angle,

$$\mathcal{H}_D = (\gamma_I \gamma_s \hbar^2 / r^3)(A + B + C + D + E + F), \tag{1.3.8}$$

where

$$A = I_z S_z (1 - 3\cos^2\theta), \tag{1.3.8a}$$

$$B = -\tfrac{1}{4}[(I_x - iI_y)(S_x + iS_y) + (I_x + iI_y)(S_x - iS_y)](1 - 3\cos^2\theta), \tag{1.3.8b}$$

$$C = \tfrac{3}{2}[(I_x + iI_z)S_z + (S_x + iS_y)I_z](\sin\theta\cos\theta\, e^{-i\phi}), \tag{1.3.8c}$$

$$D = -\tfrac{3}{2}[(I_x - iI_y)S_z + (S_x - iS_y)I_z](\sin\theta\cos\theta\, e^{i\phi}), \tag{1.3.8d}$$

$$E = -\tfrac{3}{4}[(I_x + iI_y)(S_x + iS_y)]\sin^2\theta\, e^{-2i\phi}, \tag{1.3.8e}$$

$$F = -\tfrac{3}{4}[(I_x - iI_y)(S_x - iS_y)]\sin^2\theta\, e^{2i\phi}. \tag{1.3.8f}$$

For this Hamiltonian, it can be shown that the dipolar contribution to the longitudinal relaxation rate for nucleus I is

$$R_1^I = 1/T_1^I = \gamma_I^2 \gamma_s^2 \hbar^2 S(S+1)[\tfrac{1}{12}J^{(0)}(\omega_I - \omega_S) + \tfrac{3}{2}J^{(1)}(\omega_I) + \tfrac{3}{4}J^{(2)}(\omega_I + \omega_S)], \tag{1.3.9}$$

while for transverse relaxation the rate is

$$R_2^I = 1/T_2^I = \gamma_I^2 \gamma_s^2 \hbar^2 S(S+1)[\tfrac{1}{6}J^{(0)}(0) + \tfrac{1}{24}J^{(0)}(\omega_I - \omega_S) + \tfrac{3}{4}J^{(1)}(\omega_I) + \tfrac{3}{2}J^{(1)}(\omega_S) + \tfrac{3}{8}J^{(2)}(\omega_I + \omega_S)]; \tag{1.3.10}$$

[14] C. P. Slichter, "Principles of Magnetic Resonance." Springer-Verlag, Berlin and New York, 1978.

$J^{(0)}$, $J^{(1)}$, and $J^{(2)}$ are the spectral density functions associated with the fluctuations in

$$F^{(0)} = (1 - 3\cos^2\theta)r^{-3},$$

$$F^{(1)} = \sin\theta\cos\theta\, e^{i\phi}r^{-3}, \qquad (1.3.11)$$

$$F^{(2)} = \sin^2\theta\, e^{2i\phi}r^{-3}.$$

For intramolecular dipolar relaxation, where $\mathbf{r}$ is fixed, the spectral density functions for Eqs. (1.3.9) and (1.3.10) can be shown to be

$$J^{(0)}(\omega) = \tfrac{24}{15}r^{-6}[\tau_c/(1 + \omega^2\tau_c^2)],$$

$$J^{(1)}(\omega) = \tfrac{4}{15}r^{-6}[\tau_c/(1 + \omega^2\tau_c^2)], \qquad (1.3.12)$$

$$J^{(2)}(\omega) = \tfrac{16}{15}r^{-6}[\tau_c/(1 + \omega^2\tau_c^2)].$$

It is generally assumed that molecular reorientation in the liquid state satisfies the Stokes–Einstein relationship; that is,

$$\tau_c = \tau_v = 4\pi\eta a^3/3kT, \qquad (1.3.13)$$

where η is the viscosity of the medium in which the molecule is embedded and a is the radius of a hypothetical sphere used to approximate the molecule. Frequently, $\omega_0\tau_c \ll 1$. This is the so-called extreme narrowing condition, a situation which is not always satisfied for large biological macromolecules. In any case, when the extreme narrowing condition obtains, $T_1 = T_2$ and the intramolecular dipolar relaxation rates are given by

$$(R_1^I)_{\text{intra}} = (R_2^I)_{\text{intra}} = \tfrac{4}{3}\gamma_I^2\gamma_S^2\hbar^2 S(S + 1)\tau_c r^{-6}. \qquad (1.3.14)$$

From this result we may gain some insight into those factors that influence the intramolecular dipole–dipole relaxation rate. For example, since the spectral density functions are proportional to r^{-6}, the dipole–dipole interaction will be maximal when I and S are in closest proximity, as, for example, when they are covalently bonded to each other. It can also be noted that the dipolar relaxation rate will be highly dependent upon the gyromagnetic ratios of the interacting nuclei. This is, of course, to be expected since γ is proportional to the nuclear magnetic moment μ, whose magnitude determines the strength of the microscopic magnetic field acting at the resonant nucleus.

In order to illustrate how these ideas may be exploited we have selected some results from the work of Allerhand and co-workers[15] on the T_1s for the ^{13}C nuclei of cholesteryl chloride dissolved in CCl_4. If we may assume that any rotational anisotropy in the molecular motions is small, all the

---

[15] A. Allerhand, D. Doddrell, and R. Komoroski, *J. Chem. Phys.*, **55**, 189 (1971).

carbons located in the rigid ring structure will have the same correlation time τ_v. The T_1s for these nuclei will then be a function of structural factors only, namely, the number and the gyromagnetic ratio of the nuclei coupled magnetically to the carbon atoms and the respective internuclear distances. Because of the low natural abundance of ^{13}C (about 1%) and the strong r^{-6} dependence of the relaxation rate, the only significant dipolar contribution to the ^{13}C T_1 should come from directly bonded *protons*, in which case the r values will be similar as well. Accordingly, we predict that T_1 for the ring carbons should decrease with the number of protons bonded to them. This expectation is clearly borne out in Fig. 6, which shows that carbons with 0, 1, and 2 protons have average T_1 values of 3.4, 0.49, and 0.26 sec, respectively. An apparent anomaly is the T_1 value for the ring-bonded methyl carbons, which despite the three directly bonded protons is 1.5 sec. However, here the increased rotational freedom of the methyl rotor results in a faster effective τ_c than τ_v. This effect of τ_c on the relaxation rate is even more pronounced in the case of the methyl carbons attached to the terminal end of the aliphatic tail segment; the T_1 values are about 2.0 sec for these carbons.

Finally, for intermolecular dipole–dipole interactions the interspin distance r is also time dependent, in additional to θ and ϕ. The situation is therefore somewhat more complex, but if the extreme narrowing condition is applicable and the spins are of the same kind, it can be shown that

$$(R_1^I)_{\text{inter}} = (\tfrac{8}{15}\pi)\gamma_I^4\hbar^2 I(I + 1)N/dD, \tag{1.3.15}$$

where N is the density of spins per cubic centimeter, d is the distance of closest approach, and D is the translational diffusion coefficient. Within the Stokes formulation D and d are related by

$$D = kT/6\pi a\eta, \qquad d = 2a. \tag{1.3.16}$$

It should be evident that the intermolecular contribution to the relaxation rate can be suppressed by diluting the molecule in a magnetically inert solvent.

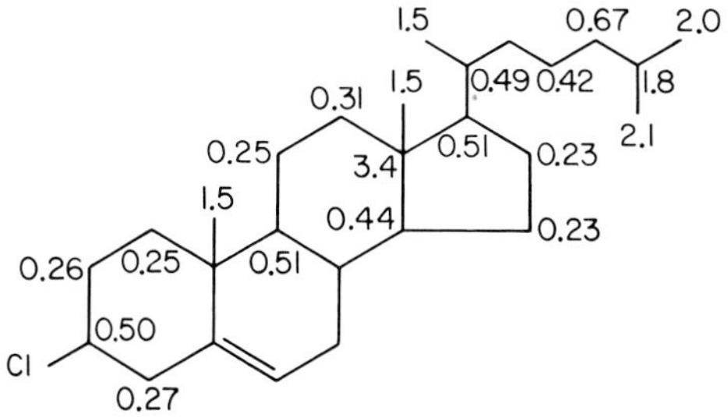

FIG. 6. ^{13}C spin–lattice relaxation times (sec) for cholesteryl chloride in CCl_4. (From Allerhand *et al.*[15])

1.3.4.2. Spin–Spin Coupling. In addition to the direct dipolar coupling of nuclear magnetic moments, there can also be indirect interactions between nuclei which are mediated via the bonding electrons. This effect is called spin–spin or scalar coupling and can lead to fine structure in the resonance signals. If the spin coupling interaction fluctuates in time it can also act as a mechanism for relaxation.

The spin Hamiltonian that describes this interaction between two spins is given by

$$\mathscr{H}_J = \mathbf{I} \cdot \mathbf{J} \cdot \mathbf{S}, \tag{1.3.17}$$

where $\mathbf{J}$ is the tensor describing the bilinear coupling of nuclei I and S through interactions with their electrons. In solution, only the trace of $\mathbf{J}$ can be measured and the spin Hamiltonian simplifies to

$$\mathscr{H}_J = J(\mathbf{I} \cdot \mathbf{S}), \tag{1.3.18}$$

where J is the well-known scalar spin–spin coupling constant.

An example of how spin–spin coupling manifests itself in the frequency spectrum is illustrated in Fig. 7, where we show the ^{13}C spectrum of the choline methyl resonance from a solution of phosphatidylcholine dissolved in a mixed solvent of chloroform, methanol, and water (50:40:15).[16] Here, ^{13}C is chemically bonded to ^{14}N, and the two nuclei are spin coupled with J/h of 3.7 Hz. Since $I = 1$ for ^{14}N, the effect of this coupling is to split the ^{13}C resonance into three lines, corresponding to the three possible orientations of the ^{14}N spin in a magnetic field, namely, $m_I = -1, 0, +1$.

It should be evident that $\mathscr{H}_J$ can become time dependent when either one of the spins undergoes rapid nuclear relaxation, or if the coupling between I and S is modulated by rapid chemical exchange for one of the spins. Therefore J splittings will only be manifested in the NMR spectrum when J/h is greater than either the nuclear relaxation rate or the frequency of dissociation between I and S. Of course, when this condition does not obtain, spectral splitting vanishes and the time-dependent scalar coupling affords another mechanism for relaxation.

Scalar relaxation of the first kind occurs when the chemical exchange rate of a coupled nucleus is greater than both J and the nuclear relaxation rate. For this mechanism only a single correlation function is involved, describing the probability that S and I are still coupled at $t + \tau$. Assuming that I is always bound to some S spin, one can show that the relaxation rates for I are

$$R_1^I = \tfrac{2}{3}J^2[\tau_e/\{1 + (\omega_I - \omega_S)^2\tau_e^2\}]S(S + 1) \tag{1.3.19}$$

and

$$R_2^I = \tfrac{1}{3}J^2[\tau_e + (\tau_e/\{1 + (\omega_I - \omega_S)^2\tau_e^2\})]S(S + 1), \tag{1.3.20}$$

where τ_e is the chemical exchange time constant.

[16] R. Murari and W. J. Baumann, *J. Am. Chem. Soc.* **103**, 1238 (1981).

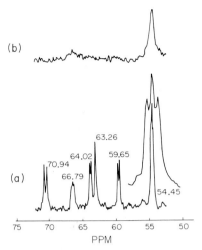

FIG. 7. Proton-decoupled 20-MHz ^{13}C NMR spectra of egg yolk phosphatidylcholine in (a) $CDCl_3/CD_3OD/D_2O$ (50:50:15), and (b) D_2O. Peak assignments in (a) are given for: $N(CH_3)_3$ (54.45 ppm), $P-O-CH_2$ (59.65 ppm), sn-1 CH_2-O-R_1 (63.26 ppm), sn-3 CH_2-O-P (64.02 ppm), CH_2-N (66.79 ppm), and sn-2 $CH-O-R_2$ (70.94 ppm). Insert is an expansion of the $N(CH_3)_3$ peak. Reprinted with permission from Murari and Baumann.[16] Copyright © 1981 American Chemical Society.

A second type of scalar relaxation can exist when the relaxation rate for one of the nuclei is much greater than both J and/or the exchange rate $1/\tau_e$. This may occur when a different relaxation mechanism, e.g., quadrupolar (see Section 1.3.4.3) is predominant for the second nucleus. Under these conditions the coupling between the two spins is rapidly modulated by fluctuations in the S spin as it makes frequent transitions among its energy states.

For spin–lattice relaxation, only the time-dependent components of the spin–spin interaction having frequency ω_I will contribute to T_1^I, and in fact only those components of the S magnetization not aligned with the Zeeman field will be important. The contribution to the spin–lattice relaxation rate for I is simply

$$R_1^I = 2J^{(1)}(\omega_I), \tag{1.3.21}$$

where $J^{(1)}(\omega)$ is the Fourier spectrum of the autocorrelation function

$$G^{(1)}(\tau) \cong \tfrac{1}{6}J^2 S(S+1)e^{i\omega_S\tau}\,e^{-\tau/T_2^S}. \tag{1.3.22}$$

The result is

$$R_1^I = (\tfrac{2}{3}J^2)S(S+1)[T_2^S/\{1 + (\omega_I - \omega_S)^2(T_2^S)^2\}]. \tag{1.3.23}$$

For the transverse relaxation rate, where stationary fields also can contribute, it is necessary to include a second autocorrelation function involving the z components of S. Therefore

$$R_2^I = J^{(1)}(\omega_I) + \tfrac{1}{2}J^{(0)}(0), \tag{1.3.24}$$

where $J^{(0)}(\omega)$ is the Fourier transform of the correlation function

$$G^{(0)}(\tau) = [\tfrac{1}{3}J^2 S(S + 1)]e^{-\tau/T_1^S}. \qquad (1.3.25)$$

Thus the expression for the transverse relaxation rate is

$$R_2^I = [\tfrac{1}{3}J^2 S(S + 1)][T_2^S/\{1 + (\omega_I - \omega_S)^2(T_2^S)^2\} + T_1^S]. \qquad (1.3.26)$$

Scalar relaxation of the second type can be illustrated from the phospholipid studies of Murari and Baumann,[16] which we cited earlier in this section (Fig. 7). The spin–lattice and spin–spin relaxation rates for the ^{14}N nucleus in phosphatidylcholine are controlled by the rotational rate of the CH_2—N segment,[17,18] which, in turn, depends upon the aggregation state of the phospholipids in solution. In accordance with this expectation, the ^{14}N T_1 and T_2 values for phosphatidylcholine liposomes in D_2O are 0.06 and 0.0006 sec, respectively, while in chloroform, methanol, and water (50:50:15), a medium where phospholipids do not aggregate, the corresponding values are 0.15 and 0.177 sec. Concomitant with the increased relaxation rates of the ^{14}N in the liposomes, the ^{14}N splittings in the choline methyl resonance of the ^{13}C spectrum are collapsed.

1.3.4.3. Quadrupolar Interactions. Atomic nuclei having $I \geq 1$ possess an electric quadrupole moment because the charge on the nucleus is distributed asymmetrically. The interaction of this asymmetric nuclear charge distribution with the electron cloud can partially lift the $2I + 1$ spatial degeneracy of the nuclear spin, even in the absence of an external magnetic field. This coupling is usually referred to as the electric quadrupole interaction.

The Hamiltonian operator for this problem is given by

$$\mathscr{H}_Q = \mathbf{I} \cdot \mathbf{Q} \cdot \mathbf{I}, \qquad (1.3.27)$$

where $\mathbf{Q}$ is the quadrupole energy tensor and is related to the electric field gradient tensor at the nucleus, $\mathbf{V}$, by

$$\mathbf{Q} = [eQ/2I(2I - 1)]\mathbf{V}, \qquad (1.3.28)$$

where e is the electron charge and Q the electric quadrupole moment of the nucleus. Expanding $\mathscr{H}_Q$ in the principal axis system (a, b, c) of $\mathbf{Q}$ or $\mathbf{V}$, we get

$$\mathscr{H}_Q = [eQ/4I(2I - 1)][V_{cc}(3I_c^2 - I^2) + (V_{aa} - V_{bb})(I_a^2 - I_b^2)], \qquad (1.3.29)$$

which may be simplified to

$$\mathscr{H}_Q = [e^2 qQ/4I(2I - 1)][(3I_c^2 - I^2) + \eta(I_a^2 - I_b^2)] \qquad (1.3.30)$$

[17] J. P. Behr and J. M. Lehn, *Biochem. Biophys. Res. Commun.* **49**, 1573 (1972).
[18] R. E. London, T. E. Walker, D. M. Wilson, and N. A. Matwiyoff, *Chem. Phys. Lipids* **25**, 7 (1979).

by taking advantage of Tr $\mathbf{V} = 0$ and introducing $V_{cc} = eq$ and $\eta = (V_{aa} - V_{bb})/V_{cc}$.

The importance of the electric quadrupole interaction is determined by the magnitudes of Q and q. Generally, the greater the nuclear charge and the degree to which this charge distribution deviates from spherical symmetry, the larger Q is. For example, Q for ^{2}H, ^{14}N, and ^{79}Br are 0.0028, 0.016, and 0.33 respectively, in units of $e \times 10^{-24}$ cm^2. Similarly, q, the electric field gradient, measures the deviation of the electron charge distribution from spherical symmetry at the nucleus. For the ^{35}Cl anion, $q = 0$; however, when the chlorine is covalently bound, as in CH$_3$Cl, the electric field gradient has a value of $38.4 \times 10^{-21}/V$ m^2. In biological systems, the most frequently studied quadrupolar nuclei are ^{14}N, ^{2}H, ^{23}Na, and ^{35}Cl.

When a quadrupolar nucleus is placed in a magnetic field, there will be a competition between the Zeeman interaction to align the magnetic moment of the nucleus with the magnetic field and the electric quadrupole interaction to align the nuclear quadrupole with the electron cloud of the molecule. Obviously, reorientation of the electron cloud *vis-à-vis* the magnetic field via rotational tumbling of the molecule will "confuse" the nucleus, cause mixing of the nuclear spin states, broaden their energy levels, and promote nuclear spin relaxation. This mechanism is, in fact, the most important spin relaxation pathway for quadrupolar nuclei. If the rotational reorientation of the molecule is sufficiently fast, such that $\omega_0 \tau_c \ll 1$, the contribution of electric quadrupolar interaction to the relaxation rates can be shown to be

$$R_1 = R_2 = \frac{3}{40} \frac{(2I + 3)}{I^2(2I - 1)} \left(1 + \frac{\eta^2}{3}\right)\left(\frac{e^2qQ}{\hbar}\right)^2 \tau_c. \qquad (1.3.31)$$

1.3.4.4. Chemical Shift Anisotropy. In the presence of an external magnetic field, the effective field seen by a nucleus will differ from $\mathbf{H}_0$ by a small amount due to magnetic screening by the surrounding electron cloud. This screening effect arises from the electronic currents induced in the molecule by $\mathbf{H}_0$. Since the electronic environment is usually asymmetric about the nucleus, the magnitude of the shielding will depend on the orientation of the molecule relative to the external field. For solid samples this orientational dependence gives rise to the well-known "chemical shift anisotropy."

With magnetic shielding, the Zeeman Hamiltonian is modified to

$$\mathscr{H}_z = \mathbf{\mu} \cdot (1 - \mathbf{\sigma}) \cdot \mathbf{H}_0, \qquad (1.3.32)$$

where $\mathbf{\sigma}$ is the anisotropic chemical shielding tensor. In terms of the principal values of $\mathbf{\sigma}$, the effective magnetic shielding σ_z along $\mathbf{H}_0$ exhibits the following angular dependence:

$$\sigma_z = \sigma_{aa} \sin^2 \theta \cos^2 \phi + \sigma_{bb} \sin^2 \theta \sin^2 \phi + \sigma_{cc} \cos^2 \theta, \qquad (1.3.33)$$

where θ and ϕ define the polar and azimuthal angles of $\mathbf{H_0}$ relative to the principal axes of $\boldsymbol{\sigma}$.

In solution, σ_z is time dependent because of the rapid reorientation of the molecule; this modulation of the shielding field can contribute to spin–lattice and transverse relaxation. If we assume for sake of simplicity that $\boldsymbol{\sigma}$ is axially symmetric, i.e., $\sigma_{cc} = \sigma_\parallel$ and $\sigma_{aa} = \sigma_{bb} = \sigma_\perp$, and the condition of extreme narrowing holds, the relaxation rates for this mechanism are

$$R_1 = \tfrac{2}{15}\gamma^2 H_0^2 (\Delta\sigma)^2 \tau_c \qquad (1.3.34)$$

and

$$R_2 = \tfrac{7}{45}\gamma^2 H_0^2 (\Delta\sigma)^2 \tau_c, \qquad (1.3.35)$$

where $\Delta\sigma = \sigma_\parallel - \sigma_\perp$. Observations of spectral width and T_1 dependence on H_0^2 having been reported for ^{13}C, ^{19}F, and ^{31}P nuclei in a number of compounds,[19,20] indicating that anisotropic shielding can indeed contribute to nuclear relaxation in some systems. With the advent of superconducting solenoid NMR spectrometers capable of NMR observations at magnetic fields greater than 10 T, anisotropic shielding effects are likely to become more prevalent.

1.3.5. Effects of Anisotropic and Restricted Motions

1.3.5.1. Anisotropic Motions.   We have assumed in our discussion of relaxation mechanisms that the molecular motions that modulate the various interactions are rapid *and isotropic*. In studies of biological systems, however, the approximation of isotropic motion is often a gross oversimplification. More likely, we expect that molecules will undergo anisotropic motions, that is, motions which take the molecule over all directions in space with equal probability, but with quite different rates of reorientation about the various molecular axes. In less rigid systems, a part of the molecule may also undergo segmental motion at rates quite distinct from those motions of the whole system.

The characterization of a molecule undergoing anisotropic motion requires at least two rotational correlation times. For example, if we consider the motions of a rod-shaped molecule in solution, it can be seen that rotation about the long axis will be much more rapid than molecular rotations perpendicular to this axis. The inclusion of such motional details can have a profound effect on the NMR relaxation times. Although we do not treat anisotropic motion here, it is important to emphasize that, in the interpretation of relaxation data, treatments based on single correlation times are

[19] H. M. McConnell and C. H. Holm, *J. Chem. Phys.*, **25**, 1289 (1956).
[20] H. S. Gutowsky and D. E. Woessner, *Phys. Rev.* **104**, 843 (1956).

not applicable for systems where the molecules are undergoing anisotropic motions.

1.3.5.2. Restricted Motions. In studying complex biological systems one is often interested in molecules that are found in large, macromolecular assemblies and arrays and exhibit a high degree of order. In such an environment these molecules may undergo motions that are not only anisotropic but also restricted in the angular spatial orientations that can be assumed. Such motional restrictions lead to first-order dispersions in the NMR spectrum due to chemical shift anisotropy and incompletely averaged dipolar and quadrupolar interactions. While these effects are not difficult to treat, space limitations preclude the discussion of these details in the present chapter. However, we present in Section 1.5.3 an example to outline some features of the NMR of motionally restricted systems.

1.4. Experimental Methods

In this section a number of experimental techniques are described. Some of these methods were chosen because they are of general utility, while others have proven especially useful in studies of biological systems. In addition some recent advances in solid-state NMR have been included. These techniques hold great promise for studies of complex biological phenomena, such as those involving cell membranes, where motional restrictions prevent the averaging of certain spin interactions.

1.4.1. Pulse Techniques

1.4.1.1. Measurement of T_1, T_2, and T_{1_ρ}

T_1. It was previously noted that one of the most useful ways to detect magnetic resonance is by measuring the free-induction decay signal following a 90° pulse. Since T_1 characterizes the time required for the M_z component to return to equilibrium, it cannot be directly measured from the FID, which reflects only the decay of M_{xy}. However, T_1 can be determined by utilizing a two-pulse sequence. One such procedure, often referred to as the 180°-τ-90° method, calls for an initial pulse which rotates the magnetization vector **M** 180° from its position along the Zeeman axis (see Fig. 8a). Following this pulse, M_z will begin to relax back toward equilibrium along $+z$. If, however, after time τ a second pulse is applied, rotating **M** by 90° and thereby placing it in the xy plane, a signal will be induced in the receiver coil. The initial height of this signal is a measure of the magnetization along the z axis at time τ after the first pulse. By repeating this sequence for different values of

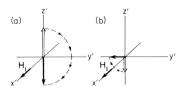

FIG. 8a. Measurement of spin–lattice relaxation by the 180°-τ-90° method: (a) H_1 is applied along x', rotating M_z to M_{-z}, (b) after a time τ, H_1 is applied long enough to rotate the net magnetization along $-y'$.[12]

τ, one obtains a measure of the rate at which M_z returns to equilibrium. Integrating

$$dM_z/d\tau = -(M_z - M_0)/T_1 \qquad (1.4.1)$$

with the initial condition $M_z = -M_0$ at $\tau = 0$ gives the following expression for the magnetization recovery:

$$M_z = M_0(1 - 2\,e^{-\tau/T_1}). \qquad (1.4.2)$$

For a heterogeneous spin system containing a number of distinct populations of resonant nuclei, the T_1 values of any particular species can be obtained by a varient of the scheme described above if their signals are spectrally resolved in the frequency spectrum. The method calls for Fourier-transforming the FID observed after the 90° pulse to obtain the so-called partially relaxed frequency spectrum at each time τ following the magnetization inversion. If the height of the ith signal in the partially relaxed spectrum is A^i_τ, then

$$\ln(A^i_\infty - A^i_\tau) = \ln(2A^i_\infty) - \tau/T^i_1, \qquad (1.4.3)$$

and its spin–lattice relaxation rate $1/T^i_1$, may be determined by plotting $\ln(A'_\infty - A^i_\tau)$ versus τ.

T_2. The transverse relaxation time T_2 is often measured directly from the linewidth of a resonance signal. This procedure assumes that the line shape of the resonance is Lorentzian, wherein T_2 is related to the linewidth at half height by $\Delta v = 1/(\pi T_2)$; and that the linewidth contribution from magnetic field inhomogeneity is not significant, a condition not normally attained in high-resolution NMR.

The intrinsic T_2 is best determined by the Carr–Purcell spin-echo techniques.[21] This method employs a 90°, τ, 180° pulse sequence and is illustrated in Fig. 8b. Following a 90° pulse, **M** is rotated onto the xy plane. After a period τ there will be a loss in the phase coherence of the magnetic moments, but if at this time a 180° pulse is applied, *inhomogeneous* T_2 processes will

[21] H. Y. Carr and E. M. Purcell, *Phys. Rev.* **94**, 630 (1954).

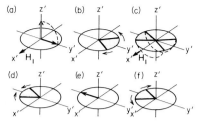

FIG. 8b. Spin-echo method in the rotating frame: (a) 90° pulse H_1 along x' rotates M_z to the y' axis, (b) magnetic field inhomogeneity causes nuclei to lose phase for a time τ, (c) 180° pulse along x' rotates dephasing magnetization into the $-y'$ half of the $x'y'$ plane, (d) nuclei rephasing towards $-y'$, (e) maximum magnetization along $-y'$, (f) dephasing of spin echo.[12]

bring the moments back in phase after another period τ. However, the maximum signal induced in the receiver coil at this time will have decayed by $e^{-2\tau/T_2}$ because of homogeneous T_2 processes. The cycle can be repeated by allowing the spins to dephase for another time τ, and applying a second 180° pulse to refocus the spins and obtain nuclear induction at the receiver coil. It is evident that the amplitude of the induced signal decays exponentially with a time constant equal to half the homogeneous T_2. For systems having more than one spin species, the peak heights in the frequency spectrum may be measured after Fourier transformation of the FID at 2τ. However, this is rarely done because of the difficulties in obtaining a properly phased spectrum, particularly from complex systems with coupled spins.

$T_{1\rho}$. Although T_1 measurements provide a convenient parameter to sample motion in biological systems, they are limited in that spin-lattice relaxation is only effected by relatively rapid motions, occurring near 10^{-8}–10^{-10} sec. Although T_2 measurements do permit one to sample slower motions, the sensitivity of the method does not extend beyond times slower than about 10^{-6} sec. However, there exists a third relaxation time $T_{1\rho}$ (referred to as T_1 in the rotating frame) which allows one to sample motions as slow as 10^4 per second.

$T_{1\rho}$ is measured by switching the phase of H_1 following a 90° pulse so that it becomes aligned with M_{xy}. $H_{eff} \cong H_1$ at resonance and the spins become locked to H_1 in the rotating frame, very much like the alignment of M along H_0 in the laboratory frame. The decay of M_{xy} under the spin-locked condition yields $T_{1\rho}$. When the spins are locked by H_1, the efficiency of relaxation will depend upon the amplitude of local fields fluctuating at γH_1 rather than γH_0. Thus by appropriate choice of H_1, relatively slow molecular motions can be sampled.

1.4.1.2. Measurement of Diffusion Coefficients. The diffusion of molecules can introduce significant errors in the measurements of T_2 because of H_0 inhomogeneity across the sample volume. However, under appropriate

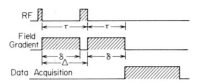

FIG. 9a. Pulse sequence and data acquisition period for a pulsed gradient experiment. Echo development occurs in the presence of a gradient which is then turned off to sample the echo decay envelope.[22]

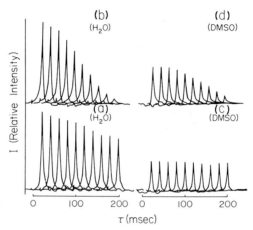

FIG. 9b. Decay of the proton signals from a dimethyl sulfoxide–H_2O (1:1) solution as a function of the echo development time τ shown in (a). Spectra were obtained in the presence [(b) and (d)] and absence [(a) and (c)] of a pulsed field gradient.[22]

conditions one can take advantage of this effect and use it to measure the molecular diffusion coefficient D (cm^2/sec). For a sample in the presence of magnetic field gradient G, the spin-echo amplitude at 2τ, $M_{(2\tau)}$, is given by

$$M_{(2\tau)} = M_0\, e^{(-2\tau/T_2) - 2/3\gamma^2 G^2 D\tau^3}. \qquad (1.4.4)$$

The self-diffusion D can be measured by comparing $M_{(2\tau)}$ determined in the presence and absence of G at various values of τ.

A significant improvement in this method can be realized by utilizing a pulsed gradient imposed on the sample following each rf pulse (see Fig. 9a). Fourier transformation of the resulting FID allows one to calculate the diffusion coefficient from any spectral peak that can be resolved (see Fig. 9b).[22]

1.4.1.3. Multiple-Pulse Methods. The angular dependence of dipolar interactions is of no consequence for molecules undergoing rapid, random reorientation since the dipolar interactions are averaged to zero by such

[22] T. L. James and G. G. McDonald, *J. Magn. Reson.* **11**, 58 (1973).

motions. However, in systems where this condition does not obtain, such as solids and membranes, this averaging does not occur and the signal is broadened. Unfortunately, this effect masks many of the weaker spin interactions, such as chemical shifts and spin–spin couplings, which also contribute to the spectrum. Since these parameters hold important information concerning molecular structure, it is of great interest to have a method which allows them to be observed.

In 1966 it was reported that dipolar effects in solids could be reduced by a modification of the Carr–Purcell pulse sequence.[23,24] Unfortunately, there was a concomitant loss in the chemical shift and spin coupling effects. However, this led Waugh and co-workers to formulate a number of multiple-pulse sequences that allow for the coherent averaging of dipolar effects in solids; these sequences permit unmasking of the weaker interactions.[25,26] To illustrate this, we will briefly discuss the initial four-pulse sequence developed by Waugh, Huber, and Haeberlen (WHH),[27] although better averaging is now possible by utilizing eight- and sixteen-pulse sequences. Details about these further developments may be found in Haeberlen,[25] Mehring,[26] and Vaughan.[28]

For dipolar broadening in a two-spin system consisting of $I = S = \frac{1}{2}$, we can disregard the terms C, D, E, and F in Eq. (1.3.8) since they serve only to induce relatively weak absorptions at frequencies ω_0 and $2\omega_0$. Rewriting the simplified Hamiltonian, we get

$$\mathscr{H}_D = [\gamma_I\gamma_s\hbar^2/2r^3(1 - 3\cos^2\theta)](3I_zS_z - \mathbf{I}\cdot\mathbf{S}). \qquad (1.4.5)$$

From this result, it is evident that $\mathscr{H}_D$ can be averaged to zero both in real and spin space. The purpose of the WHH four-pulse cycle is to average-out $\mathscr{H}_D$ through an appropriate choice of pulses so that

$$\langle 3I_zS_z - \mathbf{I}\cdot\mathbf{S}\rangle = 0. \qquad (1.4.6)$$

This averaging in spin space may be accomplished by imposing the rf sequence

$$90^\circ_x - 2\tau - 90^\circ_{-x} - \tau - 90^\circ_y - 2\tau - 90^\circ_{-y} - \tau, \qquad (1.4.7)$$

where the subscripts denote the axes along which the pulses are applied. Inasmuch as this pulse sequence over the period 6τ of the cycle aligns the magnetization along all three axes x, y, z for durations of 2τ each, then

[23] P. Mansfield and D. Ware, *Phys. Lett.* **22**, 133 (1966).
[24] E. D. Ostroff and J. S. Waugh, *Phys. Rev. Lett.* **16**, 1096 (1966).
[25] U. Haeberlen, *Adv. Magn. Reson., Suppl.* **1** (1976).
[26] M. Mehring, *NMR: Basic Princ. Prog.* **11** (1976).
[27] J. S. Waugh, L. M. Huber, and U. Haeberlen, *Phys. Rev. Lett.* **20**, 180 (1968).
[28] R. W. Vaughan, *Annu. Rev. Phys. Chem.* **29**, 397 (1978).

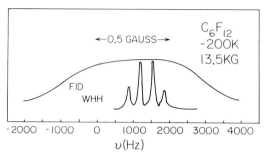

FIG. 10. Resolution of the ^{19}F multiplet from polycrystalline C_6F_{12} by application of a WHH four-pulse sequence.[26]

$I_z S_z = \frac{1}{3} \mathbf{I} \cdot \mathbf{S}$ and Eq. (1.4.6) is satisfied. Clearly, to minimize dephasing effects τ must be chosen to be much smaller than T_2.

In a typical multiple-pulse experiment, the magnetization is sampled once every cycle at a 6τ interval, and the averaged spectrum is subsequently obtained by Fourier transformation of the accumulated decay signals. We illustrate in Fig. 10 the striking effect of coherent averaging in spin space on the ^{19}F spectrum of polycrystalline C_6F_{12}.

1.4.2. Magic-Angle Spinning

In the preceding discussion, it was shown how dipolar broadening could be removed by coherent averaging of the spin operators in spin space. In this section we describe the technique of magic-angle spinning, which averages the lattice operators in real space.

The spin Hamiltonian for an ensemble of interacting nuclei in a solid undergoing rapid rotation about an axis can be conveniently written as

$$\mathscr{H} = \overline{\mathscr{H}} + \mathscr{H}'(t), \tag{1.4.8}$$

where $\overline{\mathscr{H}}$ is the time-averaged value and $\mathscr{H}'(t)$ is the time-dependent Hamiltonian having a mean value of zero. If the axis of sample rotation forms angle β with $\mathbf{H}_0$, the time-averaged dipolar contribution to this system is

$$\frac{\overline{\mathscr{H}}_D}{h} = \frac{1}{2} (3 \cos^2 \beta - 1) \sum_{i<j} \frac{h}{8\pi^2} \gamma_i \gamma_j \mathbf{r}_{ij}^{-3} (3 \cos^2 \beta'_{ij} - 1)(\mathbf{I}_i \cdot \mathbf{I}_j - 3 I_{iz} I_{jz}). \tag{1.4.9}$$

It can be seen from this equation that when β is set at the "magic angle" of $54°44'$ the value of $\overline{\mathscr{H}}_D$ goes to zero. Similarly, for the time-dependent

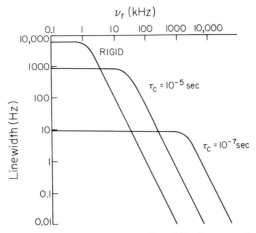

FIG. 11. Effect of magic-angle spinning rate v_r on linewidths from samples having correlation times of ∞, 10^{-5} sec, and 10^{-7} sec. Reprinted from Waugh,[30] p. 207, by courtesy of Marcel Dekker, Inc.

dipolar term

$$\frac{\mathscr{H}'_D(t)}{h} = \sum_{i<j} \frac{3h}{16\pi^2} \gamma_i\gamma_j r_{ij}^{-3}(\mathbf{I}_i \cdot \mathbf{I}_j - 3I_{iz}I_{jz})$$

$$\times [\sin 2\beta \sin 2\beta'_{ij} \cos(\omega_r t + \phi_{ij}) + \sin^2 \beta \sin^2 \beta'_{ij} \cos 2(\omega_r t + \phi_{ij})], \quad (1.4.10)$$

the value will average out to zero when the frequency of sample rotation ω_r is greater than the linewidth Δ of the rigid lattice. In both expressions, β'_{ij} is the angle between internuclear vector r_{ij} and the axis of sample rotation.

Thus, under appropriate conditions of sample spinning, one can effectively remove dipolar (as well as quadrupolar) interactions from the spectrum of a solid. Rapid spinning also removes the chemical shift anisotropy, resulting in the loss of information about these tensors. Accordingly, magic-angle spinning experiments result in high-resolution spectra such as those observed in liquids and solutions.[29]

At the present time, magic-angle spinning has not received the attention it deserves in NMR studies of biological systems. The major reason is probably that biological systems always display some motion. If the frequency of such motions is on the order of $1/\tau_c$, then it can be shown[30] that, in order to obtain further narrowing by sample spinning, the condition $\omega_r > 1/\tau_c > \Delta$

[29] E. R. Andrew, *Prog. Nucl. Magn. Reson.* **8**, 1 (1971).
[30] J. S. Waugh, *in* "NMR and Biochemistry" (S. J. Opella and P. Lu, eds.), p. 203. Dekker, New York, 1979.

must be met. This effect is illustrated in Fig. 11, where the theoretical line-width for a dipolar broadenined sample is depicted as a function of sample spinning in the presence and absence of internal motions. Of course, for systems displaying $1/\tau_c \approx \Delta$, such as a phospholipid multilayer, a significant reduction in linewidths may be observed when samples are spun at the magic angle (see Fig. 12).[31]

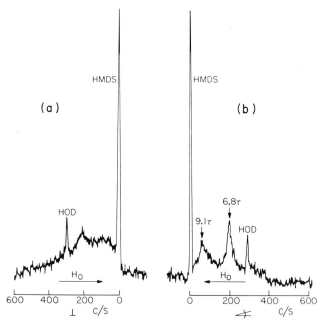

FIG. 12. Proton NMR spectrum of a sample containing 33 wt. % egg yolk lecithin in D_2O at 25° and rotating at 500 Hz about an axis perpendicular (a) or at the magic angle (b) to the external field $\mathbf{H_0}$.[31] The resolved peaks in (b) at 9.1τ and 6.6τ are assigned to the fatty acid terminal methyl group protons and the choline methyl group protons, respectively. HMDS is an external reference.

1.4.3. Double Resonance

Numerous methods have been devised to take advantage of the coupling interactions (dipolar and scalar) that exist between nuclei. Double-resonance techniques employ a second rf field which acts on one of the coupled nuclei while the other is being sampled. Under appropriate conditions these methods can remove the spin–spin splitting between scalar coupled nuclei (spin decoupling), change the spin-state distribution for one type of spin by

[31] D. Chapman, E. Oldfield, D. Doskocilova, and B. Schneider, *FEBS Lett.* **25**, 261 (1972).

equalizing the spin-state populations of a coupled species (saturation transfer or nuclear Overhauser effects, and increase the signal intensity of a dilute spin species by transfer of polarization from an abundant species (cross polarization). In this section we briefly describe these techniques and illustrate their usefulness.

1.4.3.1. Spin Decoupling. In our discussion of spin–spin coupling above we noted that the signal from one coupled nucleus will be split into a multiplet because of the different spatial orientations of the other nuclear magnetic moment. The multiplet is observed only because the rate at which transitions occur in the coupled nucleus causing the splitting is slow. If, however, we impose a second rf field H_2 at the resonance frequency of the coupled nucleus such that $\gamma H_2 \gg \pi J$, and simultaneously sample the species of interest, we now observe only a single peak at the center of the multiplet of the undecoupled spectrum. What has happened is that the second rf field induces rapid transitions between the various Zeeman states of the coupled nucleus which cause the splitting to begin with, and the spin–spin splitting is collapsed.

The usefulness of spin decoupling is apparent. Since the area under the spectral peaks is proportional to the number of nuclei, the collapse of spin–spin splitting can result in considerable signal enhancement. In addition, spin decoupling may be used to simplify complex spectra, facilitating spectral assignments. An interesting example is presented in Fig. 13. Here the ^{13}C spectrum for the aromatic region of tryptophan is complicated by interference between signals from proton-bonded and nonprotonated carbon atoms.[32] The peaks from the proton-bonded carbons all appear as singlets due to proton decoupling. However the linewidths $\Delta \nu$ of these carbons are all slightly broadened due to the strong J coupling (≥ 120 Hz). Since $\Delta \nu \propto J^2/H_2^2$, any proton-bonded carbon resonance can be effectively broadened by reducing H_2, for example by setting H_2 off resonance from the proton frequency. The effect of this incomplete proton decoupling is shown in Fig. 13b. Under these conditions the peaks arising from nonprotonated carbon atoms, which experience only weak long-range coupling (≤ 15 Hz), are easily resolved and unambiguously assigned.

1.4.3.2. Saturation Transfer. The technique of double resonance can be taken to great advantage when two spins are dipole coupled. This is because the dipolar interaction provides a mechanism for transferring magnetization from one spin species to the other. Unlike spin decoupling, where a relatively weak rf field is used to excite one of the scalar-coupled nuclei, the magnetization transfer experiment employs a second field which is strong enough to equalize the spin populations (saturate) of the irradiated species. Since the

[32] A. Allerhand, R. F. Childers, and E. Oldfield, *Biochemistry* **12**, 1335 (1973).

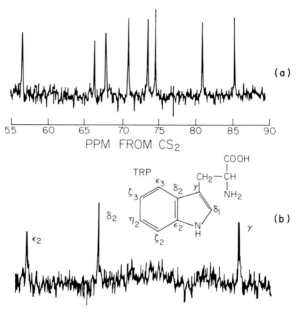

FIG. 13. (a) Fully proton-decoupled ^{13}C NMR spectrum of the aromatic resonances of tryptophan in aqueous solution. (b) Off-resonance proton-decoupled spectrum of tryptophan. Reprinted with permission from Allerhand et al.[32] Copyright © 1973 American Chemical Society.

Zeeman states of the two nuclei are coupled by dipolar interactions, the resulting increase in the relaxation rate of the saturated spins, brought about by the equalized populations, will cause a redistribution in the Zeeman populations of the nonirradiated species. This effect, called the nuclear Overhauser effect (NOE), is manifested by a change in the intensity of the NMR signal from the unsaturated spins when the latter are sampled.[33]

For the sake of discussion we shall consider a simple two-spin system consisting of $I = S = \frac{1}{2}$. The coupled Zeeman levels are shown in Fig. 14, where α and β denote the high- and low-energy spin orientations, respectively. Also shown are the transition probabilities W_i, which couple the spin states to the lattice degrees of freedom, as modulated by the dominant spin interactions. Under these conditions, the equation of motion for the observed I spin magnetization $\langle I_z \rangle$ is described by[34]

$$d\langle I_z \rangle / dt = -(W_0 + 2W_{1I} + W_2)(\langle I_z \rangle - I_0) - (W_2 - W_0)(\langle S_z \rangle - S_0);$$

$$(1.4.11)$$

[33] J. H. Noggle and R. E. Schirmer, "The Nuclear Overhauser Effect." Academic Press, New York, 1971.
[34] D. Doddrell, V. Glushko, and A. Allerhand, J. Chem. Phys. **56**, 3683 (1972).

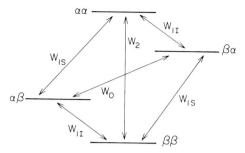

FIG. 14. Energy level diagram for coupled spins $I = S = \frac{1}{2}$. Transition probabilities between α and β (high and low) energy levels are denoted W, with subscripts 0, 1, or 2 representing the quantum of energy absorbed or emitted by I or S or both.

I_0 and S_0 are the equilibrium values for $\langle I_z \rangle$ and $\langle S_z \rangle$. If we now impose a saturating rf field at the S spin frequency, so that $\langle S_z \rangle = 0$, Eq. (1.4.11) becomes

$$d\langle I_z \rangle/dt = -(W_0 + 2W_{1I} + W_2)[\langle I_z \rangle - I_0(1 + f_I(S))]. \qquad (1.4.12)$$

The effect of saturation is therefore to change the steady-state value of $\langle I_z \rangle$ from its value at thermal equilibrium by the factor $1 + f_I(S)$, where

$$f_I(S) = [(W_2 - W_0)/(W_0 + 2W_{1I} + W_2)]\gamma_S/\gamma_I. \qquad (1.4.13)$$

For $I = S = \frac{1}{2}$ the transition probabilities are

$$W_0 = \tfrac{1}{20}K^2 J^{(0)}(\omega_S - \omega_I), \qquad W_{1I} = \tfrac{3}{40}K^2 J^{(1)}(\omega_I),$$
$$W_2 = \tfrac{3}{10}K^2 J^{(2)}(\omega_I + \omega_S), \qquad (1.4.14)$$

where $K = \hbar\gamma_S\gamma_I$ and $J^{(i)}(\omega)$ are the spectral density functions discussed in Section 1.3.4.1. From these relationships, it can be seen that the net NOE is determined not only by the relative signs and magnitudes of the nuclear gyromagnetic ratios, but also by the relative probabilities of the various relaxation pathways. Thus for dipolar relaxation of small molecules undergoing rapid motions, $W_2 \gg W_0$ and the fractional enhancement reduces to $\gamma_S/2\gamma_I$. However, the NOE is also influenced by the onset of nuclear spin relaxation mechanisms other than dipole–dipole interactions. For nuclei exhibiting a significant scalar contribution to the relaxation rate, the zero quantum transition probability will also be important and the NOE modified accordingly. In fact, because of the dependence of $f_I(S)$ on the relaxation mechanism, Overhauser effects have been used to sort out the relative relaxation contributions in several well-defined systems.[35]

[35] K. F. Kuhlmann, D. M. Grant, and R. K. Harris, J. Chem. Phys. 52, 3439 (1970).

The interpretation of saturation transfer experiments on biological molecules is complicated by both the size and intrinsic complexity of the macromolecules involved. For this reason, these studies usually take advantage of the fact that spin interactions are distance dependent; any measurable NOE can then be taken to infer that two nuclei are in reasonably close proximity. The usefulness of this approach is illustrated by the work of Balaram et al.,[36-38] who used proton NMR to characterize the hormone binding site on the protein neurophysin II. Based on preliminary studies showing a differential broadening of certain aromatic protons in position 2 or a tripeptide hormone analog, these workers attempted to further investigate the binding interaction by measuring the proton NOE of the peptide while irradiating chemically shifted protons on the protein. Their data, obtained by monitoring the aromatic protons of the peptides S-Me-Cys-Phe-Ile-NH_2, show a decrease for the resonance intensities of the 2, 3, and 4 phenyl group protons in the presence of a saturating rf field at a frequency corresponding to the ortho ring protons on the lone tyrosine of neurophyrin II. This result unambiguously places the protein tyrosine residue at the hormone binding site and further suggests that a specific interaction occurs with the peptide phenyl group.

The negative Overhauser effect observed in the above hormone/neurophyrin II study is due to the long correlation times of the protons associated with the protein. For a system of two dipolar-coupled protons, where $(\omega_I - \omega_S)^2 \tau_c^2 \ll 1$, the dependence of the fractional enhancement on τ_c is given by

$$f_I(S) = (5 + \omega^2 \tau_c^2 - 4\omega^4 \tau_c^4)/(10 + 23\omega^2 \tau_c^2 + 4\omega^4 \tau_c^4), \quad (1.4.15)$$

which for rapid motions ($\omega_I \tau_c \ll 1$) results in $f_I(S) = 0.5$, but for long correlation times, $f_I(S) = -1$. The physical basis for the negative Overhauser effect associated with slow motions is the predominance of W_0 over W_2.

1.4.3.3. Cross Polarization of Dilute Spins. In 1962 Hartmann and Hahn[39] proposed a double-resonance method for observing the magnetic properties of an isotopically or chemically rare spin S coupled to an abundant spin species I. In this experiment the abundant spins are irradiated with an rf field of amplitude H_1^I at the Larmor frequency of I. These spins are then brought into contact with the rare S spin system by imposing a second field H_1^S at the S frequency. Under these conditions both spin magnetizations may

[36] P. Balaram, A. A. Bothner-By, and J. Dadok, J. Am. Chem. Soc. 94, 4015 (1972).

[37] P. Balaram, A. A. Bothner-By, and E. Breslow, J. Am. Chem. Soc. 94, 4017 (1972).

[38] P. Balaram, A. A. Bothner-By, and E. Breslow, Biochemistry 12, 4695 (1973).

[39] S. R. Hartmann and E. L. Hahn, Phys. Rev. 128, 2042 (1962).

be considered as vectors in the xy planes of their respective rotating frames, precessing about the imposed $\mathbf{H}_1$ fields with frequencies

$$\Omega_I = \gamma_I H_1^I \qquad (1.4.16)$$

and

$$\Omega_S = \gamma_S H_1^S. \qquad (1.4.17)$$

By adjusting the amplitudes of the $\mathbf{H}_1$ fields so that

$$\gamma_I H_1^I = \gamma_S H_1^S, \qquad (1.4.18)$$

one can effectively match the Zeeman splittings of the two nuclei along their respective $\mathbf{H}_1$ fields. In the laboratory frame, of course, the interactions of the xy components of the precessing spin magnetizations are negligible, since $\gamma_I \mathbf{H}_0 \neq \gamma_S \mathbf{H}_0$. However, the z components, which are generated as the spins precess about their respective $\mathbf{H}_1$ fields in the rotating frame, will have the same frequency. This allows energy exchange to occur through the z components of the dipolar interaction, as described in Eq. (1.3.8a). As the two spin systems reach internal equilibrium, there will be a net transfer of magnetic polarization from the abundant I spins to the rare S spin system. This exchange, however, only results in a small decrease in the I magnetization. A further decrease in I magnetization occurs if the S spins are now allowed to reach equilibrium with the lattice (for example by turning off $\mathbf{H}_1^S$) and then are brought back into contact with the I spins. Repetition of this cycle ultimately results in a measurable decrease in the I magnetization, thus allowing one to sample the rare S spins indirectly.

Although the advent of FT-NMR and signal averaging techniques have alleviated many of the difficulties associated with investigations of rare spins, the observation of a dilute spin signal still presents difficulties for solid samples, where dipolar interactions are not averaged out. Similarly, for studies of labile systems or transient effects, the use of lengthy signal averaging is impractical. In an effort to mitigate this problem a technique based on the aforementioned Hartmann–Hahn experiment was introduced in 1973 by Pines, Gibby, and Waugh.[40] In this experiment, the abundant I spins are polarized and brought into contact with the S spin system, as previously described. However, in this case, it is the rare S spins that are sampled directly. Early applications of the technique utilized isotopically abundant protons (99.98 %) to enhance the signal from relatively rare spins, such as ^{13}C (1.1 %).

---

[40] A. Pines, M. G. Gibby, and J. S. Waugh, *J. Chem. Phys.* **59**, 569 (1973).

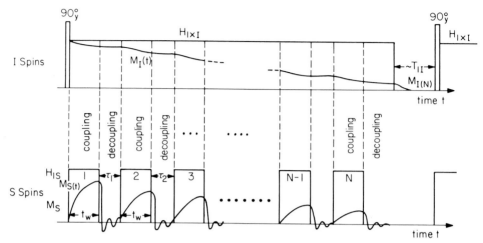

FIG. 15. Pulse sequence for a cross polarization experiment in which the magnetization from the abundant spins (M_I) is transferred to the rare spins (M_S). The FID of M_S is sampled during the periods denoted by τ_N.[26]

As a result, this method is commonly known as proton-enhanced nuclear induction spectroscopy.

One version of the proton-enhanced nuclear induction spectroscopy experiment is shown in Fig. 15. Following a 90° pulse, the I magnetization is locked along H_1^I so that it decays in a time $T_{1\rho}^I$. The two systems are next brought into contact so that energy can be exchanged via dipolar cross relaxation. This is done, as in the Hartmann–Hahn experiment, by imposing H_1^S such that the conditions of Eq. (1.4.18) are satisfied. During this period, denoted t_w in Fig. 15, the energy conserved from the longitudinal relaxation (in the rotating frame) of the I spins causes an increase in the S spin magnetization along H_1^S in the rotating frame. The rare spins are then sampled by turning H_1^S off and recording the FID while continuously irradiating the abundant I spins in order to decouple the remaining dipolar interaction. This cycle is then repeated until the rotating frame magnetization of the I spin system becomes depleted. Finally, the accumulated FIDs are averaged and transformed to yield a high-resolution spectrum of the rare S spins.

A recent application of this technique included measurement of the angular dependence of the ^{31}P chemical shift for oriented phospholipid bilayers.[41] In this experiment the enhancement by cross polarization was necessary because of the small amount of sample (about 4 μmole PO_4) that could be oriented between glass slides. Only in this way was it possible to define the chemical shift tensor in relation to the plane of the membrane.

[41] R. G. Griffin, L. Powers, and P. S. Pershan, *Biochemistry* **17**, 2718 (1978).

1.4.4. Two-Dimensional FT-NMR

Many of the difficulties that arise in NMR spectroscopy occur because the method characterizes energy absorption as a dispersion along a single variable, either time or frequency. However, as we have noted, a sample can actually be characterized by many useful variables which act on chemically identical nuclei to split and spread their signal in frequency space. Most of the techniques that have been developed to simplify complex spectra do so by suppressing some of these effects; thus much important information is lost. Another approach is to disperse the signal along different dimensions in frequency space, each direction displaying the effect of a single variable (e.g., chemical shift, spin–spin coupling). This is the rationale for the development of multidimensional NMR spectroscopy,[42] a new method which holds great promise for unraveling the complex spectra of biological macromolecules. In its simplest form, two-dimensional (2D) NMR, spectra are obtained by Fourier transformation of a signal possessing more than one independent time variable. For the sake of illustration, we consider the case in which a complex spectrum is resolved by effectively spreading it in two dimensions, based on the spin–spin coupling and chemical shift (a technique referred to as 2D, J-resolved spectroscopy[43,44]).

Although the mathematical development of this method is beyond the scope of this chapter (see Ref. 42), the basis for it can be outlined. The experimental scheme for obtaining a 2D, J-resolved spectrum, shown in Fig. 16,

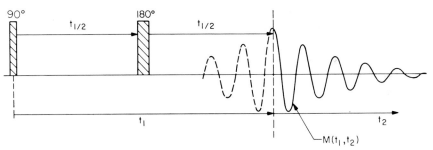

FIG. 16. Pulse sequence for obtaining a 2D, J-resolved spectrum.[43]

is simply the spin-echo method repeated at different values of t_1, where t_1 is the echo development time. If the coupling constants J are small relative to the chemical shifts between coupled nuclei, then the spin-echo amplitude will be a function of J.[45] This is due to the fact that dephasing effects from

[42] W. P. Aue, E. Bartholdi, and R. R. Ernst, *J. Chem. Phys.* **64**, 2229 (1976).
[43] W. P. Aue, J. Karhan, and R. R. Ernst, *J. Chem. Phys.* **64**, 4226 (1976).
[44] K. Nagayama, P. Bachman, K. Wüthrich, and R. R. Ernst, *J. Magn. Reson.* **31**, 133 (1978).
[45] E. L. Hahn and D. E. Maxwell, *Phys. Rev.* **88**, 1070 (1952).

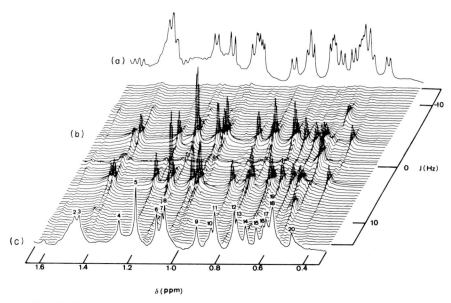

FIG. 17. (a) Standard proton NMR spectrum of bovine pancreatic trypsin inhibitor in the high-field region, displaying the unresolved multiplets arising from 19 of the 20 methyl groups present. (b) 2D, J-resolved spectrum of the same sample with the individual multiplets displayed along the J axis and the chemically shifted, decoupled peaks along δ, projected in trace (c).[44]

chemical shifts and magnetic field inhomogeneity are reversed following the 180° pulse. Thus by changing the spin-echo development time one can obtain a measure of the spin–spin couplings in the absence of chemical shifts. The decaying echo signal is then sampled for time t_2, which is a running variable of constant length. Repetition of this experiment using different values for the parameter t_1 results in a data matrix of the transverse magnetization components as a function of t_1 and t_2. The contribution each resonance line k in the multiplet of nucleus j makes to the matrix is given by

$$M_{jk}(t_1, t_2) = M_{jk}(0, 0) \cos(v_{jk} t_1 - \omega_{jk} t_2) e^{-t_1/T_{2jk} - t_2/T_{2jk}^*}. \quad (1.4.19)$$

This data matrix can then be Fourier transformed in two dimensions to give a corresponding matrix in frequency space. The contribution arising from the resonance described in Eq. (1.4.19) can be shown to be

$$|S|_{jk}(\omega_1, \omega_2) = \tfrac{1}{2} M_{jk}(0, 0) [1/T_{2jk}^2 + (\omega_1 - v_{jk})^2]^{-1/2} \\ \times [1/T_{2jk}^{*2} + (\omega_2 - \omega_{jk})^2]^{-1/2}, \quad (1.4.20)$$

where

$$\omega_{jk} = \Omega_j + v_{jk} \quad \text{and} \quad v_{jk} = 2\pi \sum_l J_{jl} m_{lk};$$

ω_j is the Larmor frequency of j, J_{jl} is the spin coupling constant, and m_{lk} are the magnetic quantum numbers of the coupled nucleus l. Thus by displaying $S(\omega_1, \omega_2)$ along the ω_2 axis one obtains the chemical shift information, while the spectrum projected along ω_1 shows the spin coupling multiplets.

The effectiveness of this technique in resolving complex spectra is illustrated in Fig. 17, where the proton spectrum of bovine pancreatic trypsin inhibitor is shown. In this case the one-dimensional spectrum contains overlapping spin multiplets from nineteen chemically shifted methyl resonances. However, in two dimensions the data are well resolved, allowing investigators to directly observe the effects of spin decoupling on each resonance.[46] This type of information can be used for the unambiguous assignment of resonances from similar groups located in different parts of the macromolecule.

1.5. Selected Studies on Biological Systems

In this section we discuss a limited number of published studies on biological applications of NMR in order to illustrate how the technique can be uniquely useful for investigating biological systems. Unfortunately, many excellent and important examples must necessarily be omitted in this limited treatise. It is hoped, however, that the few described here will stimulate researchers to seek out and develop other applications in order to capitalize on the vast wealth of knowledge that NMR techniques hold for biology. We have also decided to forgo any discussion of NMR zeugmatography since NMR imaging studies are highly technical and require specialized equipment. Readers wishing to learn more of this most interesting application of NMR to medical imaging are referred to a recent paper by Hoult and Lauterbur[47] and the literature cited therein.

1.5.1. Proteins: Probing the Active Site of an Enzyme

Practically every chemical reaction occurring within a cell is catalyzed by an enzyme. For many years biological chemists have worked at unraveling the molecular mechanisms responsible for enzyme catalysis. In these efforts much information has been derived from the investigation of enzyme reaction rates and substrate binding constants. More recent studies employing both x-ray crystallography and NMR have also allowed the definition

[46] K. Nagayama, P. Bachman, R. R. Ernst, and K. Wüthrich, *Biochem. Biophys. Res. Commun.* **86**, 218 (1979).

[47] D. I. Hoult and P. C. Lauterbur, *J. Magn. Reson.* **34**, 425 (1979).

of the active-site residues of enzymes and the orientation of the bound substrates. While both x-ray crystallography and NMR provide detailed information at the molecular level, the NMR approach has an additional advantage in that it can be done under physiological conditions, with enzymes and substrates remaining in solution.

In the following section we describe two approaches for using NMR to study enzyme mechanisms. The first technique allows one to characterize the protein residues at the active site, whereas the second approach can be used to reveal the orientation of the substrate when it is bound to the active site of the enzyme. Of course, a host of other NMR techniques have been successfully applied to the study of proteins. For discussions of these methods, the reader should consult the biological NMR texts cited in Chapter 1.1.

1.5.1.1. Isotopic Labeling. Two of the major problems associated with using NMR techniques in biology are the low natural abundance of many nuclei of interest and the difficulty of assigning resonances in a complex spectrum. In some cases, both of these problems can be obviated by incorporating labeled precursor molecules into a biosynthetic system which produces the macromolecule to be studied. This approach has been used to investigate α-lytic protease isolated from cultures of *Myxobacter*.[48,49] NMR studies of this enzyme have recently been reported using histidine enriched in either ^{13}C at the C_2 ring position[48] or ^{15}N for the imidazole nitrogens.[49] Since ^{13}C and ^{15}N are normally present in low isotopic abundance (1.11% and 0.365%, respectively), specific enrichment allows for greater sensitivity. In addition, because this enzyme has only one histidine residue, the enrichment provides an unambiguous assignment of the resonances arising from this residue.

The mechanism of action of α-lytic protease is of interest because it is representative of a ubiquitous group of enzymes called serine proteases. In addition to their similar functions, these proteases share a common type of active site, containing the same triad of amino acid residues (aspartic acid, histidine, and serine) in close proximity. Two schemes have been proposed for the involvement of this triad in the proteolytic cleavage (Fig. 18a). Based on the pH dependence of the enzymatic activity (maximum at ~ 6.9), the charge-relay mechanism depicted in Scheme I is possible only if the pK_a of the aspartyl residue is unusually high (~ 6.9) and then only if the histidyl residue has an unusually low pK_a value. On the other hand, if Scheme II is operative, the pK_a for the histidyl group should be normal, ~ 6.9. In the elegant work of Bachovchin and Roberts,[49] these investigators exploited the sensitivity of the ^{15}N chemical shift of the imidazole ^{15}N

[48] M. W. Hunkapiller, S. H. Smallcombe, D. R. Whitaker, and J. H. Richards, *Biochemistry* **12**, 4732 (1973).

[49] W. W. Bachovchin and J. D. Roberts, *J. Am. Chem. Soc.* **100**, 8041 (1978).

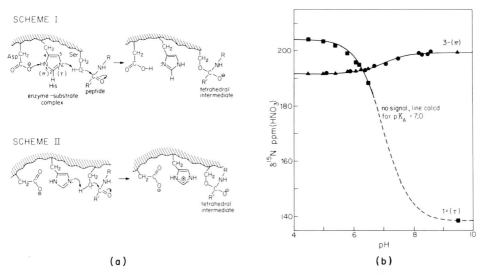

(a)   (b)

FIG. 18. (a) Proposed mechanisms for the catalytic activity of the *serine protease* family of enzymes. See text for details. (b) Effect of pH on the ^{15}N chemical shifts of ^{15}N-enriched histidine nitrogens in α-lytic protease: ●, ^{15}N-enriched at N3; ■ and ▲, enriched at N1 and N3, respectively. Reprinted with permission from Bachovchin and Roberts.[49] Copyright © 1978 American Chemical Society.

resonances to their state of protonation[50] to examine the pH titration behavior of this residue. The results of these ^{15}N experiments, shown in Fig. 18b, indicate that the histidyl residue does have a $pK_a \sim 7$, thus disproving the popular charge-relay mechanism.

1.5.1.2. Paramagnetic Probes. The approach outlined in this section requires that an unpaired electron, from either a paramagnetic ion or a spin label, be present at the active site where the substrate binds, as well as the extensive application of NMR relaxation theory[51,52] for paramagnetic systems. Because of the large magnetic moment associated with unpaired electrons (on the order of 10^3 times larger than that for protons) the relaxation for any ligand nucleus with $I = \frac{1}{2}$ at the active site will be dominated by the paramagnetic interactions. Two interactions can contribute to the nuclear spin relaxation: (a) through-space magnetic dipolar interaction between the nuclear spin and electron spin; and (b) isotropic nuclear hyperfine coupling of the unpaired electron at the nucleus under consideration. The longitudinal relaxation rate is primarily determined by five parameters: (i) the ratio of the concentration of paramagnetic species to ligand (f), (ii) the binding stoichiometry or coordination number of the electron-nuclear complex (q), (iii) the

[50] F. Blomberg, W. Maurer, and H. Ruterjans, *J. Am. Chem. Soc.* **99**, 8149 (1977).

[51] Z. Luz and S. Meiboom, *J. Chem. Phys.* **40**, 2686 (1964).

[52] T. J. Swift and R. E. Connick, *J. Chem. Phys.* **37**, 307 (1962).

lifetime of binding at the active site (τ_m), (iv) the correlation time modulating the dipolar interaction between the nuclear and electron spins (τ_c), and (v) the distance between the paramagnetic center and the nucleus (r). In this case the longitudinal relaxation rate has been shown to be

$$\frac{1}{T_{1M}} = \frac{2}{15} \frac{S(S+1)\gamma_I^2 g^2 \beta^2}{r^6} \left(\frac{3\tau_c}{1 + \omega_I^2 \tau_c^2} + \frac{7\tau_c}{1 + \omega_s^2 \tau_c^2} \right)$$
$$+ \frac{2}{3} \frac{S(S+1)A^2}{\hbar^2} \frac{\tau_e}{1 + \omega_s^2 \tau_e^2}. \tag{1.5.1}$$

Here S is the electron spin, g is the electronic "g" factor, β is the Bohr magneton, A is the nuclear hyperfine coupling constant, and τ_e is the effective correlation time for the scalar interaction. For an enzyme-bound paramagnetic center, τ_e is usually large enough so that $\omega_s^2 \tau_e^2 \gg 1$ and the contribution from modulation of scalar nuclear hyperfine interaction in Eq. (1.5.1) may be neglected. The nuclear spin relaxation is then dominated by magnetic dipolar interaction between the nuclear spin and the paramagnetic center.

In a typical experiment, one does not measure T_{1M} directly. Instead, one determines $1/T_{1p}$, the paramagnetic contribution to the longitudinal relaxation rate, from the difference between the relaxation rate measured in the presence and absence of the paramagnetic probe and takes advantage of the following relationship between T_{1p} and T_{1M}:

$$1/f T_{1p} = q/(T_{1M} + \tau_M). \tag{1.5.2}$$

Of course, τ_M must still be known; however, frequently $T_{1M} \gg \tau_M$, or the experiments can be done at temperatures where this condition is fulfilled.

The dipolar contribution to T_{1M} provides a means of determining r, the distance between the nuclear spin and the paramagnetic center, assuming that τ_c can be obtained. The value of τ_c is probably best determined by measuring T_{1p} at two or more frequencies.[53] Note that the ratio of T_{1p} at two frequencies is a function of τ_c only, provided that the dipolar mechanism dominates the relaxation. This procedure may be repeated for several substrate nuclei, and in this manner the orientation of the substrate relative to the paramagnetic probe at the active site of the enzyme can be ascertained.

1.5.2. Nucleic Acids: The Dynamic Structure of tRNA

Transfer RNAs (tRNAs), unlike proteins, all display a similar structure since they all perform an identical function in the process of translating messenger RNA. The general structure of tRNA is illustrated in Fig. 19a.

[53] A. S. Mildvan and R. K. Gupta, *Methods Enzymol.* **44G**, 322 (1978).

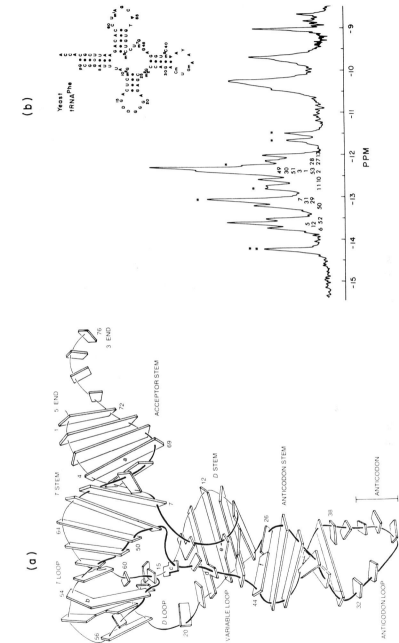

(a)

(b)

Yeast
tRNA^Phe

Fig. 19. (a) Schematic representation of the three-dimensional structure of tRNA. Reprinted from A. Rich and S. H. Kim, *Sci. Am.* **238**, 52 (1978). Copyright © 1978 by Scientific American, Inc. All rights reserved. (b) 360-MHz proton NMR spectrum of yeast tRNA^Phe (Reprinted from Reid *et al.*,[54] p. 100, by courtesy of Marcel Dekker, Inc.) The numbered assignments for the cloverleaf (secondary) base pairs correspond to the diagram (inset), while the resonances from the tertiary base pairs are marked with asterisks.

It is an L-shaped molecule with an acceptor site for binding the amino acid to be transcribed to the polypeptide during protein synthesis and a recognition site (in the anticodon loop) for binding the complex to the appropriate sequence of messenger RNA. The cross rungs represent hydrogen bonds between complementary base pairs. These hydrogen bonds resonate at very low fields in the ^{1}H NMR spectrum (-11 to -15 ppm from DSS). However, since the nucleic acid bases consist of unsaturated ring molecules which can induce ring-current shifts in adjacent spins, the hydrogen-bonded protons of a particular kind are spectrally dispersed according to base proximity relationships determined by the sequence. Accordingly, the ^{1}H NMR spectrum provides a fingerprint for the secondary structure of the tRNA, although features characteristic of the tertiary folding are often discerned[54] (see Fig. 19b).

The assignment of the individual hydrogen-bond resonances in the low-field spectrum has received considerable attention in recent years. Generally these studies have confirmed the cloverleaf model for the secondary structure of tRNAs. Some important insights into the tertiary folding of these molecules in solution have also emerged. In addition, there have been attempts to measure the rates of proton exchange between tRNA hydrogen bonds at specific sites and the solvent. These proton exchange rates can be an extremely sensitive probe of conformational fluctuations in tRNAs and offer some insights into the pathway of conformational changes that might be followed in tRNA binding interactions in solution or on the ribosome surface. Initial studies have utilized a technique called saturation recovery to measure these hydrogen-bond proton exchange rates.[55] In this method, a saturating rf field is imposed at a specific frequency corresponding to the resonance of the specific hydrogen bond to be sampled. The signal intensity is then measured from the frequency spectrum taken after the saturating field has been turned off for various time intervals τ. By varying τ, the exponential recovery of the saturated signal can be followed. The rate of this recovery measures the rate of replacement of the hydrogen-bonded proton by the unsaturated water proton, provided, of course, that the rate of exchange is greater than the intrinsic longitudinal relaxation rate of the resonance monitored. For hydrogen-bonded protons which exchange very slowly compared with T_1, the rate may be determined from deuterium exchange studies, following the decrease in signal intensity when D_2O is substituted for H_2O in the medium.[56]

[54] B. R. Reid, E. Azhderian, and R. E. Hurd, in "NMR and Biochemistry," (S. J. Opella and P. Lu, eds.), p. 91. Dekker, New York, 1979.

[55] P. D. Johnston and A. G. Redfield, Nucleic Acids Res. 4, 3599 (1977).

[56] C. K. Woodward and B. D. Hilton, Annu. Rev. Biophys. Bioeng. 8, 99 (1979).

1.5.3. Membranes: The Restricted Motions of Lipid Chains

The important role played by cell membranes in controlling biological functions is firmly established. They serve as highly specialized barriers, not only between the cytoplasm and the extracellular medium, but also in compartmentalizing many activities within the cell (e.g., nuclear and lysosomal membranes) while providing a highly specialized surface to promote certain biological reactions. The current dogma of membrane structure is the fluid mosaic model, with proteins intercalated into a lipid bilayer matrix.[57] This model raises many questions about the role of lipids in cell structure, particularly the effects which lipids might exert on the proteins and vice versa.

To answer some of these questions, there has been considerable interest in the structure and motional state of phospholipids in bilayer membranes, in order to provide benchmark information on this system. A number of NMR techniques have been applied to probe the orientational order and the molecular mobility of the phospholipid molecules in bilayers. Of these, the most informative has been deuterium NMR studies on phospholipids with perdeuterated or specifically deuterated methyl and methylene groups. These ^2H NMR studies take advantage of the quadrupolar interaction of ^2H and the manner in which these interactions are manifested in the NMR spectrum of a motionally restricted system.

The energy levels of the combined Zeeman–quadrupolar Hamiltonian, when the Zeeman interaction dominates over the quadrupolar interaction, are given by

$$E_{(m_I)} = -\gamma\hbar H_0 m_I + [e^2qQ/8I(2I-1)][3m_I^2 - I(I+1)] \\ \times [(3\cos^2\theta - 1) + \eta\sin^2\theta\cos 2\phi], \qquad (1.5.3)$$

where the angles θ and ϕ define the orientation of the major and minor axes of the electric field gradient tensor relative to the external field. The results for the case of ^2H ($I = 1$) are shown in Fig. 20. The resultant NMR spectrum can be seen to be a doublet displaced symmetrically about the Zeeman frequency by

$$v - v_0 = \pm\tfrac{3}{8}(e^2qQ/h)[(3\cos^2\theta - 1) + \eta\sin^2\theta\cos 2\phi]. \qquad (1.5.4)$$

For a C—D bond in an alkyl chain, e^2qQ/h is typically 170 kHz, and η is near zero; that is, the electric field gradient is essentially axially symmetric about the C—D bond. The ^2H NMR spectrum of a specifically deuterated

[57] S. J. Singer and G. L. Nicholson, *Science* (*Washington, D.C.*) **175**, 720 (1972).

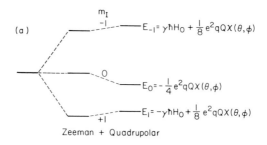

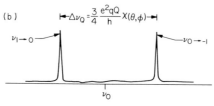

FIG. 20. (a) Nuclear energy levels of an $I = 1$ nucleus in a strong magnetic field. The angular function $\chi(\theta, \phi)$ is defined in Eq. (1.5.3). The relative contributions to the splitting from the nuclear Zeeman effect and the nuclear quadrupolar interactions are not drawn to scale. (b) The NMR spectrum expected for a single crystal of a molecule with an isolated $I = 1$ nucleus at an arbitrary orientation in the magnetic field. From "Membrane Spectroscopy" (E. Grell, ed.), p. 12. Springer-Verlag, Heidelberg.

long-chain alcohol (1,1-dideuterooctanol) intercalated into hydrated sodium octanoate and oriented at various angles with respect to the external magnetic field is shown in Fig. 21a.[58] A quadrupolar splitting is indeed observed, but it is significantly smaller than the value of 170 kHz expected for motionless acyl chains. The origin of the reduced splitting is, of course, motional averaging. The ^2H NMR spectrum of a powder sample of dipalmitoyl lecithin specifically deuterated at carbon 5 in both acyl chains is shown in Fig. 21b.[59] Note that one frequently deals with unoriented multilayers in membrane studies and only a powder spectrum is observed.

The effect of motional averaging on the electric quadrupole interaction is usually expressed in terms of an order parameter. In a phospholipid bilayer, the chain motion takes a molecular axis symmetrically about a director perpendicular to the bilayer surface, and the ^2H quadrupolar splitting may be rewritten as

$$\Delta\nu_Q = \tfrac{3}{4}(e^2qQ/h)S_{CD}(3\cos^2\theta' - 1), \tag{1.5.5}$$

where θ' is the angle between the applied magnetic field and the director and S_{CD} is the order parameter for the C—D bond relative to the director. S_{CD}s have been measured for methylene segments located at different

[58] J. Seelig and W. Niederberger, *J. Am. Chem. Soc.* **96**, 2069 (1974).
[59] J. Seelig and A. Seelig, *Biochem. Biophys. Res. Commun.* **57**, 406 (1974).

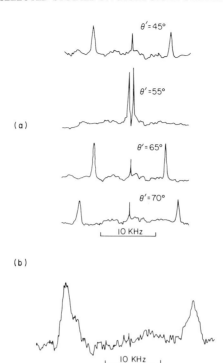

FIG. 21. (a) ^{2}H NMR spectra (13.8 MHz) of specifically deuterated fatty alcohol intercalated into hydrated sodium octanoate and oriented at various angles with respect to the
magnetic field. The angle θ' is between the magnetic field and the bilayer normal. The sharp
signal in the center of the spectra arises from a small fraction of molecules which form an isotropic
phase when the oriented multiplayers are prepared. Reprinted with permission from Seelig and
Niederberger, *J. Am. Chem. Soc.* **96**, 2069 (1974). Copyright 1974 American Chemical Society.
(b) ^{2}H NMR spectrum (13.8 MHz) of a powder sample of dipalmitoyllecithin specifically
deuterated at carbon atom 5 in both chains. Temperature 60°C. (From Seelig and Seelig.[59])

positions along the hydrocarbon chains of a phospholipid bilayer. From these
data the motional state of the phospholipid bilayer membrane can be quantified and expressed in terms of the flexibility profile, as depicted in Fig. 22.[60]

^{2}H NMR is now widely used in biophysical studies of biological membranes. The interested reader is referred to a review by Seelig[61] and to the
Specialist Periodical Reports on NMR published by the Chemical Society.

1.5.4. Intact Cells: Metabolism Studied *in Vivo*

Despite what may have appeared to be insurmountable difficulties involved in utilizing NMR as a tool for biological studies, investigators have

[60] G. W. Stockton, C. F. Polnaszek, A. P. Tullock, F. Hasan, and I. C. P. Smith, *Biochemistry*
15, 954 (1976).
[61] J. Seelig, *Q. Rev. Biophys.* **10**, 353 (1977).

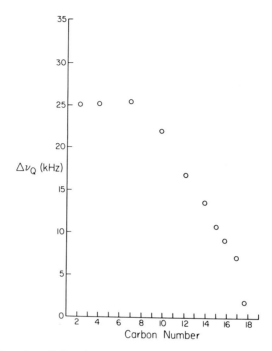

FIG. 22. Quadrupolar splittings from specifically deuterated stearic acid in egg yolk lecithin dispersions plotted as a function of the hydrocarbon chain position. Data taken from Stockton et al.[60]

persevered to the point where today living tissues and whole animals are finding their way into the spectroscopists' magnetic field. Such work involves not only the immense technical challenge of maintaining the sample under conditions compatible with life, but also presents interpretational hurdles, even when the data are successfully gathered. However, many problems, such as those involving magnetic field inhomogeneity caused by air bubbles, interference from resonant nuclei located in organelles, and tissues and organs not of interest, are gradually, if only slowly, being overcome. We view the outlook for exciting breakthroughs and giant strides in this field with some degree of optimism. In this section we briefly review two recently published studies in this area. Both of these studies have used NMR to characterize the metabolism of living cells, but they illustrate the diverse range of problems the technique is capable of tackling.

1.5.4.1. ^{31}P-NMR. The great usefulness of phosphorus NMR in studying cell metabolism arises from the fortuitous combination of its nuclear magnetic properties and its presence in high-energy substrate molecules. ^{31}P is an abundant nucleus (natural abundance of 100%), has

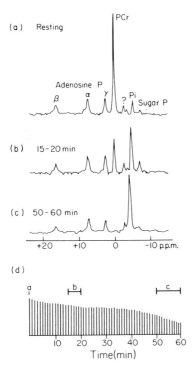

FIG. 23. (a)–(c) ^{31}P NMR spectra from anaerobic frog gastrocnemius muscles, stimulated for 1 sec every 60 sec and recorded at the time intervals indicated. (d) Decline in the isometric force development of the stimulated muscle during the time period of the experiment. From Dawson *et al.*[64] Reprinted by permission from *Nature* **274**, p. 862. Copyright © 1978 Macmillan Journals Limited.

a spin of $\frac{1}{2}$, and exhibits a large chemical shift range. These properties, together with the fact that metabolically important molecules bearing ^{31}P are usually small, with short correlation times and concomitantly long relaxation times, render ^{31}P NMR an ideal tool for the study of cell metabolism. In addition, inorganic phosphate displays a characteristic pH-dependent ^{31}P chemical shift[62] which can be used to simultaneously determined pH changes within the cell medium during metabolic studies.

To illustrate the utility of ^{31}P NMR for monitoring metabolic changes, we cite the recent studies of Dawson and co-workers.[63,64] These investigators were interested in delineating the chemical changes occurring in frog muscle as it fatigues due to repeated contractions under anaerobic conditions. In

[62] R. B. Moon and J. H. Richards, *J. Biol. Chem.* **248**, 7276 (1973).

[63] M. J. Dawson, D. G. Gadian, and D. R. Wilkie, *J. Physiol.* (*London*) **267**, 703 (1977).

[64] M. J. Dawson, D. G. Gadian, and D. R. Wilkie, *Nature* (*London*) **274**, 861 (1978).

their pioneering experiment, they suspended the muscles in a specially designed chamber which allowed them to perfuse, stimulate, and determine the strength of muscle contraction, while simultaneously using ^{31}P NMR to record the level of various metabolities as well as the cellular pH. Some of their data are reproduced in Fig. 23. Although the NMR experiments permitted only the direct measurement of ATP, phosphocreatine (PCr), and inorganic phosphate (Pi) levels from the observed peak areas in the ^{31}P NMR spectrum and the cellular pH from the inorganic phosphate chemical shift, it was also possible to infer indirectly the concentration of creatine, free ADP, and lactic acid, based on their quantitative relationships to the directly measured metabolites.[64] Taken together these results established an unambiguous correlation between muscle fatigue and the energy state of the cells.

1.5.4.2. ^{13}C-Labeled Substrates. In Section 1.5.1.1 we noted the great advantages of enriching an isotopically rare nucleus at a well-defined position in a protein. Similarly, this method can be useful in following the fate of a labeled substrate molecule as it is metabolized within a cell. Thus by exploiting NMR as a noninvasive probe one can investigate reaction kinetics and the stereospecificity of enzymes in their natural environment.

This approach has recently been utilized by den Hollander et al.[65] to investigate anaerobic glycolysis in yeast cells, using [1-^{13}C] glucose and [6-^{13}C] glucose as substrates. Because of the isotopic enrichment they were able to obtain well-resolved spectra at one-minute intervals. This capability allowed them to directly measure the rates of consumption of glucose, the production and consumption of the metabolic intermediate fructose 1,6-bisphosphate, and the production of the fermentation end products ethanol and glycerol. In addition, the use of ^{13}C-glucose labeled at the 1 or 6 carbon made it possible to follow the knetics of label scrambling caused by the enzymes aldolase and triose phosphate isomerase.

The successful application of NMR in studies of this type shows great promise for future investigations into the direct effects of various metabolites and antimetabolites *in vivo*. This approach may also prove fruitful in unraveling the origins of various metabolic diseases.

Acknowledgment

This work was supported by USPHS NIH Grant GM-22432 and American Cancer Society Grant PF-1628.

[65] J. A. den Hollander, T. R. Brown, K. Ugurbil, and R. G. Shulman, *Proc. Natl. Acad. Sci. U. S. A.* **76**, 6096 (1979).

2. NITROXIDE SPIN LABELS

By Gillian M. K. Humphries and
Harden M. McConnell

2.1. Introduction

Systematic natural science has advanced by means of two interdependent processes, specialization and generalization; the former being concerned with how things differ from one another and the latter with how they are the same. The recent course of science has been remarkable not only for its tremendous growth, but also for the increasing difficulty with which we construct boundaries between "different" scientific disciplines. This latter phenomenon should not be surprising because many such boundaries are not natural but exist as a convenience to scientists and librarians. Boundaries are no longer convenient if they are allowed to become barriers to creative science. Those who engage in highly interdisciplinary science are usually searching for unifying principles rather than specifics. They aspire to the eclectic and are especially dependent upon the lines of communication between themselves and the specialists in areas of their interest; there is no room for esoteric science. If biophysical chemistry is to be effective and intellectually stimulating, it is vital that information be communicated in a manner which is palatable to biologists, physicists, and chemists alike. Guided by this belief, we have taken the unusual step of presenting a physical method having great realized and unrealized potential in biophysical applications, at two levels.

In Chapter 2.2 spin-label theory is presented in a simplified form with the aid of copious diagrams rather than mathematical equations. This method has been used many times by Humphries to present spin-label theory to medical and biological scientists, but has not been published previously. We hope that Chapter 2.2 will enable any scientist to follow Chapter 2.4, which is a review of recent work in the McConnell laboratory concerned with using spin labels as antigenic determinants, with a view to discovering selection rules which control immune responses to membranes. Our thesis is that the physical state of antigenic determinants associated with membranes must have a marked effect on afferent and efferent immune responses to those membranes and that the use of determinants capable of providing direct information regarding their physical state, as do spin labels, is essential in

METHODS OF EXPERIMENTAL PHYSICS, VOL. 20

any study designed to test this proposition. The choice of this topic allows us to illustrate the main features of electron paramagnetic resonance (EPR) spectra of nitroxide spin labels. In brief, EPR enables us to reach conclusions regarding the distribution, orientation, rate of motion, and dielectric environment of spin-label antigenic determinants.

We also hope that Chapter 2.2 will provide an easily read introduction to spin-label theory for those who wish to pursue the subject in greater depth by reading the mathematical treatment written by McConnell and presented in Chapter 2.3.

We do not endeavor to provide an encyclopedic review of the use of spin labels, either with respect to work from this laboratory or from others, and refer the reader to other recent monographs or reviews which accomplish that purpose.[1-10] However, our discussion will touch upon certain of the most common uses of spin labels.

2.2. Spin-Label Theory: A Descriptive Treatment

2.2.1. Fundamentals Including the Resonance Condition

The essential structure of the type of spin label discussed here is a nitroxide group adjacent to two tertiary carbon atoms. Such a molecule is host to an unpaired electron: it is a stable free radical and therefore paramagnetic. Examples are shown below.

2,2,6,6-Tetramethyl-
piperidine-1-oxyl
(TEMPO)

2-(3-Carboxypropyl)-4,4-
dimethyl-2-tridecyl-
3-oxazolidinyloxyl

[1] L. J. Berliner, ed., "Spin Labeling, Theory and Applications." Academic Press, New York, 1976.

[2] G. I. Likhtenstein (P. F. Shelnitz, trans.), "Spin Labeling Methods in Molecular Biology." Wiley, New York, 1976.

[3] H. M. Swartz, J. R. Bolton, and D. C. Borg, eds., "Biological Applications of Electron Spin Resonance." New York, 1972.

[4] B. J. Gaffney, Methods Enzymol. 32B, 161 (1974).

[5] L. J. Berliner, Methods Enzymol. 49G, 418 (1977).

[6] P. C. Jost, A. S. Waggoner, and O. H. Griffith, in "Structure and Function of Biological Membranes" (L. I. Rothfield, ed.), p. 83. Academic Press, New York, 1971.

[7] J. Seelig, Biomembranes 3, 267 (1972).

[8] M. Cohn and J. Reuben, Acc. Chem. Res. 4, 214 (1971).

[9] R. A. Dwek, Adv. Mol. Relaxation Processes 4, 1 (1972).

[10] R. A. Dwek, "Nuclear Magnetic Resonance (N. M. R.) in Biochemistry." Oxford Univ. Press, London and New York, 1973.

Such molecules are stable in the pH range 3–10. If the carbon atoms adjacent to the nitrogen atom are not tertiary, disproportionation takes place as shown below.

The products are no longer spin labels because they do not contain unpaired electrons. Similarly, chemical reduction by appropriate agents such as ascorbate or glutathione destroys the paramagnetic properties of spin labels as shown below.

Certain oxidizing agents and photoinduced addition reactions can also destroy the paramagnetism.

It is helpful to look at a structural representation of the important end of TEMPO in order to understand why it is a free radical (see Fig. 1 and legend). The figure is drawn to indicate that the unpaired electron is solely associated with the nitrogen atom. This is an oversimplification; in fact, it is also significantly associated, but to a lesser extent, with the oxygen atom. This sharing has a concomitant effect on the extent of charge separation (also indicated by the structural representation); it is partly governed by the

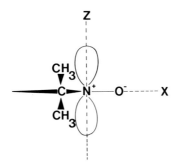

FIG. 1. Diagrammatic representation of the nitroxide group in a typical spin-label molecule. In simple terms one may consider that two of the five nitrogen valence electrons are required for the formation of the two CN bonds (only one of which is shown here) and two form a coordinate bond with the oxygen, so generating the charge separation indicated in the diagram. This leaves a single unpaired electron, which is predominantly associated with the nitrogen atom, occupying a 2p π orbital, also indicated in the diagram. The conventional assignment of z and x axes is given; the y axis is normal to the page.

dielectric properties of the environment and is an important factor leading to certain spectral changes which we discuss in Section 2.2.4.

The unpaired electrons exist in one of two possible "spin states," designated $+\frac{1}{2}$ and $-\frac{1}{2}$. The energy levels of these two spin states are not identical in the presence of an applied magnetic field ($+\frac{1}{2}$ being higher than $-\frac{1}{2}$). As a result, for a sample of spin-label molecules at equilibrium, there will be a population difference, unpaired electrons in the $+\frac{1}{2}$ state being slightly in the minority with respect to those in the $-\frac{1}{2}$ state. (The population difference at equilibrium is controlled by the temperature and the difference between the energy levels of the two spin states. This is covered in detail in Section 2.3.2.) In a manner analogous to the induction of a magnetic field by an electrical current passing through a coiled wire, spinning electrons are each associated with their own magnetic moment; they can be thought of as little bar magnets. If an electron is not paired to another of equal and opposite spin in the same orbital, it demonstrates appreciable interaction with any magnetic field to which it is subjected. One consequence is that the *difference* ΔE in energy levels between the $+\frac{1}{2}$ and $-\frac{1}{2}$ spin states increases with the strength of a magnetic field in which the molecules are placed (see Fig. 2).

If an electron in the $-\frac{1}{2}$ state is supplied with a quantum of energy exactly equal to the energy level difference between the two spin states, it may absorb it and assume the $+\frac{1}{2}$ state. This is the condition of *resonance* and is the *essential step* for generation of EPR spectra of the type which we consider here. The energy of a quantum of electromagnetic radiation is given by its frequency v multiplied by Planck's constant h. (So long as the EPR spectrometer is adjusted correctly, with the microwave power set low enough, the rate at which $+\frac{1}{2}$ state electrons revert to $-\frac{1}{2}$ electrons by "spin–lattice

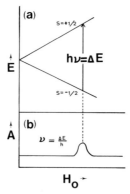

FIG. 2. (a) Effect of external magnetic field strength H_0 on the energy level E of the permitted orbitals of unpaired electrons having spin states S of $+\frac{1}{2}$ and $-\frac{1}{2}$. (b) Diagrammatic representation of resonance: absorption A of microwaves of frequency v as a function of field strength H_0 by unpaired electrons not influenced by nuclear hyperfine splitting.

relaxation" is fast enough that the system does not "saturate." This is covered in detail in Section 2.3.2.)

The principal features of an EPR spectrometer, a machine which is designed to detect the resonance condition, are illustrated in Fig. 3. If the spacially fixed magnetic field is sufficiently strong, quanta of energy susceptible to absorption by the unpaired electrons of spin labels are associated with

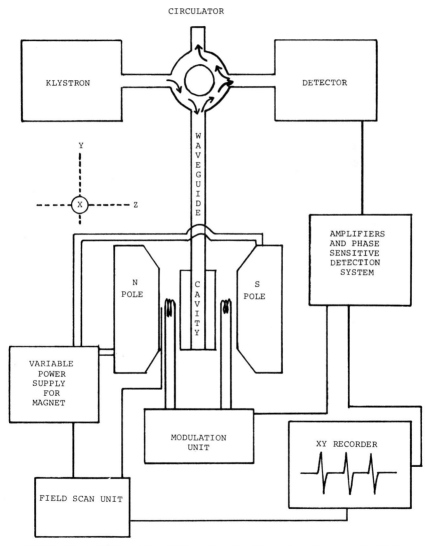

FIG. 3. Principal features of an EPR spectrometer. The axes are the same as in Fig. 9.

electromagnetic radiation of a convenient wavelength. Typically, for the spin-label work, the strength of the spacially fixed magnetic field is varied at around 3000 G, and the wavelength of electromagnetic radiation (supplied by a klystron) is kept constant at approximately 3 cm (i.e., in the "microwave" range). This is experimentally convenient because the radiation is supplied as standing waves in the "cavity," a carefully engineered structure suspended between the poles of the magnet with dimensions appropriate to the wavelength. The spin-label sample is placed within the cavity and its absorption of a portion of the microwave radiation which would otherwise be allowed to leave the cavity is responsible for the signal detected by the spectrometer.

In summary, EPR spectrometers are usually designed to measure the absorption of microwaves (supplied at a fixed frequency) as a function of increasing (or decreasing) magnetic field strength; a measurable net absorption of microwaves by the sample can occur when this allows electrons associated with spin labels to move from the lower permitted energy state to the higher. If the only relevant facts were those which have already been presented, we would expect a field versus absorption spectrum showing a single narrow absorption peak or "line": a very boring bump, which would vary little in shape or position with changes in motion, distribution, orientation, or dielectric environment of the spin labels (see Fig. 2b). The reason why nitroxide spin-label spectra are interesting is that the unpaired electrons are strongly influenced by the nitrogen nuclei, which have three possible

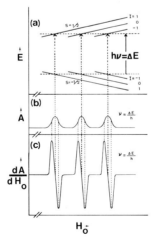

FIG. 4. (a) Effect of external magnetic field H_0 and spin state of nitrogen nuclei ($I = +1$, 0, -1) on the energy level E of permitted orbitals of unpaired electrons (spin states $S = +\frac{1}{2}$, $-\frac{1}{2}$). (b) Absorption A of microwaves of frequency v as a function of field strength H_0 by unpaired electrons: an example of nuclear hyperfine splitting. (c) First-derivative spectrum taken from (b).

spin states $(+1, 0, -1)$, and therefore exhibit magnetic properties themselves. Because the populations of nitrogen nuclei in each spin state are approximately equal under normal experimental conditions, one-third of all the unpaired electrons in a spin-label sample are associated with nitrogen nuclei in each of the three spin states. The effect on the permitted energy levels of the electrons is illustrated by Fig. 4. This shows that resonance conditions are obtained at three different positions of the field sweep, corresponding to the $+1, 0, -1$ spin states of the nitrogen nuclei. This phenomenon is an example of *nuclear hyperfine splitting*. In actual practice, EPR spectra are usually not recorded as microwave absorbtion A versus magnetic field strength H_0 (as shown in Fig. 4b), but as the first derivative of the absorption with respect to field strength dA/dH_0 versus H_0 (as shown in Fig. 4c). This is convenient for the instrument's design and because first-derivative spectra provide information in a more measurable form than absorption spectra do. However, absorption spectra are easier to understand theoretically, and we therefore continue to use them as models in the following discussion.

2.2.2. Effect of Orientation

Figure 1 illustrates the disposition of the $2p\,\pi$ orbital of the nitrogen atom, with which the unpaired electron is predominantly associated, and the conventional manner of assigning Cartesian coordinates to the molecule. As a simplification, the N, O, and two tertiary C atoms are placed in a single plane in the diagram although this is not usually strictly correct, the CNO angle being a function of the particular nitroxide molecule considered. Because of the anisotropic disposition of the unpaired electron (i.e., in the p orbital) its interaction with the large external applied magnetic field H_0 is governed by the orientation of the spin-label group with respect to the direction of H_0. Spin labels have been oriented in single host crystals so that all the spin-label axes are aligned the same way with respect to the crystallographic axes. If the orientation of such a crystal in the cavity of the EPR spectrometer is rearranged in many different ways, the observed hyperfine splitting (the distance between the absorption peaks) changes such that it is maximal when the external field H_0 is parallel to the z axis of the nitroxide group and minimal when it is perpendicular to the z axis. This is a consequence of the fact that the extent of splitting is positively correlated with the influence of the nitrogen nucleus on the electron; this is maximal when H_0 and z (the long axis of the $2p\,\pi$ orbital) are parallel. Therefore, the maximum value for the hyperfine splitting between adjacent peaks is known as $T_{\parallel}$ and the minimum as $T_{\perp}$. [In fact, because the electron is partially shared by the oxygen atom, there is a small difference between the splittings when H_0 is parallel to x and when it is parallel to y, but usually it is sufficient to consider an average of the values obtained when H_0 is parallel to any line drawn in the xy plane. It is also true that there is another shift in the position of the set of all three lines (as a group) when the crystal is rotated. This is due to a change in the "g value," which is discussed in Chapter 2.3 but need not be of concern for the purpose of the present discussion.] Figure 5a illustrates the true change in hyperfine splittings and line positions as a crystal is rotated about the spatially fixed external magnetic field axis H_0. Figure 5b is a simplified version which approximates to Fig. 5a and allows us to present a version of spin-label theory without recourse to mathematics.

For orientations of spin-labeled crystals other than $H_0 \parallel z$ or $H_0 \perp z$, the hyperfine splitting values are intermediate. If we were to take the carefully prepared crystal that we have been discussing and crush it to a powder, the signal generated by such a sample would be a composite of contributions from the randomly oriented labels. If we assume that the various positions indicated by the lines in Fig. 5b are all occupied by spin labels but, for further simplification, we consider labels with similar positions as groups or "packets"

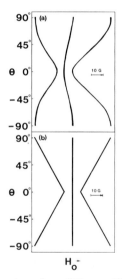

FIG. 5. (a) True change in nuclear hyperfine splittings and line positions as a crystal containing oriented nitroxide spin labels is rotated about a spatially fixed external magnetic field of variable strength H_0. θ is the angle between the xy plane of the nitroxide groups (see Fig. 1) and H_0; i.e., when $\theta = 90°$ or $-90°$ the z axis of the labels is parallel to the magnetic field, and when $\theta = 0°$ it is perpendicular. Note that the center line is equidistant from the two outer lines at all values of θ. (b) Simplified version of (a). Two simplifications have been made. (1) The "g factor," is shown to be constant (see Section 2.3.5). This is not the true case because orbital magnetic contributions by the electrons cause variation of g with orientation and give rise to the nonlinearity of the center line in (a), with concomitant shifts of the outer lines. (2) The curve associated with the outer lines has been removed.

which generate absorption spectra of fairly narrow width, we arrive at the prediction illustrated by Fig. 6. This shows that a powder spectrum is related in a very interesting way to spectra derived from single crystals rotated relative to H_0. If we consider the left-hand (low-field) side of the composite absorption spectrum (Fig. 6c) we see that the packet of labels that includes those with $z \parallel H_0$ is responsible for the field positions at which $\sum A$ commences to (i) increase increasingly rapidly and (ii) increase less rapidly. These field positions are those at which A for that particular packet (i) commences to increase and (ii) is at its maximum value (see Fig. 6b). However, the EPR spectrometer records spectra not as field versus $\sum A$, but as field versus the first derivative of A (i.e., H_0 versus $d \sum A/dH_0$), as shown in Fig. 6d. The first-derivative spectrum (Fig. 6d) of the low-field shoulder shown in the composite absorption spectrum (Fig. 6c) has the same shape as the absorption spectrum of the packet of labels with $z \sim \parallel H_0$ (Fig. 6b); that is to say, both commence at the same low-field position and achieve their maximum value at the same position. This fact is also derived mathematically in Section 2.3.8. The composite first-derivative spectrum may similarly be used to detect contributions from the packet of labels having their z axes perpendicular to H_0; the spectrum may, therefore, be used to determine both $T_{\parallel}$ and $T_{\perp}$.

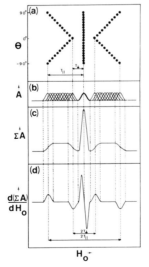

FIG. 6. Simplified version of the generation of a powder spectrum. See Fig. 4 for use of symbols, etc. (a) As Fig. 5b but indicating a noncontinuous distribution of labels as an aid to interpretation. (b) Superimposed absorption spectra of all packets of labels shown in (a). (c) Composite absorption spectrum of all packets of labels, obtained by adding the individual spectra shown in (b). (d) First-derivative spectrum taken from (c).

It should be borne in mind that the spectrum shown in Fig. 6d is derived using several simplifications which are described in the legends to Figs. 5 and 6. A comparison of Figs. 6d and 13 should be made, to observe similarities and differences between the true and synthetic spectra.

2.2.3. Effect of Motion

The individual spin labels in the powder sample discussed above are essentially motionless. The EPR spectrometer has a characteristic time interval for receiving data which can be likened to the shutter speed on a camera. Even if labels are moving, a "powder spectrum" will be obtained if the motion is very slow relative to the instrument's data gathering interval

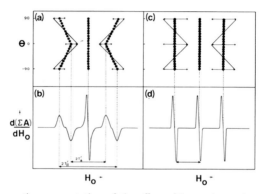

FIG. 7. Diagrammatic representation of the effect of isotropic motion on the spectra of isotropic dispersions of spin labels. (a) The tilted V-shaped solid lines to either side are the nuclear hyperfine splittings appropriate to zero or very slow motion (see Figs. 5 and 6), and the ordinate, labeled $\theta = +90$ to -90, relates to these; θ being the angle between the xy plane of nitroxide groups and H_0. Because the instrument has a fixed, characteristic data gathering interval ($\sim 10^{-9}$ sec), labels which alter their orientation appreciably during this time interval are detected via their absorption at an average H_0, appropriate to their orientation excursion during $\sim 10^{-9}$ sec. The "strings of beads" represent packets of labels that absorb at approximately equal average values of H_0 and were, at some time during the data gathering interval, at the orientation θ. (b) The first-derivative spectrum of the microwave absorption of the packets of labels, shown in (a) as a function of H_0. The values for maximal and minimal splitting ($T_{\parallel}'$ and $T_{\perp}'$) are respectively smaller and larger than those observed for powder spectra ($T_{\parallel}$ and $T_{\perp}$) and may be compared with $T_{\parallel}$ and $T_{\perp}$ for calculations regarding the motion of spin-labeled molecules. (c) and (d) are analogous to (a) and (b) respectively and illustrate the case when motion is so rapid that the orientation excursion during 10^{-9} sec is sufficiently complete as to nullify all contributions due to orientation detectable by this method. We then say that the rotational correlation time of the molecule is $\leq 10^{-9}$ sec. [It should be noted that for an isotropic distribution there are, at any time, effectively twice as many labels oriented with $z \perp H_0$ as there are labels oriented with $z \parallel H_0$. Therefore the value for the observed splitting in the true case of rapid motion is $\frac{1}{3}(T_{\parallel} + 2T_{\perp})$, not $(T_{\parallel} + T_{\perp})$ as suggested by this diagram which does not take this "weighting" into account.]

($\sim 10^{-9}$ sec). If, however, spin labels are moving so fast that their rotational correlation time is equal to, or shorter than, 10^{-9} sec, a single value for hyperfine splitting is observed, this being the weighted average of splittings appropriate to the various orientations of z with respect to H_0 [i.e., $\frac{1}{3}(T_\parallel + 2T_\perp)$]. Such spectra are termed isotropic or "fast tumbling" (see Figs. 7 and 11 and also Section 2.3.4.). For intermediate cases, various degrees of mixing of splittings become apparent; for example, at the beginning of a data gathering interval, a packet of labels might be oriented with $z \parallel H_0$ but will have rotated sufficiently by the end of the interval that the recorded hyperfine splittings of those labels are less than the maximum value observed in the absence of motion. Similarly, labels oriented with $z \perp H_0$ at the beginning or at some other time will give rise during the data gathering interval to splittings greater than the minimum value observed in the absence of motion. Changes in observed maximum and minimum splittings, relative to those obtained from powder spectra, have been used to obtain absolute or relative values for motion.

Well-known applications include determination of the rotational correlation times of spin-labeled proteins[11] and the measurement of the "order parameter" of membranes using spin-labeled fatty acids or phospholipids labeled in a fatty acid side chain.[12] Frequently, motion is not sufficiently fast to give rise to a spectrum in which the splittings are completely averaged, but yet it is impossible to observe measurable maximum and minimum splittings. In such cases one can take advantage of the relative amplitudes of the various peaks in order to obtain information regarding motion[13]; for a true fast motion, isotropic spectrum the amplitudes of all three peaks are almost equal (see Fig. 11); with decreasing motion the amplitudes of the high- and low-field peaks decrease relative to the center peak. This phenomenon is responsible for the effect discussed in Section 2.4.7, Fig. 18, when spin-label lipid haptens report the "melting" of lipid bilayers.

2.2.4. Effect of the Electrostatic Environment

In environments having high dielectric constants (i.e., polar or "hydrophilic" environments) the polar character of the NO bond is encouraged; charge separation is favored, the bond lengthens, and the nitrogen nucleus's share of the unpaired electron increases relative to that when the spin label finds itself in an apolar, "hydrophobic" environment. This phenomenon has a marked effect on EPR spectra. It is sufficient for the present discussion simply to say that the higher the electron density on the nitrogen nucleus,

[11] E. J. Shimshick and H. M. McConnell, *Biochem. Biophys. Res. Commun.* **46**, 321 (1972).
[12] W. L. Hubbell and H. M. McConnell, *J. Am. Chem. Soc.* **93**, 314 (1971).
[13] J. D. Morrisett and C. A. Broomfield, *J. Am. Chem. Soc.* **93**, 7297 (1971).

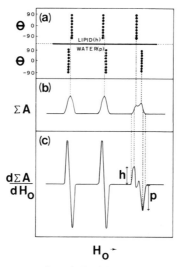

FIG. 8. Diagrammatic representation, derived in a manner analogous to that shown in Figs. 6 and 7, of a phenomenon which forms the basis of a well-known method for determining lipid phase transitions. (a) The time-averaged nuclear hyperfine splitting (see Fig. 7c) of TEMPO in the lipid, or hydrophobic, phase and the aqueous, or polar, phase. (b) Total absorption, $\sum A$, of all packets of labels shown in (a) as a function of H_0. (c) First derivative spectrum taken from (b). Note that the signal h is derived from TEMPO in the hydrophobic phase and the signal p is derived from TEMPO in the polar phase.

the greater is the influence of the nucleus and therefore the larger the hyperfine splittings. In other words, polar environments increase splittings; apolar environments decrease splittings. (For further discussion see Chapter 2.3) This is the basis of a phenomenon described in Section 2.4.6 and illustrated by Fig. 17. It is also the basis of a well-known method for detecting temperature-induced phase transitions in lipid bilayers or cell membranes,[14,15] illustrated by Fig. 8. The compound TEMPO partitions between aqueous and lipid phases with a coefficient which is a function of the fluidity of the lipid phase. The observed EPR spectrum is a composite of that generated by TEMPO dissolved in the aqueous (polar) phase and that generated by TEMPO dissolved in the lipid (hydrophobic) phase. In both environments, the spin labels are in rapid isotropic motion and the amplitude of the first-derivative signals ($2h$ and $2p$) are approximately proportional to concentration. The high-field lines of these two spectra are sufficiently well resolved that they can be used to calculate the relative amounts of TEMPO

[14] E. J. Shimshick and H. M. McConnell, *Biochemistry* **12**, 2351 (1973).

[15] C. Linden, K. Wright, H. M. McConnell, and C. F. Fox, *Proc. Natl. Acad. Sci. U. S. A.* **70**, 2271 (1973).

in the two phases. The ratio may change dramatically at temperature-induced phase transitions. (The resolution is aided by the fact that there is a "*g* value" change affecting the position of the set of three lines. This is covered in more detail in Section 2.3.5. The *g* shift is down field for apolar environments, so that the low-field lines of the two spectra overlap extensively but the high-field lines are better resolved than they would otherwise be.)

2.2.5. Effect of Distribution or Concentration of Spin Labels: Spin-Exchange Broadening

There are two ways in which EPR spectra are affected by the distance between individual spin labels; we consider spin-exchange broadening first. If labels collide, the orbitals containing unpaired electrons overlap sufficiently that there is a high probability that the electrons will exchange orientations, thus averaging the effects of the two nuclei. If the collision frequency is sufficiently high, as it will be in the case of low-molecular-weight spin labels in low-viscosity solvents at concentrations ≥ 1 mM, this spin exchange has a marked effect on the spectra. All three lines are broadened to a similar extent and eventually a single absorption peak is observed when the concentration is raised to permit such rapid exchange that the effects of the three different nitrogen populations are totally averaged out. As is the case when the lines are broadened by a decrease in motion, broadening by spin exchange also results in a decrease in amplitude of the first-derivative spectrum. An example of this effect is illustrated by Fig. 22. (See also section 2.3.12.)

2.2.6. Effect of Distribution or Concentration of Spin Labels: Dipole–Dipole Interactions

If a certain spin-label molecule is close to another but spin exchange does not occur, it may still be affected by this second label in another way. If the second label is tumbling very rapidly and isotropically around the first, the direction of its magnetic field is totally averaged and cannot affect the first spin label. If, however, the second label is moving slowly, or nonisotropically, its magnetic field will affect that of the first spin label, again broadening the peaks of the spectrum and decreasing their amplitude because an element of inhomogeneity has been introduced into the magnetic environment.

2.3. Spin-Label Theory: A Mathematical Treatment

2.3.1. The Resonance Condition

The term "magnetic resonance spectroscopy" has usually been applied to those forms of spectroscopy in which the frequency of radiation that is absorbed or emitted is nearly or exactly proportional to the strength of a laboratory magnetic field acting on the sample. The absorption of radiation by a sample placed in a magnetic field can be discussed in terms of the equation

$$hv = \Delta E, \tag{2.3.1}$$

where h is Planck's constant, v is the frequency of the radiation that is absorbed (or emitted), and ΔE is the separation of two energy levels. ΔE is exactly or nearly proportional to the strength of the applied laboratory magnetic field $\mathbf{H}_0$, at least for large fields (see Fig. 2a). In electron magnetic resonance spectroscopy, the applied fields are usually in the 3000–10,000 G range, and the radiation frequency is in the 9000–35,000 MHz range. Electron magnetic resonance spectroscopy is also called paramagnetic resonance spectroscopy. (Electron paramagnetic resonance was discovered by Zavoisky in the U.S.S.R. in 1945.[16,17])

An electron is a negatively charged particle. A circulating motion of the electron gives rise to a magnetic moment. An electron can have two kinds of circulating motion, one giving an *orbital* magnetic moment, and the other giving a *spin* magnetic moment. In organic free radicals† the orbital magnetic moment is largely quenched and makes only a small contribution to the total magnetic moment. The electron spin magnetic moment is a vector and is denoted by $\boldsymbol{\mu}$. This magnetic moment is parallel to the vector spin angular momentum $\mathbf{S}\hbar$. Here $\hbar$ is Planck's constant divided by 2π and $\mathbf{S}$ is the spin angular momentum in units of $\hbar$.

The electronic magnetic moment $\boldsymbol{\mu}$ and the spin angular momentum $\mathbf{S}\hbar$ are parallel to one another.

$$\boldsymbol{\mu} = g(e/2mc)\mathbf{S}\hbar. \tag{2.3.2}$$

The charge on the electron is e ($e = -4.8 \times 10^{-10}$ esu), m is the mass of the electron, and c is the velocity of light. The quantity g is the "spectroscopic splitting factor"; $g = 2.0023$ when the orbital contribution to the electronic magnetic moment is zero (complete orbital quenching). Equation (2.3.2) is often written in terms of the Bohr magneton β, where

$$\beta = e\hbar/2mc. \tag{2.3.3}$$

We shall write Eq. (2.3.2) in terms of the absolute value of β, $|\beta|$, in order to avoid any uncertainty in algebraic sign:

$$\boldsymbol{\mu} = -g|\beta|\mathbf{S}. \tag{2.3.4}$$

[16] E. Zavoisky, *Fiziol. Zh.* **9**, 245 (1945).
[17] E. Zavoisky, *Fiziol. Zh.* **9**, 211 (1945).

---

† Except for diatomic CH.

The negative sign arises because the charge on the electron e, is negative. The numerical value of $|\beta|$ is 0.92731×10^{-20} erg/G.

A magnetic dipole moment $\boldsymbol{\mu}$ interacts with an applied field vector $\mathbf{H}_0$ with an energy E, where

$$E = -\boldsymbol{\mu} \cdot \mathbf{H}_0; \tag{2.3.5}$$

thus,

$$E = g|\beta|\mathbf{S} \cdot \mathbf{H}_0. \tag{2.3.6}$$

If the applied field is in the z direction, then the energy E can be expressed in terms of S_z; $S_z = \pm\frac{1}{2}$, and thus there are two energy levels, as sketched in Fig. 2.

Transitions between these levels can be induced by radiation (e.g., microwaves) of the proper frequency and polarization. When a sample containing unpaired electrons is at thermal equilibrium there are more electrons in the lower energy state $(S_z = -\frac{1}{2})$ than in the upper energy state $(S_z = +\frac{1}{2})$. Incident radiation of the proper polarization and frequency then stimulates more upward $(S_z = -\frac{1}{2} \rightarrow S_z = +\frac{1}{2})$ than downward $(S_z = +\frac{1}{2} \rightarrow S_z = -\frac{1}{2})$ transitions, resulting in a net absorption of radiation. This is electron magnetic resonance.

According to Eq. 2.3.1, the frequency of the radiation v that can produce transitions between two energy levels $E(S_z = -\frac{1}{2})$ and $E(S_z = +\frac{1}{2})$ is

$$hv = E(S_z = +\tfrac{1}{2}) - E(S_z = -\tfrac{1}{2}); \tag{2.3.7}$$

$$hv = g|\beta|H_0. \tag{2.3.8}$$

Most organic free radicals have g factors that are approximately equal to 2. From Eq. (2.3.8) it follows that for radiation of frequency $v = 9.5$ kMHz, paramagnetic resonance absorption is observed at an applied field H_0 of approximately 3370 G. This frequency region is often termed "X band" by microwave spectroscopists. Most commercial paramagnetic resonance spectrometers operate in this frequency region.

The conditions for resonance given in Eq. (2.3.8) is also often expressed in terms of the magnetogyric ratio γ:

$$2\pi v = \gamma H_0 \tag{2.3.9}$$

or

$$\omega = \gamma H_0, \tag{2.3.10}$$

where

$$\gamma = g|\beta|/\hbar. \tag{2.3.11}$$

The magnetogyric ratio is the ratio of the magnetic moment of the electron $(g|\beta|S)$ to its "mechanical moment" or spin angular momentum $(S\hbar)$. The resonant frequency v can be regarded as a classical Larmor frequency of the electron magnetic moment, as discussed below in terms of the Bloch equations (Section 2.3.3.).

In free space, radiation having a frequency $v = 9.5$ GHz has a wavelength $\lambda = c/v = 3.16$ cm. Radiation of this wavelength can be conveniently handled by means of microwave tubing and resonance cavities, whose internal dimensions are usually of the order of magnitude of this wavelength.

The principal features of an EPR spectrometer are shown in Fig. 3. In the simplest possible terms, the resonance cavity can be thought of as functioning in the following way. A paramagnetic sample whose resonance is to be studied is placed in an appropriate holder in the center of the cavity and the cavity placed between the poles of an electromagnet. In the case of the commonly used reflection cavity, microwave energy from a source passes down the waveguide, through an iris, and into the cavity. This particular type of resonance cavity has a length which is equal to the wavelength of the microwave radiation. In this case the radiation is reflected back and forth many times in the cavity before either being absorbed by the sample or the walls, or being lost by going out through the iris and back up the waveguide. The radiation can thus be thought of as being "stored" in the cavity. Paramagnetic resonance absorption is observed when the field of the electromagnet is changed until the resonance condition [Eq. (2.3.8)] is met. At resonance, radiation is absorbed by the sample; this absorption is detected by a decrease in power leaving the cavity. Note that in an experiment of this type, the frequency of the radiation is normally kept fixed and the resonance absorption detected by changing the strength of the applied field.

2.3.2. Absorption Probabilities and Paramagnetic Relaxation

In the above discussion we have considered briefly the field–frequency condition for paramagnetic resonance absorption. Let us now discuss the intensity of the resonance absorption, that is, the number of photons taken from the microwave field by the paramagnetic sample when the field–frequency ratio corresponds to the resonance condition.

Consider a system of N electrons, where the electron spins interact with each other only very weakly compared to the thermal energy kT. In the presence of a uniform magnetic field of strength H_0, some of the electrons $(n\downarrow)$ will be in the lower ("spin-down") energy level, and the remaining electrons $(n\uparrow)$ will be in the upper energy level:

$$N = n\downarrow + n\uparrow. \tag{2.3.12}$$

The relative numbers of electrons in the lower and upper spin states can be calculated from Boltzmann's law:

$$n{\uparrow}/n{\downarrow} = e^{-h\nu/kT}, \tag{2.3.13}$$

where $h\nu$ is the transition energy, equal to $g|\beta|H_0$. The wavelength of X-band microwave radiation is ~ 3 cm, so that the transition energy measured in wave numbers is ~ 0.33 cm^{-1}. In the same units, room-temperature thermal energy is ~ 200 cm^{-1}. Thus $n{\uparrow}$ and $n{\downarrow}$ differ only by a few tenths of a percent. Let $n = n{\downarrow} - n{\uparrow}$ be the population difference. Under these conditions a good approximation to the difference in population n is then

$$n_{eq} \simeq N(h\nu/kT). \tag{2.3.14}$$

Here n_{eq} designates the difference in population of these two states when the electron spin magnetic energy levels are in thermal equilibrium with a "bath" at temperature T. The probabilities for absorption and stimulated emission are the same. That is, if P is the probability (in sec^{-1}) that an electron in the spin-down state absorbs a photon, then P is also the probability that an electron in the upper spin state will be stimulated to emit. (The probabilities of *spontaneous* emission of radiation can usually be neglected in magnetic resonance experiments.) We shall see later that P depends on the photon density or energy density in the radiation field.

The time rate of change of the population of the upper spin state $n{\uparrow}$ is thus

$$dn{\uparrow}/dt = -n{\uparrow}P + n{\downarrow}P. \tag{2.3.15}$$

The first term in Eq. (2.3.15) is the rate of loss of electrons in the upper spin state due to stimulated emission, and the second term is the gain of electrons in the upper spin state due to absorption. The population difference then follows the equation

$$dn/dt = -2nP. \tag{2.3.16}$$

From this equation we see that a system of spins initially in thermal equilibrium at time $t = 0$ would have an exponentially decreasing population difference if a suitable radiation field were applied at time $t = 0$:

$$n(t) = n(0)e^{-2Pt} \tag{2.3.17}$$

According to this result the spin system saturates with a time constant of $\frac{1}{2}P$ and the absorption of radiation decreases at this rate. Fortunately, this is not usually the case because of paramagnetic relaxation.

Paramagnetic relaxation describes the processes whereby the spin magnetic energy (sometimes called "Zeeman energy") comes to thermal equilibrium with its molecular or crystalline environment (usually called the "lattice" or "bath"). This relaxation is often a simple first-order process, so

that in the *absence* of an applied radiation field, but in the presence of a
thermal bath with which the spins can exchange energy, we have

$$dn/dt = -(1/T_1)(n - n_{eq}).$$ (2.3.18)

Here T_1 is the electron paramagnetic relaxation time. In the presence of both
the radiation field that induces transitions between the two spin states and
paramagnetic relaxation that tends to bring the spin state populations to
thermal equilibrium, the net time dependence of the population difference is
governed by the differential equation

$$dn/dt = -2Pn - (1/T_1)(n - n_{eq}).$$ (2.3.19)

Under steady-state conditions where $dn/dt = 0$ we find for the steady-state
difference in population n_{ss}

$$n_{ss} = n_{eq}/(1 + 2PT).$$ (2.3.20)

From this equation one draws a physically obvious conclusion: as long as
the incident power (radiation energy density) is low enough, the rate of
radiation-induced transitions is small compared to the rate of paramagnetic
relaxation, and the resonance absorption of radiation does not appreciably
disturb the difference in population of the two spin states. The power absorbed
by the sample under steady-state conditions is $n_{ss} Ph\nu$. This can also be written
in the following form:

$$\text{power absorption} = [NP(h\nu)^2/kT]/(1 + 2PT_1).$$ (2.3.21)

We now discuss the absorption probability P. As already mentioned, most
paramagnetic resonance experiments are carried out using a resonant
microwave cavity. The electromagnetic radiation that is stored in this
cavity has both electric and magnetic components that oscillate at the
resonant frequency of the cavity. The lines of force corresponding to one
phase of the microwave magnetic field are shown in Fig. 9. The shaded
region is, in fact, the region of the microwave cavity where the aqueous
samples are normally placed.

The axis systems in Figs. 3 and 9 are the same. The resonance cavity is
placed between the pole faces of the electromagnet so that the applied field
H_0 is parallel to the cavity axis z. The oscillatory magnetic component of the
microwave radiation field is then in the x direction. Near the center of the
cavity this field is linearly polarized and uniform. Let the strength and
direction of this linearly polarized field be represented by

$$H_i' = H_i' \cos(\omega t + \delta).$$ (2.3.22)

Here $\mathbf{i}$ is a unit vector in the x direction, H_1 is the amplitude of the oscillatory
field, ω is 2π times the microwave frequency, and δ is a phase factor that can

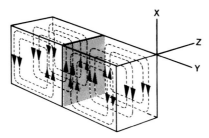

FIG. 9. Lines of force of the oscillatory magnetic field associated with the microwave radiation within the cavity. The sample is placed in the central shaded area. The axis system is the same as that used in Fig. 3.

be set equal to zero. For convenience we write the amplitude of this field as $2H_1$ instead of H_1':

$$2H_1 \equiv H_1', \qquad (2.3.23)$$

$$H_1' = \mathbf{i}(2H_1) \cos(\omega t + \delta). \qquad (2.3.24)$$

An elementary quantum-mechanical calculation yields the following result for the transition probability for an electron[18]:

$$P = \tfrac{1}{4}(\gamma H_1)^2 g(v). \qquad (2.3.25)$$

Here $g(v)$ is the absorption line-shape function, which is normalized as

$$\int_0^\infty g(v)\,dv = 1. \qquad (2.3.26)$$

We do not give a quantum-mechanical derivation of Eq. (2.3.25) here since essentially the same result is obtained later from the Bloch equations. The derivation of Eq. (2.3.25) does involve the assumption that the source of (microwave) radiation has a frequency distribution that is narrow compared to the width of the absorption curve, described by $g(v)$. This condition is always valid for the paramagnetic resonance experiments of interest here, where typical linewidths are never much less than 1 MHz, and the microwave radiation has a frequency stability of better than one part in 10^7. The line-shape function $g(v)$ can be thought of as describing the distribution of energy levels between which transitions can take place. The distribution of energy levels within the sample can arise from local magnetic fields within the sample that are necessarily compounded with the external field and define a distribution of resonance frequencies. If the energy levels between which

[18] A. Carrington and A. D. McLachlan, "Introduction to Magnetic Resonance," p. 21. Harper & Row, New York, 1967.

transitions take place have a sufficiently short lifetime, then this can also contribute to the width of the distribution of frequencies given by $g(v)$.

2.3.3. The Bloch Equations

Consider an electron i with spin magnetic moment $\boldsymbol{\mu}_i$ and spin angular momentum $\hbar \mathbf{S}_i$. A magnetic field $\mathbf{H}$ exerts a torque of $\boldsymbol{\mu}_i \times \mathbf{H}$ on the spin magnetic moment, with a consequent change in the direction of the angular momentum:

$$\frac{d}{dt}(\hbar \mathbf{S}_i) = \boldsymbol{\mu}_i \times \mathbf{H}. \tag{2.3.27}$$

If we multiply through by the magnetogyric ratio γ, we obtain

$$\frac{d}{dt}\boldsymbol{\mu}_i = \gamma \boldsymbol{\mu}_i \times \mathbf{H}. \tag{2.3.28}$$

Then we may sum over all the electrons i in the sample and obtain

$$\frac{d\mathbf{M}}{dt} = \gamma \mathbf{M} \times \mathbf{H}, \tag{2.3.29}$$

where $\mathbf{M}$ is the total electron magnetic moment

$$\mathbf{M} = \sum \boldsymbol{\mu}_i. \tag{2.3.30}$$

Equation (2.3.29) gives the time rate change of the electron magnetization due to the externally applied fields, but does not take into account changes in the magnetization due to internal fields in the sample that produce electron relaxation. If we take the z axis to be in the direction of the large, steady applied field, then the contributions of relaxation to the time rate of change of M are assumed to have the following form:

$$dM_z/dt = -(M_z - M_0)/T_1, \tag{2.3.31}$$

$$dM_x/dt = -M_x/T_2, \tag{2.3.32}$$

$$dM_y/dt = -M_y/T_2. \tag{2.3.33}$$

According to Eq. (2.3.31), the electron magnetization relaxes towards its equilibrium value M_0 with the relaxation time T_1. The equilibrium value of the electron magnetization in the field direction is easily shown to be

$$M = N\hbar^2 I(I + 1)/3kT. \tag{2.3.34}$$

Here N is the number of electrons in the sample. Equations (2.3.32) and and (2.3.33) assume that the components of electron magnetization perpendicular to the steady large field $\mathbf{H}_0$ relax to their equilibrium value, zero, with time constant T_2. If we add together the rate of change of the electron magnetization due to the externally applied fields H [Eq. (2.3.29)] and due to relaxation [Eqs. (2.3.31)–(2.3.33)] we obtain the Bloch equations. (These were originally devised to describe nuclear magnetic resonance.[19]) The Bloch equations can be written in the form of a single vector equation

$$\frac{d\mathbf{M}}{dt} = \gamma\mathbf{M} \times \mathbf{H} - \frac{k(M_z - M_0)}{T_1} - \frac{(iM_x + jM_y)}{T_2}. \qquad (2.3.35)$$

Here $\mathbf{i}, \mathbf{j}, \mathbf{k}$ are unit vectors in the x, y, z directions, respectively, and

$$\mathbf{H} = \mathbf{H}_0 + \mathbf{H}_1'; \qquad (2.3.36)$$

$\mathbf{H}_0$ is the steady large field applied in the z direction, and H_1' is a (usually weak) oscillatory field perpendicular to H_0. In both microwave electron paramagnetic resonance instruments and radio-frequency nuclear magnetic resonance instruments, the applied oscillatory field (with angular frequency ω) is usually linearly polarized. Let this linearly polarized field be in the x direction and of amplitude $2H_1$,

$$\mathbf{H}_1' = 2H_1\mathbf{i}(\cos \omega t). \qquad (2.3.37)$$

This linearly polarized field can be decomposed into two circularly polarized counter-rotating components:

$$\mathbf{H}_1' = \mathbf{H}_1^+ + \mathbf{H}_1^-, \qquad (2.3.38)$$

$$\mathbf{H}_1^- = H_1[\mathbf{i}(\cos \omega t) - \mathbf{j}(\sin \omega t)], \qquad (2.3.39)$$

$$\mathbf{H}_1^+ = H_1[\mathbf{i}(\cos \omega t) + \mathbf{j}(\sin \omega t)]. \qquad (2.3.40)$$

When the frequency ω is close to the resonance frequency $\omega_0 = \gamma H_0$, one of these components ($\mathbf{H}_1^+$ or $\mathbf{H}_1^-$) will be nearly "in phase" with the precessing electron magnetization, and the other component ($\mathbf{H}_1^-$ or $\mathbf{H}_1^+$) will be "out of phase" with the precessing magnetization. The out-of-phase component of $\mathbf{H}_1'$ has a negligible effect on the motion of the electron magnetization and can be neglected from Eq. (2.3.35). Therefore we include in Eq. (2.3.35) only the cirularly polarized component that is in phase with the electron magnetization at resonance. The magnetogyric ratio γ is negative for electrons. As can be seen from Eq. (2.3.29) and Fig. 10, the magnetization (in the absence of any $\mathbf{H}_1$ field) precesses clockwise about z, when viewed from below the xy plane in the positive z direction. Thus when γ is negative, $\mathbf{H}_1^+$ also rotates clockwise and is the in-phase component.

[19] F. Bloch, *Phys. Rev.* **70**, 460 (1946).

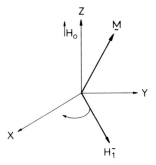

Fig. 10. Illustration of the axis system used in the Bloch equations, Section 2.3.3.

The Bloch equations are readily solved by a transformation to the rotating coordinate system $\mathbf{i}'$, $\mathbf{j}'$, $\mathbf{k}'$. Here $\mathbf{i}'$ is a unit vector in the direction of $\mathbf{H}_1$ ($= H_1^+$ when γ is negative), $\mathbf{j}'$ is a unit vector perpendicular to $\mathbf{i}'$ which lies in the xy plane, and $\mathbf{k}' = \mathbf{k}$. Let

$$\mathbf{M} = \mathbf{i}'u + \mathbf{j}'v + \mathbf{k}'M_z, \qquad (2.3.41)$$

where u is the in-phase electron magnetization, v is the out-of-phase electron magnetization, and M_z is, as before, the electron magnetization in the external field direction. In terms of u, v, and M_z, the Bloch equations are

$$du/dt = (\omega_0 - \omega)v - u/T_2, \qquad (2.3.42)$$

$$dv/dt = -(\omega_0 - \omega)u + \gamma H_1 M_z - v/T_2, \qquad (2.3.43)$$

$$dM_z/dt = -\gamma H_1 v - (M_z - M_0)/T_1. \qquad (2.3.44)$$

Under steady-state, or "slow passage" conditions where $du/dt = dv/dt = dM_z/dt = 0$, Eqs. (2.3.42)–(2.3.44) are readily solved for u, v, and M_z:

$$u = \gamma H_1 T_2^2 (\omega_0 - \omega)M/[1 + T_2^2(\omega_0 - \omega)^2 + \gamma^2 H_1^2 T_1 T_2], \qquad (2.3.45)$$

$$v = \gamma H_1 T_2 M/[1 + T_2^2(\omega_0 - \omega)^2 + \gamma^2 H_1^2 T_1 T_2], \qquad (2.3.46)$$

$$M_z = [1 + T_2^2(\omega_0 - \omega)^2]/[1 + T_2^2(\omega_0 - \omega)^2 + \gamma^2 H_1^2 T_1 T_2]. \qquad (2.3.47)$$

The instantaneous power absorbed by the sample is $\mathbf{H} \cdot d\mathbf{M}/dt$, and the average power absorbed (averaged over one cycle of the microwave field) is

$$P(\omega) = \frac{\omega}{2\pi} \int_0^{2\pi/\omega} \mathbf{H} \cdot \frac{d\mathbf{M}}{dt}\, dt. \qquad (2.3.48)$$

The average power absorbed by the sample is then

$$P(\omega) = \omega \gamma H_1^2 M_0 T_2/[1 + T_2^2(\omega - \omega_0)^2 + \gamma^2 H_1^2 T_1 T_2]. \qquad (2.3.49)$$

This absorption curve has a Lorentzian line shape that is independent of the microwave power (e.g., H_1^2) when the microwave power is low enough,

$$\gamma^2 H_1^2 T_1 T_2 \ll 1. \tag{2.3.50}$$

The quantity $\gamma^2 H_1^2 T_1 T_2$ is termed the saturation factor. If we consider the condition for resonance, $\omega = \omega_0$, and compare Eqs. (2.3.49) and (2.3.21) we can obtain an expression for the absorption probability P:

$$P = \tfrac{1}{2}\gamma^2 H_1^2 T_2. \tag{2.3.51}$$

This result is the same as that obtained by direct quantum-mechanical calculation of the transition probability.

Most commercial paramagnetic resonance spectrometers are designed so that the recorded signal is proportional to the out-of-phase rotating magnetic moment v and do not record the power absorbed by the sample, $P(\omega)$. (Strictly speaking, the signal usually recorded is the derivative of v, dv/dH, in field-sweep spectra.) We shall refer to the amplitude of the out-of-phase component v as the "absorption signal."

2.3.4. Nuclear Hyperfine Structure

The most important feature of the paramagnetic resonance spectra of organic free radicals is the rich nuclear hyperfine structure. This hyperfine structure arises from the magnetic interaction between the electron spin magnetic moment and nuclear spin magnetic moments. The nuclear hyperfine interaction splits the paramagnetic resonance into a number of components. For example, Fig. 11 shows the paramagnetic resonance spectrum of 2,2,6,6-tetramethyl 1-4-oxopiperidine (TEMPONE) in water.

(TEMPONE)

The resonance spectrum consists of three sharp lines separated by 16.0 ± 0.05 G. These three lines correspond to the three possible orientations of the ^{14}N nucleus in the applied magnetic field. The ^{14}N nucleus has a nuclear spin I which is equal to one. The component of the nuclear spin angular momentum in the applied field direction is designated m. When $I = 1$, the quantitized values of m are $m = 1, 0, -1$ (see Fig. 4). Proton nuclear hyperfine splittings are not resolved in the paramagnetic resonance spectra of many aliphatic nitroxide radicals. However, these proton hyperfine interactions often contribute to the observed resonance linewidths.

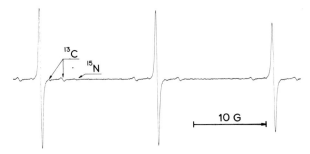

FIG. 11. EPR spectrum of a dilute aqueous solution of TEMPONE. See text for discussion.

The additional weak signals in Fig. 11 are due to a small fraction of free radicals that contain ^{13}C or ^{15}N nuclei instead of ^{14}N and ^{12}C. The natural abundances of ^{13}C and ^{15}N are 1.1 % and 0.37 %, respectively.

Another important feature of the paramagnetic resonance spectra of organic free radicals is the spectroscopic splitting factor g, which is nearly but not exactly equal to the free-spin g factor $g_0 = 2.0023$. This deviation of the g factor from the free-spin value is due to a (second-order) spin-orbit interaction. In other words, there is an electron-spin-induced electron orbital motion that creates a magnetic field which acts back on the electron spin. The deviation of the free-radical g factor from the free-electron g factor means that the electron paramagnetic resonance spectrum is displaced from a free-electron resonance spectrum. For example, the spectroscopic splitting factor g for the resonance spectrum of TEMPONE in Fig. 11 is $g = 2.00562 \pm 0.00002$.

It is sometimes convenient to consider the effect of the spin–orbit interaction (g-factor effect) and the hyperfine interaction on paramagnetic resonance spectra from the following point of view. A free electron with spin $+\frac{1}{2}$ has two energy levels in an applied field of strength $|\mathbf{H}|$, and the separation of these energy levels is

$$hv = g_0|\beta||\mathbf{H}|. \tag{2.3.52}$$

Paramagnetic resonance transitions between these levels then occur at the transition frequency v. The field $\mathbf{H}$ acting on the "free" electron in a molecule such as a nitroxide free radical is equal to the externally applied field $\mathbf{H}_0$ plus local fields $\mathbf{H}_{loc}$ originating within the molecule. Thus

$$hv = g_0|\beta||\mathbf{H}_0 + \mathbf{H}_{loc}|. \tag{2.3.53}$$

The local magnetic field acting on an electron in a nitroxide free radical has two components, the hyperfine field due to the ^{14}N nucleus, and the spin–orbit field due to the (spin–induced) orbital motion of the electron itself.

Thus

$$hv = g_0|\beta||\mathbf{H}_0 + H_{hf} + \mathbf{H}_g|. \qquad (2.3.54)$$

Here $\mathbf{H}_{hf}$ is the local field acting at the electron due to the magnetic moment of the ^{14}N nucleus and $\mathbf{H}_g$ is the local field acting at the electron due to the orbital motion of the electron. Since ^{14}N has a nuclear spin $I = 1$ which can take on three orientations in a magnetic field, the local hyperfine field $\mathbf{H}_{hf}$ has three possible values. On the other hand, for organic free radicals in general, and for nitroxide free radicals in particular, the effective field $\mathbf{H}_g$ due to spin–orbit interaction that acts on the "free electron" spin is, to a good approximation, in the same direction as $\mathbf{H}_0$ and proportional to $\mathbf{H}_0$:

$$\mathbf{H}_g = (\delta_g/g_0)\mathbf{H}_0. \qquad (2.3.55)$$

Here $g = g_0 + \delta g$. Also, to a good approximation, the only effective component of $\mathbf{H}_{hf}$ is the component in the direction of $\mathbf{H}_0$, namely $\mathbf{H}'_{hf}$. Thus Eq. (2.3.54) becomes

$$hv = g|\beta|H_0 + g_0|\beta|H'_{hf}. \qquad (2.3.56)$$

Since the second term in this equation is small compared to the first, we see that the paramagnetic resonance spectrum of a nitroxide free radical is a three-line spectrum, centered at the field H_0, where $hv = g|\beta|H_0$, with two hyperfine lines centered about this resonance position, corresponding to applied fields $\mathbf{H}_0 + |\mathbf{H}'_{hf}|$ and $\mathbf{H}_0 - |\mathbf{H}'_{hf}|$. The three possible values of H'_{hf}, namely $-|H'_{hf}|, 0, +|H'_{hf}|$, correspond to $m = -1, 0, 1$. Since these values of m are very nearly equal in probability, the three hyperfine lines have essentially the same intensity. In the paramagnetic resonance spectrum exhibited in Fig. 11, the local hyperfine field acting at the odd electron is ± 16 (± 0.05) G.

In general, the hyperfine field acting at the electron due to the nitrogen nucleus, $\mathbf{H}_{hf}$, has two components:

$$H_{hf} = \mathbf{H}_{hf}\ \text{isotr} + \mathbf{H}_{hf}\ \text{anisotr}. \qquad (2.3.57)$$

In the theory of the effect of molecular *motion* on the magnetic resonance spectra, it is shown that the *anisotropic* contribution to the hyperfine interaction is averaged to zero by the rapid tumbling motions of low-molecular-weight free radicals in liquids of low viscosity, such as water, at room temperature. That is, under these conditions only the isotropic interaction contributes to the splitting, and this isotropic hyperfine field is H'_{hf} isotr. As we shall discuss later (Section 2.3.9), the anisotropic part of the hyperfine interaction contributes to the widths of the resonance lines even when the molecular motion is very fast. The anisotropic part of the hyperfine interaction makes a large contribution to the hyperfine splittings in the paramagnetic resonance spectra of oriented nitroxide radicals, as discussed in Section 2.3.7.

The theory of the *origin* of the nuclear hyperfine interaction in organic free radicals has been developed extensively. An understanding of this theory is not necessary for most of the biophysical applications of spin labels. We include a brief outline of this theory here for the sake of completeness and because this theory leads in a straightforward way to the *spin Hamiltonian* for nitroxide free radicals. This spin Hamiltonian provides a general and convenient quantitative starting point for the calculation and interpretation of magnetic resonance spectra. For an alternative discussion, see Abragam and Bleaney.[20]

Let us first consider the magnetic interaction between a nucleus located at the origin of a Cartesian coordinate system $x = y = z = 0$ and an electron located at the point $\mathbf{r} = \mathbf{i}x + \mathbf{j}y + \mathbf{k}z$. The energy of the magnetic dipole–dipole interaction is

$$\mathcal{H}_d = r^{-3}[\boldsymbol{\mu}_e \cdot \boldsymbol{\mu}_n - 2(\boldsymbol{\mu}_e \cdot \mathbf{r})(\boldsymbol{\mu}_n \cdot \mathbf{r})/r^2]. \tag{2.3.58}$$

Here r is the distance between the electron with spin magnetic moment $\boldsymbol{\mu}_e$ and the nucleus with spin magnetic moment $\boldsymbol{\mu}_n$. As discussed above [cf. Eq. (2.3.53)], the free-electron spin magnetic moment $\boldsymbol{\mu}_e$ is

$$\boldsymbol{\mu}_e = -g_0|\beta|\mathbf{S}. \tag{2.3.59}$$

The corresponding expression for the nuclear spin magnetic moment is

$$\boldsymbol{\mu}_n = g_N\beta_N\mathbf{I}, \tag{2.3.60}$$

where g_N is the nuclear g factor, β_N is the nuclear magneton, and $\mathbf{I}$ is the nuclear spin angular momentum in units of $\hbar$. The energy of this point dipole–dipole hyperfine interaction can then be written

$$\mathcal{H}_d = h\mathbf{S} \cdot \mathbf{T}_p \cdot \mathbf{I}, \tag{2.3.61}$$

where

$$\mathbf{T}_p = -h^{-1}g_0|\beta|g_N\beta_N r^{-3}(\mathbf{U} - 3\mathbf{rr}/r^2). \tag{2.3.62}$$

Here $\mathbf{U}$ is a unit dyadic:

$$\mathbf{U} = \mathbf{ii} + \mathbf{jj} + \mathbf{kk}. \tag{2.3.63}$$

For a discussion of dyads and dyadics, see Goldstein.[21]

The reader can show for himself that by inserting an electron–nuclear distance of the order of, say, 1 Å in these equations, one obtains hyperfine

[20] A. Abragam and B. Bleaney, "Electron Paramagnetic Resonance of Transition Ions." Oxford Univ. Press (Clarendon), London and New York, 1970.

[21] H. Goldstein, "Classical Mechanics," p. 147. Addison-Wesley, Reading, Massachusetts, 1950.

interaction energies of the same order of magnitude as the observed hyperfine splittings.

For quantitative work, the above calculation needs to be improved by taking into account the facts that (i) there is no single distinguishable odd electron in a nitroxide free radical, and (ii) the electron spin magnetic moment is not concentrated at a fixed point in space, but rather is distributed in space according to the molecular electronic wave function. The spin distribution in nitroxide free radicals can be discussed by reference to Fig. 1. We assume that the oxygen, nitrogen, and two bonded carbon atoms lie in a plane. This is precisely true in one nitroxide free radical whose structure has been determined.[22] For determinations of the structure of other nitroxides, see Berliner[23] and Lajzerowicz-Bonneteau.[24] The "odd" electron is localized in a 2p π atomic orbital centered on the nitrogen atom. The spin distribution doubtless also has some π-orbital amplitude on the oxygen atom as well. Since the odd electron is described by a quantum-mechanical distribution function, it is clear that the net dipole–dipole interaction must take into account this distribution of the spin. Let $\mathbf{m}(r)\, dV$ be the contribution to the total electron spin magnetic moment arising from the element of volume dV.

$$\boldsymbol{\mu}_e = \int \mathbf{m}(r)\, dV. \tag{2.3.64}$$

It is convenient to define a spin density function $\rho(r)$ such that[25]

$$\mathbf{m}(r) = -g_0|\beta|\mathbf{S}\rho(r). \tag{2.3.65}$$

The spin density function is normalized to 1:

$$\int \rho(r)\, dV = 1. \tag{2.3.66}$$

One can readily generalize the *point dipole* interaction in Eq. (2.3.60) to a *distributed dipole* interaction as follows.

$$\mathscr{H}_d = h\mathbf{S} \cdot \mathbf{T}_d \cdot I, \tag{2.3.67}$$

where

$$\mathbf{T}_d = \int_\varepsilon \mathbf{T}_p \rho(\mathbf{r})\, dv. \tag{2.3.68}$$

[22] J. C. A. Boeyens and G. J. Kruger, *Acta Crystallogr. Sect. B* **B26**, 668 (1970).

[23] L. J. Berliner, *Acta Crystallogr. Sect. B* **B26**, 1198 (1970).

[24] J. Lajzerowicz-Bonneteau, *in* "Spin Labeling, Theory and Applications" (L. J. Berliner, ed.), p. 239. Academic Press, New York, 1976.

[25] H. M. McConnell and J. Strathder, *Mol. Phys.* **2**, 129 (1959).

The ε in Eq. (2.3.68) denotes a small region surrounding the nucleus which is excluded from the integration because of the fact that the dipolar interaction formula $\mathbf{T}_p$ becomes infinite near the origin. This difficulty will be considered later in this section.

The point dipole dyadic $\mathbf{T}_p$ is *symmetric* in the sense that the coefficient of **ij** is equal to the coefficient of **ji**, $(T_d)_{xy} = (T_d)_{yx}$, and so forth. Thus the distributed dipole dyadic has this symmetry as well. In molecules where there is significant contribution of orbital magnetic moment to the nuclear hyperfine interaction, as in the case of paramagnetic transition-metal ions, the effective electron-spin, nuclear-spin dyadic need not be symmetric. In the case of nitroxide radicals the antisymmetric components in the hyperfine coupling between the electron spin and nuclear spins are negligible in magnitude, and in their effects on the observed resonance spectra.

The spin density distribution function $\rho(\mathbf{r})$ can be readily evaluated from an approximate molecular electronic wave function. Irrespective of the detailed form of this spin density distribution function, the dipolar hyperfine tensor $\mathbf{T}_d$ conforms to the following trace equation;

$$\mathrm{Tr}\{\mathbf{T}_d\} = 0, \tag{2.3.69}$$

$$(T_d)_{xx} + (T_d)_{yy} + (T_d)_{zz} = 0. \tag{2.3.70}$$

This result follows from the fact that the point dipolar interaction itself obeys the same trace equations, as one can easily show by expanding the vector $\mathbf{r}$ in Eq. (2.3.62) in Cartesian coordinates. A second point to note regarding the dipolar tensor $\mathbf{T}_d$ is that if the spin density in a nitroxide free radical is approximated by a single odd electron in the $2p\,\pi$ orbital sketched in Fig. 1, then the dipolar tensor is axially symmetric. If z is taken to be the symmetry axis, then $(T_d)_{xx} = (T_d)_{yy}$. An explicit calculation of the dipolar tensor $\mathbf{T}_d$ for a $2p\,\pi$-orbital spin distribution yields the following result:

$$(T_d)_{zz} = \tfrac{4}{5}h^{-1}g|\beta|g_N\beta_N\langle r^{-3}\rangle. \tag{2.3.71}$$

Here $\langle r^{-3}\rangle$ is the average value of $\langle r^{-3}\rangle$, where the average is taken over the radial probability distribution for the odd electron. From the above discussion it follows that

$$(T_d)_{xx} = (T_d)_{yy} = \tfrac{1}{2}(T_d)_{zz}. \tag{2.3.72}$$

One can make semiquantitative estimates of these dipolar terms using 2p Slater atomic orbitals. Estimated values for nitrogen are $(T_d)_{xx} = (T_d)_{zz} = -38$ MHz, $(T_d)_{zz} = +76$ MHz.[25,26] For an estimate of $\langle r^{-3}\rangle$ derived from self-consistent field Hartree functions for atomic nitrogen, see Dousmanis.[27]

[26] A. Carrington and H. C. Longuet-Higgins, *Mol. Phys.* **5**, 447 (1962).
[27] G. C. Dousmanis, *Phys. Rev.* **77**, 967 (1955).

Since the elements of the dyadic $\mathbf{T}_d$ are symmetric, this dyadic can be written in diagonal form (all nondiagonal elements equal to zero). The axes x, y, z for which $\mathbf{T}_d$ is diagonal are then a set of principal axes. When x, y, z are a set of principal axes, we shall use the notation $(T_d)_x$, $(T_d)_y$, and $(T_d)_z$ in place of $(T_d)_{xx}$, $(T_d)_{yy}$, and $(T_d)_{zz}$.

We now return to the expression for the dipolar interaction given in Eq. (2.3.68) and consider the small region ε that is excluded from the integration. If the spin distribution $\rho(\mathbf{r})$ is due exclusively to spin distributions that can be described in terms of $p, d, f, \ldots$, atomic wave functions, then the small region ε need not be excluded from the integration because the divergence of $\mathbf{T}_p$ does not produce any problem in the integration, since for these atomic functions the spin density goes to zero at the origin. On the other hand, if the odd-electron spin distribution has contributions from s orbitals, which is the case for nitroxide radicals, then it is necessary to consider this small region about the nucleus in detail.

The physical problem concerning integration near the nucleus is that the simple dipole–dipole interaction formula, Eq. (2.3.59) or (2.3.62), is only adequate when the two magnetic dipoles are far apart, relative to their intrinsic spatial extent, or size. The problem of calculating the magnetic hyperfine spin–spin interaction between a nucleus and an electron in an s orbital was first treated by Fermi[28] using the Dirac relativistic theory of the electron spin (see also Breit and Doerman[29]). The following semiclassical derivation of the "contact" isotropic interaction has been taken in part from a treatment of this problem by Ramsey.[30]

Draw a hypothetical sphere of radius R_s about the ^{14}N nucleus as center. This corresponds to the small volume excluded in the distributed dipole integration in Eq. (2.3.68). Let R_s be large enough that the magnetic interaction between the nucleus and the electron magnetization outside the sphere can be treated by the distributed dipole formula. This means that R_s must be large compared to the "size" of the nucleus. Let us assume that the effective size of the nucleus is R_n, of the order of 10^{-5} Å. Now consider the magnetic interaction between the nucleus and the magnetization inside the sphere with radius R_s. We assume that the spin density $\rho(\mathbf{r})$ is uniform across the radius R_s so that the sphere is small enough to have a uniform spin density due to the electron. The magnetic interaction of the nucleus with the electron spin magnetization inside the sphere is

$$\mathcal{H}_i = -\boldsymbol{\mu}_N \cdot \mathbf{B} \tag{2.3.73}$$

$$= -g_N \beta_N \mathbf{I} \cdot \mathbf{B}, \tag{2.3.74}$$

[28] E. Fermi, Z. Phys. **60**, 320 (1930).

[29] G. Breit and F. W. Doerman, Phys. Rev. **36**, 1732 (1930).

[30] N. F. Ramsey, "Molecular Beams." Oxford Univ. Press (Clarendon), London and New York, 1956.

where $\mathbf{B}$ is the (uniform) magnetization inside the sphere. From classical magnetostatics we know that

$$\mathbf{B} = \tfrac{8}{3}\pi\mathbf{m}(0) \tag{2.3.75}$$

$$= \tfrac{8}{3}\pi g_0 |\beta| \mathbf{S}\rho(0). \tag{2.3.76}$$

This leads to the following expression for the isotropic hyperfine interaction:

$$\mathscr{H}_i = g_0 |\beta| g_N \beta_N \tfrac{8}{3}\pi\rho(0)\mathbf{S} \cdot \mathbf{I} \tag{2.3.77}$$

$$= ha\mathbf{S} \cdot \mathbf{I}, \tag{2.3.78}$$

where the isotropic hyperfine splitting constant a is

$$a = h^{-1} g_0 |\beta| g_N \beta_N \tfrac{8}{3}\pi\rho(0). \tag{2.3.79}$$

Theoretical calculations of s-orbital spin density at the ^{14}N nucleus in planar NH_3^+ and also in diatomic NO give a spin density $\rho(0)$ that is *positive*. The theoretical calculations yield isotropic ^{14}N hyperfine coupling constants from Eq. (2.3.79) that are equal to $+58.5$ MHz[31] and $+40$ MHz,[27] respectively. These calculations should be applicable at least approximately to aliphatic nitroxide free radicals, which have isotropic hyperfine interactions equal to ~ 45 MHz. To a good approximation, nitroxide radicals are examples of "π-electron radicals"; that is, for a single electron the spin density at the ^{14}N nucleus is zero. The observed spin density at the nucleus is due to electron spin correlation effects (configuration interaction), which have been discussed at length in the literature for various nuclei (see Ref. 32 p. 39).

The total nuclear hyperfine interaction $\mathscr{H}_{hf}$ is the sum of the anisotropic dipolar interaction between the electron spin and the nucleus when the electron is outside the sphere and the isotropic contact interaction between the electron and the nucleus when the electron spin is inside the sphere:

$$\mathscr{H}_{hf} = \mathscr{H}_d + \mathscr{H}_i, \tag{2.3.80}$$

$$\mathscr{H}_{hf} = h\mathbf{S} \cdot \mathbf{T} \cdot \mathbf{I}, \tag{2.3.81}$$

$$\mathbf{T} = \mathbf{T}_d + \mathbf{U}a. \tag{2.3.82}$$

As we have stated earlier in this section, the anisotropic dipolar interaction is averaged to zero by rapid isotropic tumbling motions. This occurs when the tumbling rate, or inverse correlation time, is large compared to the anisotropy of the dipolar terms. This is the case for low-molecular-weight nitroxide free radicals in solutions of low viscosity, such as TEMPONE in

[31] G. Giaconnetti and P. L. Nordio, *Mol. Phys.* **6**, 301 (1963).
[32] E. G. Rozantzev, "Free Nitroxyl Radicals" (B. J. Hazzard, transl.), p. 39. Plenum, New York, 1970.

water. In such cases the observed (high-field) splittings are equal to the isotropic hyperfine interactions, as is the case for the spectrum in Fig. 11. Here the hyperfine splittings are 16.0 ± 0.05 G, corresponding to an isotropic hyperfine coupling constant

$$a = (2.8025 \text{ MHz/G}) \left(\frac{2.0023}{2.0056} \right) (16.0) \qquad (2.3.83)$$

$$= 44.8 \pm 0.14 \text{ MHz.} \qquad (2.3.84)$$

If we combine our previous theoretical estimates for the elements of the dipolar tensor $\mathbf{T}_d$ with the experimental value for the isotropic hyperfine coupling constant, we obtain

$$T_z \sim 76 + 45 = 121 \text{ MHz,} \qquad (2.3.85)$$

$$T_x = T_y \sim -38 + 45 = 7 \text{ MHz.} \qquad (2.3.86)$$

These theoretical estimates are compared with experimental values of $\mathbf{T}$ in the next section. Note that Eqs. (2.3.85) and (2.3.86) are based on the theoretical evidence that the spin density $\rho(0)$ at the ^{15}N nucleus is positive, and thus a is positive. Additional evidence for the validity of this conclusion will be evident from subsequent discussions. Note that since $\text{Tr}\{\mathbf{T}_d\} = 0$ [cf. Eq. (2.3.69)], it follows that

$$\tfrac{1}{3} \text{Tr}\{\mathbf{T}\} = a. \qquad (2.3.87)$$

2.3.5. The Spectroscopic Splitting Factor g

The theory of the origin of the g-factor term in organic free radicals is more complex than the theory of the nuclear hyperfine interactions and will not be described here except for a few qualitative remarks.

Nonlinear polyatomic free radicals have no first-order electronic orbital angular momentum; that is, the electron orbital motion is "quenched." The spin–orbit interaction can induce a small orbital magnetic moment in the molecule. The magnitude of this induced moment is expected to be proportional to the magnitude and sign of $\mathbf{S}$, but the induced moment need not be in the same direction as $\mathbf{S}$. Thus we expect the total magnetic moment (orbital magnetic moment plus spin magnetic moment) to have the form

$$\boldsymbol{\mu}_e = -|\beta| \mathbf{S} \cdot \mathbf{g}. \qquad (2.3.88)$$

The energy of interaction of this magnetic moment with the externally applied field H_0 is

$$\mathscr{H}_z = -\boldsymbol{\mu}_e \cdot \mathbf{H}_0 \qquad (2.3.89)$$

$$= |\beta| \mathbf{S} \cdot \mathbf{g} \cdot \mathbf{H}_0. \qquad (2.3.90)$$

The term $\mathcal{H}_z$ is often referred to as the electron Zeeman Hamiltonian. For molecules such as nitroxide radicals, **g** can be assumed to be a symmetric tensor.[33]

The deviations of the **g** tensor from the isotropic free-electron spin **g** tensor $g_0\,\mathbf{U}$ can be understood in terms of spin-orbit-induced (virtual) transitions of electrons between different one-electron orbital states. (This is the mixing of lower-energy occupied molecular orbitals with higher-energy unoccupied orbitals). Transitions between localized atomic 2p orbitals such as the following are "allowed": $p_x \rightarrow p_y$, $p_y \rightarrow p_z$, $p_z \rightarrow p_x$. The lowest-energy transitions contribute most significantly to the g-factor deviation; these transitions involve the half-empty 2p π orbital, which is a p_z orbital in our axis system. From this discussion it follows that g_z must be close to the free-electron g factor, since the only contribution to g_z comes from high-energy $p_x \rightleftharpoons p_y$ virtual transitions. The g_x and g_y terms are expected to differ much more from the free-electron g factor. These qualitative expectations have been verified for a number of organic free radicals, including nitroxide free radicals. For an early note discussing these ideas, see McConnell and Robertson.[34]

2.3.6. The Spin Hamiltonian

From the above discussion, we arrive at the following expression for the spin Hamiltonian:

$$\mathcal{H} = |\beta|\mathbf{S} \cdot \mathbf{g} \cdot \mathbf{H} + h\mathbf{S} \cdot \mathbf{T} \cdot \mathbf{I} - \beta_N g_N \mathbf{I} \cdot \mathbf{H}. \qquad (2.3.91)$$

We have included the nuclear Zeeman term $\beta_N g_N \mathbf{I} \cdot \mathbf{H}$.

In this discussion of the nuclear hyperfine interaction, the magnetic interaction between the spin-orbit-induced orbital magnetic moment and the nucleus has been neglected. This contribution is expected to be the order of magnitude of $\Delta g/g$ smaller than electron–spin interaction itself, where Δg is of the order of magnitude of the deviation of the g factor from the isotropic g factor. For nitroxide free radicals, the orbital contribution to nuclear hyperfine interaction is negligible since $\Delta g/g$ is $\gtrsim 3 \times 10^{-3}$.

Let us now return to the paramagnetic resonance spectrum of TEMPONE in water, shown in Fig. 11. To understand this spectrum in terms of the spin Hamiltonian equation (2.3.91) we recall first that rapid tumbling averages out the anisotropic terms, so we are left with an isotropic, time-average Hamiltonian

$$\mathcal{H} = |\beta|g\mathbf{S} \cdot \mathbf{H} + ha\mathbf{S} \cdot \mathbf{I} - \beta_N g_N \mathbf{I} \cdot \mathbf{H}. \qquad (2.3.92)$$

[33] H. M. McConnell, *Proc. Natl. Acad. Sci. U. S. A.* **44**, 766 (1958).
[34] H. M. McConnell and R. E. Robertson, *J. Phys. Chem.* **61**, 1018 (1957).

The resonance spectrum can be obtained from the spin Hamiltonian equation (2.3.92) by quantum-mechanical calculation. However, this spectrum can also be obtained by means of elementary physical arguments.

The interaction between an electron spin and an externally applied magnetic field can be thought of classically in terms of a precession of the spin angular momentum about the field direction. The precession frequency v is $h|\beta|gH_0 \simeq 9.5$ MHz when $H_0 \simeq 3300$ G. This precession frequency is orders of magnitude larger than the precession frequency of the nucleus in the external field, or in the nuclear hyperfine field $(ha\mathbf{I})/(g|\beta)$. Thus, if $\mathbf{h}$ is the direction of the external field, the components of $\mathbf{S}$ and $\mathbf{I}$ perpendicular to $\mathbf{h}$ are "decoupled" from one another since they precess about the external field direction at such different frequencies. Thus the secular or time-average interaction between the electron and nuclear spin is

$$\mathcal{H} = |\beta|gH_0 S_h + haS_h I_h + \beta_N g_N H_0 I_h. \tag{2.3.93}$$

In strong applied fields (e.g., 3000 G and above) the components S_h and I_h of the electron and nuclear spins in the applied field direction are good quantum numbers.

$$S_h = \pm \tfrac{1}{2}, \tag{2.3.94}$$

$$I_h = +1, 0, -1. \tag{2.3.95}$$

When we apply the selection rules $\Delta S_h = \pm\tfrac{1}{2}$ (an allowed magnetic resonance dipole transition) and $\Delta I_h = 0$, then in a field-swept spectrum resonance signals are observed at values of the applied field H_0 that satisfy the equation

$$hv = |\beta|gH_0 + haI_h, \tag{2.3.96}$$

where v is the fixed microwave frequency. Thus

$$H_0 = h(v + aI_n)/g|\beta|. \tag{2.3.97}$$

A field-sweep spectrum thus consists of three signals at applied fields equal to $h(v + a)/g|\beta|$, $hv/g|\beta|$, and $h(v - a)/g|\beta|$. The corresponding energy levels and resonance transitions are shown schematically in Fig. 4. These energy levels are only accurate for strong fields. For calculations of energy levels in low applied fields, see Rozantzev[32] and Ryzhkov and Stepanov.[35]

2.3.7. Effect of Orientation: Single-Crystal Spectra

It is sometimes possible to include a nitroxide free radical in a diamagnetic host single crystal as a substitutional impurity.[36] The free radical often takes

[35] V. M. Ryzhkov and A. P. Stepanov, *Geofiz. Priborostr.* **12**, 35 (1962).
[36] O. H. Griffith, D. W. Cornell, and H. M. McConnell, *J. Chem. Phys.* **43**, 2909 (1965).

on one or more well-defined, fixed orientations relative to the axes of the host crystal. It is then found that the paramagnetic resonance spectrum of the free radical depends strongly on the orientation of the single crystal (and thus the free radical) relative to the direction of the externally applied field. As one example of this type of experiment, Fig. 12 shows paramagnetic resonance spectra of the N-oxyl-4',4'-dimethyloxazolidine derivative of 5-α-cholestan-3-one (**I**) when this substance is incorporated as a substitutional impurity (at a concentration of the order of one percent) in single crystals of cholesteryl chloride.[12]

I

It is seen that the hyperfine splittings depend strongly on the orientation of the applied magnetic field when the field direction lies in the **ab** plane of the crystal. The maximum hyperfine splitting is 31.91 ± 0.02 G when the applied field is parallel to the crystallographic axis **b**; the minimum hyperfine splitting is 6.0 ± 0.1 G when the applied field is perpendicular to **b**. A plot of the hyperfine splittings as a function of the angle θ between the applied field direction and crystallographic **b** axis is given in Fig. 5a. Plots such as this may be used to determine the directions of the principal axes of the nuclear hyperfine interaction, the magnitudes of the principal hyperfine interactions, the directions of the principal axes of the **g** tensor, and the magnitudes of the

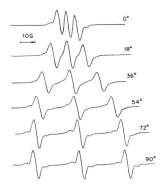

FIG. 12. Paramagnetic resonance spectra of the cholesterol spin label incorporated at a concentration of the order of one percent in single crystals of cholesteryl chloride. The angle $\theta = 0°$ represents a field direction perpendicular to the twofold monoclinic symmetry axis **b**. The angle $\theta = 90°$ represents a field direction parallel to **b**, and to the z axis of the spin label (indicated in Fig. 1).

elements of this tensor. In other words, plots such as those given in Fig. 5a can be used to determine the elements of the spin Hamiltonian discussed in Section 2.3.6. Many biophysical applications of anisotropic paramagnetic resonance spectra do not require a detailed quantitative analysis in terms of a spin Hamiltonian. Therefore we first discuss the significance of data such as are given in Figs. 12 and 5a from an empirical, utilitarian point of view, and then subsequently justify this discussion in terms of the spin Hamiltonian.

When the applied field H_0 is parallel to a principal axis of the nuclear hyperfine interaction, the nuclear hyperfine splitting is an *extremum*. That is, there is no first-order change in the hyperfine splitting for any small deviation in the direction of H_0 from the principal axis direction. For nitroxide free radicals with a fixed, strongly immobilized orientation in a single crystal, the principal axis direction that is simplest to find experimentally is the principal axis corresponding to the largest hyperfine splitting. We denote this principal axis direction z. The principal axis z is easiest to locate experimentally, because the hyperfine splitting is not only an extremum, but is an absolute maximum and is much larger than the other hyperfine splittings. In the case of the free radical (I) dissolved in cholesteryl chloride, the largest hyperfine splitting, 31.91 ± 0.02 G, is observed where the applied field H_0 is parallel to the crystallographic axis **b**. The other two principal axes are necessarily perpendicular to z, and to one another. These principal axis directions are denoted x and y. When the applied field H_0 is rotated in a plane perpendicular to the principal axis z the minimum splitting corresponds to the principal axis direction with the smallest nuclear hyperfine interaction, and the direction perpendicular to this gives the third principal axis direction with intermediate nuclear hyperfine splitting. In the case of label (I), when the applied field is rotated in the **ac** plane (percendicular to **b** and to z), one observes a maximum splitting of 6.31 ± 0.1 G, and a minimum splitting of 5.83 ± 0.1 G.

The principal axis directions of the **g** tensor and the components of the **g** tensor can be found in much the same way. If one plots the resonance position for the central hyperfine line, the highest field position of this resonance defines the minimum principal value of the **g** tensor. The corresponding field direction gives the direction of this principal axis. It can be seen from the data for the resonance line positions of the label (I) when the applied field direction is parallel to the principal nuclear hyperfine axis z. This maximum field position is an absolute maximum for this direction, so these principal axes of the **g** tensor and **T** tensor coincide.

This principal value of **g**, g_z, can be calculated from the equation $g_z = hv/|\beta|H_z$ where H_z is the maximum field strength. It is found that $g_z = 2.0024 \pm 0.0001$.

The other two values of **g**, g_x and g_y, as well as the corresponding principal axis directions, x and y, can be determined from the maximum and minimum values of the applied magnetic field that give the central ($m = 0$) resonance line when the applied field direction is perpendicular to z. It is found that $g_x = 2.0090 \pm 0.0001$ and $g_y = 2.0060 \pm 0.001$. The directions of these principal axes deviate from those of the nuclear hyperfine interaction by a rotation around z of about $10°$. Because the nuclear hyperfine anisotropy is so small in directions perpendicular to z the directions of the principal hyperfine axes x and y can only be determined to within $\pm 5°$; there is a similar uncertainty in the directions of the x and y principal axes of the **g** tensor.

The earliest study of nitroxide free radicals in single crystals was made by Griffith et al.[36] This study and others, such as that of Libertini and Griffith,[37] permitted determination of T and g values for several simple nitroxides.

Anisotropic paramagnetic resonance spectra can also be observed when nitroxide spin labels are attached to protein molecules in single crystals. For example, Berliner and McConnell[38] have shown that the active site of the enzyme α-chymotrypsin can be acylated in solution, and also in single crystals,[39] according to the following reaction:

Here E—OH designates the enzyme α-chymotrypsin, and the OH refers to the hydroxyl group of the active-site serine. Under the experimental conditions employed by Berliner and McConnell[39] the enzyme deacylates at a negligible rate in the single crystals, and paramagnetic resonance spectra can be observed from these crystals. More extensive studies of this reaction in solution and in single crystals have been carried out.[40,41]

[37] L. J. Libertini and O. H. Griffith, J. Chem. Phys. **53**, 1359 (1970).
[38] L. J. Berliner and H. M. McConnell, Proc. Natl. Acad. Sci. U. S. A. **55**, 708 (1966).
[39] L. J. Berliner and H. M. McConnell, Biochem. Biophys. Res. Commun. **43**, 651 (1971).
[40] L. J. Berliner and R. S. Bauer, J. Mol. Biol. **129**, 165 (1979).
[41] R. S. Bauer and L. J. Berliner, J. Mol. Biol. **128**, 1 (1979).

We now return to a brief discussion of the resonance spectra of oriented spin labels in single crystals, in terms of the spin Hamiltonian derived in Section 2.3.4 [cf. Eq. (2.3.91)],

$$\mathcal{H} = |\beta| \mathbf{S} \cdot \mathbf{g} \cdot \mathbf{H} + h\mathbf{S} \cdot \mathbf{T} \cdot \mathbf{I} + \beta_N g_N \mathbf{I} \cdot \mathbf{H}. \tag{2.3.98}$$

The energy levels giving rise to the angle-dependent spectra shown in (I) can be determined using the spin Hamiltonian by a straightforward quantum-mechanical calculation. Instead, we give simple physical arguments using (2.3.98) that lead to a good approximation to the observed energy levels and spectra.

The last two terms in the spin Hamiltonian are linear in the nuclear spin. Thus the nuclear spin magnetic moment $\beta_N g_N \mathbf{I}$ "sees" a magnetic field equal to

$$h\mathbf{S} \cdot \mathbf{T}/\beta_N g_N + \mathbf{H}, \tag{2.3.99}$$

where the first term arises from the hyperfine field due to the unpaired electron and the second term arises from the externally applied field. Note that in general the hyperfine field is not in the same direction as the externally applied field, and that the *direction* of the hyperfine field depends on the direction of the electron spin angular momentum $\mathbf{S}$. The smallest value of $|\mathbf{T}|$ is ~ -16 MHz, which corresponds to a field of the order of 5.2×10^4 G acting on an ^{14}N nucleus. Applied fields H of the order to 3300 G can then be neglected compared to this hyperfine field, at least as a first approximation. Under these conditions the simplified approximation spin Hamiltonian is simply

$$\mathcal{H} = |\beta| \mathbf{S} \cdot \mathbf{g} \cdot \mathbf{H} + h\mathbf{S} \cdot \mathbf{T} \cdot \mathbf{I}. \tag{2.3.100}$$

The second term is of the order of the nuclear hyperfine interaction, $\gtrsim 100$ MHz, and is small compared to the first term (~ 9500 MHz when $H \sim 3200$ G). To a good approximation the electron spin is quantized in a direction defined by the first term since $\mathbf{g}$ is very close to $g_0 \mathbf{U}$; the electron spin is quantized in the external field direction, defined by the unit vector $\mathbf{h}$:

$$\mathcal{H} = |\beta| S_h H \mathbf{h} \cdot \mathbf{g} \cdot \mathbf{h} + h\mathbf{S} \cdot \mathbf{T} \cdot \mathbf{I} \tag{2.3.101}$$

The ^{14}N nuclear spin "sees" a local field parallel to $h\mathbf{S} \cdot \mathbf{T}$; the secular component of this field is $hS_h \mathbf{h} \cdot \mathbf{T}$. If m is the quantized component of the nuclear spin in the local field direction, then the spin Hamiltonian becomes

$$\mathcal{H} = |\beta| S_h H \mathbf{h} \cdot \mathbf{g} \cdot \mathbf{h} + hS_h m (\mathbf{h} \cdot \mathbf{T} \cdot \mathbf{T} \cdot \mathbf{h})^{1/2}, \tag{2.3.102}$$

$$\mathcal{H} = |\beta| HS_h (g_z c_z^2 + g_x c_x^2 + g_y c_y^2) + hS_h m (T_z^2 c_z^2 + T_x^2 c_x^2 + T_y^2 c_y^2)^{1/2}. \tag{2.3.103}$$

Here c_z, c_x, c_y are the direction cosines of the field in the z, x, y principal axis system of the spin Hamiltonian. In terms of the polar and azimuthal angles θ and ϕ, these direction cosines are

$$c_z = \cos \theta, \qquad (2.3.104)$$

$$c_x = \sin \theta \cos \phi, \qquad (2.3.105)$$

$$c_y = \sin \theta \sin \phi. \qquad (2.3.106)$$

For brevity, let

$$g = g_z c_z^2 + g_x c_x^2 + g_y c_y^2 \qquad (2.3.107)$$

and

$$T = (T_z^2 c_z^2 + T_x^2 c_x^2 + T_y^2 c_y^2)^{1/2}. \qquad (2.3.108)$$

Again using the selection rule for allowed electron spin magnetic dipolar transitions in strong applied fields, $\Delta S_h = \pm\frac{1}{2}$, $\Delta m = 0$, we find that observed resonance transitions are observed at applied fields H for which

$$h\nu = |\beta| gH + hTm. \qquad (2.3.109)$$

The applied fields giving resonance transitions are therefore

$$H^{\text{res}}(m) = (h/(\nu - Tm)/g|\beta|) \qquad (2.3.110)$$

where $m = -1, 0, +1$.

2.3.8. Isotropic Distributions of Strongly Immobilized Labels: Powder Spectra

In many biophysical applications of spin labels one encounters resonance spectra arising from isotropic distributions of strongly immobilized labels. Such spectra can arise, for example, from labels attached to proteins in powdered crystals, or from labels rigidly attached to large, slowly tumbling proteins in solution.

The paramagnetic resonance spectrum of an isotropic distribution of strongly immobilized nitroxide free radicals is illustrated in Fig. 13. Certain qualitative features of this spectrum are easily understood and have been covered by Section 2.2.2 and illustrated by Fig. 6.

In isotropic distributions of nitroxide radicals, some radicals have orientations for which the principal axis z is perpendicular, or nearly perpendicular, to the applied field direction. These radicals will then exhibit hyperfine splittings that range between T_x and T_y, with the triplet hyperfine pattern corresponding to the x direction being centered about an applied field direction determined by g_x and the triplet hyperfine pattern corresponding

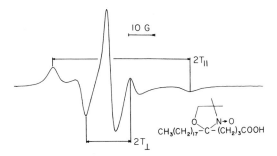

FIG. 13. Paramagnetic resonance spectrum of a fatty acid spin label, whose formula is shown, strongly immobilized in phosphatidylcholine bilayers.

to the y direction being determined by g_y. At X band these two fields are separated by ~ 3 G. Centered about each of these field positions are hyperfine splittings of T_x and T_y. These six resonance signals are then overlapped with the $m = 0$ component of the hyperfine triplet corresponding to the principal z direction. It is therefore not surprising that the "perpendicular" hyperfine structure is not well resolved. As we see below, however, inner hyperfine structure *is* well resolved when there is axial symmetry, $T_x = T_y$, $g_x = g_y$, and when the resonance linewidths are narrow.

The outer hyperfine signals in isotropic distributions have been very useful in many biophysical application of spin labels. We shall therefore discuss these outer signals in some detail.

The apparent relatively high intensity of the outer hyperfine signals in Fig. 13 reflects the fact that these signals correspond to the $H_0 \parallel z$ extremum in the hyperfine splitting pattern shown in Fig. 5a. Since the resonance position does not change rapidly with the angle θ between the applied field H_0 and z when θ is near $90°$, a relatively large number of molecules contribute resonance absorption at positions of extrema.

An interesting and very useful feature of these outer hyperfine extrema is their shape. The paramagnetic resonance spectrum seen in Fig. 13 is the *derivative* of the absorption spectrum of an isotropic distribution of radical orientations. As we show below, to a good approximation the outer signals of this derivative-curve spectrum have the *shape* of the *absorption curve* itself, for the special case that the applied field H_0 is exactly parallel to z.[12] Moreover, the separation of the peaks of these two outer "absorption curves" is to a very good approximation exactly equal to T_z.

Let the paramagnetic resonance absorption spectrum due to the hyperfine state m be $a_m(\xi)$, where

$$\xi = H = H_r^m(\theta, \phi), \qquad (2.3.11)$$

and $H_r^m(\theta, \phi)$ is the field position of the resonance line for hyperfine component m when the applied field direction in the principal hyperfine axis system is described by the usual polar and aximuthal angles θ, ϕ.

The contribution of hyperfine component m to the total absorption spectrum is

$$A_m(H) = \int_{\theta=0}^{\theta=\pi/2} a_m(\xi) \sin\theta \, d\theta. \qquad (2.3.112)$$

The corresponding contribution to the derivative-curve spectrum is

$$\frac{\partial Am}{\partial H} = \int_{\theta=0}^{\theta=\pi/2} \frac{\partial a_m(\xi)}{\partial \xi} \sin\theta \, d\theta. \qquad (2.3.113)$$

This derivative-curve line-shape expression can now be greatly simplified by using the following approximation to the resonance line position $H_r^m(\theta, \phi)$:

$$H_r^m(\theta, \phi) \simeq H_r^m(0, 0) + \lambda_m(1 - \cos\theta). \qquad (2.3.114)$$

This approximation is very important and needs to be considered in detail. Two approximations are in fact involved. First, the ϕ dependence of the line position is neglected, and the θ dependence is taken to vary as $\cos\theta$. The quantitative aspects of these approximations are illustrated in Fig. 14, which shows the resonance line positions $H_r^m(\theta, \phi)$ for the cholesterol spin label (2.3.I). The data in Fig. 14 are the same as the data in Fig. 5a, except that in the upper part of Fig. 14 a plot is given of $H_r^m(0, 0) + \lambda_m(1 - \cos)$ for $m = \pm 1, 0° \gtrsim \theta \gtrsim 70°$. In the lower half of this figure is given a rough plot of the total range of $H_r^m(\theta, \phi)$ for any possible value of ϕ. This range is due to

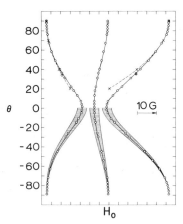

FIG. 14. Hyperfine splittings as a function of position of oriented spin labels. See Section 2.3.8 for further discussion.

deviations of the spin Hamiltonian from axial symmetry. From these plots it will be seen that the approximation in Eq. (2.3.14) is an adequate representation of the resonance line positions in the range $45° \gtrsim \theta \gtrsim 90°$ to an accuracy of the order of the linewidth itself (3–5 G). In the plot in the upper half of Fig. 14, $\lambda_1 = 18.6$ G (low-field line), and $\lambda_1 = -47.1$ G (high-field line).

The approximation in Eq. (2.3.114) can now be inserted in Eq. (2.3.113), with the understanding that only the derivative-curve spectrum arising from radicals for which $\theta \gtrsim 45°$ will be accurate. The indicated integration in Eq. (2.3.113) is readily carried out, using the approximation in Eq. (2.3.114), with the result

$$\partial Am/\partial H = (\lambda_1/m)a_m(H - H_r^m(0, 0)) + \cdots . \qquad (2.3.115)$$

Here $+ \cdots$ indicates the term arising from the $\theta = 0$ lower limit of the integration, which is dropped because Eq. (2.3.114) is invalid in this region. The function $a_m(H - H_r^m(0, 0))$ is of course just the absorption curve, which is centered at the position it would have in a perfectly oriented crystal for which the applied field $\mathbf{H}_0$ is parallel to the principal axis z. Note in Fig. 13 that the low-field signal is positive since λ_1 is positive, and that the high-field signal is negative since λ_{-1} is negative.

From this calculation it is clear that from the resonance spectrum of an isotropic distribution of strongly immobilized nitroxide spin labels one can deduce the hyperfine splitting parameter T_z. The observed separation of the outer hyperfine signals is equal to $2T_z$. The g factor g_z can be determined from the center point of these outer extrema. Further quantitative description of the strongly immobilized spectra may be obtained using computer calculations.[1,42,43]

2.3.9. Effects of Isotropic Motion on Spectra

The paramagnetic resonance spectrum of TEMPO shown in Fig. 11 exhibits three sharp hyperfine components, as discussed in Section 2.3.4. Since the probabilities for the three ^{14}N nuclear spin quantum states $m = 1$, 0, -1 are very nearly equal, the integrated intensities for the three hyperfine components are expected to be equal. But the derivative-curve peak heights are evidently not equal. This difference in the linewidths of the three hyperfine components can easily be shown to be related to the motional freedom of the nitroxide free radical.

In Section 2.3.7 we saw that the paramagnetic resonance spectrum of a nitroxide free radical depends on its orientation in space (relative to the

[42] D. D. Thomas and H. M. McConnell, *Chem. Phys. Lett.* **25**, 470 (1974).
[43] B. J. Gaffney and H. M. McConnell, *J. Magn. Reson.* **16**, 1 (1974).

direction of an applied magnetic field). Of the three resonance lines, the position of the central line ($m = 0$) changes least with angle of rotation, and the position of the high-field ($m = -1$) line changes most with angle (see Fig. 5a). In the presence of the rapid motion considered here, there ensues the phenomenon of *motional averaging*. That is, when the rapid, isotropic diffusion rotation of the spin label becomes fast enough, the microwaves are absorbed at the *average* spacing of the hyperfine energy levels, rather than the instantaneous spacing. This gives rise to the sharp three-line spectra seen in Fig. 11 rather than the orientation-dependent spectra, or strongly immobilized spectra, considered in Section 2.3.7. On the other hand, this rotational averaging is never perfect. Understandably, then, the largest residual linewidths even with very fast isotropic motion are found for the high-field line ($m = -1$), and the smallest linewidth is usually found for the central hyperfine line ($m = 0$).

The theory of hyperfine multiplet line shapes has been developed extensively. See Berliner[1] and additional references contained therein. The basic idea is that in the presence of molecular motion the spin Hamiltonian (2.3.91) becomes time dependent. This is because the axes of the hyperfine and g-factor dyadics $\mathbf{T}$ and $\mathbf{g}$ are molecule fixed, whereas the spin vectors $\mathbf{S}$ and $\mathbf{I}$ are effectively fixed in the laboratory frame. That is, in the presence of rapid motion the components S_h and I_h of $\mathbf{S}$ and $\mathbf{I}$ in the field direction are "good quantum numbers."

We now turn to a quantitative treatment of paramagnetic resonance line shapes of nitroxide spin labels for the case of *slow* isotropic rotation.[44] The simplest approach is to use the Bloch equations, modified to include the effect of diffusion:

$$\frac{d\mathbf{M}}{dt} = \gamma\mathbf{M} \times \mathbf{H} - \frac{1}{T_2}(u\mathbf{i} + v\mathbf{j}) - \frac{1}{T_1}(M_z - M_0)\mathbf{k} + D\nabla^2\mathbf{M}. \quad (2.3.116)$$

This modified Bloch equation was first introduced by Torrey[45] to describe the effects of diffusion in a liquid on nuclear magnetic resonance spectra. Here $\mathbf{M}$ is the spin magnetization per unit volume (or per unit area on a sphere; see below) and D is the diffusion constant. The term $D\nabla^2\mathbf{M}$ gives the change in this magnetization per unit time due to diffusion of spin magnetization into and out of this unit volume. For rotational diffusion the magnetization is represented as a function of θ and ϕ, $M = M(\theta, \phi, t)$. Here

$$M(\theta, \phi, t) \sin\theta \, d\theta \, d\phi$$

[44] R. C. McCalley, E. J. Shimshick, and H. M. McConnell, *Chem. Phys. Lett.* **13**, 115 (1972).
[45] H. C. Torrey, *Phys. Rev.* **104**, 563 (1956).

gives the spin paramagnetism due to radicals so oriented that the large applied field lies between θ and $\theta + d\theta$ and between ϕ and $\phi + d\phi$ in the principal axis system of the spin Hamiltonian of these radicals. For the simple case of a spin Hamiltonian with axial symmetry, we need only consider the dependence of $\mathbf{M}$ on θ, $\mathbf{M} = M(\theta, t)$. For this special case ∇^2 in Eq. (2.3.116) is

$$\nabla_\theta^2 = \frac{1}{\sin\theta}\left[\frac{d}{d\theta}\left(\sin\theta\,\frac{d}{d\theta}\right)\right]. \tag{2.3.117}$$

From differential calculus the quantity $\nabla^2 M(\theta)$ can be written as a limit,

$$\nabla^2\mathbf{M}(\theta) = \lim_{\Delta\to 0}\{(1/\Delta^2\sin\theta)[-2\sin\theta\cos\tfrac{1}{2}\Delta\mathbf{M}(\theta)$$
$$+ \sin(\theta + \tfrac{1}{2}\Delta)\mathbf{M}(\theta + \Delta) + \sin(\theta - \tfrac{1}{2}\Delta)\mathbf{M}(\theta - \Delta)]\} \tag{2.3.118}$$

The form of this equation suggests how the change of magnetization due to rotational diffusion can be approximated by discrete jumps of the magnetization between adjacent angular zones on the surface of a unit sphere. This approximation is described below.

Divide a unit sphere into N zones between $\theta = 0$ and $\theta = \tfrac{1}{2}\pi$, so that the angular width Δ of each of the zones is the same,

$$\Delta = \pi/2N. \tag{2.3.119}$$

(When the direction $\theta = 0$ is taken to be the direction of the applied strong magnetic field $\mathbf{H}$, only molecules with principal axis directions $0 \le \theta \le \tfrac{1}{2}\pi$ need be considered explicitly since the overall spectrum of this set of molecules is identical to the overall spectrum of the molecules for which $\tfrac{1}{2}\pi \le \theta \le \pi$.) The polar angle θ_n corresponding to the center of each zone is

$$\theta_n = (n - \tfrac{1}{2})\Delta, \tag{2.3.120}$$

and the zones are numbered $n = 1, 2, \ldots, n$.

The total electron magnetization in each zone is

$$\mathbf{M}_n = \int_{\theta_n - \Delta/2}^{\theta_n + \Delta/2} 2\pi r\sin\theta\mathbf{M}(\theta)\,d\theta. \tag{2.3.121}$$

For small Δ (and a unit sphere, $r = 1$),

$$\mathbf{M}_n \simeq 2\pi\sin\theta_n\mathbf{M}(\theta_n). \tag{2.3.122}$$

We may now substitute $\theta = \theta_n$ in Eq. (2.3.116) and replace $M(\theta_n)$ by $[1/(2\pi \sin \theta_n)]\mathbf{M}_n$, where $\mathbf{M}_n$ is the magnetization in the nth zone. When this substitution is made, one obtains the following differential equation for $\mathbf{M}_n$:

$$\frac{d\mathbf{M}_n}{dt} = \gamma\mathbf{M}_n \times \mathbf{H} - \frac{1}{T_2}(u_n\mathbf{i} + v_n\mathbf{j}) - \frac{1}{T_1}(M_{nz} - M_n^0)\mathbf{k}$$

$$- \frac{1}{t_n}\mathbf{M}_n + \pi(n+1 \to n)\mathbf{M}_{n+1} + \pi(n-1 \to n)\mathbf{M}_{n-1}. \qquad (2.3.123)$$

Here

$$\pi(n \pm 1 \to n) = (D/\Delta^2)\sin(\theta_n \pm \tfrac{1}{2}\Delta)/\sin(\theta_n \pm \Delta), \qquad (2.3.124)$$

for $1 \leq n, n + 1 \leq N$;

$$t_n^{-1} = (D/\Delta^2)\cos\tfrac{1}{2}\Delta, \qquad 1 \leq n \leq N - 1, \qquad (2.3.125)$$

$$= (D/\Delta^2)\cos\Delta/\cos\tfrac{1}{2}\Delta, \qquad n = N. \qquad (2.3.126)$$

The quantity t_n^{-1} gives the rate of spin magnetization leaving the nth zone due to rotational diffusion, and the quantities $\pi(n \pm 1 \to n)$ give the corresponding expressions for the rates of magnetization entering the nth zone from the adjacent zones. Equation (2.3.123) is a suitable form for computer program calculations; these equations have the same general form as do the Bloch equations when modified to include chemical exchange effects on nuclear resonance spectra.

We now must consider the fact that the paramagnetic resonance spectrum of a nitroxide free radical depends on the nitrogen nuclear spin quantum number m, and on the orientation of the strong applied field $\mathbf{H}_0$ relative to the principal hyperfine axis of the nitroxide spin Hamiltonian. When this spin Hamiltonian has axial symmetry, as assumed for the present calculation, we need only consider the angle θ, the angle between $\mathbf{H}_0$ and $\mathbf{k}$, the principal axis corresponding to the largest ^{14}N hyperfine splitting. The modified Bloch equations (2.3.123) include the effect of the anisotropic hyperfine interaction by regarding the effective field acting on the free electron as depending on θ_n, the angle describing the zone in which the principal axis of the radical lies. Under steady-state conditions, in the limit of no saturation, and in a coordinate system rotating at the microwave frequency, the Bloch equations are

$$0 = \left(\frac{-1}{T_2} - \frac{1}{t_n}\right)u_n + \gamma_e H_n' v_n + \pi(n+1 \to n)u_{n+1} + \pi(n-1 \to n)u_{n+1},$$

$$(2.3.127)$$

$$0 = -\gamma_e H_n' U_n + \left(\frac{-1}{T_2} - \frac{1}{t_n}\right)v_n + \gamma_e H_1 M_{0_n}$$

$$+ \pi(n+1 \to n)v_{n+1} + \pi(n-1 \to n)v_{n-1}. \qquad (2.3.128)$$

In these equations $\mathbf{H}$ is separated into its two components, $\mathbf{H} = H_0 \mathbf{k} + 2H_1 \mathbf{i} \cos(2\pi v_e)$. $H'_n = H_0 - H^{res}(n, m)$, where $H^{res}(n, m)$ is the resonance field for an electron in zone n, subject to a nuclear hyperfine field due to a nitrogen nucleus with spin component m.

A convenient approximate expression for $H^{res}(n, m)$ can be obtained directly from Eq. (2.3.110):

$$H^{res}(n, m) = (2\pi v/|\gamma_e|)[1 + (\Delta g/g_{\parallel}) \sin^2 \theta_n]$$
$$+ m(T_{\parallel}^2 \cos^2 \theta_n + T_{\perp}^2 \sin^2 \theta_n)^{1/2} \qquad (2.3.129)$$

where $\Delta g = g_{\parallel} - g_{\perp}$, and $m = 1, 0, -1$.

The above set of $2N$ simultaneous equations was solved by McCalley et al.[44] using a computer and employing a 3° angular mesh ($N = 30$ zones) and a one-gauss field mesh. A set of reference curves for obtaining correlation times τ_2 from the inward shifts due to slow motion of the outer hyperfine extrema was determined and compared with authentic spectra.

The Bloch equation calculations assume that the rotational motion is sufficiently slow that it is adiabatic—the local field acting at the nitrogen nucleus changes direction so slowly that no nuclear (or electron) spin transitions are induced. McCalley et al.[44] estimate that this approximation breaks down when the rotational motion moves the hyperfine field through an angular displacement of the order of a radian in a time of the order of the reciprocal of the anisotropy of the hyperfine interaction. This time is of the order of 5×10^{-7} sec, so this approximation introduces no significant error for approximating rotational correlation times in the range 2×10^{-8}–3×10^{-7} sec.

In the case of globular proteins we expect from the Debye equation

$$\tau_2 = 4\pi\eta a^3/3kT \qquad (2.3.130)$$

that the derived correlation times should depend linearly on the viscosity. Shimshick and McConnell[11] have used a particular extrapolation method in which the field shifts are plotted versus $(T/\eta)^{2/3}$. With a knowledge of the strongly immobilized hyperfine separation, the plot of correlation time τ_2 versus η/T was obtained. The data follow the Debye equation to within experimental accuracy. In cases where the same proteins have been studied using fluorescence spectroscopy the results of the two methods are in excellent agreement. A general theoretical treatment of the effects of slow motion on spin-label paramagnetic resonance spectra has been given by Freed (see Ref. 1, p. 53).

A novel technique, termed saturation transfer spectroscopy, has been developed by Hyde,[46] Thomas,[47] and Dalton[48] to determine very slow

[46] J. S. Hyde, Methods Enzymol. **49**, 480 (1977).
[47] D. D. Thomas, Biophys. J. **24**, 439 (1978).
[48] L. R. Dalton, Adv. Magn. Reson. **8**, 149 (1976).

molecular motions, with correlation times as long as a millisecond. Both the experimental and theoretical aspects of this method are too involved to permit discussion here. Thomas and McConnell[42] have given a comparatively elementary treatment of this problem using the Bloch equations.

2.3.10. Line Shapes with Fast Anisotropic Motion and Partial Orientation—Frequency Amplitude Order Parameters

High-frequency anisotropic motion can arise, for example, when a spin label is attached to a macromolecule through one or more bonds about which rapid rotation or rapid partial rotation (oscillation) can take place. This high-frequency anisotropic motion is also often found in membranes or in liquid crystals when molecules with anisotropic shapes are bound in anisotropic liquid environments by noncovalent bonds.

In the presence of molecular motion, the spin Hamiltonian $\mathcal{H}$ [cf. Eq. (2.3.91)] becomes time dependent. This is because (i) the dyadics $\mathbf{g}$ and $\mathbf{T}$ are anisotropic, (ii) the principal axes of these dyadics are fixed in the free-radical molecule, and (iii) the spin angular momentum vectors $\mathbf{S}$ and $\mathbf{I}$ are often fixed in the laboratory reference frame because they are quantized in the direction of the strong applied field. ($\mathbf{S}$ and $\mathbf{I}$ may sometimes be fixed to the laboratory reference frame simply because of conservation of angular momentum.) Consider the time-dependent hyperfine interaction $\mathcal{H}_{hf}(t)$,

$$\mathcal{H}_{hf}(t) = h\mathbf{S} \cdot \mathbf{T} \cdot \mathbf{I}. \tag{2.3.131}$$

Imagine that a magnetic field is applied in the direction $\mathbf{h}$, the field being so strong that both S_h and I_h are good quantum numbers. Since the components of $\mathbf{S}$ and $\mathbf{I}$ perpendicular to $\mathbf{h}$ precess at very different frequencies, their interaction averages to zero. Thus, to a good approximation in high fields, the hyperfine energy is

$$\mathcal{H}_{hf} = hS_h I_h \mathbf{h} \cdot \mathbf{T} \cdot \mathbf{h} = hS_h I_h(\gamma_{hx}^2 T_{xx} + \gamma_{hy}^2 T_{yy} + \gamma_{hz}^2 T_{zz}). \tag{2.3.132}$$

Here $\gamma_{hx} = \gamma_{hx}(t)$, $\gamma_{hy} = \gamma_{hy}(t)$, $h_{hz} = \gamma_{hz}(t)$ are the time-dependent direction cosines of the field direction $\mathbf{h}$ in the principal axis system x, y, z of the hyperfine Hamiltonian $\mathcal{H}_{hf}$, Eq. (2.3.131). If the molecular motion is sufficiently rapid, the observed interaction energy is the time average of $\mathcal{H}_{hf}$, namely $\mathcal{H}'_{hf}$:

$$\mathcal{H}'_{hf} = hS_h I_h(\overline{\gamma_{hx}^2} T_{xx} + \overline{\gamma_{hy}^2} T_{yy} + \overline{\gamma_{hz}^2} T_{zz}). \tag{2.3.133}$$

If the rapid molecular motion is *isotropic*, $\overline{\gamma_{hx}^2} = \overline{\gamma_{hy}^2} = \overline{\gamma_{hz}^2} = \frac{1}{3}$, we obtain the simple isotropic hyperfine spin Hamiltonian discussed above in connection with Eq. (2.3.93), since $T_{xx} + T_{yy} + T_{zz} = 3a$, where a is the isotropic hyperfine coupling constant. If the rapid molecular motion is *not* isotropic in the sense that $\overline{\gamma_{hx}^2} = \overline{\gamma_{hy}^2} = \overline{\gamma_{hz}^2} = \frac{1}{3}$, then there is a residual hyperfine interaction $\mathcal{H}'_{hf}$ which is necessarily anisotropic. The residual or effective hyperfine interaction can be represented by the hyperfine dyadic $\mathbf{T}'$:

$$\mathcal{H}'_{hf} = h\mathbf{S} \cdot \mathbf{T}' \cdot \mathbf{I}. \tag{2.3.134}$$

A quantitative relation between the elements of $\mathbf{T}'$ and $\mathbf{T}$ is easily obtained. Let the principal axes of $\mathbf{T}'$ be $\mathbf{i}', \mathbf{j}', \mathbf{k}'$, and let the strong applied field be in the direction of $\mathbf{k}'$. Then from Eq. (2.3.134),

$$\mathcal{H}'_{hf} = hS_{z'} I_{z'} T'_{z'z'}, \tag{2.3.135}$$

and from Eq. (2.3.133),

$$\mathcal{H}'_{hf} = hS_{z'} I_{z'} (\overline{\gamma_{z'x}^2} T_{xx} + \overline{\gamma_{z'y}^2} T_{yy} + \overline{\gamma_{z'z}^2} T_{zz}). \tag{2.3.136}$$

We obtain simply

$$T'_{z'z'} = \overline{\gamma_{z'x}^2} T_{xx} + \overline{\gamma_{z'y}^2} T_{yy} + \overline{\gamma_{z'z}^2} T_{zz}, \tag{2.3.137}$$

where $\overline{\gamma_{z'x}^2}$ and $\overline{\gamma_{z'y}^2}$ and $\overline{\gamma_{z'z}^2}$ are time averages of the direction cosines of z' in the x, y, z principal axis system of $\mathbf{T}$. Similar equations hold for the elements $T'_{y'y'}$ and $T'_{x'y'}$.

The nine quantities $\overline{\gamma_{ij}^2}$, $i = x', y', z', j = x, y, z$, are not all independent of one another. Thus there are three equations (one each for z', x', y') such as

$$\overline{\gamma_{z'x}^2} + \overline{\gamma_{z'y}^2} + \overline{\gamma_{z'z}^2} = 1, \tag{2.3.138}$$

and also three equations (one each for z, x, y) such as

$$\overline{\gamma_{x'z}^2} + \overline{\gamma_{y'z}^2} + \overline{\gamma_{z'z}^2} = 1. \tag{2.3.139}$$

The effective Hamiltonian is

$$\mathcal{H}' = |\beta| \mathbf{S} \cdot \mathbf{g}' \cdot \mathbf{H} + h\mathbf{S} \cdot \mathbf{T}' \cdot \mathbf{I} + \beta_N g_N \mathbf{I} \cdot \mathbf{H}. \tag{2.3.140}$$

The elements of $\mathbf{g}'$ are related to the elements of $\mathbf{g}$ by equations similar to those relating the elements of $\mathbf{T}'$ and $\mathbf{T}$ [e.g., Eq. (2.3.137)]. Since the principal axes of $\mathbf{g}$ and $\mathbf{T}$ nearly coincide for nitroxide free radicals, so do the principal axes of $\mathbf{g}'$ and $\mathbf{T}'$.

Instead of using the direction cosines $\overline{\gamma_{ij}^2}$ to describe the motional averaging of one axis system relative to the other, it is conventional to use a set of equivalent numbers, the order parameters S_{ij}, defined by the equations

$$S_{ij} = \frac{1}{2}(3\overline{\gamma_{ij}^2} - 1). \tag{2.3.141}$$

In terms of these order parameters the sum rules in Eqs. (2.3.138) and (2.3.139) have the equivalent forms

$$\sum_i S_{ij} = 0, \tag{2.3.142}$$

$$\sum_j S_{ij} = 0. \tag{2.3.143}$$

In the frequently studied liquid-crystal-like systems (phospholipid bilayers and biological membranes), motional averaging very often leads to effective Hamiltonians $\mathscr{H}'$ which account for observed spectra to within the experimental accuracy. In such cases, two independent order parameters can be determined. The relevant equations for this case are given below

$$T'_{z'z'} = a(1 - S_{z'z}) + S_{z'z} T_{zz} + \tfrac{1}{3}(S_{z'x} - S_{z'y})(T_{xx} - T_{yy}). \tag{2.3.144}$$

A similar equation holds for $g'_{z'z}$:

$$g'_{z'z'} = \bar{g}(1 - S_{z'z}) + S_{z'z} g_{zz} + \tfrac{1}{3}(S_{z'x} - S_{z'y})(T_{xx} - T_{yy}). \tag{2.3.145}$$

The principal elements of T' and g' can often be determined from experimental spectra using methods analogous to those described in Section 2.3.8. From observed values of $\mathbf{T}'$, $\mathbf{g}'$, $\mathbf{T}$, and $\mathbf{g}$ one can, in principle, determine all the elements of the order-parameter matrix. For example, observed values of T'_{zz} and $g'_{z'z'}$ enable one to solve Eqs. (2.3.144) and (2.3.145) for $S_{z'z}$ and $S_{z'x} - S_{z'y}$. These two quantities then permit the calculation of $S_{z'x}$ and $S_{z'y}$ since $S_{z'x} + S_{z'y} = 1 - S_{z'z}$. In a similar way, a determination of $T'_{x'x'}$ and $g'_{x'x'}$ yields the values of $S_{x'x}$, $S_{x'y}$, and $S_{x'z}$. The other order parameters $S_{y'x}$, $S_{y'y}$, and $S_{y'z}$ are obtained from the above order parameters and Eqs. (2.3.142) and (2.3.143).

Let $T'_{\parallel}$ and $T'_{\perp}$ ($g'_{\parallel}$ and $g'_{\perp}$) be the elements of $\mathbf{T}'$ (and $\mathbf{g}'$), where the effective Hamiltonian has axial symmetry and the symmetry axis is k'. From the foregoing equations one can obtain the following convenient set of equations:

$$\frac{T'_{\parallel} - T'_{\perp}}{T_{zz} - \tfrac{1}{2}(T_{xx} + T_{yy})} = S_{k'k} + \frac{1}{2} \frac{(T_{xx} - T_{yy})(2S_{k'i} + S_{k'k})}{T_{zz} - \tfrac{1}{2}(T_{xx} - T_{yy})}, \tag{2.3.146}$$

$$\frac{g'_{\parallel} - g'_{\perp}}{g_{zz} - \tfrac{1}{2}(g_{xx} + g_{yy})} = S_{k'k} + \frac{1}{2} \frac{(g_{xx} - g_{yy})(2S_{k'i} + S_{k'k})}{g_{zz} - \tfrac{1}{2}(g_{xx} + g_{yy})}. \tag{2.3.147}$$

These equations have been used by Gaffney and McConnell[43] for determinations of order parameters $S_{k'k}$ and $S_{k'i}$ (and hence $S_{k'j}$) for phospholipid spin labels incorporated into phospholipid bilayers. In this work it was found experimentally that $S_{k'i} \simeq S_{k'j} \simeq -\tfrac{1}{2}S_{k'k}$.

In view of the fact that T_{xx}, T_{yy}, and particularly $|T_{xx} - T_{yy}|$ are small for nitroxide free radicals, the second member of the right-hand side of Eq.

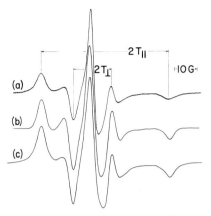

FIG. 15. Paramagnetic resonance spectra of a phospholipid spin label in an aqueous dispersion of egg lecithin–cholesterol (2:1 mole ratio), at room temperature. Applied field is ~ 3300 G. (a) Observed spectrum, (b) spectra calculated using Gaussian line shapes, (c) spectra calculated using Lorentzian line shapes. (Reprinted with permission from Hubbell and McConnell, *J. Am. Chem. Soc.* **93**, 314 (1971). Copyright by the American Chemical Society.)

(2.3.146) can often be neglected for fatty acid and phospholipid labels in bilayers and membranes, leading to the convenient expression

$$S_{k'k} = (T'_{\parallel} - T'_{\perp})/[T_{zz} - \tfrac{1}{2}(T_{xx} + T_{yy})]. \qquad (2.3.148)$$

When "order parameters" for nitroxide spin labels are referred to in the literature without further definition, they generally signify the order parameter $S_{k'k}$ given in Eq. (2.3.148), where $\mathbf{k}'$ is the axially symmetric symmetry axis, and z is the principal axis of the largest hyperfine splitting (π-orbital axis). Before discussing rapid, anisotropic molecular motion in more detail, let us illustrate the development up to this point with some early spectra and calculations by Hubbell and McConnell.[12]

Figure 15 shows the paramagnetic resonance spectra of the following phospholipid spin label incorporated in lipid bilayers of egg lecithin and cholesterol (2:1 mole ratio).

Values of $T'_{\parallel}$ are obtained from the observed spectra as indicated. First approximations to $T'_{\perp}$ are obtained as indicated, as in the first approximation to $g'_{\perp}$ (center point of the inner hyperfine extrema). The value of $g'_{\parallel}$ is obtained from the center point of the outer hyperfine extrema. These parameters were then used as the first step in the computation of spectra, using either isotropic Lorentzian or isotropic Gaussian line shapes. This process was then repeated until the best overall agreement between observed and calculated spectra was obtained. It will be seen in Fig. 15 that the comparison between observed and calculated spectra is good, but not perfect. For a guide to more general treatments of relaxation and resonance spectra, see Freed.[49] See also Lin and Freed,[50] Smith,[51] and Libertini et al.[52]

2.3.11. Effect of Electron–Electron Spin–Spin Interaction: Dipolar Interactions in Biradicals with Fixed Distances between Spins

The paramagnetic resonance spectra of organic free radicals are sometimes strongly affected by interactions between unpaired electrons. There are two interactions that affect the resonance spectra. The first is the coulomb-exchange (or simply "exchange") interaction. The second is the (electron spin)–(electron spin) magnetic dipolar interaction, analogous to the (electron spin)–(nuclear spin) magnetic dipolar interaction responsible for the aniso-tropic hyperfine interaction discussed in Section 2.3.4. The exchange inter-action is analogous to the contact hyperfine interaction only in that both are isotropic. The physical origins are entirely different: the isotropic hyperfine interaction is a magnetic interaction, and the isotropic electron spin exchange is coulombic. The *origin* of this interaction is discussed in textbooks on quantum mechanics and is not discussed further here.

An extreme case of a strong dipolar interaction is found in the case of a nitroxide biradical prepared by Keana and Dinerstein[53]:

II

[49] J. H. Freed, *J. Chem. Phys.* **66**, 4183 (1977).
[50] W.-J. Lin and J. H. Freed, *J. Phys. Chem.* **83**, 379 (1979).
[51] I. C. P. Smith, *Can. J. Biochem.* **57**, 1 (1979).
[52] L. J. Libertini, C. A. Burke, P. C. Jost, and O. H. Griffith, *J. Magn. Reson.* **15**, 460 (1974).
[53] J. F. W. Keana and R. J. Dinerstein, *J. Am. Chem. Soc.* **93**, 2808 (1971).

The two unpaired electrons in this molecule are quite close to one another. Molecular models suggest that the electrons are 3.8–5.2 Å apart, the uncertainty reflecting various plausible low-energy conformations of the piperdine ring.

Two unpaired electrons have a magnetic dipolar spin–spin interaction with one another that is analogous to the electron-nuclear spin–spin dipolar hyperfine interaction discussed in Section 2.3.4. For the case of two electrons, the equations analogous to the dipolar equations (2.3.57) and (2.3.61) are

$$\mathcal{H}_d = r^{-3}[\boldsymbol{\mu}_e \cdot \boldsymbol{\mu}'_e - 3(\boldsymbol{\mu}_e \cdot \mathbf{r})(\boldsymbol{\mu}'_e \cdot \mathbf{r})/r^2], \tag{2.3.149}$$

$$\boldsymbol{\mu}_e = -g_0|\beta|\mathbf{s}, \tag{2.3.150}$$

$$\boldsymbol{\mu}'_e = -g_0|\beta|\mathbf{s}', \tag{2.3.151}$$

$$\mathcal{H}_d = h\mathbf{s} \cdot \mathbf{T} \cdot \mathbf{s}'. \tag{2.3.152}$$

The paramagnetic resonance spectrum has been observed in solution, in rigid glasses, and in oriented multilayers. Although the nuclear hyperfine structure in these resonance spectra is not well resolved, large splittings due to the electron–electron dipolar interaction are evident in the observed spectra. Keana and Dinerstein have analyzed the observed spectra under the assumption that the colomb-exchange interaction is large compared to the nuclear hyperfine interaction, in which case the total electron spin S is a good quantum number, where

$$\mathbf{S} = \mathbf{s} + \mathbf{s}'. \tag{2.3.153}$$

In this case, the dipolar or "fine structure" contribution spin Hamiltonian can be written

$$\mathcal{H}_d = DS_z^2 + E(S_x^2 - S_y^2), \tag{2.3.154}$$

where

$$D = \tfrac{3}{4}T_z, \tag{2.3.155}$$

$$E = \tfrac{1}{4}(T_x - T_y). \tag{2.3.156}$$

The observed spectra can be accounted for with $D = 777$ MHz and $E \simeq 0$. The corresponding distance estimated for two point dipoles ($D = 80.6 \times 10R^{-3}$) is $\hat{R} = 4.7$ Å.

Dipolar interactions between the electron spins on different spin labels are often manifest as broadenings of the paramagnetic resonance lines. In some cases the relative distance and homogeneous orientations of pairs of spin labels are such that (electron spin)–(electron spin) dipolar fine-structure splittings due to spins on different spin-label molecules can be clearly resolved. A particularly interesting example can be found in the work of Marsh and Smith, who have observed dipolar splittings due to neighboring pairs of cholestane spin labels of (I) when incorporated into phosphatidylcholine-cholesterol bilayers. Marsh and Smith have used such data to study the "condensing effect" of the cholesterol bilayer structure.[54] Dipolar interactions in general have been discussed so extensively in the literature that there is no need to present a more elaborate discussion here.

Spin–spin dipolar interactions between spin labels and paramagnetic metal ions have often been observed in studies of metal-containing enzymes in solution. In the frequently encountered case where the electron paramagnetic relaxation rate of the metal ion is large compared to the dipolar interaction itself (in frequency units), this interaction may manifest itself in a most unusual way, as first described by Taylor et al.[55] and by Leigh.[56] For many enzymes, bound spin labels can exhibit a "strongly immobilized," powder-type spectrum since the rotational correlation times of the enzymes are large compared to the reciprocal of the hyperfine splittings, in frequency units. In the presence of an intense, fluctuating magnetic field, the resonance signal from the nitroxide group is essentially obliterated, except for those (nitroxide spin)–(metal ion spin) pairs whose orientational vectors make angles close to the magic angle, 54°, to the external magnetic field. For such molecules the dominant part of the dipolar interaction, proportional to $1 - 3\cos^2\theta$, vanishes. The net effect of this dipolar interaction is found both experimentally and in theoretically simulated spectra to produce an apparent reduction in the *amplitude* of the spin-label signal, with comparatively little change in the characteristic shape of the strongly immobilized spectrum.

2.3.12. Effect of Electron–Electron Spin–Spin Interaction: Spin–Spin Interaction Determined by Translational Motion and Resulting from Colliding Spins

In the foregoing discussion we have considered the magnetic resonance spectra of interacting pairs of paramagnetic centers held a fixed distance apart. Another important situation arises in liquids, or liquidlike systems, where paramagnetic spin labels are normally far apart and noninteracting, but where they occasionally diffuse together and undergo strong spin-exchange and/or dipolar interactions. Such effects have often been observed for various free radicals in solutions, and have in fact been studied in some detail for nitroxide free radicals. Sometimes the major spectral effects are due to the spin-exchange interaction, with the dipolar interaction only producing a relatively small broadening of the lines with increasing radical

[54] D. Marsh and I. C. P. Smith, *Biochim. Biophys. Acta* **298** (1973).

[55] J. S. Taylor, J. S. Leigh, and M. Cohn, *Proc. Natl. Acad. Sci. U. S. A.* **64**, 219 (1969).

[56] J. S. Leigh, *J. Chem. Phys.* **52**, 2608 (1970).

concentration. An example of broadening as a result of increasing head-group spin-labeled lipid concentration in the plane of a lipid bilayer membranes is illustrated by Fig. 22.

Just as is the case for chemical bonds, the spin-exchange interaction between nitroxide radicals depends strongly on the distance between the NO groups, because the overlap of electron wave functions is required for the effect. Thus, when the radicals in solution are far apart (i.e., ≥ 10 Å) the spin-exchange interaction can be assumed to be negligible. When the radicals come closer to one another, say to within the van der Waals radius, the spin-exchange interaction becomes very strong; that is, strong relative to the nuclear hyperfine interaction and electron–electron dipolar energies.

The spin-exchange interaction between nitroxide free radicals can be thought of as equivalent to a physical exchange of two electrons, one on each radical. Since electrons are, of course, identical to one another, it is more accurate to say that the spin-exchange interaction produces an exchange of the electron spin momentum (or magnetization) on the two radicals. This can be demonstrated in a formal quantum-mechanical way through the use of Dirac spin-exchange identity:

$$\mathbf{S}_i \cdot \mathbf{S}_j = \tfrac{1}{2}P_{ij} - \tfrac{1}{4}. \tag{2.3.157}$$

In this equation P_{ij} is an operator that permutes the labels of the electrons. For example, suppose that the spin wave function for the odd electron on the first radical is $\Phi_1(1) = a_1(t)\alpha(1) + b_1(t)\beta(1)$, and that the spin wave function for the odd electron on the second radical is $\Phi_2(2) = a_2(t)\alpha(2) + b_2(t)\beta(2)$. The Hamiltonian operator $hJ\mathbf{S}_i\mathbf{S}_2$ can then give rise to transitions between the states $\Phi_1(1)\Phi_2(2)$ and $\Phi_1(2)\Phi_2(1)$. This then corresponds to an interchange of spin angular momentum between the two radicals. We can now understand the effect of spin exchange on the paramagnetic resonance spectra of nitroxide free radicals. In an applied field of strength $\mathbf{H}_0$, and in the absence of spin exchange, a solution containing nitroxide free radicals contains three groups of electron spins, those for which the nitrogen nuclear spins are $I_z = 1, 0, -1$. The corresponding precession frequencies are $v = g|\beta|H_0h^{-1} - a, g|\beta|H_0h^{-1}, g|\beta|H_0h^{-1} + a$. Consider that the high-frequency (or low-field) resonance signal is being observed. In terms of the Bloch equations we see that magnetizations in the rotating frame $u(1)$ and $v(1)$ are different from zero under these conditions, whereas $u(0) = v(0) = u(-1) = v(-1) = 0$. In the presence of the exchange interaction, the $u(1)$ and $v(1)$ magnetization is transferred to the $I_z = 0$ and $I_z = -1$ states with no loss of magnitude or phase, and this loss of magnetization is replaced by the incoming zero magnetization of the $I_z = 0$ and $I_z = -1$ states. In the presence of slow exchange this then leads to a simple lifetime broadening of the resonance lines, and in the presence of rapid exchange (compared to a in

$\sec^{-1}$) to a narrowing of the resonance to a single line. Let us now consider this line broadening in more quantitative detail.

Let τ be the "lifetime" of a nitroxide radical between collisions. That is, let τ^{-1} be the probability per unit time that a nitroxide radical has a "collision" with a second nitroxide radical. Let τ_p be the duration of the collision. Here a "collision" means that the associated exchange interaction is of the order of magnitude of the hyperfine interaction, or greater. It is assumed that $\tau \gg \tau_p$.

In considering the effect of collisions on the paramagnetic resonance spectra, we must decide first whether these collisions are *strong* or *weak*. If $J\tau_p \gg 1$ the collision is *strong*, and if $J\tau_p \ll 1$ the collision is *weak*. In a strong collision the spin magnetization has, so to speak, no recollection of which radical it was on before the collision. Thus after a strong collision the spin magnetization has a 50:50 chance of having the spin magnetization of the colliding partner. Since nitroxide groups that are in direct van der Waals contact with one another doubtless have an exchange interaction J which is at least of the order of $\sim 10^{11} \sec^{-1}$, we see that a collision will be strong if $\tau_p \gtrsim 10^{-1}$ sec.

In dilute solutions of nitroxide free radicals (cf. Fig. 22) the weak broadening of the resonance lines can be attributed to a lifetime-limited state of the radical. Let T_2^{-1} be the broadening of any given resonance line due to spin-exchange collisions. Of the collisions that take place at a rate of τ^{-1}, only $\frac{2}{3}$ of these are between radicals with different components of the nitrogen nuclear spin angular momentum, and thus only $\frac{2}{3}$ of all collisions can be effective in line broadening due to spin exchange. Of these collisions, assumed to be *strong*, only $\frac{1}{2}$ result in a net spin exchange. Thus the linewidth enhancement due to spin-exchange collisions is

$$1/T_2 = \tfrac{2}{3}\tfrac{1}{2}1/\tau = \tfrac{1}{3}\tau. \qquad (2.3.158)$$

The lifetimes τ and τ_p of course depend on the detailed molecular properties of the liquid. For our present purposes it is sufficient to imagine that the liquid has a quasicrystalline structure, so that the molecular environment of each radical has a relatively well-defined structure. If the liquid is "ideal" the lifetimes τ and τ_p are related by the equation

$$\tau/\tau_p = N_s/zN_R, \qquad (2.3.159)$$

where N_s and N_R are the concentrations of solvent and radical in the solution. In this quasi-crystalline model τ_p is the average time that the radical stays in its site, and τ_p^{-1} is its hopping frequency. The number of new neighbors encountered with each step is z. If D is the diffusion constant for the free radical in the solvent of interest, we have

$$D = \tfrac{1}{6}(1/\tau_p)\lambda^2, \qquad (2.3.160)$$

where λ is the "jump distance," the distance moved in each step. Equation (2.3.159) can then be written

$$\tau^{-1} = (zN_R/N_s)6D/\lambda^2. \tag{2.3.161}$$

Therefore, in the limit of infrequent but strong spin-exchange collisions, the the enhancement of the linewidth is

$$1/T_2 = \tfrac{1}{3}(zN_R/N_s)(6D/\lambda^2). \tag{2.3.162}$$

In this limit the enhancement of the linewidth is seen to be proportional to the free-radical concentration.

The case of *rapid* spin exchange between radicals corresponds to an exchange rate that is large compared to the hyperfine splitting a. It can be shown that, for a three-line spectrum, the strongly narrowed residual linewidth is of the form[57]

$$1/T_2 = \tfrac{2}{27}\tau[(\omega_1 - \omega_2)^2 + (\omega_2 - \omega_3)^2 + (\omega_3 - \omega_1)^2], \tag{2.3.163}$$

where ω_1, ω_2, and ω_3 are the angular resonance frequencies of the three lines:

$$(\omega_1 - \omega_2)^2 = (\omega_3 - \omega_2)^2 = (2\pi a)^2 \tag{2.3.164}$$

and

$$(\omega_1 - \omega_3)^3 = (4\pi a)^2. \tag{2.3.165}$$

Thus

$$1/T_2 = \tfrac{4}{3}a^2\tau. \tag{2.3.166}$$

Miller et al.[58] have carried out a study of the paramagnetic resonance spectra of di-t-butyl nitroxide in the solvent dimethylformamide. These authors found that in the slow-exchange region the hyperfine linewidths are proportional to the radical concentration. These investigators find for the enhanced linewidth

$$1/T_2 = 3.7 \times 10^9(N_R/N_s). \tag{2.3.167}$$

If we return to Eq. (2.3.162) and make the plausible assumptions that $z \sim 6$ and $\lambda \sim 3.5 \times 10^{-8}$, then Eq. (2.3.167) leads to an experimental diffusion constant of $D \sim 3.8 \times 10^{-6}$ cm^2/sec.

In the above discussion, we have considered the effect on paramagnetic resonance line shapes of spin-exchange collisions between spin labels for two

[57] A. Hudson and G. K. Luckhurst, *Chem. Rev.* **69**, 191 (1969).
[58] T. A. Miller, R. N. Adams, and P. M. Richards, *J. Chem. Phys.* **44**, 4022 (1966).

limiting cases. One case corresponds to strong but very frequency collisions. Spectra in the intermediate range can be calculated using the Bloch equations, modified to include the effect of spin-exchange collisions. We have already used modified Bloch equations to treat the effect of slow motion on spin-label spectra. Thus we give here only a very brief discussion of the Bloch equations to treat the spin-exchange problem.

Assume for the moment that a nitroxide spin label gives three Lorentzian resonance lines with corresponding transverse relaxation times $T_2(m)$, where the nitrogen nuclear spin quantum numbers m are $m = 1, 0, -1$. Let u_m and v_m be the out-of-phase and in-phase electron spin magnetizations, and let ω_m be the three corresponding angular resonance frequencies in a frequency-sweep spectrum. The Bloch equations that allow for the effects of strong, random exchange collisions on u_m and v_m are

$$u_m + (\omega - \omega_m)v_m = -u_m\left(\frac{1}{T_2(m)} + \sum_{m'}\frac{1}{\tau_{mm'}}\right) + \sum_{m'}\frac{u_{m'}}{\tau_{m'm}}, \qquad (2.3.168)$$

$$v_m - (\omega - \omega_m)u_m = -v_m\left(\frac{1}{T_2(m)} + \sum_{m'}\frac{1}{\tau_{mm'}}\right) + \sum_{m'}\frac{v_{m'}}{\tau_{m'm}} - \gamma H_1 M_z m. \qquad (2.3.169)$$

In these equations $\tau_{mm'}^{-1}$ is the probability per unit time that electron spin magnetization with Larmor frequency ω_m will undergo a transition (due to spin-exchange collisions) to a state where the Larmor frequency is $\omega_{m'}$ (where $m \neq m'$). Since the populations of the three m states are essentially the same, and the spin-exchange interaction affects each of these states equally, one can use the parameter $\tau^{-1} = 2\tau_{mm'}^{-1}$ to measure the first-order decay of electron spin magnetization from each of the states m. The quantity $1/\tau$ can be referred to conveniently as the "spin transfer rate," the probability per unit time that an electron spin moment will undergo a transition, due to spin-exchange collisions, such that it experiences a new nuclear hyperfine field. From our earlier discussion we see that only $\frac{1}{3}$ of all the molecular collisions are effective for spin transfer, so

$$1/\tau = \tfrac{1}{3}K, \qquad (2.3.170)$$

where K is the frequency of the collisions; that is, the probability per unit time that one spin label will encounter another in a "strong" collision.

The Bloch equations with spin-exchange terms have been used to determine the rates of lateral diffusion of lipid spin labels in phospholipid bilayers[59] and in biological membranes.[60]

[59] P. Devaux, C. J. Scandella, and H. M. McConnell, *J. Magn. Reson.* **9**, 474 (1973).
[60] P. Devaux and H. M. McConnell, *Ann. N. Y. Acad. Sci.* **222**, 56 (1973).

For example, Devaux, Scandella, and McConnell[59] incorporated the phospholipid spin label (1,14) in phospholipid bilayers of egg lecithin and

$$
\begin{array}{c}
\text{(structure of phospholipid spin label (1,14))}
\end{array}
$$

other phospholipids, and also into membranes derived from the sarcoplasmic reticulum of rabbit skeletal muscle. The incorporation of this label up to concentrations of the order of a few percent of all the phospholipids has a relatively small effect on the rate of Ca^{2+} ion uptake by these vesicles, in the presence of ATP.[60] This particular label was chosen since the paramagnetic group is nearest the most flexible end of the chain, and thus, in an isotropic sample, the resonance signals are nearly isotropic and sharp (cf. Section 2.3.10). Through comparisons of observed phospholipid spin-label spectra in the membranes as a function of label concentration and calculated spin-label spectra as a function of concentration it was possible to deduce the rate of lateral diffusion of the above phospholipid spin label in these various membranes. The lateral diffusion coefficients so obtained were of the order of $D \simeq 2 \times 10^{-8}$ cm^2/sec. A similar approach was used by Träuble and Sackmann in an early study of the lateral diffusion of phospholipid spin labels in model bilayer membranes.[61] An alternative approach to this measurement of lateral diffusion using spin-labeled lipids is discussed later (Section 2.4.4.)

2.3.13. Enhancement of Nuclear Relaxation by Spin Labels

Because the spin magnetic moment of the electron is so much larger than the magnetic moments of nuclei, nitroxide spin labels introduced into biological systems can have large effects in enhancing nuclear relaxation, compared to relaxation due to the interactions of the nuclei among themselves. For example, Kornberg and McConnell[62] and Brûlet and McConnell[63] used spin-label-phospholipid-enhanced nuclear relaxation of nonlabeled lipids to study lateral diffusion of phospholipids in model membranes. Many elegant studies of paramagnetic enhanced nuclear relaxation (a field pioneered by M. Cohn) have been carried out to provide significant structural and kinetic information on enzymes and other macromolecular systems.

[61] H. Träuble and E. Sackmann, *J. Am. Chem. Soc.* **94**, 4499 (1972).
[62] R. D. Kornberg and H. M. McConnell, *Proc. Natl. Acad. Sci. U. S. A.* **68**, 2564 (1971).
[63] P. Brûlet and H. M. McConnell, *Proc. Natl. Acad. Sci. U. S. A.* **73**, 2977 (1976).

Available space makes it impossible to summarize this large body of work here. For leading references, the reader is referred to the books by Berliner[1] and Dwek.[10]

2.4. The Use of Spin Labels as Antigenic Determinants Capable of Reporting Their Physical State

2.4.1. Background

Our interest in spin-label lipid haptens centers on the control of immune responses to hapten-bearing synthetic lipid membranes by physical factors pertaining to such membranes. For example, is a particular immune response helped or hindered by rapid motion of antigenic determinants in the plane of the membrane? Is there an optimal spacing for such determinants and an optimal distance they should extend from the surface of the membrane? Do different immune responses have different requirements with regard to the physical state of membrane-bound determinants?

With answers to such questions we hope to be in a position to elucidate molecular mechanisms controlling immune function at the level of the cell membrane. Specific examples of immune responses and our experimental approach towards their elucidation are referred to in Section 2.4.8. These examples only concern antibody-dependent efferent responses. We have commenced work in other areas of immune function, but we have not yet employed spin labels in these endeavors.

We would like to make two important points regarding the use of stable nitroxides as antigenic determinants. The first is that this concept represents a significant departure from that which led to their more common use as molecular labels or probes of systems in which they have no part except as "reporters." When used in that way it is important that the experimental design and interpretation take account of possible inappropriate perturbation of the system by the presence of these rather large reporters. When spin labels are used as antigenic determinants they are *an essential participant in the action*: their function is to perturb, or control, the system in whatever way is natural. The second important point is that, to our knowledge, these studies represent the first attempt to use antigenic determinants capable of directly reporting their physical state to answer very important questions regarding the control of immune function by physical factors concerning antigen presentation.

However, we consider that the examples presented in this section serve to illustrate the various basic types of information that may be obtained by using spin labels, whether they be used as bystander reporters or as central components of the system of interest. Among these examples are several that serve to illustrate a characteristic of spin labels which has been under-exploited, their susceptibility to agents which chemically (or photochemically) destroy their paramagnetic properties. Such techniques are particularly useful for studying systems composed of several compartments that differ in the ease with which they are penetrated by spin labels and/or reagents.

2.4.2. Spin-Label Lipid Haptens

Three different spin-label lipid haptens have been used successfully to study the physical properties of membranes and certain relatively simple immune responses to membranes. The structures of these are shown below.

III

IV

V

All are soluble in organic solvents such as ethanol and chloroform, but form ordered aggregates in aqueous dispersion. When haptens are mixed with other suitable amphipathic lipids, such as phosphatidylcholine, and introduced to an aqueous medium by an appropriate method, continuous lipid bilayer structures are formed. These include the lipid haptens in the manner illustrated by Fig. 16.[64] Further details regarding particular types of lipid structures are given, where relevant, below. (Part 11 is devoted to lipid bilayers.)

[64] P. Brûlet, G. M. K. Humphries, and H. M. McConnell, *in* "Structure in Biological Membranes" (S. Abrahamsson and I. Pascher, eds.), pp. 321–329. Plenum, New York, 1977.

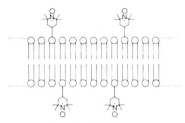

FIG. 16. Diagrammatic representation of a lipid bilayer containing a spin-label lipid hapten. (From Brûlet *et al.*[64])

2.4.3. Determination of the Rate of Inside–Outside Transitions of Phospholipids in Vesicle Membranes

Compound **III** was used by Kornberg and McConnell[65] to demonstrate that the rate of inside–outside transitions (flip-flop) of phospholipids in vesicle membranes is slow. This phenomenon has important consequences with respect to maintenance of the asymmetry of biological membranes,[66] the major structural feature responsible for the vectorial nature of membrane function. We feel safe in assuming that asymmetry is of fundamental importance for membrane-dependent cellular immune function; signals received on the external side are believed to be modulated by the membrane to give appropriate messages to the cell.

The experimental procedure used to determine the rate of flip-flop involved preparation of small (~ 250 Å diameter) single-compartmented egg phosphatidylcholine vesicles by sonication. At 0°C, such vesicles protect trapped water-soluble spin labels from chemical reduction, and consequential loss of paramagnetic properties, by sodium ascorbate (see Section 2.2.1). Therefore the membranes of similar vesicles lacking water-soluble spin labels but including compound **III** in the manner illustrated by Fig. 16 become asymmetric when treated with ascorbate at 0°C, spin labels being exclusively on the internal face of the vesicle membranes. The rate at which equilibrium is approached, at different temperatures, was used to determine rate constants and activation energies for the process. A typical result was that the asymmetry decayed with a half-time of 6.5 hr at 30°C. Using other lipids, even slower rates for flip-flop have been reported since these measurements were made. (See Ref. 66.)

2.4.4. Determination of Lateral Diffusion of Spin-Label Lipid Haptens in Membranes

The first measurements of the rate of lateral diffusion of spin-labeled lipids in membranes were made by Devaux and McConnell[67] and Träuble and

[65] R. D. Kornberg and H. M. McConnell, *Biochemistry* **10**, 1111 (1971).

[66] J. E. Rothman and J. Lenard, *Science (Washington, D.C.)* **195**, 743 (1977).

[67] P. Devaux and H. M. McConnell, *J. Am. Chem. Soc.* **94**, 4475 (1972).

Sackmann.[61] Devaux and McConnell created a nonequilibrium distribution of labeled and nonlabeled lipids in multibilayer membranes, then followed the time dependence of the spin–spin-mediated line broadening to deduce the lateral diffusion coefficient. (An important early clue as to the rapid lateral diffusion of lipids in vesicles was provided by Kornberg and McConnell,[62] who studied the dipolar effect of spin labels on the proton nuclear resonance spectra of nonlabeled lipids to infer the rapid lateral motion of lipids.)

The lateral diffusion of spin-labeled lipid haptens in hydrated planar phospholipids multilayers has been measured by Sheats and McConnell using low probe concentrations and a photochemical reaction to generate an initial concentration gradient.[68] The equation for this reaction is given below. The product is stable and nonparamagnetic.

$$[Co^{III}(CN)_5(CH_2CO_2-)]^{4-} \ + \ R-\overset{hv}{\longrightarrow}$$

$$Co^{II}(CN)_6^{3-} \ + \ R-\boxed{\;\;\;}\;N-O-CH_2CO_2^-$$

For this work the EPR spectrometer is interfaced with a computer programmed for double integration of spectra. Sequential exposures of partially masked multilayer samples to light from an argon ion laser (3511 and 3534 Å) are followed by determination of residual spin-label signal by EPR spectroscopy and computer integration. Diffusion constants calculated from such data are similar to those obtained using other methods (cited in Ref. 68). For example, at 48°C, the diffusion coefficient for dipalmitoyl phosphatidylcholine is found to be 9.9×10^{-8} cm^2/sec.

2.4.5. Determination of the Surface Area of Lipid Membranes

As continuous lipid bilayer preparations are rarely unilamellar, their external surface area cannot be determined by simple calculation. It is frequently important to know this quantity, or, for work involving immune responses to spin-label lipid haptens, to know the number of such molecules accessible to the exterior of lipid structures. A technique similar to that discussed in Section 2.4.4 has therefore been used by Schwartz and McConnell to determine these quantities.[69]

[68] J. R. Sheats and H. M. McConnell, *Proc. Natl. Acad. Sci. U. S. A.* **75**, 4661 (1978).
[69] M. A. Schwartz and H. M. McConnell, *Biochemistry* **17**, 837 (1978).

The number of externally exposed spin-label haptens is a measure of the external surface area of membrane structures containing a low concentration of such molecules. Photochemical reduction of exposed haptens is induced by an argon ion laser in the presence of excess 4-(N,N,N-trimethylamino) benzyl pentacyanocobaltate ions. This ion pentrates phospholipid bilayers very slowly, is stable in light and air for several days, and is highly specific for photolysis of nitroxides even in the presence of air and phospholipids. These properties make it more suitable for this application than the ion used by Sheats and McConnell.[68] The reaction proceeds via an alkyl free radical split off from the pentacyanocobaltate complex. The rate of reaction is first rapid and then slow. This latter rate is thought to result from slow leakage of radicals through the membrane. In order to determine the proportion of external labels, reduction of spin-label EPR signal intensity is, therefore, measured at various times after illumination. For the commonly used multi-compartmented "liposomes" (prepared by hand-shaking dried lipid films with aqueous solutions after hydration for a suitable length of time and at an appropriate temperature) the external signal is between 5.0 and 5.5 % for all preparations of dimyristoylphosphatidylcholine or dipalmitoylphosphatidylcholine. This figure is also obtained when such bilayers include up to 50 mole % cholesterol.

2.4.6. Antibodies Specific for Spin-Label Determinants

Antibodies specific for TEMPO have been prepared by immunizing rabbits either with spin-labeled hemocyanin, to prepare IgG,[70] or with spin-labeled bacterial flagellin, to prepare IgM.[71] They have been purified with respect to immunoglobulin class and specificity for spin labels.[70-73] We are currently producing monoclonal antibodies with specificity for TEMPO.

As described in Sections 2.2.3 and 2.3.6, when spin labels are in rapid isotropic motion their hyperfine splittings are averaged. Therefore the amplitude of the peaks (or lines), particularly the low- and high-field lines, is relatively high (due to superimposition of the contributions from many labels). Reference to Figs. 6 and 7 will, we hope, suffice to convince the reader that immobilization of spin labels, resulting as it does in the representation of very many different splittings, can be correctly predicted to produce a

[70] G. M. K. Humphries and H. M. McConnell, *Biophys. J.* **16**, 275 (1976).

[71] G. M. K. Humphries and H. M. McConnell, *Proc. Natl. Acad. Sci. U. S. A.* **74**, 3537 (1977).

[72] P. Rey and H. M. McConnell, *Biochem. Biophys. Res. Commun.* **73**, 248 (1976).

[73] D. G. Hafeman, J. W. Parce, and H. M. McConnell, *Biochem. Biophys. Res. Commun.* **86**, 522 (1979).

decrease in the amplitude of the lines, particularly the high- and low-field lines.

When specific antibodies bind to TEMPO groups, either presented at the surface of a membranes or as constituents of a low-molecular-weight, water-soluble species,[70] the EPR spectrum changes dramatically, indicating a marked reduction in motion. The new spectrum is a typical "powder spectrum," except that the hyperfine splittings are remarkably broad $(2T_{\parallel} = 78-79$ G, in contrast to the usual maximum $2T_{\parallel}$ of approximately 69 G). It is probable that the charge separation in the NO bond is enhanced by an electrostatic interaction within the antibody combining site, such as a hydrogen bond to the oxygen atom. This would increase the unpaired electron density on the nitrogen atom, relative to that previously obtained from samples giving rise to powder spectra, and would lead to increased splittings as discussed in Section 2.2.4.

These effects are illustrated well by Fig. 17, which shows the effect of mixing immunoglobulins, with or without specificity for TEMPO, with TEMPO-bearing bilayers. As the signal intensity of the "rapid motion" spectral component is proportional to the concentration of unbound spin labels, this principle has been used to determine the affinity of antibodies specific for TEMPO.[72] Univalent haptens are employed for such measurements. An unexpected result is that the affinity of such antibodies for the reduced form of a TEMPO derivative is very close to that for the TEMPO derivative itself. Although this cross reactivity surprises us, it probably accounts for the fact that anti-TEMPO activity can be obtained; there is a high probability that the spin-label immunogen is reduced *in vivo* before stimulating the antibody response.

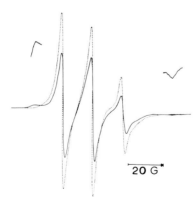

FIG. 17. EPR spectra of large single-compartmented dimyristoylphosphatidylcholine vesicles containing 2.5% of spin-label hapten III together with IgG prepared from rabbit anti-TEMPO hemocyanin serum (—) or IgG prepared from rabbit antihuman erythrocyte serum (---). The vesicles were prepared by dialysis of a detergent solution of lipids. (From Brûlet et al.[64])

2.4.7. Physical Factors Affecting the Binding of Spin-Label Specific Antibodies to Hapten-Bearing Bilayers

Changes in (i) the length of the hydrophilic moiety of spin-label haptens, (ii) the fluidity of their host bilayers, (iii) the cholesterol content of those bilayers, and (iv) the two-dimensional concentration of haptens in bilayers, have all been shown to affect the EPR spectra of spin-label hapten-bearing model membranes. All four of these types of variation are also accompanied by changes in the ability of such membranes to bind TEMPO-specific antibodies and to interact with complement in an antibody-dependent manner. Our view is that such events are likely to parallel physiologically important events concerning antibody-dependent immune effector mechanisms such as complement-mediated cell lysis and certain types of cell-mediated lysis.

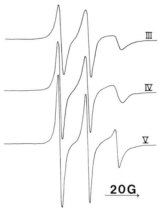

FIG. 18. Dipalmitoylphosphatidylcholine bilayers containing 0.5 mole % of each of the three haptens (Structures **III–V**); see Section 2.4.2. Spectra were recorded at 32°C and stored, integrated, normalized, and replotted by means of a Digital PDP 8/e computer interfaced with a Varian E12 EPR spectrometer. For these three labels, the relative heights of the spectral lines indicate increased motion of the nitroxide head group as the length of the hydrophilic moiety of the molecule is increased.

The EPR spectra shown in Fig. 18 indicate that the mobility of the head group increases with the length of the hydrophilic moiety for the case of membrane-bound spin-label haptens **III–V**. Antibody binding has also been shown to be positively correlated with the length of the hydrophilic moiety for the case of these three haptens (in bilayers of phosphatidylcholine or phosphatidylcholine/cholesterol).[74]

[74] P. Brûlet and H. M. McConnell, *Biochemistry* **16**, 1209 (1977).

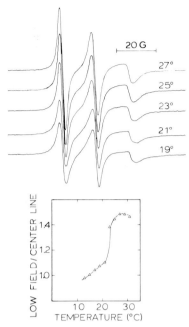

FIG. 19. Electron paramagnetic resonance spectra of dimyristoylphosphatidylcholine liposomes containing 0.5 mole % spin-label hapten **IV**, together with a plot of temperature versus the ratio of the amplitude of the low-field line to that of the center line. The principal phase transition temperature of dimyristoylphosphatidylcholine is 23°C. (It should be noted that melting the bilayers both sharpens the spectrum, by increasing the rate of motion, and broadens it, by increasing the collision rate of spin-label haptens. In this instance, because the hapten is present at a low concentration in the membrane, the dominant effect is that due to increasing the rate of motion. When haptens are already sufficiently concentrated to show significant spin-exchange broadening in the solid state, melting is accompanied by an appreciable increase in broadening. (From Humphries.[75])

Figure 19 illustrates the spectral consequences of melting bilayers containing a spin-label lipid hapten.[75] The predictable increase in mobility of the head group is indicated. Agglutination of haptenated liposomes or vesicles by amounts of IgM or IgG antibodies which do not saturate the available binding sites is strongly favored by the fluid state of the membranes. This effect can be readily observed when membranes and antibodies are mixed at a temperature below the principal phase transition temperature and then warmed slowly through it; a marked increase in agglutination commences at the phase transition (G. M. K. Humphries, unpublished observation). Figure 20 shows the binding of IgG antibodies to liposomes composed of two

[75] G. M. K. Humphries, in "Liposomes in Biological Systems" (P. Gregoriadis and A. Allison, eds.), pp. 345–375. Wiley, New York, 1980.

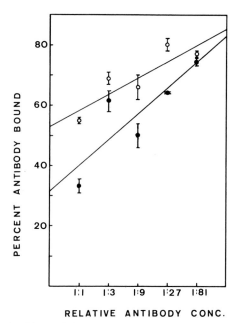

RELATIVE ANTIBODY CONC.

Fɪɢ. 20. Binding of rabbit antinitroxide IgG to dimyristoylphosphatidylcholine (○) or dipalmitoylphosphatidylcholine (●) liposomes, containing spin-label hapten **IV**, as a function of antibody dilution. The 1:1 specific antibody concentration was estimated to be of the order of $10^{-7}M$ and the liposomes were present at a concentration of $2 \times 10^{-4}M$ total phospholipid. Hapten was 1 mole % of total lipid. Since ca. 5% of the total hapten on these liposomes is on the outermost membrane surface, the total exposed hapten concentration was of the order of $10^{-7}M$. After 15 min incubation at 37°C, lipsomes were removed by ultracentrifugation, and residual antibody was assayed by a Wasserman–Levine complement fixation assay[76] in which the antigen was liposomes consisting of cholesterol, dimyristoylphosphatidylcholine, and spin-label hapten **IV** in the molar ratio 50:45:5. (Previously unpublished data, G. M. K. Humphries.)

different phosphatidylcholines at a temperature which is above the principal thermotropic phase transition temperature of one lipid but not of the other. In both cases it is apparent that the percentage of antibody bound increases as the ratio of antibody/liposomes decreases. It is also evident that "fluid" liposomes bind more antibody than "solid" liposomes under these experimental conditions, but that the differences in absolute amounts bound to the two types of liposomes become less significant as antibody/lipid decreases.

The cholesterol content of phosphatidylcholine membranes has a marked effect on the spectra of spin-label haptens included with these lipids. This is illustrated by Fig. 21. The observed spectral changes indicate that the rate of motion of the head group is increased by increasing the cholesterol content.

[76] E. Wasserman and L. Levine, *J. Immunol.* **87**, 290 (1961).

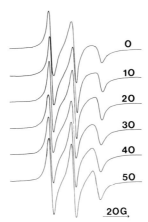

FIG. 21. EPR spectra of liposomes composed of dipalmitoylphosphatidylcholine and the various molar proportions of cholesterol, in the range 0–50 mole %, indicated on the figure. Liposomes contained 3 mole % spin-label hapten **IV**. Spectra were stored, integrated, normalized, and replotted by means of a Digital PDP 8/e computer interfaced with a Varian E12 EPR spectrometer. Similar spectral changes are observed when dimyristoylphosphatidylcholine bilayers are used. The temperature was 35°C.

This phenomenon has been studied recently by Rubenstein *et al.*[77] In brief, it is evident that the effect of cholesterol on the physical state of lipid haptens in phosphatidylcholine bilayers is complex. In addition to its positive or negative effect on lateral mobility of membrane molecules,[78–80] EPR spectra indicate that cholesterol, at ~20–50 mole %, permits increased mobility of spin-label hapten head groups, possibly because of the increased rotational freedom resulting from a decrease in their interaction with phosphatidylcholine head groups. Antibody binding is favored by inclusion of cholesterol at 20–50 mole % in such bilayers.[74]

We have discussed three examples of physical changes which control antibody binding, and in each case increased mobility of the spin-label head group, as indicated by EPR spectroscopy, appears to favor binding. A fourth type of physical change which affects antibody binding is the two-dimensional concentration of haptens in the membranes. Figure 22 shows the spectral changes observed when the distance between adjacent haptens is decreased. The broadening of the lines is typical of spin exchange, and this result is consistent with a homogeneous distribution of spin labels in the plane of the membranes.

[77] J. L. R. Rubenstein, J. C. Owicki, and H. M. McConnell, *Biochemistry* **19**, 569 (1980).

[78] J. L. R. Rubenstein, B. A. Smith, and H. M. McConnell, *Proc. Natl. Acad. Sci. U. S. A.* **76**, 15 (1979).

[79] B. R. Copeland and H. M. McConnell, *Biochim. Biophys. Acta* **599**, 95 (1980).

[80] J. C. Owicki and H. M. McConnell, *Biophys. J.* **30**, 383 (1980).

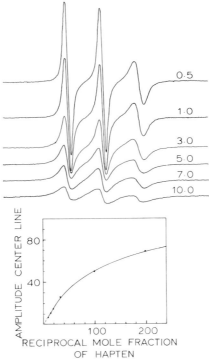

FIG. 22. Spectral consequence of decreasing the spacing of spin-label hapten **IV** in di-palmitoylphosphatidylcholine membranes with 50 mole % cholesterol; EPR spectra of lipo-somes at 32°C containing *equal amounts* of hapten at the various molar percentages (with respect to total lipid) indicated. Also shown is a plot of the amplitude of the center line (in arbitrary units) versus the reciprocal of the mole fraction of hapten present in the bilayers. This latter parameter is proportional to the average area of membrane available to each hapten. The data indicate a random, monomolecular dispersion of this hapten in cholesterol-dipalmitoylphosphatidyl-choline bilayers of the chosen composition. Spectra were obtained with a Varian E12 EPR spectrometer interfaced with a Digital PDP 8/e computer. Samples were prepared as specified but such that all had the same *overall* concentration of nitroxide (i.e., total lipid per milliliter of liposome suspension was varied). Even so, because of the heterogeneous nature of lipid sus-pensions, spectra were integrated and normalized to the same spin concentration and then plotted at these new normalized values in order to measure comparable spectra parameters. (From Humphries and McConnell.[71])

Binding of spin-label-specific IgM is favored by decreasing the distance between adjacent haptens, even for the case of membranes for which the rate of lateral diffusion is very high.[71] This presumably relates to the ease with which multipoint binding is maintained.

A systematic study of the effect of hapten density on IgG binding has not yet been attempted. However, some early work using a monoclonal antibody suggested that for some IgG molecules, hapten density may be important even in the range where the distance between adjacent haptens is less than that between the two binding sites of an IgG molecule. It is our suspicion that the ease with which antibodies bind to *single sites* on membranes may, in some cases, be far less than that with which they bind to the same determinants in soluble form. We anticipate that the use of spin-label haptens and monoclonal antibodies will be particularly useful for elucidation of this important point.

It would be of interest to measure off rates for antibodies bound to lipid haptens on membranes. Schwartz et al.[81] have approached this problem by studying selective hydrogen exchange between nitroxides and reduced nitroxides. It is apparent that different nitroxide species differ greatly in their affinity for hydrogen atoms. Hydrogen exchange between an appropriate nitroxide-reduced nitroxide couple can only occur when neither species is bound to antibodies. Oxidation of the reduced form of a suitable water-soluble nitroxide (i.e., one which does not cross react, in the immunological sense, with TEMPO and which is a hydrogen donor relative to TEMPO) by lipid haptens, as they are released from the protection of TEMPO-specific antibodies, has been used in some preliminary studies designed to measure antibody off rates from membranes.

2.4.8. Efferent Immune Responses Controlled by Antibodies Bound to Antigen-Bearing Membranes

Our thesis is that the many responses initiated by the binding of antibodies to cell membranes will be elucidated by the study of analogous systems in which the target membranes are synthetic. Antibodies are polyvalent, IgG and IgE having two antigen-binding sites and the secreted (serum) form of IgM having ten. Of central importance is the question of the significance of multipoint versus single-site binding of antibodies to membranes. A full discussion of this subject is clearly not appropriate for this book, but a few points of general interest to biophysical chemists will be mentioned.

When antibodies bind to membranes their concentration (or time spent) in the region of the membrane increases and their orientation in that region is no longer isotropic. This increase in concentration and anisotropy is favored by multivalent rather than monovalent binding. To what extent, and how, do these properties of local concentration and orientation of antibodies control antibody-triggered responses? It has been suggested that, in addition to concentration and orientation, antibody conformational changes are induced by binding to antigens and are required for various functions such as activation of complement. This topic is controversial. For a discussion see Ref. 82.

[81] M. A. Schwartz, J. W. Parce, and H. M. McConnell, *J. Am. Chem. Soc.* **101**, 3592 (1979).
[82] H. Metzger, *Adv. Immunol.* **18**, 169 (1974).

Regarding the physical state of membrane-bound antibodies, Henry, Parce, and McConnell,[83] using freeze-fracture methods, have observed no significant aggregation of IgG bound to haptenated bilayers. Such IgG appeared to be aggregated by subsequent addition of Clq. More recently Smith et al.[84] have shown that IgG bound to lipid haptens in bilayers diffuses as rapidly as do the unbound lipids.

We have undertaken several studies of the control of antibody-dependent complement activation (or depletion of activity) by spin-label lipid hapten-bearing synthetic membranes. These have concerned the effect of changing (i) the length of the hydrophilic moiety of the hapten,[74] (ii) membrane fluidity,[64,71,75,85] (iii) cholesterol content,[71,74] and (iv) two-dimensional hapten concentration.[71] In most cases IgG antibodies have been used, but Ref. 71 concerns IgM. As is true for the case of antibody binding all these types of physical variation do, under certain circumstances, affect complement activation. To what extent this is due to quantitative differences in antibody binding and to what extent other differences (related to the behavior of bound antibodies or the way in which they are bound) play a part in not always clear and is the subject of continuing investigation.

Another related area of study has been the binding of antibody-coated synthetic membrane structures by macrophages and neutrophils, and the induction of phagocytosis, superoxide production, and a respiratory burst which are frequently a consequence of such binding. See Refs. 86–89.

Acknowledgments

We are greatly indebted to Laurie K. Doepel for skillful assistance in preparation of this manuscript. This work has been supported by National Institutes of Health Grants No. 5RO1 AI13587 (H.M.McC.), and 1R23 AI14813 (G.M.K.H.), and by National Science Foundation Grant No. PCM 77-23586 (H.M.McC.).

[83] N. Henry, J. W. Parce, and H. M. McConnell, *Proc. Natl. Acad. Sci. U. S. A.* **75**, 3933 (1978).

[84] L. M. Smith, J. W. Parce, B. A. Smith, and H. M. McConnell, *Proc. Natl. Acad. Sci. U. S. A.* **76**, 4177 (1979).

[85] A. F. Esser, R. M. Bartholomew, J. W. Parce, and H. M. McConnell, *J. Biol. Chem.* **254**, 1768 (1979).

[86] D. G. Hafeman, J. W. Parce, and H. M. McConnell, *Biochem. Biophys. Res. Commun.* **86**, 522 (1979).

[87] J. T. Lewis, D. G. Hafeman, and H. M. McConnell, *Biochemistry* **19**, 5376 (1980).

[88] D. G. Hafeman, J. T. Lewis, and H. M. McConnell, *Biochemistry* **19**, 5387 (1980).

[89] J. T. Lewis, D. G. Hafeman, and H. M. McConnell, *in* "Liposomes and Immunobiology" (B. Tom and H. Six, eds.), pp. 179–191. Am. Elsevier, New York, 1980.

3. RAMAN SPECTROSCOPY

By Margaret R. Bunow

3.1. Introduction

Raman vibrational spectroscopy, in technological development the younger sibling of infrared spectroscopy, surpasses its elder in immense versatility and applicability to studies of biological molecules. These advantages over infrared derive largely from the fact that water, the natural biological solvent, is a weak Raman scatterer although highly opaque over much of the infrared spectrum, and the fact that molecules can be studied in any state, whether in crystalline array or aqueous phase or biological cell fraction. Furthermore, sample size is economical, typically being of the order of several microliters, and the sample usually survives its examination unaltered and intact. As in infrared spectroscopy, no perturbing molecules need be added to the sample.

The Raman spectrum is a record of vibrational modes for all groups of atoms in the scattering molecules whose concerted atomic motions result in a change of polarizability. The information in this record concerning molecular structure and conformation constitutes an abundance of riches. The Raman spectrum is obtained by spectral analysis of the light scattered from a sample illuminated by a laser source; the spectral record has the form of the intensity I of scattered light versus the displacement v_v of the frequency of scattered light from the incident laser frequency v_L. The intensities of vibrational modes which are vibronically coupled to electronic transitions may be enhanced by illumination at frequencies within the electronic absorption band; this method permits selective study of the neighborhood of biological chromophores. The high intensity of such resonance-enhanced bands makes possible the study of much lower concentrations, which in many cases better represent the biological norm. Typical concentrations of biomolecules studied in solution are, in conventional (nonresonance) Raman spectroscopy, 0.1 M or the maximum soluble; this figure may drop to several micromolars in resonance Raman work. Both conventional and resonance Raman biological studies burgeoned in the 1970s, when sufficiently powerful laser sources (> 100 mW for conventional scanning, < 100 mW for resonance) became readily available.

METHODS OF EXPERIMENTAL PHYSICS, VOL. 20

This article reviews the principles of measurement of Raman scattering and then presents a guide to the application of conventional and resonance Raman spectral analysis to several classes of biological molecules. The emphasis is placed on the strategic choice of problem and its handling in light of the body of spectroscopic data already gathered for biophysical applications.

3.2. Origin of the Raman Spectrum

If a beam of light at visible frequency v_L illuminates a sample, most of the scattered light is unshifted in frequency, that is, elastically scattered. Upon spectral analysis, many weak bands, representing inelastic scattering, are found at both higher and lower frequencies about the central Rayleigh band. The very near neighbors are Brillouin bands resulting from the interaction of light with coherent fluctuations in the sample medium. In addition, the array of vibrational bands composing the Raman spectrum is displaced by from 3 to 3500 cm^{-1} from the central band. These displacements result from the interaction of light with the changing electrical field of the electronic clouds, which closely follow motion of the oscillating atoms of the sample molecules.

In the Raman phenomenon, the molecule in the ground state interacts with an incident photon of energy hv_L, excurses to a "virtual" (nonstationary) state, and exits to a second intermediate state by creating a vibrational quantum hv_v. Finally, upon emitting a photon of energy $h(v_L - v_v)$, the molecule returns to its ground electronic state but to the first excited vibrational level rather than to the ground vibrational state. (see Fig. 1). The process is essentially instantaneous ($\sim 10^{-15}$ sec) and involves two photons, one incident, and one emitted. The emitted line thus occurs at a frequency close to v_L, that is, usually in the visible range, but displaced from v_L by the vibrational frequency v_v, which is usually reported in wave numbers (cm^{-1}) of displacement. The interaction of many photons with the sample molecules produces the complete Raman vibrational spectrum. The spectrum is very complex, limited to fewer than $3N - 6$ lines for an N-atom nonlinear molecule only by quantum-mechanical selection rules and by the detectable intensity of the allowed lines.

The process just described is known as Raman Stokes scattering, and the Raman spectrum consists of bands shifted to frequencies lower than those of the incident line. An anti-Stokes spectrum is seen at frequencies greater than v_L; this is produced by incident photons interacting with the smaller number of molecules which populate the vibrational levels above the ground-state level in proportion to the Boltzmann factor $e^{-hv_v/kT}$ (Fig. 1).

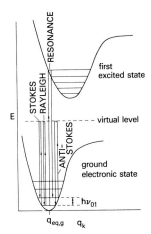

FIG. 1. Relationship between Stokes, anti-Stokes, and resonance Raman effects, where $h\nu_{01}$ is the energy difference between the first excited and lowest vibrational levels of the ground electronic state, and $q_{eq,g}$ is the equilibrium position of the vibrational coordinate q_k in the ground electronic state.

This spectrum is too weak to scan more than 500 wave numbers (cm^{-1}) from ν_L. Transitions involving a jump of more than one vibrational level are also weak.

When ν_L falls within or near molecular absorption bands, the molecule may transit to an excited electronic state, and sets of vibrational bands of the proper symmetry and overlap relative to the electronic transition may become resonance enhanced (Fig. 1).

3.3. Analysis of the Raman Spectrum

Each Raman vibrational band is characterized by its frequency ν_v, intensity, bandwidth, and depolarization ratio. These quantities are defined and interpreted as follows.

Characteristic vibrational frequencies can be calculated by normal coordinate analysis of the set of atoms of a molecule or chemical grouping connected by "springs" for which the force constants are related to bond strength. The atoms vibrate with small changes in atom–atom distances and angles about their equilibrium positions. The vibrational frequencies are clearly sensitive to alterations of normal coordinates or of bond strengths. This sensitivity is the foundation for Raman spectral investigations not only of molecular conformation but also of its mutability in differing environments.

The induced dipole moment $\mathbf{P}$ of a molecule is related to the electric field vector $\mathbf{E}$ of the impinging light by the polarizability tensor $\boldsymbol{\alpha}$:

$$\begin{Bmatrix} P_x \\ P_y \\ P_z \end{Bmatrix} = \begin{pmatrix} \alpha_{xx} & \alpha_{xy} & \alpha_{xz} \\ \alpha_{yx} & \alpha_{yy} & \alpha_{yz} \\ \alpha_{zx} & \alpha_{zy} & \alpha_{zz} \end{pmatrix} \begin{pmatrix} E_x \\ E_y \\ E_z \end{pmatrix}. \tag{3.3.1}$$

The molecular polarizability $\boldsymbol{\alpha}$ can be expanded in terms of linear deviations from the equilibrium coordinates of the molecule

$$\boldsymbol{\alpha} = \boldsymbol{\alpha}_0 + \sum_i \left(\frac{\partial \boldsymbol{\alpha}}{\partial q_i} \right)_0 q_i + \cdots, \tag{3.3.2}$$

where q_i is a normal coordinate describing a molecular vibration and higher (anharmonic) terms are neglected. The $\boldsymbol{\alpha}_0$ term gives rise to Rayleigh scattering, and the $[(\partial \boldsymbol{\alpha}/\partial q_i)]_0 q_i$ term expresses the dependence of the induced dipole moment on the $3N - 6$ molecular vibrations which may contribute to the Raman spectrum. The matrix of these coefficients, which is symmetric in the conventional vibrational Raman effect, composes the derived polarizability tensor. These coefficients are evaluated in terms of the probabilities of transitions between the initial and final states and contain information on the magnitude and direction of the induced dipole moment.

The intensity of a Raman line scattered by randomly oriented molecules can now be written as dependent upon the sum of the products of cofficients α_{jk}, where the coefficients α_{jk} for the ith vibration are evaluated by quantum mechanical theory, as well as proportional to the concentration c of scatterers, the incident light intensity I_0, and the fourth power of the frequency of scattered light, as in classical scattering processes:

$$I \propto (v_L - v_v)^4 \sum_{j,k} |\alpha_{jk}|^2 c I_0. \tag{3.3.3}$$

The indices j, k run over x, y, and z.

The symmetry of small molecules and of chemical groupings in large molecules can be characterized in terms of the set of symmetry operations consisting of rotations, reflections, and inversions which transform the molecule into itself. Using group-theoretical rules, one can count and classify by symmetry the number of Raman lines expected for a molecule. Totally symmetric vibrations, for which the periodic fluctuation in polarizability has the full symmetry of the molecular skeleton, contribute only diagonal elements to the polarizability tensor. Non-totally-symmetric vibrations have nonzero off-diagonal elements. It follows that measurement of the polarization of the scattered light for each Raman line gives direct

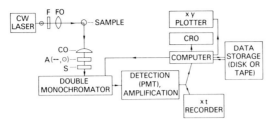

Fig. 2. Schematic diagram of instrumentation for measurement of conventional or resonance Raman scattering: F, dispersion or interference filter of laser output; FO, focusing optics; CO, collection optics; A, analyzer; S, polarization scrambler. Typically, the incident laser beam is polarized out of the plane of the paper (⊙); the Raman spectrum is recorded first with the analyzer rotated in the plane defined by the incident and scattered beams (↔), and then normal to the plane so defined (⊙). The ratio of peak intensities for each vibration, measured in the respective spectra, is the depolarization ratio ρ.

information on the symmetry, and helps one to deduce the assignment of the associated vibration.

The depolarization ratio, a measure of the rotation of the electric field vector of the scattered light, may be measured in one of several related scattering geometries. In Fig. 2, the electric field vector of the polarized laser beam is chosen to be normal to the plane containing incident and observed scattering directions. The scattered light is analyzed by transmitting light with the field vector first parallel, then normal to this plane; the ratio of intensities thus recorded for each vibration is the depolarization ratio ρ.[1] The polarization scrambler placed between the analyzer and slit (Fig. 2) eliminates unequal grating response to different polarizations.

For molecules oriented randomly in solution, this measurement gives $0 < \rho < 0.75$ for totally symmetric vibrations, the exact value reflecting the symmetry group, and $\rho = 0.75$ for non-totally-symmetric vibrations. The ratio ρ can be measured accurately only for liquids or clear solutions; multiple reflections of incident and scattered light in turbid solutions destroy the polarization. If a sample is a single crystal or oriented polymer, it is possible to choose the orientations of the sample axes with respect to the incident beam so as to measure individual elements of the tensor α and assign vibrational modes more precisely.

In practice most mode assignments in biophysical studies are made from comparative scans of model compounds or by reference to tables of chemical group frequencies compiled for Raman and infrared work. These results are refined by calculation. Raman and infrared vibrational spectra are largely complementary. Infrared activity requires that a vibration change one of the three components of the dipole moment of a molecule;

[1] H. H. Claessen, H. Selig, and J. Shamir, *J. Appl. Spectrosc.* **23**, 8 (1969).

in general, highly polar chemical groups have strong infrared bands while chemical groups of low polarity and high polarizability are strongly Raman active. If a molecule has a center of symmetry, it can be shown that Raman- and infrared-active modes are mutually exclusive; as the symmetry of the molecule is lowered, more bands are found to be allowed in both types of spectra. When a mode has both Raman and infrared activity, the frequencies in both spectroscopic appearances coincide, allowing use of infrared spectro- scopic compilations.

The fourth useful spectroscopic variable is the linewidth. Raman spectra of crystalline substances at low temperatures exhibit sharp lines. These lines are broadened when the molecule is placed in solution, reflecting the increased molecular conformational freedom and range of solute–solvent interactions. Observation of line splitting for dissolved or crystalline samples can be interpreted in terms of inequivalence of molecular orientations, bonding, or environment.

Several classical works[2,3] and review articles[4–8] may be consulted for details of theory. A variety of introductions to theory and modern instru- mentation are available.[5,7,9–11]

3.4. Resonance Raman Scattering

When a molecule is illuminated by light at frequency ν_L near or beneath an electronic absorption band, the mixing of electronic and vibrational states allows preresonance or resonance enhancement of the intensities of Raman

[2] G. Herzberg, "Infrared and Raman Spectra of Polyatomic Molecules." Van Nostrand-Reinhold, New York, 1945.

[3] E. B. Wilson, Jr., J. C. Decius, and P. C. Cross, "Molecular Vibrations." McGraw-Hill, New York, 1955.

[4] B. P. Stoicheff, in "Methods in Experimental Physics: Molecular Physics" (D. Williams, ed.), Vol. 3, 2nd ed., p. 111. Academic Press, New York, 1962.

[5] G. Placzek, in "Handbuch der Radiologie" (E. Marx, ed.), Vol. 6, Part 2, p. 205. Akad. Verlagsges., Leipzig, 1934.

[6] J. Tang and A. C. Albrecht, in "Raman Spectroscopy" (H. A. Szymanski, ed.), Vol. 2, p. 33. Plenum, New York, 1970.

[7] J. R. Durig and W. C. Harris, in "Physical Methods of Chemistry" (A. Weissberger and B. Rossiter, eds.), Vol. 1, Part 3b, p. 85. Wiley, New York, 1972.

[8] I. W. Shepherd, in "Advances in Infrared and Raman Spectroscopy" (R. J. H. Clark and R. E. Hester, eds.), Vol. 3, p. 127. Heyden, London, 1977.

[9] M. C. Tobin, "Laser Raman Spectroscopy," Vol. 35 of "Chemical Analysis" (P. J. Elving and I. M. Kolthoff, eds.), p. 1. Wiley (Interscience), New York, 1971.

[10] S. K. Freeman, "Applications of Laser Raman Spectroscopy." Wiley, New York, 1974.

[11] N.-T. Yu, CRC Crit. Rev. Biochem. 4, 229 (1977).

vibrational modes associated with the chromophore. The preresonance and resonance regimes are theoretically distinguishable.[12]

The polarizability can now be expressed, according to Albrecht's formulation,[13] as a function of two types of terms, called A and B terms. Higher terms of smaller magnitude can usually be neglected. The reader is referred to reviews of the theory.[6,12–16]

3.4.1. A-Term Scattering

A-term scattering involves vibrational interaction with a single excited electronic state. This type is usually dominant in resonance Raman scattering, but it is nonzero only for totally symmetrical vibrational modes. A-term scattering depends upon Franck–Condon overlaps, which require shifts in the excited-state equilibrium coordinates or changes in the shape of the excited-state potential well relative to the ground state. Accordingly, a useful rule for identifying vibrational modes involved in A-term resonance was promulgated by Hirakawa and Tsuboi[17]: when the laser frequency ν_L approaches an absorption band, the vibrational modes most strongly enhanced are those whose normal coordinates lie along the direction describing the distortion of the molecule upon transition to the excited electronic state. Vibrational stretching modes aligned with bond weakening such as that associated with $\pi–\pi^*$ transitions receive strong A-term resonance enhancement. Examples occurring in biologically interesting polyenes and porphyrins are given below (Sections 3.7.9 and 3.7.6). Vibrational modes associated with charge-transfer transitions of ligand-metal complexes in proteins are similarly enhanced (Section 3.7.7).[14,18]

The magnitude of A-term scattering is a function of the square of the transition moment of the electronic state in resonance such that it becomes proportional to the square of the extinction maximum of the absorption band. The total A-term scattering for each vibrational mode is found by summing over terms describing all contributing excited states. An analogous summation is made for B-term scattering, except that two excited states are included in each term.[13,19]

[12] B. B. Johnson and W. L. Peticolas, *Annu. Rev. Phys. Chem.* **27**, 465 (1976).

[13] A. C. Albrecht, *J. Chem. Phys.* **34**, 1476 (1961).

[14] T. G. Spiro and P. Stein, *Annu. Rev. Phys. Chem.* **28**, 501 (1977).

[15] A. Warshel, *Annu. Rev. Biophys. Bioeng.* **6**, 273 (1977).

[16] J. Behringer, *Mol. Spectrosc.* (*Chem. Soc. London*) **2**, 100 (1974).

[17] A. Y. Hirakawa and M. Tsuboi, *Science* (*Washington, D.C.*) **188**, 359 (1975).

[18] T. G. Spiro and B. P. Gaber, *Annu. Rev. Biochem.* **46**, 553 (1977).

[19] F. Inagaki, M. Tasumi, and T. Miyazawa, *J. Mol. Spectrosc.* **50**, 286 (1974).

3.4.2. B-Term Scattering

B-term scattering involves vibronic mixing of two excited states; vibrations with B-term activity are allowed any symmetry contained in the direct product of the representations of the two electronic transitions. The magnitude of B-term scattering depends on the product of the transition dipole moments of the two electronic states being mixed. Transitions that have become allowed in the absorption spectrum by vibronic mixing are generally reflected in the excitation profiles of the contributing vibrational modes in the resonance Raman spectrum.[6]

Vibrational modes that have rotational symmetry are particularly effective in vibronic mixing in producing anomalously or inversely polarized bands. The mixing of in-plane vibrational modes with π–π^* transitions in porphyrins provided the first example of inverse polarization in vibrational Raman scattering.[20]

3.4.3. The Polarizability Tensor

When vibronic mixing with electronic states is admitted, the polarizability tensor is no longer necessarily symmetric. As a result, the depolarization ratio ρ may assume a value in the following ranges: (i) $0 < \rho < 0.75$, the vibration is totally symmetric, and the Raman mode is polarized; (ii) $\rho = 0.75$, the vibration is non-totally-symmetric, and the mode is depolarized; (iii) $0.75 < \rho < \infty$, α is unsymmetric, and the polarization is termed anomalous; (iv) $\rho = \infty$, α is antisymmetric, and the polarization is termed inverse.

A-term scattering, as noted previously, requires total symmetry of the mode; B-term scattering may fall into any of the above categories. A- and B-term scattering can hence essentially be distinguished by the values of the measured depolarization ratios. The two types can also be distinguished, less easily, by the shape of the excitation profiles in the region of preresonance enhancement.[12,21–23]

The excitation profile is a plot of the measured intensity, and of the depolarization ratio, for each vibrational mode against the variable v_L.[24,25] The intensity profile plots are corrected for the usual v^4 dependence of scattering [see Eq. (3.3.3)] and for instrumental spectral response, and normalized with respect to a nonresonant reference mode. These plots

[20] T. G. Spiro and T. C. Strekas, *Proc. Natl. Acad. Sci. U. S. A.* **69**, 2622 (1972).

[21] A. C. Albrecht and M. C. Hutley, *J. Chem. Phys.* **55**, 4438 (1971).

[22] T. C. Strekas, A. J. Packer, and T. G. Spiro, *J. Raman Spectrosc.* **1**, 197 (1973).

[23] A. G. Doukas, B. Aton, R. H. Callender, and B. Honig, *Chem. Phys. Lett.* **56**, 248 (1978).

[24] L. Rimai, R. G. Kilponen, and D. Gill, *J. Am. Chem. Soc.* **92**, 3824 (1970).

[25] B. B. Johnson, L. A. Nafie, and W. L. Peticolas, *Chem. Phys.* **19**, 303 (1977).

contain information on the relation of the resonance scattering to the absorption spectrum, as well as on vibrational symmetry and its alteration by environment.[12,14,25-27]

3.5. Instrumentation

The basic Raman experimental configuration provides for spectral analysis of scattered light (Fig. 2). A continuous-wave (CW) laser beam is focused on a sample, and the light scattered at 90° is collected and directed into the slit of a double monochromator. The double monochromator arrangement nearly eliminates the intense scattering of laser light unshifted in frequency; currently available holographic gratings have impressively high stray-light rejection. The gratings of the scanning monochromaters are preferably driven by a stepping motor to allow flexibility for computer interfacing.

The spectrally analyzed signal is detected at the photomultiplier cathode and amplified. High-intensity signals are processed by direct current amplification or synchronous detection and can be recorded on an xt recorder or converted to digital form for computer processing. Weak signals are optimally processed by pulse counting circuitry; the data can be recorded in digital form or integrated for recording on an xt recorder.[28] The weakness of many bands in the typically complex Raman spectra of large biological molecules makes the addition of computer-controlled data acquisition and data manipulation facilities very desirable.[29,30]

The basic Raman instrumentation can be purchased commercially as a "package" which allows easy interchange of laser source. The CW argon ion laser is the workhorse of Raman spectroscopy. Its strongest lasing lines lie at 488.0 (blue) and 514.5 nm (green). Substitution of a krypton ion laser supplies several lines toward the red, the strongest at 647.1 nm; the helium-neon laser supplies a 632.8-nm line. Alternatively, a range of longer-wavelength lines may be supplied by adding a dye laser pumped by an argon laser or flash lamp.

For near-ultraviolet excitation, two weak argon laser lines at 351.1 and 363.8 nm are widely used. Frequency-doubled pulsed argon laser lines lie at even shorter wavelengths. Ultraviolet work requires quartz optics and

[26] T. C. Strekas and T. G. Spiro, *J. Raman Spectrosc.* **1**, 387 (1973).

[27] T. G. Spiro, *Biochim. Biophys. Acta* **416**, 169 (1975).

[28] J. K. Nakamura and S. E. Schwartz, *Appl. Opt.* **7**, 1073 (1978).

[29] W. F. Edgell, E. Schmidlin, T. J. Kuriakose, and P. Lurix, *Appl. Spectrosc.* **30**, 428 (1976).

[30] J. W. Arthur and D. J. Lockwood, *J. Raman Spectrosc.* **2**, 53 (1974).

may require exchange of gratings for those more efficient toward the ultraviolet. The electronics for gating of pulsed signals and other modifications are generally acquired separately from "packaged" instrumentation.

3.5.1. Sample Handling

The standard sample container is a melting-point capillary holding either a solid or liquid sample. Small glass or quartz cuvettes are also used. Stationary samples may be thermostatted at temperatures from that of liquid nitrogen to 100°C, or may be held in vacuum, controlled atmosphere, or humidity, using a variety of sample chambers.[9,31]

The molecular order and spectral reproducibility of solids is improved by annealing the sample by repeated immersion of the sealed capillary in liquid nitrogen. Solid samples should be well tamped to maximize scattering and conduction of heat away from the sample area heated by the laser beam. Suspended samples (such as subcellular fractions or crystals in mother liquor) can be centrifuged to concentrate the sediment.

Absorbing (colored) samples suffer heating in the focused laser beam; resonance Raman studies as a rule require precautions against heating, photoreaction, and photodecomposition. These effects are reduced by moving the sample rapidly in the beam path. Solids are usually pressed into an annular groove on a (motor- or air-driven) spinning plate.[10,32] Absorbing samples in solution can be pumped in a closed circuit through a capillary segment or through a jet orifice past the laser beam.[33,34] Other precautions include beam defocusing and use of lower laser power.

3.5.2. Optimization and Standardization of Signal

Attainment of a good signal-to-noise ratio in conventional Raman spectroscopy requires optimization of the factors v_L and I_0 of Eq. (3.3.3), and an adequate, usually maximal, sample concentration c. The signal generally increases nearly tenfold (for equal laser intensities) as the laser line is moved from the red to the blue, a result of the v_L^4 scattering dependence [Eq. (3.3.3)] multiplied by the photomultiplier and grating responses.

To optimize resonance Raman signals, the sample concentration is adjusted in successive trials so that absorbance does not surpass scattering.[35]

[31] I. W. Levin, in "Human Responses to Environmental Odors" (A. Turk, J. W. Johnston, Jr., and D. Moulton, ed.), p. 45. Academic Press, New York, 1974.

[32] W. Kiefer and H. J. Bernstein, *Appl. Spectrosc.* **25**, 609 (1971).

[33] R. H. Callender, A. Doukas, B. Crouch, and K. Nakanishi, *Biochemistry* **15**, 1621 (1976).

[34] R. Mathies, A. R. Oseroff, and L. Stryer, *Proc. Natl. Acad. Sci. U. S. A.* **73**, 1 (1976).

[35] T. C. Strekas, D. H. Adams, and T. G. Spiro, *Appl. Spectrosc.* **28**, 324 (1974).

Slow drifts in incident power, sample turbidity, and background fluorescence make small intensity changes hard to quantitate. For this reason, computer averaging of multiple fast scans is preferable to slow scanning. Spectra accumulated under varying conditions may be computer subtracted to yield difference spectra and thus detect small spectral changes or cancel solvent or fluorescence background. Difference spectra may be obtained directly by gated, two-channel accumulation of the two signals scattered from a rotating, two-compartment cell[36]; such an arrangement has been used to compare cytochrome spectra.[37]

Absolute intensity measurements are uncommon in biological studies; usually spectral intensity changes are measured with respect either to a spectral band of the sample assumed to be invariant or to an added, chemically noninteractive intensity standard such as the 983-cm^{-1} band of sulfate or 610-cm^{-1} band of cacodylate.

3.5.3. Management of Fluorescence

Biological molecules tend to exhibit a fluorescence background, due sometimes to intrinsic chromophores and more often to impurities, which overwhelms the weak Raman signal. The fluorescence can be substantially reduced by a combination of the following procedures.

1. Purification by repeated recrystallization or fractionation. Dissolved samples may be passed through activated charcoal or alumina, then a small-pore filter. Proteins may be purified by ion-exchange chromatography or gel electrophoresis, lipids by silica gel chromatography.

2. Optimization of signal/fluorescence. Moving the laser line to the red to skirt the absorption bands associated with the fluorescence, or alternatively as far to the blue as possible to skirt the longer-wave length fluorescence envelope frequently improves the spectrum dramatically. If the Raman signal can be detected above the fluorescence background in a single scan, multiple spectral accumulation plus subtraction of fluorescence background will result in good-quality spectra. Because the fluorescence background fluctuates, rapid, repeated scanning is preferred. The background fluctuation is partly due to convective motion of photoquenched and unquenched molecules in and out of the beam; a narrow-walled capillary may reduce this circulation.

3. Photoquenching by exposing the sample for a period of hours to as intense a laser beam as it can tolerate. The sample should be kept cool,

[36] W. Kiefer, in "Advances in Infrared and Raman Spectroscopy" (R. J. H. Clark and R. E. Hester, eds.), Vol. 3, p. 30. Heyden, London, 1977.

[37] D. L. Rousseau, J. A. Shelnutt, and J. M. Friedman, *Biophys. J.* 21, 68a (1978).

most easily by directing a stream of chilled air or nitrogen on it. Even momentary occlusion of the beam reverses the photoquenching effect.

4. Energy transfer to an acceptor molecule, which must have close access to the fluorescent impurity sites. Iodide ion is frequently used for water-soluble samples; the free iodine level must be kept low by addition of a reducing agent.

5. Modulation spectroscopy, assuming the background is not a function of time or small shifts in laser frequency.[38]

6. Rejection of fluorescence (of 1–10 nsec lifetime, typically, for biological molecules) by gated detection of the (nearly instantaneous) Raman signal scattered by illumination with nanosecond or shorter laser pulses. Using nanosecond pulses generated by a mode-locked argon ion laser, the signal-to-noise ratio has been improved less than tenfold.[8] The use of vidicon or photodiode array detectors should yield further improvement.

7. Coherent anti-Stokes resonance scattering (see Chapter 3.9).

Of all these methods, those requiring modified instrumentation are expensive and time consuming. Sample purification is in nearly all circumstances the critical procedure; photoquenching is also routinely utilized.

3.6. Strategy of Raman Spectroscopic Applications in Biology

Raman modes strong enough to be followed as indicators of molecular conformation usually belong to highly polarizable chemical groups or to groups which exist in high concentration in a small number of conformations. Thus highly polarizable groups such as $-C=C-$ and $-S-S-$ groups give strong, useful Raman bands. Short group frequency tables are found in introductory texts.[9,10] Similarly, the backbone configuration of biological polymers that undertake a regular conformation is apparent from the typical Raman modes. When a conformational marker mode happens to be juxtaposed with a conglomerate of other modes, either it or the interfering modes may be shifted into an emptier spectral "window" by selective deuteration or other isotopic substitution of the molecule, by chemical modification, or by the use of deuterium oxide instead of water as solvent. The identity of the mode can at the same time verified by its frequency shift upon such treatment.

Since the frequency, intensity, and depolarization ratio of a Raman mode may change in response to a change in molecular conformation,

[38] J. Funfschilling and D. F. Williams, *Appl. Spectrosc.* **30**, 443 (1976).

TABLE I. Applications of Raman Spectroscopic Analysis to Biomolecules

Useful vibrational modes	Structural information	Biological insight	Example
		Lipids (Section 3.7.1)	
Stretching modes of polyethylene chains	Chain conformation and interchain interaction	Membrane fluidity; its perturbability	Miscibility of different lecithins[a]
		Nucleic acids (Section 3.7.2)	
Preresonance-enhanced ring modes of bases	Base pairing, base stacking	Conformation of nucleic acids in nuclei, viruses, ribosomes, etc.	Structure of rRNA in ribosomes[b]
Phosphate diester stretching modes	Helix tilt		
		Polysaccharides (Section 3.7.3)	
Modes associated with sugar linkages, side groups of homopolymers and copolymers	Sugar linkage, sugar chain, and side-group configuration	Structure of cartilage Possible monitor of helix–coil transitions in plant polysaccharides	Structure of chondroitin sulfate[c]
		Proteins (Sections 3.7.4, 3.7.5)	
Amide I, III; backbone C–C stretching modes	Polypeptide conformation (e.g., α-helical content)	Conformation, especially of noncrystalline proteins in native environment	Structure of membrane-bound Ca^{2+}-ATPase, calsequestrin[d]
Conventional or preresonance-enhanced modes associated with side groups	Side-chain configuration and environment	Interaction of local region of protein with solvent, substrate, and other agents	Location of tryptophan residues in calsequestrin[d]
		Porphyrins and hemoproteins (Section 3.7.6)	
Resonance-enhanced vibrations in plane of porphyrin ring	Porphyrin symmetry, conformation, bonding energetics	Electron transport or O_2-carrying mechanisms in cytochromes, hemoglobin, chlorophyll	Hemoglobin structure and dynamics (see text)
		Nonheme metalloproteins (Section 3.7.7)	
Resonance-enhanced modes involving liganded metal ion	Metal ion situation, bonding scheme		
Resonance-enhanced modes involving liganded metal ions	Metal-ligand geometry, bonding scheme	Structure of redox or transport sites	O_2 binding in hemocyanin[e]

(continued)

TABLE I (*continued*)

Useful vibrational modes	Structural information	Biological insight	Example
	Extrinsic chromophores in proteins (Section 3.7.8)		
Resonance-enhanced modes involving labeled substrate or active site of enzyme	Configuration at site of bound chromophore	Mechanisms of enzyme activity	Chemistry, kinetics of papain-labeled substrate complex[f]
	Biological polyenes (Section 3.7.9)		
Preresonance- or resonance-enhanced stretching modes of polyene backbone	Configuration of polyene, polarity of environment	Polyenes as photopigments; membrane fluidity	Photopigments of lobster carapace[g]
	Kinetics (Chapter 3.8)		
Resonance-enhanced modes for most kinetics ($\tau < 10$ sec); nonresonance modes nonresonance modes for slower processes ($\tau > 10$ sec)	Dependent on choice of vibrational modes	Kinetics of hemoproteins, photoreceptors, enzymes; conformational transitions in biopolymers	Photocycle of bacteriorhodopsin (see text)
	Nonlinear phenomena (Chapter 3.9)		
Spectral information from resonance CARS similar to that from resonance Raman, but fluorescence free	Dependent upon molecule and chromophore under study	Same insights as for resonance Raman studies; potential for kinetic studies	Flavin adenine dinucleotide–glucose oxidase binding[h]

[a] R. Mendelsohn and J. Maisano, *Biochim. Biophys. Acta* **506**, 192 (1978).

[b] M. G. Hamilton, B. Prescott, and G. J. Thomas, Jr., *Biophys. J.* **25**, 224a (1979).

[c] R. Bansil, I. V. Yannas, and H. E. Stanley, *Biochim. Biophys. Acta* **541**, 535 (1978).

[d] J. L. Lippert, R. M. Lindsay, R. I. Schultz, and D. J. Osterhout, *Biophys. J.* **25**, 108a (1979).

[e] T. B. Freedman, J. S. Loehr, and T. M. Loehr, *J. Am. Chem. Soc.* **98**, 2809 (1976).

[f] P. R. Carey, R. G. Carriere, D. J. Phelps, and H. Schneider, *Biochemistry* **17**, 1081 (1978).

[g] V. R. Salares, N. M. Young, H. J. Bernstein, and P. R. Carey, *Biochemistry* **16**, 4751 (1977).

[h] P. K. Dutta, J. R. Nestor, and T. G. Spiro, *Proc. Natl. Acad. Sci. U. S. A.* **74**, 4146 (1977).

bond energy and polarizability, local electrical field and symmetry, and/or intermolecular order, the correlation of changes in Raman spectra with conformation has required a great deal of comparative spectral observation and computation. As might be expected, Raman spectral correlations are imperfect for heteropolymers, such as globular proteins, which have a distribution of backbone configurations; however, the ordered secondary structures, α-helix and β-structure, are readily recognized in the Raman spectrum.

In Chapter 3.7, the information available from Raman spectroscopy is surveyed for several categories of biological molecules. The sensitivity of the Raman profile to conformation in each case delimits the types of biological questions that may be posed using this technique. A summary of these Raman spectral applications is given in Table I.

3.7. Conformational Studies

3.7.1. Conformation of Lipids in Biological Membranes

The simplicity and redundancy of the nonpolar lipid chains endows the lipid array of natural and model membranes with a simple, strong Raman spectrum which is exquisitely conformation sensitive. Raman bands reflecting hydrocarbon chain order can be roughly classified as those reflecting primarily only intrachain conformation (skeletal optical modes associated with C–C stretching, 1150–1050 cm^{-1}); and intrachain plus interchain interactions (methylene C–H bending modes at 1450 cm^{-1} and C–H stretching modes, 3000–2800 cm^{-1}). Vibrations associated with the chains are listed in Table II, along with the direction of change of frequency and peak height with increasing conformational disorder. The reader is alerted that some of the assignments in Table II supersede several erroneous ones found in the literature.

The spectral pattern in the skeletal optical region can be correlated with the populations of *trans* and *gauche* conformers of the chains.[39-42] The profile of the C–H stretching region, analogously, is a function of lipid "fluidity."[42-44] Peak height ratios h_{2880}/h_{2850}, h_{2930}/h_{2850}; and h_{1090}/h_{1130} and h_{1090}/h_{1060} are most commonly used to monitor changes in interchain

[39] R. G. Snyder, *J. Chem. Phys.* **47**, 1316 (1967).

[40] J. L. Lippert and W. L. Peticolas, *Proc. Natl. Acad. Sci. U. S. A.* **68**, 1572 (1971).

[41] N. Yellin and I. W. Levin, *Biochemistry* **16**, 642 (1977).

[42] N. Yellin and I. W. Levin, *Biochim. Biophys. Acta* **489**, 177 (1977).

[43] K. Larsson, *Chem. Phys. Lipids* **10**, 165 (1973).

[44] M. R. Bunow and I. W. Levin, *Biochim. Biophys. Acta* **487**, 388 (1977).

TABLE II.   Selected Conformation-Sensitive Modes of Hydrocarbon Chains of Lipids[a]

$\bar{v}$ (cm^{-1})	Assignment	Change with increasing lipid disorder	
		Frequency	Peak height
Methylene modes [b–e]			
2925 (2200)[f]	C–H asym stretch, infrared active	+ (+)	+ (−)
2880 (2180)[f]	C–H asym stretch	+ (0)	− (0)
2850 (2103)	C–H sym stretch	+ (+)	− (−)
1450 (985–960)	CH$_2$ bending deformations	− (−)	− (−)
1295 (940–920)	CH$_2$ twisting deformations	+ (+)	− (−)
C–C skeletal stretching modes[b,g,h]			
1130 (1249)	All-trans conformers	− (−)	− (−)
1090 (?)	Gauche conformers	−	+
1060 (1145)	All-trans conformers	+ (−)	− (−)
Terminal methyl group C–H stretching modes[b,d,e,i,j]			
2960 (2217)	C–H asym stretch	w (w)	w (w)
2935, 2070 (2126, 2075)	C–H sym stretch, split by Fermi resonance	w (0)	w (−)
"Reference" mode[d,e]			
720 (720)	Choline group sym stretch (plus CD$_2$ modes)	0 (0)	0 (+)

[a] Frequencies and behavior of perdeuterated chains are in parentheses. Asym, asymmetric; sym, symmetric; w, mode too weak to monitor.

[b] R. C. Spiker, Jr. and I. W. Levin, *Biochim. Biophys. Acta* **388**, 361 (1975).

[c] M. R. Bunow and I. W. Levin, *Biochim. Biophys. Acta* **487**, 388 (1977).

[d] M. R. Bunow and I. W. Levin, *Biochim. Biophys. Acta* **489**, 191 (1977).

[e] B. P. Gaber, P. Yager, and W. L. Peticolas, *Biophys. J.* **22**, 191 (1978).

[f] These two assignments may be reversed [see M. R. Bunow and I. W. Levin, *Biochim. Biophys. Acta* **489**, 191 (1977)].

[g] J. L. Lippert and W. L. Peticolas, *Biochim. Biophys. Acta* **282**, 8 (1972).

[h] R. C. Spiker, Jr., and I. W. Levin, *Biochim. Biophys. Acta* **433**, 457. (1976).

[i] S. Sunder, R. Mendelsohn, and H. J. Bernstein, *Chem. Phys. Lipids* **17**, 456 (1976).

[j] I. R. Hill and I. W. Levin, *J. Chem. Phys.* **70**, 842 (1979).

plus intrachain order and intrachain order, respectively[42,44–46]; alternatively, peak heights may be normalized to the height of a stable band, such as the 720-cm^{-1} choline mode in fully hydrated lecithin,[47] or to a mode of an external standard.[48] Several frequency shifts[42,49–51] and bandwidths,[52,53] also associated with the hydrocarbon chains, are equally good monitors of fluidity.

The conformation and packing of lipids in the membrane bilayer is influenced by chain length and unsaturation, and head group bulk, charge, and hydrogen bonding capability. Figure 3 illustrates the sensitivity of the Raman spectrum to the conformation of the alkyl chains in two different, fully hydrated glycolipids, each containing two fatty chains and one sugar residue (galactose in cerebroside and inositol in phosphatidylinositol) attached to the head group. The high degree of chain order in the cerebrosides (note the relatively narrower peaks), relative to phosphatidylinositol reflects the behavior of the almost totally saturated chains (compare the intensities of C=C bands at 1665 cm^{-1}), the presence of one very long (24 carbon) acyl chain, and stabilization by intermolecular hydrogen bonding in the head group of the cerebroside (observable in the amide I region) which is absent in phosphatidylinositol.

Normalized lateral and intrachain "order parameters" have been introduced to interpret the peak height ratios of Table I for phospholipid arrays.[47] These are not absolutely defined or necessarily linear with increasing chain disorder unless additional care is taken to choose well-defined reference states for the solid state, and reference modes which themselves do not vary significantly. A more refractory problem arises in considering the C–H stretching region. The peak height h_{2880} of the lateral order parameter rests on a background produced by Fermi resonance of the methylene C–H symmetric stretching modes, which peak at 2850 cm^{-1}, with overtone and combination modes of the 1450-cm^{-1} methylene deformation modes. (Fermi resonance interaction of a fundamental with an overtone or combination mode of compatible symmetry and similar energy usually results in mode splitting. A broad dispersion band is seen in the present case.) The Fermi resonance background has substantially different contributions

[45] R. Mendelsohn, *Biochim. Biophys. Acta* **290**, 15 (1972).
[46] M. R. Bunow and I. W. Levin, *Biochim. Biophys. Acta* **464**, 202 (1978).
[47] B. P. Gaber and W. L. Peticolas, *Biochim. Biophys. Acta* **465**, 260 (1977).
[48] R. C. Spiker, Jr. and I. W. Levin, *Biochim. Biophys. Acta* **455**, 560 (1976).
[49] R. C. Spiker and I. W. Levin, *Biochim. Biophys. Acta* **433**, 457 (1976).
[50] M. R. Bunow and I. W. Levin, *Biochim. Biophys. Acta* **489**, 191 (1977).
[51] B. P. Gaber, P. Yager, and W. L. Peticolas, *Biophys. J.* **22**, 191 (1978).
[52] R. Mendelsohn, S. Sunder, and H. J. Bernstein, *Biochim. Biophys. Acta* **413**, 329 (1075).
[53] R. Mendelsohn, S. Sunder, and H. J. Bernstein, *Biochim. Biophys. Acta* **443**, 613 (1976).

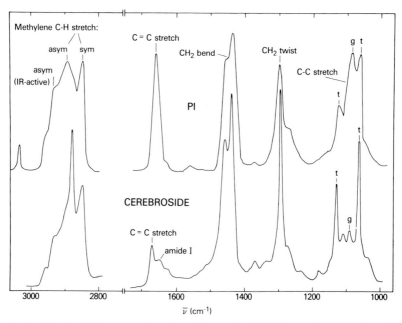

FIG. 3. Vibrational Raman spectra of two different glycolipids illustrating spectral sensitivity to conformation of alkyl chains. Top, plant phosphatidyl inositol (PI), liquid crystalline phase, 15°C. Bottom, brain cerebrosides, gel phase, 15°C. The band heights may be compared to the parameters of Table I: note the peak height ratios of methylene C–H asymmetric stretching modes at 2935 cm^{-1} and at 2880 cm^{-1} to the symmetric stretching modes at 2850 cm^{-1}, and the band profiles in the skeletal optical region (1150–1050 cm^{-1}), with respect to the high gauche conformer (g) population of PI (top spectrum) versus the high trans conformer population (t) for cerebrosides (bottom spectrum). The more disordered PI shows broader bands (M. R. Bunow and I. W. Levin, unpublished).

in the lipid gel and liquid-crystalline states, as well as temperature dependence within these states. Since the 2880-cm^{-1} peak height is the sum of methylene C–H asymmetric stretching modes, which broaden upon increasing chain disorder, plus the variable background, the 2880-cm^{-1} peak height itself cannot be expected to act as a linear parameter of intrachain order, nor can it easily be deconvoluted.[54] The peak height ratios in the 3000–2800 cm^{-1} region serve actually as phenomenological parameters of lipid order.

Modes in the C–C stretching region (1150–1050 cm^{-1}) can be treated in a more analytical manner. A fairly conservative approach allows the direct computation, from the gauche/trans peak height ratio (h_{1090}/h_{1130}) and an estimate of the gauche/trans conformer energy difference, of the

[54] R. G. Snyder, S. L. Hsu, and S. Krimm, *Spectrochim. Acta, Part A* **34A**, 395 (1978).

number of gauche conformers formed at a given temperature in lecithins with saturated chains. In this case, the reference state (all-trans chains) was defined as that at $-180°C$.[41]

Using these two sets of peak height ratios for the 3000–2800 and 1150–1050 cm^{-1} regions, one can follow the main gel to liquid-crystalline phase transition in saturated phospholipids,[40,42,47,52,55] the pretransition in saturated lecithins,[42,47,55,56] and the increased disorder imposed by high curvature in small lipid vesicles.[48] Since the frequency of the 1130 cm^{-1} all-trans skeletal mode is sensitive to chain length for short chains, it has been possible to monitor the melting of chain segments on each side of the double bond in unsaturated phospholipids[57] and in the unsaturated acyl chains of cerebrosides.[57a]

Modes associated with the polar head groups of lipids are also observable. Conformation of the glycerophosphocholine head group of lecithin[58] and sensitivity of glyceryl, phosphate, and choline group vibrational frequencies to hydration of lecithin[50] have been studied. The hydration-induced frequency shifts of the amide I and C=C vibrational modes of the head-group region of different cerebrosides, combined with Raman spectral parameters of chain packing, allow one to distinguish different molecular packings in hydrated cerebroside bilayers.[57a]

The response of lipid lamellar fluidity to the incorporation of additional components including cholesterol,[40,45,46,49] antibiotics,[46,59] polypeptides,[60] and proteins[61,62] has been studied by Raman spectroscopy. The depth of penetration of the added component can be judged by the degree of disruption of order in the original matrix,[46,60,63] and sometimes by the observation in the Raman spectrum of a second peak around 2880 cm^{-1}, which may be interpreted to indicate the formation of two populations of lipid, that segregated by the added compound and that of independent bulk lipid.[62] The lipid may melt in a biphasic manner.[62]

[55] N. Yellin and I. W. Levin, *Biochim. Biophys. Acta* **468**, 490 (1977).

[56] B. P. Gaber, P. Yager, and W. L. Peticolas, *Biophys. J.* **21**, 161 (1978).

[57] J. L. Lippert and W. L. Peticolas, *Biochim. Biophys. Acta* **282**, 8 (1972).

[57a] M. R. Bunow and I. W. Levin, *Biophys. J.* **32**, 1007 (1980).

[58] Y. Koyama, S. Toda, and Y. Kyogoku, *Chem. Phys. Lipids* **19**, 74 (1977).

[59] E. Wiedekamm, E. Bamberg, D. Brdiczka, G. Wildermuth, F. Macco, W. Lehmann, and R. Weber, *Biochim. Biophys. Acta* **464**, 442 (1977).

[60] S. P. Verma and D. F. H. Wallach, *Biochim. Biophys. Acta* **426**, 616 (1976).

[61] K. Larsson and R. P. Rand, *Biochim. Biophys. Acta* **326**, 245 (1973).

[62] W. Curatolo, S. P. Verma, J. D. Sakura, D. M. Small, G. G. Shipley, and D. F. H. Wallach, *Biochemistry* **17**, 1802 (1978).

[63] M. R. Bunow and I. W. Levin, *in* "Membrane Transport Processes" (D. C. Tosteson, Yu. A. Ovchinnikov, and R. Latorre, eds.), Vol. 2, p. 1. Raven, New York, 1978.

In these studies thermal phase transition curves, constructed from Raman spectra accumulated at a number of temperatures, are usually compared for different samples of reconstituted model membranes. If the lipid or added components are absorbing (as reflected also in a high fluorescence level), local heating in the laser beam of the compacted, fatty samples may vary from sample to sample, invalidating the thermal comparisons. This effect may be detected by testing for the dependence of spectral profile on laser beam power, or independent determinations of lipid phase transition temperatures may be made using a calorimeter or a microscope equipped with heating stage, and corrections then made.

In order to monitor separately the conformation of a lipid species in a multicomponent membrane containing either other lipids or a large amount of protein, one lipid species may be enriched in or substituted by its analog containing perdeuterated chains. The methylene C–D stretching modes for this species can then be observed free of interference in the 2300–2050 cm^{-1} region[50,51,53,64]; other modes of the perdeuterated chains also lie in spectrally "clean" regions[50,51] (Table I). Using this approach, one should be able to identify a species of deuterated lipid bound to intrinsic membrane proteins by its isolable, differential behavior compared to that of the bulk lipid.

The fluidity of natural membranes, e.g., of red blood cells,[65,66] sarcoplasmic reticulum,[67] and thymocytes[68] has been assessed by Raman spectroscopy. Increased fluidity of membranes of transformed and lectin-treated cells has been reported.[69–71] Supplemental analysis of resonance-enhanced Raman spectra of lipid-embedded polyene probes has confirmed the interpretation of the behavior of lipid chains; polyenes act as rather useful probes of fluidity (see Section 3.7.9).

[64] R. Mendelsohn and J. Maisano, *Biochim. Biophys. Acta* **506**, 192 (1978).

[65] S. P. Verma and D. F. H. Wallach, *Proc. Natl. Acad. Sci. U. S. A.* **73**, 3558 (1976).

[66] F. P. Milanovich, B. Shore, R. C. Harvey, and A. T. Tu, *Chem. Phys. Lipids* **17**, 79 (1976).

[67] F. P. Milanovich, Y. Yeh, R. J. Baskin, and R. C. Harney, *Biochim. Biophys. Acta* **419**, 243 (1976).

[68] S. P. Verma, D. F. H. Wallach, and R. Schmidt-Ullrich, *Biochim. Biophys. Acta* **394**, 633 (1975).

[69] E. B. Carew, H. W. Kaufman, P. W. Robbins, and H. E. Stanley, *Biophys. J.* **16**, 106a (1976).

[70] S. P. Verma, R. Schmidt-Ullrich, W. S. Thompson, and D. F. H. Wallach, *Cancer Res.* **37**, 3490 (1977).

[71] R. Schmidt-Ullrich, S. P. Verma, and D. F. H. Wallach, *Biochim. Biophys. Acta* **426**, 477 (1976).

3.7.2. Conformation of Nucleic Acids

Nucleic acids have proven to be superb polymers for Raman spectroscopic studies. The conformational states of these polymers are represented by the spectral profiles of three classes of vibrational modes: numerous modes assigned to in-plane ring stretching of the nucleic acid bases, stretching modes of the carbonyl groups involved in hydrogen bonding and base pairing, and phosphate diester stretching modes associated with the sugarphosphate backbone (see Table III). Spectra superior to those from proteins and lipids are obtained with only about 15 mg/ml of polynucleotides.

TABLE III.   Spectral Regions Containing Conformational Markers for Hydrated Nucleic Acids

$\bar{\nu}$ (cm^{-1})	Assignment	Conformational sensitivity[a–c]
(1684–1620)[d]	C=O stretching, base carbonyls	H bonding or base pairing; also base stacking
1579–1186	In-plane ring modes of the different bases	Base pairing and/or stacking
1100–1091	O–P–O$^-$ symmetric stretch	Reference mode
814–795	O–P–O diester symmetric stretch	Helical conformation of backbone and its disruption[e]
814–807	A form of helix	
795	Disordered backbone	
787	B form of helix	
786–656	In-plane ring modes for the different bases	Base stacking

[a] G. J. Thomas, Jr., *Appl. Spectrosc.* **30**, 483 (1976).
[b] S. C. Erfurth and W. L. Peticolas, *Biopolymers* **14**, 247 (1975).
[c] M. C. Chen, R. Giege, R. C. Lord, and A. Rich, *Biochemistry* **14**, 4385 (1975).
[d] Parentheses indicate frequencies observed in D$_2$O.
[e] S. C. Erfurth, P. J. Bond, and W. L. Peticolas, *Biopolymers* **14**, 1245 (1975).

Hypochromic and hyperchromic effects seen in the ultraviolet absorption spectra of polynucleotides are largely reproduced in the behavior of the in-plane ring vibrational modes in the Raman spectrum; as would be suspected, some in-plane modes in the Raman spectrum obtained with visible irradiation are enhanced by preresonance effects involving the 260-nm absorption bands.[72–76]

[72] S. C. Erfurth and W. L. Peticolas, *Biopolymers* **14**, 247 (1975).
[73] P. C. Painter and J. L. Keonig, *Biopolymers* **15**, 241 (1976).
[74] M. Pézolet, T.-J. Yu, and W. L. Peticolas, *J. Raman Spectrosc.* **3**, 55 (1975).
[75] M. Tsuboi, A. Y. Hirakawa, and Y. Nishimura, *J. Raman Spectrosc.* **2**, 609 (1974).
[76] D. C. Blazej and W. L. Peticolas, *Proc. Natl. Acad. Sci. U. S. A.* **74**, 2639 (1977).

Denaturation of deoxyribonucleic acid and ribonucleic acid (DNA and RNA), with concurrent loss of base stacking and base pairing interactions, is reflected primarily in large Raman spectral intensity changes, as well as in several frequency shifts of vibrational modes assigned to the different bases; these have been tabulated along with other mode parameters for DNA[72] and RNA.[77-80]

The intensities and frequencies of the carbonyl modes of the bases respond to rupture of hydrogen bonds occurring in the premelting and melting phenomena in DNA and in the melting of RNA; these modes participate to some extent in the hyper/hypochromic shifts involving the bases; that is, they also reflect the degree of base stacking.[72] The carbonyl modes are observed for nucleic acids in deuterium oxide, so that the interfering background from water modes at 1650 cm^{-1} is removed. The consequent proton–deuteron exchange of the nucleic acids has been carefully evaluated.[77]

The conformation of the sugar-phosphate backbone has been very successfully correlated by experiment and calculation with the frequency of the phosphate diester symmetric stretching mode (807–787 cm^{-1}). The frequency of this vibration is sensitive to the torsional angles about the P–O bonds as well as to ribose conformation[81-83]; it is an extremely valuable conformational marker for both DNA and RNA. It has been used to categorize conformations of RNAs and DNAs in solution in terms of the A and B helices (at 75 and 98% relative humidity, respectively) of DNA fibers.[84]

Overall, Raman spectroscopy is the best technique for determining secondary structure of RNAs and DNAs in association with virus proteins or with histones. Reference spectral states for fully stacked and paired polynucleic acids are supplied by pure double-stranded DNA or highly base-paired RNA. The wholly denatured state is modeled by thermally denatured nucleic acids. Decrements in base stacking, hydrogen bonding, and changes in backbone configuration upon association with protein can then be followed by monitoring the score of Raman modes described above, with the proper normalizations for base composition. Raman spectral

[77] G. J. Thomas, Jr., *Appl. Spectrosc.* **30**, 483 (1976).

[78] K. A. Hartman, R. C. Lord, and G. J. Thomas, Jr., in "Physico-Chemical Properties of Nucleic Acids" (J. Duchesne, ed.), Vol. 2, p. 1. Academic Press, New York, 1973.

[79] G. J. Thomas, Jr. and Y. Kyogoku, in "Practical Spectroscopy" (E. G. Brame, Jr., and J. G. Grasseli, eds.), Vol. 1, Part C, p. 717. Dekker, New York, 1977.

[80] R. C. Lord and G. J. Thomas, Jr., *Spectrochim. Acta, Part A* **23A**, 2551 (1967).

[81] E. B. Brown and W. L. Peticolas, *Biopolymers* **14**, 1259 (1975).

[82] S. C. Erfurth, J. K. Kiser, and W. L. Peticolas, *Proc. Natl. Acad. Sci. U. S. A.* **69**, 938 (1972).

[83] G. Forrest and R. C. Lord, *J. Raman Spectrosc.* **6**, 32 (1977).

[84] S. C. Erfurth, P. J. Bond, and W. L. Peticolas, *Biopolymers* **14**, 1245 (1975).

studies have thus shown, for example, the apparent preservation of B-helical structure upon binding of DNA to poly-L-lysine and poly-L-arginine, contrasted to the introduction of some disorder into some A-type RNAs upon similar binding[77,85]; noncooperative thermal rupture of base pairs, as contrasted to cooperative unstacking of bases in the RNA of MS2 virus, and stabilization of the RNA by the presence of MS2 coat protein[86]; and the state of DNA in chromatin,[87,88] including specific guanine–arginine attachment.[89] The major conformations of histones or virus proteins can be simultaneously determined in association with the nucleic acids, as well as separately.[77,86,87,90] Raman studies of details of DNA structure, such as the effects of cross linking[91] or alkylation[92] are also highly rewarding.

Recent Raman studies of nucleic acids using ultraviolet excitation[74,76,93,94] demonstrate that these resonance-enhanced spectra can now be isolated where the nucleic acids occur as a minor (<10%) biological component.

3.7.3. Raman Spectral Analysis of Polysaccharides

Polysaccharides that lack repeating structure, such as the branched oligomers attached to glycoproteins, are poor candidates for Raman analysis: their spectra are weak but complex. For homopolymers such as cellulose[95] and amylose,[96] and copolymers such as hyaluronic acid and chondroitin sulfates,[97] one can determine the configuration of the glycosidic linkages[97] and of side groups such as the sulfate groups in chondroitin sulfate[97] and differentiate polymorphic forms, as has been done for amylose.[96]

[85] B. Prescott, C. H. Chou, and G. J. Thomas, Jr., *J. Phys. Chem.* **80**, 1164 (1976).

[86] G. J. Thomas, Jr., B. Prescott, P. E. McDonald-Ordzie, and K. A. Hartman, *J. Mol. Biol.* **102**, 103 (1976).

[87] G. J. Thomas, Jr., B. Prescott, and D. E. Olins, *Science* (*Washington, D.C.*) **197**, 385 (1977).

[88] D. C. Goodwin and J. Brahms, *Nucleic Acids Res.* **5**, 835 (1978).

[89] S. Mansy, S. K. Engstrom, and W. L. Peticolas, *Biochem. Biophys. Res. Commun.* **68**, 1242 (1976).

[90] J.-G. Guillot, M. Pézolet, and D. Pallotta, *Biochim. Biophys. Acta* **491**, 423 (1977).

[91] R. Herbeck, T.-J. Yu, and W. L. Peticolas, *Biochemistry* **15**, 2656 (1976).

[92] S. Mansy and W. L. Peticolas, *Biochemistry* **15**, 2650 (1976).

[93] Y. Nishimura, A. H. Hirakawa, M. Tsuboi, and S. Nishimura, *Nature* (*London*) **260**, 173 (1976).

[94] L. Chinsky, P. Y. Turpin, M. Duquesne, and J. Brahms, *Biochem. Biophys. Res. Commun.* **75**, 766 (1977).

[95] J. J. Cael, K. H. Gardener, J. L. Koenig, and J. Blackwell, *J. Chem. Phys.* **62**, 1145 (1975).

[96] J. J. Cael, J. L. Koenig, and J. Blackwell, *Biopolymers* **14**, 1885 (1975).

[97] R. Bansil, I. V. Yannas, and H. E. Stanley, *Biochim. Biophys. Acta* **541**, 535 (1978).

3.7.4. Raman Vibrational Analysis of Protein Conformation

The polypeptide backbone of proteins is revealed in the Raman spectrum as a superposition of vibrations from each planar peptide unit. The strongest of these vibrational bands are the amide I (primarily C=O stretching plus C–N stretching) and amide III (C–N stretching plus N–H in-plane bending) modes. Extensive model studies and normal-coordinate calculations on homopolypeptides and simple copolymers have established the frequency-secondary structure correlations shown in Table IV. These have been refined by studies of natural proteins, which as heteropolymers have lower symmetry, diverse side-chain populations, and different charged-residue distributions, thus imposing further spectral qualifications. The reader is referred to several review articles.[18,98–101]

The uniformity of the backbone C–C–N conformation and the well-defined intrachain and interchain hydrogen bonding patterns in the α helix and antiparallel β sheet, respectively, result in fairly sharp amide I and amide III bands. The threefold helix assumed by polyglycine II has also been analyzed.[102] The so-called random-coil or disordered structure encompasses a variety of conformations and hydrogen bonding and solvent contacts. A general correlation between the torsional angle ψ for rotation about the C^{α}–C′ peptide bond and the amide III band frequency has been proposed[101]; this correlation must be modified by considering additional factors including the angle ϕ of rotation about the C^{α}–N bond and the range of hydrogen bonding strengths.[101,103] The resulting amide I and III bands are relatively broad, and the frequency for the amide III band in particular is relatively variable; the amide I band frequency is suggested to be a more dependable marker for the disordered state.[103]

The amide I and amide III bands are used to estimate the amounts of each type of secondary structure in a protein. The amide I band overlaps a broad water band which must be subtracted,[104] or, more commonly, spectra of proteins in solution must also be obtained in D_2O. Upon hydrogen-deuterium exchange on the amide nitrogen, the amide I band shifts to slightly lower frequencies; the amide III band shifts to 1110–950 cm^{-1}, a region which contains a substantial number of other vibrational modes. The observed translocation of peaks upon deuteration helps confirm their

[98] B. G. Frushour and J. L. Koenig, *in* "Advances in Infrared and Raman Spectroscopy" (R. J. H. Clark and R. E. Hestor, ed.), Vol. 1, p. 35. Heyden, London, 1975.

[99] N.-T. Yu, *CRC Crit. Rev. Biochem.* **4**, 229 (1977).

[100] H. E. Van Wart and H. A. Scheraga, *Methods Enzymol.* **49G**, 67 (1977).

[101] R. C. Lord, *Appl. Spectrosc.* **31**, 187 (1977).

[102] E. W. Small, B. Fanconi, and W. L. Peticolas, *J. Chem. Phys.* **52**, 4369 (1970).

[103] S. L. Hsu, W. H. Moore, and S. Krimm, *Biopolymers* **15**, 1513 (1976).

[104] T. W. Barrett, W. L. Peticolas, and R. R. Robson, *Biophys. J.* **23**, 349 (1978).

TABLE IV.   Selected Conformation-Sensitive Modes of Proteins

$\bar{v}$ (cm^{-1})	Assignment	Conformational sensitivity
	Polypeptide backbone[a-e]	
1670–1655 (1660–1632)[f]	Amide I band:	Secondary structure:
1690–1685, 1645–1640 (variable)		β turn ("U turn")[e]
1670 (1660) sharp		Antiparallel β sheet
1664 (1657) broad		Disordered
1655 (1632)		α helix
1315–1235	Amide III band:	Secondary structure:
1315–1275 weak		α helix
1305		β turn
1254–1240 broad		Disordered
1239–1235 sharp		Antiparallel β sheet
995 (1010)	C–C stretch	Antiparallel β sheet
940 (950)	C–C (C–C–N) stretch	α helix[g]
	Amino acid side groups	
		Secondary, tertiary structure:
2580–2560	S–H stretch	Free S–H, its accessibility[h]
1410	COO$^-$ symmetric stretch	Intensity increases with ionization[g]
1360	Tryptophan ring	Intensity decreases with denaturation[i,j]
850, 830	Tyrosine doublet	Relative peak heights reflect ionization, H bonding strength, indirectly reflect environment[k]
745–700, 670–630	C–S stretch, methionine	Frequency ranges for *gauche*, *trans* conformers, respectively[l]
540–510	S–S stretch, cystine	Frequency sensitive to bond conformation[b,m]

[a] T. G. Spiro and B. P. Gaber, *Annu. Rev. Biochem.* **46**, 553 (1977).
[b] N.-T. Yu, *CRC Crit. Rev. Biochem.* **4**, 3 (1977).
[c] T.-J. Yu, J. L. Lippert, and W. L. Peticolas, *Biopolymers* **12**, 2161 (1973).
[d] S. L. Hsu, W. H. Moore, and S. Krimm, *Biopolymers* **15**, 1513 (1976).
[e] J. Bandekar and S. Krimm, *Proc. Natl. Acad. Sci. U. S. A.* **76**, 774 (1979).
[f] Parentheses indicate frequencies observed in D_2O solution.
[g] B. G. Frushour and J. L. Koenig, *Biopolymers* **13**, 1809 (1974).
[h] N.-T. Yu and E. J. East, *J. Biol. Chem.* **250**, 2196 (1975).
[i] M. C. Chen, R. C. Lord, and R. Mendelsohn, *J. Am. Chem. Soc.* **96**, 3038 (1974).
[j] N.-T. Yu, *J. Am. Chem. Soc.* **96**, 4664 (1974).
[k] M. N. Siamzawa, R. C. Lord, M. C. Chen, T. Takematsu, I. Harada, H. Matsuura, and T. Shimanouchi, *Biochemistry* **14**, 4870 (1975).
[l] N. Nogami, H. Sugeta, and T. Miyazawa, *Bull. Chem. Soc. Jpn.* **48**, 2417 (1975).
[m] H. Sugeta, *Spectrochim. Acta, Part A* **31A**, 1729 (1975).

assignment. The weak amide III band of the α helix at about 1300 cm^{-1} is superimposed upon methylene deformation modes; loss of 30% of the band intensity in this region upon deuteration indicates very high helix content.[99]

To quantify the amounts of each type of structure in the protein it is assumed that the intensity of each vibrational mode employed as a conformational marker is linearly related to the concentration of the conformer. The intensity of the methylene deformation modes at 1450 cm^{-1} from side chains of the protein is used as an internal reference. Several groups of workers have made such estimates.[105-107] Lippert and co-workers[107] estimated fractions of α-helix, β-sheet, and disordered structure in ten proteins in solution; the estimates fell within the range of results from x-ray diffraction and circular dichroism studies. The heterogeneity of the natural proteins and the variety of structural segments found therein probably precludes accurate estimation of any but the predominant structure. The amide I, nominally infrared-active amide II, and amide III modes in simple amides are enhanced upon ultraviolet excitation[108]; it may be possible to separate these modes in proteins from the rest of the protein spectrum by their excitation profiles.

Raman spectroscopy is uniquely qualified for comparative studies of crystalline and solution states of soluble proteins. For small proteins, the introduction of solvent may considerably alter conformational energies so that moderate changes in backbone conformation become apparent. For larger proteins, a number of side-chain vibrational modes act as additional Raman parameters of conformational alteration. These are described below (Section 3.7.5). Unlike circular dichroism and optical rotatory dispersion spectra, Raman spectra of proteins in solution are not affected by solution turbidity. Comparative Raman studies of many proteins in crystalline, lyophilized, and dissolved (native and denatured) states have been performed.[18,98-101] Structure determinations by x-ray diffraction of crystalline proteins provide the most detailed reference state for these comparisons. Raman spectral mapping of conformational variation in proteins has a broad field of applicability. Isoenzyme conformations, for example, have been distinguished.[109] Such mappings can detect irreversible changes associated with preparative procedures or protein modification for

[105] B. G. Frushour and J. L. Koenig, *Biopolymers* **13**, 1809 (1974).

[106] M. Pézolet, M. Pigeon-Gosselin, and L. Coulombe, *Biochim. Biophys. Acta* **453**, 502 (1976).

[107] J. L. Lippert, D. Tyminski, and P. J. Desmeules, *J. Am. Chem. Soc.* **98**, 7075 (1976).

[108] I. Harada, Y. Sugawara, H. Matsuura, and T. Shimanouchi, *J. Raman Spectrosc.* **4**, 91 (1975).

[109] J. Twardowski, *Bipolymers* **17**, 181 (1978).

cases where the protein cannot be monitored by an assay of biological activity.

Few nonresonance Raman studies have attempted to detect overall conformational changes in large proteins associated with biological action; noteworthy is the report of increased backbone disordering in rabbit anti-ovalbumin upon precipitation with ovalbumin.[110] This study is also noteworthy in confirming β structure not previously well documented in the protein under study.

The biologically active site of small peptide hormones and antibiotics occupies a large proportion of the molecule, so that behavior of the amide I and ester carbonyl stretching modes can be largely associated with the active site. Prototypical Raman studies include those of oxytocin,[111,112] valinomycin,[113,114] and gramicidin A'.[115] For large proteins, studies of the active site are best accomplished by exploiting resonance Raman effects that are dealt with below (Section 3.7.8).

Raman spectral resolution of the conformation of opsin[116] and red blood cell proteins[117] in the native membranes, of lens protein in the intact ocular lens,[99] and of proteins in single-catch muscle fibers[118] demonstrates the capability of the technique for whole-tissue analysis. The best objects for this approach are tissues rich in a single protein, so that the Raman spectrum does not comprehend a rather meaningless average of protein conformations.

3.7.5. Vibrational Markers of Protein Side-Groups

The frequency of the stretching mode of the S–S linkage in cystine (540–485 cm^{-1}) is a function of the CS–SC and the SS–CC dihedral angles.[99,119,120] Identification of the disulfide band can be verified by chemical modification. Structural heterogeneity can in some cases be inferred from this mode; for example, the distribution of S–S frequencies allowed inference of multiple conformers of the nonapeptide lysine vasopressin in

[110] P. C. Painter and J. L. Koenig, *Bipolymers* **14**, 457 (1975).

[111] F. R. Maxfield and H. A. Scheraga, *Biochemistry* **16**, 4443 (1977).

[112] A. T. Tu, J. B. Bjarnason, and V. J. Hruby, *Biochim. Biophys. Acta* **533**, 530 (1978).

[113] I. M. Asher, K. J. Rothschild and H. E. Stanley, *J. Mol. Biol.* **89**, 205 (1974).

[114] K. J. Rothschild, I. M. Asher, H. E. Stanley, and E. Anastassakis, *J. Am. Chem. Soc.* **99**, 2032 (1977).

[115] K. J. Rothschild and H. E. Stanley, *Science (Washington, D.C.)* **185**, 616 (1974).

[116] K. Rothschild, J. R. Andrew, W. J. DeGrip, and H. E. Stanley, *Science (Washington, D.C.)* **191**, 1176 (1976).

[117] J. L. Lippert, L. E. Gorczyca, and G. Meiklejohn, *Biochim. Biophys. Acta* **382**, 51 (1975).

[118] M. Pézolet, M. Pigeon-Gosselin, and J. P. Caille, *Biochim. Biophys. Acta* **533**, 263 (1978).

[119] H. Sugeta, *Spectrochim. Acta, Part A* **31**, 1792 (1975).

[120] H. E. Van Wart and H. A. Scheraga, *J. Phys. Chem.* **80**, 1823 (1976).

aqueous solution.[111] Similarly, the conformation of the thioether side chain of methionine is sensitively mirrored by the C–S stretching frequency $(745-630 \text{ cm}^{-1})$.[121]

Tyrosine exhibits a Fermi resonance doublet, involving ring modes, at 850 and 830 cm^{-1}. The relative heights of peaks within the doublet are sensitive to the state of ionization and the type of hydrogen bonding of the phenolic hydroxyl group; the peak height ratio may be interpreted, indirectly, in terms of tyrosine residues being buried within or exposed at the aqueous interface of the protein, and may be used to monitor protein unfolding.[122] The intensities of tryptophan ring modes at 1359 and 1337 cm^{-1} are indicative of tryptophan environment, diminishing when tryptophans are exposed to solvent.[123,124] The intensities of tryptophan modes pre-resonance enhanced by excitation at 363.8 nm have been tested as monitors of the tryptophan-containing active site of lysozyme.[125]

3.7.6. Resonance Raman Studies of Porphyrins and Hemoproteins

The most popular object of resonance Raman biophysical studies is the porphyrin macrocycle, the ubiquitous chromophoric site of redox and oxygen binding activity. Several reviews of work in this field are available.[15,18,27,99,126–128] The resonance Raman studies of hemoproteins described briefly here supply examples of the properties of A-term and B-term scattering.

The 16-atom inner ring of the porphyrin macrocycle has 18 π electrons. The lowest two of the in-plane $\pi-\pi^*$ transitions in hemoproteins and metallo-porphyrins are split by configuration interaction to form the Soret ($\sim 400 \text{ nm}$) and the α and β bands (above 500 nm) where the β band, a higher-energy sideband to the α band, results from vibronic mixing between the Soret and α transitions. Excitation at laser frequencies near the intense Soret band enhances only totally symmetric, or A-term vibrational modes.[22,27,12] The complete excitation profile of the Soret band has not been obtained

[121] N. Nogami, H. Sugeta, and T. Miyazawa, *Bull. Chem. Soc. Jpn.* **48**, 2417 (1975).

[122] M. N. Siamzawa, R. C. Lord, M. C. Chen, T. Takamatsu, I. Harada, H. Matsuura, and T. Shimanouchi, *Biochemistry* **14**, 4870 (1975).

[123] M. C. Chen, R. C. Lord, and R. Mendelsohn, *J. Am. Chem. Soc.* **96**, 3038 (1974).

[124] N.-T. Yu, *J. Am. Chem. Soc.* **96**, 4664 (1974).

[125] K. G. Brown, E. B. Brown, and W. B. Person, *J. Am. Chem. Soc.* **99**, 3128 (1977).

[126] R. H. Felton and N.-T. Yu, *in* "The Porphyrins" (D. Dolphin, ed.), Vol. 3, Part A, p. 347. Academic Press, New York, 1978.

[127] T. G. Spiro, *Proc. R. Soc. London, Ser. A* **345**, 89 (1975).

[128] P. R. Carey, *Q. Rev. Biophys.* **11**, 309 (1978).

because of the lack, until recently, of a spectrum of near-ultraviolet laser frequencies. Partial data are available for ferrihemoglobin[22] and cytochrome P450$_{CAM}$.[129]

Apparently because of the weakness of the α band, A-term scattering is difficult to observe by excitation at wavelengths near the α band. Instead, excitation near the α and β bands results in enhancement of B-term modes identified by their depolarization ratios.[26,27] The vibronic mixing of α and Soret bands which is responsible for the β band gives rise to the B-term scattering[22,27] (but see also Ref. 130). The excitation profile (mode intensity versus wavelength) for each resonance-enhanced Raman mode of vibrational frequency v_v possesses a peak coincident with the α-band maximum, and a secondary peak displaced into the β band by v_v.[27] This behavior is predicted by vibronic theory for simple B-term scattering.[5,25,27] The excitation profile for the depolarization ratio peaks midway between the two intensity profile peaks.[25,131] Far from resonance, the depolarization ratio must approach a value $\rho \leq 0.75$ characteristic of the nonresonant scattering process.

When the Raman scattering tensor is not symmetric, there are actually three tensor invariants: the trace and the symmetric and antisymmetric parts of the anisotropy.[132] The depolarization ratio ρ, which supplies just two invariants, is usually measured in resonance Raman work and may be sufficient for mode assignment and for determining the actual preservation of symmetry in the molecule. For example, the observation of several bands showing anomalous polarization where inverse polarization was expected has been interpreted in terms of reduction of porphyrin symmetry by the effects of substituents or by rhombic distortion.[133,134] The alternative interpretation of accidental degeneracies among modes of different symmetry, producing "hybrid" depolarization ratios, was considered unlikely in these cases. It is also possible to measure all three scattering-tensor invariants by means of a circularly polarized laser beam and additional optics.[5,36,133,135] Such measurements for the resonance-enhanced modes of ferrocytochrome c have allowed the conclusion that the porphyrin symmetry is in fact lower than that of point group C_{4v}.[133]

The lifetimes of resonance Raman transitions can be estimated from excitation profile bandwidths; comparisons of the perturbation of heme

[129] P. M. Champion, I. C. Gunsalus, and G. C. Wagner, *J. Am. Chem. Soc.* **100**, 3743 (1978).

[130] Y. Nishimura, A. Y. Hirakawa, and M. Tsuboi, *J. Mol. Spectrosc.* **68**, 333 (1977).

[131] D. W. Collins, D. B. Fitchen, and A. Lewis, *J. Chem. Phys.* **59**, 5714 (1973).

[132] T. G. Spiro and T. C. Strekas, *Proc. Natl. Acad. Sci. U. S. A.* **69**, 2622 (1972).

[133] J. Nestor and T. G. Spiro, *J. Raman Spectrosc.* **1**, 539 (1973).

[134] R. Mendelsohn, S. Sunder, and H. J. Bernstein, *J. Raman Spectrosc.* **3**, 303 (1975).

[135] M. Pézolet, L. A. Nafie, and W. L. Peticolas, *J. Raman Spectrosc.* **1**, 455 (1973).

TABLE V.   Selected Resonance-Enhanced Modes of Hemoproteins

$\bar{\nu}$ (cm^{-1})	Polarization[a]	Configurational sensitivity
Porphyrin ring modes (C–C, C–N stretch, methine bridge C–H bend)		
1642–1601	dp	Frequency sensitive to spin and oxidation state[b]
~1590–1570	ap	Frequency sensitive to ring center–pyrrole N distance, ring size[c]
1502–1466	p	Frequency sensitive to oxidation state[a]
1375–1344	p	Frequency sensitive to oxidation state, ligand electron donating capacity[b]
Fe–axial ligand stretching modes		
567		Fe–O stretch (oxyhemoglobin), bond strength[d]
497–413	p	Fe–axial ligand (methemoglobin), sensitive to ligand identity[e]

[a] Here dp denotes depolarized; ap, anomalously polarized; p, polarized.

[b] P. M. Champion, I. C. Gunsalus, and G. C. Wagner, *J. Am. Chem. Soc.* **100**, 3743 (1978).

[c] L. D. Spaulding, R. C. C. Chang, N.-T. Yu, and R. H. Felton, *J. Am. Chem. Soc.* **97**, 2517 (1975).

[d] H. Brunner, *Naturwissenschaften* **61**, 129 (1974).

[e] S. A. Asher, L. E. Vickery, T. M. Schuster, and K. Sauer, *Biochemistry* **16**, 5849 (1977).

symmetry in cytochromes c and b_5 have been made using such data.[136] In sum, the depolarization ratio and the excitation profile yield evidence on deviations of a molecular configuration from its expected symmetry.

Besides confirming aspects of the theory of resonance scattering, resonance Raman studies of porphyrins and hemoproteins provide detailed information on the electronic environment of porphyrin molecules in protein pockets. These pockets consist of one or more ligands to the metal ion of the porphyrin plus a protein enclosure which in some cases is covalently linked to the periphery of the porphyrin.

Several bands associated with heme ring in-plane vibrations and enhanced as described above act as markers of the oxidation and/or spin state of the heme (see Table V). The lowering of ring stretching frequencies in the Fe(II) state, compared to Fe(III), has been suggested to be due to greater back donation of electrons from the ion d_π orbitals into $\pi-\pi^*$ antibonding porphyrin orbitals and resultant bond weakening.[15,137,138] Variations in

[136] J. M. Friedman, D. L. Rousseau, and F. Adar, *Proc. Natl. Acad. Sci. U. S. A.* **74**, 2607 (1977).

[137] T. G. Spiro and T. C. Strekas, *J. Am. Chem. Soc.* **96**, 338 (1974).

[138] T. G. Spiro and J. M. Burke, *J. Am. Chem. Soc.* **98**, 5482 (1976).

the Fe(II)–ligand bonding scheme affect these frequencies.[129,139] Unusually low frequencies for band I in cytochrome P450$_{CAM}$ in both oxidized and reduced states have been taken as evidence for the identity of the strong electron-donating axial ligand of P450 as the mercaptide sulfur of cysteine.[129,140] Other bands are also sensitive to the identities of the diverse ligands of cytochromes,[141,142] while the 1590-cm^{-1} band has been correlated with ring expansion.[143]

The frequency lowering of these "marker" bands upon conversion of hemes from low to high spin has been interpreted in terms of slight doming of the heme group (downward bending of the pyrrole rings) when the iron atom moves out of the mean heme plane,[129] and, more importantly,[15] expansion of the heme ring as also reflected in increased ring nitrogen–iron distances.[11,15,143]

The Fe–axial ligand stretching modes in hemes are desirable monitors of heme catalytic activity. These stretching modes, lying approximately normal to the set of heme ring π–π* transitions, are poorly enhanced by excitation at wavelengths shorter than 500 nm (in or above the α and β bands), but have been enhanced in methemoglobin derivatives using dye laser excitation at 500–650 nm of charge-transfer bands in this region.[144]

Analysis of the Raman modes described above benefited greatly from an enormous spectroscopic literature on the structurally diverse heme proteins and crystallizable metalloproteins.[145] The sensitivity of resonance Raman spectroscopy to small structural changes around the heme group equals or surpasses that of x-ray analysis and of newer approaches such as extended x-ray absorption fine-structure studies[146]; the difficult step in resonance Raman analysis is still the precise parametrization of electronic structure in terms of Raman spectral variables.[11,15]

A sizable number of Raman investigations are now focused on the problem of the mechanism coupling oxygen binding to subsequent cooperative interplay in the hemoglobin tetramer, that is, how the electronic restructuring of the heme group that follows changes in iron coordination number and spin-state accompanying oxygenation, is sensed by the protein pocket

[139] T. Kitagawa, Y. Kyogoku, T. Iizuka, and M. I. Saito, *J. Am. Chem. Soc.* **98**, 5169 (1976).

[140] Y. Ozaki, T. Kitagawa, Y. Kyogoku, H. Shimada, T. Iizuka, and Y. Ishimura, *J. Biochem.* (*Tokyo*) **80**, 1447 (1976).

[141] T. Kitagawa, Y. Ozaki, J. Teraoka, Y. Kyogoku, and T. Yamanaka, *Biochim. Biophys. Acta* **494**, 100 (1977).

[142] T. Kitagawa, Y. Ozaki, Y. Kyogoku, and T. Horio, *Biochim. Biophys. Acta* **495** (1977).

[143] L. D. Spaulding, R. C. C. Chang, N.-T. Yu, and R. H. Felton, *J. Am. Chem. Soc.* **97**, 2517 (1975).

[144] S. A. Asher, L. E. Vickery, T. M. Schuster, and K. Sauer, *Biochemistry* **16**, 5856 (1977).

[145] J. L. Hoard, *Science* (*Washington, D.C.*) **174**, 1295 (1971).

[146] P. Eisenberger and B. M. Kincaid, *Science* (*Washington, D.C.*) **200**, 1441 (1978).

and then by the other, distant heme groups.[11,15,138,147–149] Similarly, the dynamics of heme restructuring upon substrate or oxygen ligation are being tackled for cytochromes; representative references are given for this wide field.[129,136,141,142,150,151] Studies of chlorophyll environment in intact chloroplasts are also in progress.[152]

3.7.7. Resonance Raman Parameters for Nonheme Metalloproteins

Charge-transfer transitions of ligand–metal complexes in nonheme iron-containing proteins and metalloenzymes provide (A-term) resonance-enhanced views of the active sites of these proteins. Stretching modes of the iron–sulfur linkages involving sulfide or cysteine bonds to iron can be enhanced by excitation in the visible region.[18,100,129,153] It has been hypothesized that distorted geometry of the iron–sulfur complexes—for example, the inequivalence in bond lengths in the approximately tetrahedral configuration of iron and sulfur atoms in oxidized rubredoxin—can be related to redox potential in these proteins.[154] Loss of tetrahedral symmetry is suggested by the nonzero depolarization ratio observed for a totally symmetric mode in the highly symmetric tetrahedral configuration assumed for oxidized rubredoxin.[153] Loss of symmetry, that is, inequivalence of Fe–S (cysteine) linkages, in the cubanelike Fe–S clusters in oxidized ferredoxin is demonstrated more convincingly by observation of splitting of one of the Fe–S symmetric splitting modes. This splitting is also seen in the spectrum of reduced high-potential iron protein.[155] The identification of two Fe–S (cysteine) symmetric stretching modes where one was expected provides a good example of the utility of comparison of the number of observed modes, characterized by their depolarization ratios, to the number expected according to simple group-theoretical rules.[2]

The coordination of O_2 to Fe irons in hemaerythrin and to Cu ions in hemocyanin can be inferred from the O–O stretching frequencies resonance

[147] W. H. Woodruff and S. Farquharson, *Science (Washington, D.C.)* **201**, 831 (1978).

[148] L. D. Barron and A. Szabo, *J. Am. Chem. Soc.* **97**, 660 (1975).

[149] M. F. Perutz, J. V. Kilmartin, K. Nagai, A. Szabo, and S. R. Simon, *Biochemistry* **15**, 378 (1976).

[150] I. Salmeen, L. Rimai, and G. Babcock, *Biochemistry* **17**, 800 (1978).

[151] S. A. Fairhurst and L. H. Sutcliffe, *Prog. Biophys. Mol. Biol.* **34**, 1 (1978).

[152] M. Lutz, *Biochim. Biophys. Acta* **460**, 408 (1977).

[153] T. V. Long, T. M. Loehr, J. R. Alkins, and W. Lovenberg, *J. Am. Chem. Soc.* **93**, 1809 (1971).

[154] B. L. Vallee and R. J. P. Williams, *Proc. Natl. Acad. Sci. U. S. A.* **59**, 498 (1968).

[155] S.-P. Tang, T. G. Spiro, C. Antanaitis, T. H. Moss, R. H. Holm, T. Herskovitz, and L. E. Mortenson, *Biochem. Biophys. Res. Commun.* **62**, 1 (1975).

enhanced by illumination in the charge-transfer bands of these respiratory pigments.[18,156,157] Copper-ligand modes in blue copper proteins have also been analyzed.[158] The reader is referred to recent reviews.[18,100,128]

3.7.8. Extrinsic Chromophores for Resonance Raman Studies of Proteins

Vibrations involving the active site of enzymes which contain no intrinsic chromophores can be selectively enhanced upon addition of a chromophoric substrate or inhibitor. The resonance signal intensity is compatible with kinetic studies. This approach, pioneered by Carey and co-workers, yields a wealth of detail on substrate–enzyme binding configurations and dynamics,[128,159–165] and hence is anticipated to be widely exploited. Several reviews are available.[11,18,100,128,160]

3.7.9. Resonance-Enhanced Vibrational Behavior of Biological Polyenes

The exceptionally simple resonance and preresonance Raman spectra of extended linear polyenes are dominated by C=C and C– stretching modes. These are enhanced primarily by A-type vibronic mixing with π–π^* transitions.[15,166–168] For all-trans polyenes, the C=C stretching frequency (1600–1500 cm^{-1}) varies inversely with increasing electron delocalization, that is, polyene length, and with electronic absorption band wavelength.[169] The frequency is also sensitive to solvent polarity.[170] In addition, modes between 1400 and 1100 cm^{-1}, resonance enhanced by admixture of C=C stretching, serve to fingerprint the length and isomeric identity of polyenes.[171] Heteroatoms bonded in conjugation with the polyene also possess

[156] T. J. Thamann, J. S. Loehr, and T. M. Loehr, *J. Am. Chem. Soc.* **99**, 4187 (1977).

[157] D. M. Kurtz, Jr., D. F. Shriver, and I. M. Klotz, *J. Am. Chem. Soc.* **98**, 5033 (1976).

[158] O. Silman, N. M. Young, and P. R. Carey, *J. Am. Chem. Soc.* **98**, 744 (1976).

[159] P. R. Carey and H. Schneider, *Biochem. Biophys. Res. Commun.* **57**, 831 (1973).

[160] P. R. Carey and H. Schneider, *Acc. Chem. Res.* **11**, 122 (1978).

[161] P. R. Carey and H. Schneider, *J. Mol. Biol.* **102**, 679 (1976).

[162] P. R. Carey, R. G. Carriere, K. R. Lynn, and H. Schneider, *Biochemistry* **15**, 2387 (1976).

[163] K. Kumar, R. W. King, and P. R. Carey, *Biochemistry* **15**, 2195 (1976).

[164] R. L. Petersen, T.-Y. Li, J. T. McFarland, and K. L. Watters, *Biochemistry* **16**, 726 (1977).

[165] R. K. Scheule, H. E. Van Wart, B. L. Vallee, and H. E. Scheraga, *Proc. Natl. Acad. Sci. U. S. A.* **74**, 3273 (1977).

[166] S. Sufrà, G. Dellepiane, G. Masetti, and G. Zerbi, *J. Raman Spectrosc.* **6**, 267 (1977).

[167] A. Warshel and M. Karplus, *J. Am. Chem. Soc.* **96**, 5677 (1974).

[168] A. G. Doukas, B. Aton, R. H. Callender, and B. Honig, *Chem. Phys. Lett.* **56**, 248 (1978).

[169] L. Rimai, M. E. Heyde, and D. Gill, *J. Am. Chem. Soc.* **95**, 4493 (1973).

[170] M. E. Heyde, D. Gill, R. G. Kilponen, and L. Rimai, *J. Am. Chem. Soc.* **93**, 6776 (1971).

[171] R. Callender and B. Honig, *Annu. Rev. Biophys. Bioeng.* **6**, 33 (1977).

resonance-enhanced modes.[170] These different modes have allowed very fruitful resonance Raman observations of the behavior of the retinylidene Schiff base of rhodopsin and isorhodopsin[171] (Chapter 3.8).

All-trans polyenes have been shown to act as Raman probes of membrane fluidity when they are mixed with the hydrocarbon chains of lipid membranes. Isolated cell membranes contain minute amounts of β carotene.[68,172] The intensity of the C=C stretching mode of β carotene decreases dramatically when the lipid matrix of the membrane passes, with increasing temperature, from the gel to the liquid-crystalline state.[172] The mode frequency of β-carotene in sciatic nerve tissue also responds to excitation of the nerve.[173] The latter experiment illustrates that the dynamics of live tissue may be followed using resonance Raman probes.

The polyenic antibiotic amphoterician B acts only upon cell membranes containing sterol, probably by formation of an oligomeric sterol–amphotericin B pore.[174,175] The frequency and intensity of the preresonance-enhanced C=C stretching mode of amphotericin B decrease upon the gel to liquid-crystalline phase transition in lecithin–cholesterol lamellae but are indifferent to the phase transition in cholesterol-free lamellae. It can be concluded that amphotericin B is well embedded in sterol-containing, but only surface bound in sterol-free lipid bilayers.[46] The spectral responses probably reflect changes in the fit of the polyene among the hydrocarbon chains, and the decrease in dielectric coefficient sensed by the polyene, as the lipid bilayer thins and becomes more fluid.[46]

3.8. Kinetic Studies

Raman spectral studies of photoinduced chemical and physical events occurring on the nanosecond time scale can be accomplished using a pulsed laser source, for example, a nitrogen-laser-pumped tunable dye laser, combined with a gated detection system incorporating a vidicon or photodiode array target and optical multichannel analysis. A band several hundred cm^{-1} wide of the signal dispersed by a monochromator can usually be observed per pulse.[147,176,177] Studies on this time scale require a signal

[172] S. P. Verma and D. F. H. Wallach, *Biochim. Biophys. Acta* **401**, 168 (1975).

[173] B. Szalontai, Cs. Bagyinka, and L. I. Horváth, *Biochem. Biophys. Res. Commun.* **76**, 660 (1977).

[174] B. DeKruijff and R. A. Demel, *Biochim. Biophys. Acta* **339**, 57 (1974).

[175] A. Finkelstein and R. Holz, *Membranes* **2**, 377 (1973).

[176] A. Campion, J. Terner, and M. A. El-Sayed, *Nature (London)* **265**, 659 (1977).

[177] W. H. Woodruff and G. H. Atkinson, *Anal. Chem.* **48**, 186 (1976).

intensity characteristic of a resonance Raman (Chapter 3.4) or resonance CARS (Chapter 3.9) process.

Kinetics of the photochemical bacteriorhodopsin cycle on the microsecond to millisecond time scale have been followed by imaginative substitution for the pulsed laser of the electromechanically chopped beam of a CW laser, combined again with vidicon detection.[178-181] Studies of bacteriorhodopsin kinetics on the microsecond time scale have also been tackled by flowing the sample past the beam at a controlled, variable flow rate.[182-184] In this approach, the steady-state concentration profiles are pumped by a CW laser beam and detected either with a conventional photomultiplier tube or with a vidicon and optical multichannel analysis. Steady-state concentration profiles of a photoinduced process can also be observed with conventional detection methods by means of a photolysis pulse followed after a fixed delay by a probe pulse. These pulses have also been produced by electromechanical chopping.[178,185]

Studies of macromolecular dynamics, e.g., enzyme–substrate association-dissociation (time constant $\tau < 10^{-3}$ sec) and substrate-induced conformational changes ($10^{-4} < \tau < 10^{-1}$ sec) thus become feasible. Kinetic processes may be triggered, as in other types of spectroscopy, by concentration-jump (for example, using a stopped-flow arrangement[186]) or temperature-jump impulses. Rapid scanning on a slower scale, of seconds, can be achieved by computer-controlled scanning over a limited spectral region, using a conventional grating instrument. Modification of the scan drive has been suggested for faster scanning.[187]

The bacteriorhodopsin studies mentioned above illustrate the dynamic and structural resolution gained by the combination of stationary-state and kinetic resonance Raman studies. Bacteriorhodopsin, consisting of glycoprotein plus retinal bound in Schiff base linkage, is structurally very similar to the visual pigment rhodopsin[188] for which similar studies have been performed,[15,171] but its photosynthetic function as a transmembrane

[178] J. Terner, A. Campion, and M. A. El-Sayed, *Proc. Natl. Acad. Sci. U. S. A.* **74**, 5212 (1977).

[179] A. Campion, M. A. El-Sayed, and J. Terner, *Biophys. J.* **20**, 369 (1977).

[180] A. Campion, M. A. El-Sayed, and J. Terner, *Adv. Laser Spectrosc.* **1**, 128 (1977).

[181] J. Terner, C.-L. Hsieh, A. R. Burns, and M. A. El-Sayed, *Biochemistry* **18**, 3629 (1979).

[182] M. A. Marcus and A. Lewis, *Science (Washington, D.C.)* **195**, 1328 (1977).

[183] J. Terner, C.-L. Hsieh, and M. A. El-Sayed, *Biophys. J.* **26**, 527 (1979).

[184] J. Terner, C.-L. Hsieh, A. R. Burns, and M. A. El-Sayed, *Proc. Natl. Acad. Sci. U. S. A.* **76**, 3046 (1979).

[185] A. R. Oseroff and R. H. Callender, *Biochemistry* **13**, 4243 (1974).

[186] M. Crunelle-Cras and J. C. Merlin, *J. Raman Spectrosc.* **6**, 261 (1977).

[187] J. M. Remy, B. Sombret, and F. Wallart, *J. Mol. Struct.* **45**, 349 (1978).

[188] D. Osterholt and W. Stoeckenius, *Nature (London), New Biol.* **233**, 149 (1971).

proton pump is better understood than the visual process.[182] The kinetics of proton cycling of the retinylidene chromophore have been studied by absorption spectroscopy, and isomeric retinal intermediates have been thus identified by the rates of appearance of new, visible region absorption maxima.[189–191] The cycle is shown in Fig. 4. The structures of the intermediates of the cycle are not apparent from the nearly structureless absorption bands at room temperature, but retinals can be "fingerprinted" by their resonance or preresonance Raman spectra[170,192] and matched to the appearances of bands in bacteriorhodopsin preparations.[184,193–195] The intermediate M_{412}, for example, appears to resemble 13-*cis*-retinal.[193] Substantial stationary-state concentrations in bacteriorhodopsin preparations are seen only for R_{570} and M_{412}; the other intermediates are seen only with kinetic resolution.

Using 5-nsec laser pulses at 582.1 nm, researchers have observed the C=C stretching mode of the first intermediate, K_{590}, as a shoulder arising at about 1515 cm^{-1} on the 1530-cm^{-1} band of R_{570}.[176] The lower frequency of this mode is correlated with the longer absorption maximum of K_{590} (cf. Section 3.7.9) which has been observed within picoseconds of illumination.[189] The initial rearrangement occurring in the chromophore in picoseconds (isomerization versus electron redistribution) is controversial.[189]

The step at which the Schiff base is deprotonated can be located in the time-resolved resonance Raman excitation profiles.[179,183,192] In the resonance Raman spectrum, the protonation state of the retinal chromophore is differentiated by the frequency of the Schiff base C=N stretching mode, resonance enhanced by conjugation with the retinal chain, and located at 1643 cm^{-1} when protonated or 1620 cm^{-1} unprotonated. In Fig. 4a, the C=C stretching band at 1570 cm^{-1} associated with the intermediate M_{412}, and the 1620-cm^{-1} band associated with the deprotonated C=N stretching mode are seen to increase dramatically in time when laser excitation is at 476.5 nm, a wavelength near the 412-nm absorption maximum of M_{412}. In contrast (Fig. 4b), when laser excitation lies at 514.5 nm, near the 570-nm absorption maximum of R_{570}, the prominent feature in the time-resolved spectrum is the 1530-cm^{-1} band, the C=C stretching mode asso-

[189] E. P. Ippen, C. V. Shank, A. Lewis, and M. A. Marcus, *Science (Washington, D.C.)* **200**, 1279 (1968).

[190] R. H. Lozier, R. A. Bogomolni, and W. Stoeckenius, *Biophys. J.* **15**, 955 (1974).

[191] B. Chance, M. Porte, B. Hess, and D. Osterholt, *Biophys. J.* **15**, 913 (1974).

[192] D. Gill, M. E. Heyde, and L. Rimai, *J. Am. Chem. Soc.* **93**, 6288 (1971).

[193] B. Aton, A. G. Doukas, R. H. Callender, B. Becker, and T. G. Ebrey, *Biochemistry* **16**, 2004 (1977).

[194] R. H. Callender, A. Doukas, R. Crouch, and K. Nakanishi, *Biochemistry* **15**, 1621 (1976).

[195] A. Lewis, J. Spoonhower, R. A. Bogomolni, H. Lozier, and W. Stoeckenius, *Proc. Natl. Acad. Sci. U. S. A.* **71**, 4462 (1974).

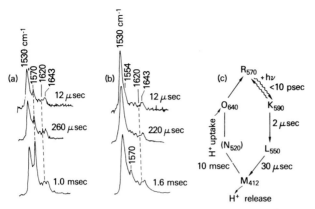

FIG. 4. Resonance-enhanced kinetically resolved Raman spectra of bacteriorhodopsin, 1700–1500 cm^{-1}, using excitation at (a) 476.5 and (b) 514.5 nm from an argon ion laser. (c) Bacteriorhodopsin photochemical cycle; the subscript on each intermediate is the wavelength (nm) of its visible absorption maximum. Spectra adapted from Campion *et al.*[179]

ciated with R_{570}, the parent compound. The 1643-cm^{-1} band visible in Fig. 4b indicates the protonated state of R_{570}. Information about the structure of the other intermediates shown in Fig. 4c has been obtained by similar analyses, primarily employing the steady-state flow technique with appropriately fast sampling by the laser beam. The combined results of these studies establishes M_{412} as the first deprotonated intermediate of the photocycle, as suggested by the data of Fig. 4.

It is hoped that the differential resonance Raman studies just described, showing the isolability of structural information by a combination of excitation profile and time analysis, presage new applications of resonance Raman techniques to other problems in biological kinetics.

3.9. Nonlinear Phenomena

Coherent anti-Stokes Raman scattering (CARS) is a four-photon process accomplished with two relatively high-powered pulsed laser beams of frequencies v_1 and v_2 passing through the sample at a small, phase-matching angle. A CARS signal beam emerges at frequency $v_3 = 2v_1 - v_2$. Whenever the difference $v_1 - v_2$ is a vibrational frequency, an anti-Stokes Raman line is seen. In order to scan the Raman spectrum, one of the incident lasers, a tunable dye laser, is tuned through the desired spectral interval.

In the CARS process, very high conversion efficiencies (10^4–10^6 times those in spontaneous Raman scattering) are obtainable, and the resultant strong anti-Stokes Raman lines occur not only at higher frequencies than most fluorescence but in a beam spatially separated from the incident beams and requiring only a single monochromator for filtering. The tremendous fluorescence rejection and signal strength means that CARS has great potential for biological studies. If a broad-band dye laser is used, a "single-shot" spectrum may be obtained, with great potential for kinetic studies.[196-199]

Both conventional and resonance Raman anti-Stokes spectra may be obtained in a CARS experimental configuration. However, the background emission arising from aqueous solvent is competitive with CARS scattering from the solute; only, therefore, under resonance conditions is a good CARS signal obtained.[200] Since resonance Raman conditions are precisely those suffering from the worst intrinsic fluorescence problems, CARS is exactly the technique needed for vibrational studies of particularly fluorescent biological chromophores. It should be noted that samples must be nonturbid; multiple scattering by sample molecules results in loss of incident beam coherence and loss of the CARS signal. A significant CARS background from glass sample holders has also been reported to be troublesome.[201]

CARS spectral profiles and depolarization ratios differ somewhat from those obtained in analogous conventional and resonance Raman studies, but essentially equivalent spectral information can be extracted from the CARS spectra.[202] A resonance CARS spectrum has been obtained for ferrocytochrome c, illustrating the similarities in spectral profile to the resonance Raman spectrum and the absence of fluorescence background[200]; depolarization ratios for ferrocytochrome c have also been measured in the CARS configuration.[202] In a resonance CARS study of highly fluorescent flavin adenine dinucleotide (FAD), assignments were suggested for some of the modes enhanced by vibronic mixing with the first $\pi-\pi^*$ transition in the plane of the flavin, allowing a frequency increment in the 1359-cm^{-1} band in the presence of glucose oxidase to be interpreted as a shift in the flavin N_3-nitrogen N–H deformation mode, reflecting hydrogen bonding of FAD to enzyme.[203]

Another Raman process, also nonlinear in response to the applied field, is inverse Raman scattering. The inverse Raman spectrum appears as a

[196] B. S. Hudson, *Annu. Rev. Biophys. Bioeng.* **6**, 135 (1977).

[197] W. M. Tolles, J. W. Nibler, J. R. McDonald, and A. B. Harvey, *Appl. Spectrosc.* **31**, 253 (1977).

[198] B. Hudson, W. Hetherington, III, S. Cramer, I. Chabay, and G. K. Klauminzer, *Proc. Natl. Acad. Sci. U. S. A.* **73**, 3798 (1976).

[199] I. Chabay, G. K. Klauminzer, and B. S. Hudson, *Appl. Phys. Lett.* **28**, 27 (1976).

[200] J. Nestor, T. G. Spiro, and G. Klauminzer, *Proc. Natl. Acad. Sci. U. S. A.* **73**, 3329 (1976).

[201] J. J. Black, T. R. Gilson, D. A. Greenhalgh, and L. C. Laycock, *Laser Focus* **14**, 84 (1978).

[202] J. R. Nestor, *J. Raman Spectrosc.* **7**, 90 (1978).

[203] P. K. Dutta, J. R. Nestor, and T. G. Spiro, *Proc. Natl. Acad. Sci. U. S. A.* **74**, 4146 (1977).

vibrational absorption spectrum on the anti-Stokes side when a pulsed laser at fixed frequency and a continuum from a dye laser simultaneously illuminate the sample. The vibrational spectrum again lies at higher frequencies than the fluorescence output.[204] Exploitation of CARS for biological studies has, however, taken precedence over applications of the inverse Raman effect. Inverse Raman effects can occur with other nonlinear processes: negative bands observed in CARS scanning of a vitamin B_{12} sample have been conjectured to be due to inverse Raman effects.[200]

3.10. The Raman Microprobe

A Raman microscope with simultaneous spatial and spectral resolution has found application in minerology[205]; a commercial model is available. Use of the microscope in biology is limited because of damage of the biological specimen by laser beam heating. If specimen damage can be controlled, in part by using a defocused illuminating beam and a well-cooled microscope stage, some interesting biological investigations can be envisaged.

3.11. Conclusions and Prognostications

Raman vibrational spectroscopists who were able in the 1970s to acquire newly developed laser sources had a glorious field day of the sort which a technological revolution customarily brings about. Further work remains to be done to quantify the spectral correlations for the behavior of biological molecules, and then to apply these correlations. Biological applications of resonance Raman spectroscopy, most especially to problems in biochemical kinetics, and also to cellular and subcellular systems including semiordered systems oriented with respect to the beam direction, are anticipated to expand in scope.

Acknowledgment

The author would like to thank Dr. Ira W. Levin for reading this article and for providing a stimulating research environment.

[204] D. A. Long, in "Advances in Raman Spectroscopy" (J. P. Mathieu, ed.), Vol. 1, p. 1. Heyden, London, 1973.
[205] M. Delhaye and P. Dhamelincourt, J. Raman Spectrosc. 3, 33 (1975).

4. PICOSECOND LASER SPECTROSCOPY

By Takayoshi Kobayashi

4.1. Introduction

The discovery of the laser[1] has led to the development of many new fields in spectroscopy. The spatial and temporal coherence of the laser makes it a unique light source for spectroscopy. New kinds of spectroscopy have been devised taking advantage of these properties. Temporal coherence is used in coherent anti-Stokes Raman scattering (CARS),[2] coherent Stokes Raman scattering (CSRS), photon echo, and other coherent transient spectroscopies, and also in high-resolution spectroscopy. Spatial coherence gives a means of producing a high photon density and is used in many fields of nonlinear spectroscopy.[2,3] In addition to the above-mentioned properties, some kinds of lasers have one more important characteristic for spectroscopic measurements—short pulse width. This property is utilized in high-time-resolution spectroscopy, which is the main subject of this chapter.

Since the environment of biological molecules is usually inhomogeneous and aperiodically organized, they have very broad electronic absorption or emission spectra. Therefore, in many cases, spectroscopy of high time resolution is of more importance to biological systems than high-spectral-resolution spectroscopy. For example, in photosynthesis, the absorption of a photon by chlorophyll leads to an excited state responsible for electron transport whose lifetime is of the order of 10 psec. Similarly, the early photoproducts of rhodopsin in visual transduction appear in a picosecond regime.

The 1960s was a time of very rapid development in short pulse generation by lasers. Only two years after the first report on laser oscillation in 1960,[1] a Q-switched laser with a pulse of several tens of nanoseconds was invented.[4] Q-switching techniques involve an artificial impairment of the optical path in a laser with the purpose of delaying the onset of laser oscillations in order

---

[1] T. H. Maiman, *Nature (London)* **187**, 493 (1960); A. Javan, W. R. Bennett, Jr., and D. R. Herriott, *Phys. Rev. Lett.* **6**, 106 (1961).

[2] M. S. Feld and V. S. Letokhov, eds., "Coherent Nonlinear Optics." Springer-Verlag, Berlin and New York, 1980.

[3] N. Bloembergen, "Nonlinear Optics," Benjamin, New York, 1965.

[4] F. J. McClung and R. W. Hellwarth, *J. Appl. Phys.* **33**, 828 (1962).

METHODS OF EXPERIMENTAL PHYSICS, VOL. 20

to increase the Q value of the laser resonator and to obtain a very short (a few tens of nanoseconds) high-power pulse called a "giant pulse."

This Q-switched laser increased the time resolution of ordinary flash photolysis[5] a 1000-fold, i.e., from a few tens of microseconds to a few tens of nanoseconds.[6] Before the appearance of mode-locked lasers, Q-switched lasers were used as powerful pulsed light sources for nanosecond spectroscopy. The advent of the mode-locked solid-state laser[7] enables one to study photophysical, photobiological, or photochemical transient species with rise times or lifetimes in the picosecond regime. As picosecond laser spectroscopy was applied to the study of semiconductor and chemical species, the techniques of picosecond laser pulse generation, detection, and time resolution were improved by many physicists and chemical physicists. These improvements led to the application of picosecond laser spectroscopy to biophysical situations. Among the biophysical studies using picosecond laser spectroscopy, photosynthetic applications are most prevalent. Other important biological systems such as rhodopsin, hemoglobin, and DNA are also being studied by many research groups.

Picosecond spectroscopy offers special advantages for the study of photobiological systems such as rhodopsin and chloroplasts. Life utilizes solar energy in two ways, in the transmission of information and in the conversion of light energy to chemical energy. To be efficient in the application of light energy in the visible region, where solar energy transmitted to the earth is most intense, biological chromophores should have large capture cross sections in this region. The light-absorbing substances are rhodopsin, in visual pigments, chlorophyll–protein complexes in photosynthetic pigments, and bacteriorhodopsin in proton pumping pigments. The chromophores in rhodopsin and bacteriorhodopsin are protonated Schiff bases of retinal and those in chlorophyll–protein complexes are chlorophylls, carotenes, and xanthenes. The chromophores have an electronic structure similar to that of many organic dye molecules. The chromophores and organic dye molecules have an intense transition between the ground state (S_0) and the lowest excited singlet state (S_1) in the visible region. The relationship between the $S_1 \rightarrow S_0$ radiative lifetime τ_r and the $S_1 \leftarrow S_0$ molar extinction coefficient ε is given by the equation[8]

$$\tau_r = (3.47 \times 10^8) \bigg/ \left(v_{max}^2 \int \varepsilon \, dv \right) \quad (\text{sec}),$$

[5] R. G. W. Norrish and G. Porter, *Nature (London)* **164**, 658 (1949); G. Porter, *Proc. R. Soc. (London), Ser. A* **200**, 284 (1950).

[6] J. R. Novak and M. W. Windsor, *J. Chem. Phys.* **47**, 3075 (1967); *Proc. R. Soc. (London), Ser. A* **308**, 95 (1968).

[7] H. W. Mocker and R. J. Collins, *Appl. Phys. Lett.* **7**, 270 (1965); A. J. DeMaria, D. A. Setser, and H. Heynau, *Appl. Phys. Lett.* **8**, 174 (1966).

[8] F. Perrin, *J. Phys. Radium* **7**, 390 (1926); G. N. Lewis and M. Kasha, *J. Am. Chem. Soc.* **67**, 994 (1945).

where v_{max} is the wave number in cm^{-1} of the $S_1 \leftarrow S_0$ absorbance maximum. Integration is over the $S_1 \leftarrow S_0$ absorption band. Since many ordinary dye molecules have a bandwidth of the order of 5×10^3 cm^{-1} and ε of the order of 10^5 M^{-1} cm^{-1} in the visible region, τ_r is calculated by the equation to be 1–10 nsec. In order for biological systems to utilize efficiently the light energy absorbed by the pigments without losing the energy by photoemission processes, the initial photobiological reaction must be completed in a much shorter time than the radiative lifetime, which is 1–10 nsec; i.e., it must be in the picosecond time scale. This is the reason why picosecond spectroscopy is invaluable for the study of the primary process of vision and photosynthesis.

In addition to the application of picosecond spectroscopy to the primary photochemical processes in photosynthetic and visual pigments, there are a number of other important applications to biological systems. Some of them are as follows.

Fluorescence in complicated biological macromolecules is emitted from the lowest excited singlet state S_1 because of the very efficient radiationless

TABLE I. List of Abbreviations

AMP	amplifier
AR	antireflection
BChl	bacteriochlorophyll
BPh	bacteriopheophytin
BS	beam splitter
CARS	coherent anti-Stokes Raman scattering
CONT CELL	picosecond continuum generation cell
CSRS	coherent Stokes Raman scattering
F	filter
II	image intensifier
IR	infrared
L	lens
M	mirror
ML	mode locked
MONO	monochromator
Nd/glass	neodimium/glass
Nd/YAG	neodimium/yttrium aluminum garnet
OMA	optical multichannel analyzer
P	polarizer
PM	photomultiplier
SC	scatterer
SHG	second-harmonic generator
SPS	single pulse selector
TEA	transversely excited atmospheric pressure
THG	third-harmonic generator
TPF	two-photon fluorescence
UV	ultraviolet
VD	variable optical delay

$S_n \rightarrow S_1$ internal conversion. Since the radiative life of S_n is generally in the nanosecond range, the time constant of the radiationless process should be in the picosecond range.

After the absorption of light by antenna chlorophyll, the light energy is transferred to a reaction center without losing the energy by fluorescence emission in the transfer process. The energy transfer should be therefore in the picosecond range. Since absolute rates of transfer are obtained by picosecond spectroscopy, we can determine the structural spacings between the energy donors and acceptors.

The motion of side chains of biological macromolecules or polymers are known to occur in 10^{-13}–10^{-14} sec from the frequency of infrared or Raman spectra. Because of the interaction between these side chains and the protein constituting pigments, the motion might slow down to the picosecond time scale. Therefore this process can also be studied by picosecond spectroscopy.

In Section 4.2.1 the comparison of techniques and equipment between nanosecond and picosecond spectroscopy are made. In Section 4.2.2 the principles of generation and the classification of picosecond light pulses are described. In Section 4.2.3 various methods of picosecond spectroscopy are described. In Chapter 4.3 examples of the application of picosecond spectroscopy to biophysical systems, photosynthesis, and vision are shown.

In the appendix we discuss several nonlinear optical phenomena useful for generating pulsed light for excitation and interrogation, or for pulse width measurements. Some of the optical elements and detectors used in picosecond spectroscopy are also described briefly in the Appendix. A list of abbreviations used in this article is given in Table I.

4.2. Nanosecond and Picosecond Spectroscopy

4.2.1. Comparison of Nanosecond and Picosecond Spectroscopy

Flash photolysis was initiated by Norrish and Porter,[5] who applied the technique to photophysical and photochemical processes of various molecules mainly in solution. The time resolution of flash photolysis was limited by the pulse width (typically a few tens of microseconds) of the excitation flash. Soon after the success of Q-switch oscillation of ruby lasers, lasers could be used for nanosecond spectroscopy. The application of Q-switch lasers improved the time resolution of photolysis experiments by a factor of a thousand. Not only Q-switch ruby lasers but also other nanosecond lasers such as N_2 lasers and Q-switch Nd/YAG lasers have been utilized for excitation light pulses. At this particular stage of development, time resolution was not limited by detectors but mainly by the excitation laser pulse width. Many sophisticated electronic devices, such as wide-band (up to 1 GHz) oscilloscopes, high-speed photomultipliers, single-photon counters, lock-in amplifiers, boxcar integrators, and transient digitizers have become commercially available in the past 20 years and have been extensively used for nanosecond to millisecond laser photolysis and spectroscopy. However, these devices are not suited for picosecond spectroscopy because they do not have time resolution on the picosecond time scale. In order to resolve times in the picosecond region, several completely new detection methods were invented.

4.2.1.1. Classification of Picosecond and Nanosecond Spectroscopy. Picosecond or nanosecond spectroscopic methods can be classified as either emission or absorption methods. In either case, there are two types of measurement: the observation of the time dependence of emission intensity or absorbance, and that of the time-resolved emission or absorption spectrum. Excitation light sources only are used for emission spectroscopy, while excitation and monitoring light sources are used for absorption spectroscopy.

There are two approaches to nanosecond or picosecond spectroscopy, a pulse technique and modulation technique. In the pulse technique, a perturbation of nanosecond or picosecond duration is given to a sample. By using perturbing pulses with a shorter pulse width than the time constants of the phenomena to be observed, one can determine the time course of the resulting process by observing the intensity of the monitoring light or the emission. In the modulation technique, the sample is periodically perturbed by excitation light with well-known phase characteristics. If an emissive species with a finite exponential decay lifetime τ is excited by a modulated light source with frequency ω, light emitted from the species is also modulated with the same frequency and a phase shift θ given by $\tan \theta = \omega\tau$. By measuring the phase shift θ, the lifetime of the species can be obtained. The modulation method can be used only for emission processes exhibiting a single exponential decay. In what follows we discuss only the pulse technique.

The classifications of nanosecond and picosecond absorption and emission spectroscopy are summarized in Table II.

4.2.1.2. Excitation Pulsed Light Source for Picosecond and Nanosecond Spectroscopy. Picosecond excitation light for the pulse method can be obtained from various kinds of mode-locked lasers[9]: induced Raman scattering of mode-locked lasers from various kinds of liquids,[10,11] the

[9] D. J. Bradley, in "Ultrashort Light Pulses" (S. L. Shapiro, ed.), p. 18. Springer-Verlag, Berlin and New York, 1977.

[10] D. von der Linde, in "Ultrashort Light Pulses" (S. L. Shapiro, ed.), p. 204. Springer-Verlag, Berlin and New York, 1977.

[11] D. H. Auston, in "Ultrashort Light Pulses" (S. L. Shapiro, ed.), p. 123. Springer-Verlag, Berlin and New York, 1977.

TABLE II.   Classification of Emission and Absorption Spectroscopy in Nanosecond
to Subpicosecond Time Region

Time domains	Excitation light sources	Monitoring light sources for absorption	Detector systems
Nanosecond (1000–1 nsec)	Q-switch ruby and Nd/YAG lasers; N_2 and H_2 lasers; excimer lasers; dye lasers pumped by the above lasers; N_2, H_2, and D_2 flash lamps	Xe flash lamp; air breakdown	Photomultiplier; wideband oscilloscope; boxcar integrator; sampling scope
Subnanosecond (1–0.1 nsec)	Mode-locked gas (He–Ne, Ar ion) lasers, mode-locked solid-state lasers (ruby, Nd/glass, Nd/YAG), TEA N_2 lasers and dye lasers pumped by the lasers	Xe flash lamp	Sampling oscilloscope; fast photodiode; very wide-band oscilloscope; streak camera
Picosecond (1000–1 psec)	Mode-locked solid-state lasers (ruby, Nd/glass, Nd/YAG); synchronously pumped dye lasers	Picosecond continuum; nanosecond laser; nanosecond flash lamp	Echelon and OMA; streak camera; variable delay and photomultiplier; Kerr gate, variable delay, and photomultiplier
Subpicosecond (1–0.1 psec)	Passively mode-locked dye lasers, synchronously pumped dye lasers	Subpicosecond continuum	Variable delay and photomultiplier; Kerr gate, variable delay, and photomultiplier

second and higher harmonics of the picosecond pulse,[11] parametric oscillation.[11] Synchrotron radiation can also provide several ten or several hundred picosecond pulses.[12]

Nanosecond flash lamps such as hydrogen, deuterium, and nitrogen lamps with a pulse width of a few nanoseconds are commercially available as excitation light sources.[13] Single-shot solid lasers, such as Q-switch ruby[14] and Nd/glass[15] lasers; repetitively pulsed lasers, such as nitrogen,[16] hydro-

[12] C. Kunz, ed. "Synchrotron Radiation Techniques and Applications." Springer-Verlag, Berlin and New York, 1979.

[13] "Handbook of TRW Fluorimetry" Manual of TRW Nanosecond Fluorimetry Equipment, Model 31A and 32A. TRW, Inc.

[14] V. Evtuhov and J. K. Neeland, Lasers 1, 1 (1966).

[15] E. Snitzer and C. G. Young, Lasers 2, 191 (1968).

[16] H. G. Heard, Bull. Am. Phys. Soc. 9, 65 (1964).

TABLE III.   Pulsed Lasers for Nanosecond Spectroscopy

Laser	λ_{osc} (nm)[a]	Typical[b] repetition rate (Hz)	Typical[b] peak power (MW)	Typical single-pulse energy[b] (J)	Pulse width (nsec)
Gas lasers					
N_2 laser	337.1	100–1	0.1–1	$(0.3–10) \times 10^{-3}$	3–10
H_2 laser	156.7–161.3	< 10	0.01	$(10–50) \times 10^{-6}$	1–5
Solid-state lasers					
Ruby laser	694.3	single shot	200–300	5	20–30
2nd harmonic	347.2	single shot	40–60	1	20–30
Nd/glass laser	∼1060	single shot	50–200	1–3	15
2nd harmonic	∼530	single shot	10–40	0.2–0.6	15
Nd/YAG laser	1064	10	10–50	0.1–1	10–20
2nd harmonic	532	10	3–15	0.03–0.3	10–20
Excimer lasers					
ArF laser	193	10–20	1–2	0.02	20
KrF laser	248	10–20	5–7	0.1	20
XeCl laser	308	10–20	2	0.05	30
XeF laser	351	10–20	2	0.04	20–30

[a] Wavelength of laser oscillation.
[b] Repetition rate, peak power, and energy are typical of commercially available lasers or homemade lasers.

gen[17] and Nd/YAG lasers[18]; and dye lasers[19] pumped by flash lamps or nanosecond lasers have been powerful light sources for nanosecond spectroscopy and laser photolysis. Dye lasers pumped by excimer lasers as well as excimer lasers[20] themselves are also useful light sources for nanosecond experiments.

The chemical composition of ruby for use in laser oscillation is Al_2O_3 with 0.05% (by weight) Cr_2O_3 where only the chromium ions participate in the laser oscillation phenomenon. Ruby laser oscillation is interpreted in terms of a three-level laser. In a three-level laser system, laser oscillation takes place between a lower level and an upper one where the latter is populated by relaxation from excited levels. Nd/YAG is yttrium aluminum garnet doped with Nd^{3+}; its laser oscillation is interpreted in terms of a four-level laser. In a four-level system, laser oscillation takes place between an upper and a

[17] P. A. Bazhulin, I. N. Knyazev, and G. G. Petrash, *Sov. Phys. JETP (Engl. Transl.)* **20**, 1068 (1965); **22**, 1260 (1960).
[18] H. G. Danielmeyer, *Lasers* **4**, 1 (1976).
[19] F. P. Schäfer, ed., "Dye Lasers." Springer-Verlag, Berlin and New York, 1973.
[20] J. J. Ewing, *Phys. Today* **31**, 32 (1978); Ch. K. Rhodes, ed., "Excimer Lasers." Springer-Verlag, Berlin and New York, 1979.

lower level, the former is populated by exciting Nd^{3+} from the ground level to levels located above the upper level. The gain of the active medium of a three-level system is more sensitive to temperature than that of a four-level system. The oscillation wavelength, peak power, pulse energy, and pulse width of these nanosecond lasers are listed in Table III.

4.2.2. Picosecond Light Pulses

4.2.2.1. Generation of Picosecond Pulse. The operating principle of the mode-locked laser can be explained in terms of either the time domain or the frequency domain (Fig. 1). The time-domain explanation is illustrated in Fig. 1b and is given as follows. Experimentally, a passively mode-locked laser is obtained by inserting a saturable dye in a cavity composed of mirrors, an active medium, and a laser gain medium. As the light beam passes through the active medium, the light intensity in the wavelength region where the gain is greater than the loss is increased while it is decreased outside this spectral region; the beam then passes through the dye, where the noise and leading edge of the pulse are absorbed more than the peak and trailing edge because of nonlinear absorption in the dye solution. In the cavity, the beam passes back and forth many times, shortening the pulse width each time it passes through the dye solution. The laser pulse increases in intensity when it passes through the active medium, with the final duration of the pulse theoretically limited by the gain spectrum width of the laser medium used (Nd/glass, Nd/YAG, ruby, Ar ion, Kr ion, or dye solution). Pulse trains from a real

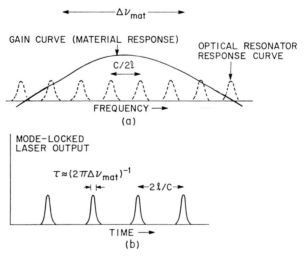

FIG. 1. Mode locking in frequency (a) and time domains (b).

resonator do not achieve the theoretical pulse width limit. This is due to spreading of the pulse width caused by laser material and cavity dispersion.

The mechanism for mode-locking a laser as visualized in the frequency domain is illustrated in Fig. 1a. Each transverse mode in a laser cavity has a set of longitudinal modes in the frequency region where the laser gain overcomes resonator loss. Each laser material (gain medium) has its own gain curve. The geometrical configuration of the laser resonator and gain medium is responsible for resonator loss. The longitudinal modes are separated in frequency by $c/2l$, where l is the optical path length between the highly reflective back mirror and the output mirror which make up the laser cavity and c is the speed of light. Without the mode locker, the output of the laser varies in the distribution of the phases and amplitude of the longitudinal frequency components. If a set of longitudinal modes is maintained in a fixed phase and fixed amplitude relationship, the output will be a well-defined function of time forming the mode-locked laser pulse train. With a lasing medium of an adequate bandwidth, a picosecond pulse train with a constant time separation of $c/2l$ can be produced.

4.2.2.2. Classification of Picosecond Light Pulses. Several typical picosecond light sources which can be used for research in biology are listed in Table IV. They can be classified as follows: (i) single-shot lasers and (ii) repetitively pulsed lasers. There are various kinds of single-shot mode-locked lasers, namely, the Nd/glass laser, the ruby laser, the flash-lamp-pumped dye laser, as well as dye lasers synchronously pumped by the fundamental or second harmonic of a ruby laser or the second or third harmonic of a Nd/glass laser. The Nd/YAG laser and passively or synchronously mode-locked dye lasers pumped by an Ar or Kr ion laser are repetitively pulsed lasers. Repetitively pulsed lasers are more appropriate for obtaining more precise data since it is easy to average data at high repetition rates. When the repetitively pulsed lasers are used for photoreactive samples, which undergo irreversible photochemical reactions, the samples must be circulated using a pump and a flow cell.

Photoreactive samples in the crystalline phases, in polymer matrices, in glass matrices of organic solvents at low temperatures, or in biological tissues must be moved by a motor attached to the sample so that the irradiated part of the sample goes out of the probing region.

Recently Shank et al.[21] succeeded in producing subpicosecond pulses of several gigawatts at a repetition rate of 10 Hz by means of an Ar-laser-pumped rhodamine 6G laser amplified by a three-stage dye amplifier pumped by the second harmonic of an amplified Q-switched Nd/YAG laser. This

[21] E. P. Ippen and C. V. Shank, in "Picosecond Phenomena" (C. V. Shank, E. P. Ippen, and S. L. Shapiro, eds.), p. 103. Springer-Verlag, Berlin and New York, 1978.

TABLE IV.   Pulsed Lasers for Picosecond Spectroscopy

Laser	λ_{osc} (nm)[a]	Typical[b] repetition rate	Typical[b] peak power (MW)	Typical[b] energy per pulse (mJ)	Pulse width (psec)
Ruby laser (amplified single pulse)	694.3	single shot	1000–5000	20–100	20
2nd harmonic	347.2	single shot	100–500	2–10	15–20
Nd/glass laser (amplified single pulse)	1060	single shot	5000–10,000	50	5–8
2nd harmonic	~530	single shot	500–1000	5	4–6
Nd/YAG laser (amplified single pulse)	1064	1–10 Hz	1000	30	30–50
2nd harmonic	532	1–10 Hz	100	3	20–40
Ar-laser-pumped passively mode-locked rhodamine 6G laser[c]	610–630	0.1 MHz	0.001	5×10^{-7}	0.5
Amplified Ar-laser-pumped passively mode-locked rhodamine 6G laser[d]	610–630	10 Hz	1000	0.5	0.5

[a] Wavelength of laser oscillation.

[b] Repetition rate, peak power, and energy are typical of commercially available lasers or homemade lasers.

[c] C. V. Shank and E. P. Ippen, *Appl. Phys. Lett.* **24**, 373 (1974).

[d] E. P. Ippen and C. V. Shank, *in* "Picosecond Phenomena" (C. V. Shank, E. P. Ippen, and S. L. Shapiro, eds.), p. 103. Springer-Verlag, Berlin and New York, 1978.

high-power repetitively pulsed laser is of great use for obtaining a subpicosecond continuum which can be used as a monitoring and reference light in absorption spectroscopy. The continuum is generated in nonlinear optical processes such as parametric four-wave mixing and/or self-phase modulation.

4.2.3. Various Methods for Picosecond Absorption and Emission Spectroscopy

Biologists may be interested both in the time-resolved absorption or emission spectrum at a fixed delay after excitation and in the time dependence of the absorbance or emission intensity at a fixed wavelength. The former is necessary for the assignment of the transient species and the latter for the determination of the rates of the corresponding processes.

In the following section, we give some details of several different experimental schemes summarized in Table V. The reason why so many different schemes have evolved is that there are various picosecond lasers, detection

TABLE V. Various Methods for Picosecond Absorption and Emission Spectroscopy

Method number, absorption or emission (A, E)	Time dependence of absorbance or emission intensity (T) or time-resolved spectrum (TRS)	Single-shot (S) or multishot (M) experiment	Time resolution method	Figure number
1, A	T	M	Variable delay	2a
2, A	T	S	Echelon and OMA	2b
3, A	TRS	S	Variable delay	4
4, E	T	M	Kerr shutter	5a
5, E	T	S	Streak camera	5b
6, E	TRS	S	Kerr shutter	6
7, A	T, TRS	S	Echelon and OMA in two-dimensional mode	2b
8, E	T, TRS	S	Streak camera and OMA in two-dimensional mode	5b

systems, and methods to resolve on a picosecond time scale. Readers interested in a particular application can skip to the relevant section.

4.2.3.1. Absorption Spectroscopy for Time Dependence Measurement Using a Repetitively Pulsed Laser. In the block diagram shown in Fig. 2a, an Ar-laser-pumped dye (rhodamine 6G) laser is amplified by a three-stage amplifier (AMP) pumped by the second harmonic of an amplified Q-switched Nd/YAG laser used as a light source. A cavity dumper attached to the dye laser is operated at 10 Hz in synchronization with the Nd/YAG laser for amplification. This is an example of a repetitively pulsed light source. The second harmonic (315 nm) of the laser output is used as an excitation light source. The second harmonic generated by SHG is for an excitation light source and is reflected by BS, VD with a stepping motor, and M1. It is focused by L3 on a sample cell. The fundamental pulse after passing BS is reflected by M2, VD, and M3 and focused by L1 into a continuum generation cell. The generated continuum is focused by L3 for use as a probe pulse and detected by an ordinary photomultiplier (PM) attached to a monochromator (MONO). The principle of this method is as follows: Monitoring light intensity is probed before and after excitation. By changing the variable delay time of 315 nm used for the excitation of a sample, the delay time dependence of the absorbance change is determined. The monitoring light signal is averaged over many laser shots; averaging is easily performed by means of a repetitive pulse, while it takes a very long time to obtain a single absorbance-change curve if a single-shot laser is used instead of a repetitively pulsed laser, and it is difficult to eliminate the effects of long-term fluctuation

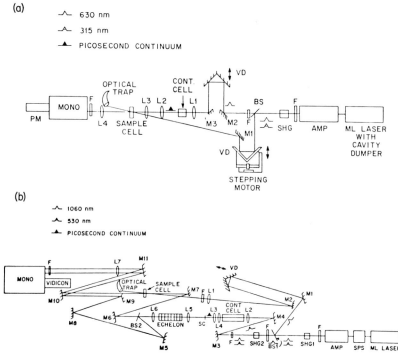

FIG. 2. Block diagrams of (a) repetitive-pulse and (b) single-shot picosecond spectroscopy equipment to measure the time dependence of absorbance of intermediates. Picosecond lasers used are (a) Ar-laser-pumped ML dye (rhodamine 6G) laser with a cavity dumper and a Nd/YAG-laser-pumped dye amplifier, and (b) ML Nd/glass laser.

of the laser intensity by averaging with this method. This is the big disadvantage of using a single-shot laser. The experimental system can be modified to be a dual-beam spectroscope in order to minimize the effects of intensity fluctuations (see p. 176).

4.2.3.2. Echelon Method for the Measurement of Time Dependence of Absorbance Using a Single-Shot Laser. This method, illustrated in Fig. 2b, uses an echelon to convert time resolution to spatial resolution.[22] In the block diagram, a mode-locked Nd/glass laser is used as a single-shot light source. The single pulse is amplified by AMP after being extracted from a pulse train by SPS. The second harmonic generated by SHG1 is for the excitation of a sample and is reflected by BS1, M1, VD, M2, and focused on the sample cell by L1. The third or fourth harmonic can also be used as the excitation light source. The second harmonic generated by SHG2 is reflected

[22] M. R. Topp, P. M. Rentzepis, and R. P. Jones, *Chem. Phys. Lett.* **9**, 1 (1971).

by M3 and M4 and focused by L2 into a continuum generation cell (CONT CELL). The generated continuum is focused by L3 on a scatterer (SC) and collimated by L5. After passing through an echelon the continuum is split into monitoring and reference beams by BS2. The monitoring beam passes through BS2 and then is reflected by M6 and M7 and focused on the sample cell by L6. Both excitation and monitoring pulses travel nearly equal optical distances. They are subsequently focused at a small angle to each other in the sample cell by lenses L1 and L6. The reference beam is reflected by BS2, M5, M8, M9, M10, and M11, and then focused by L7 together with the monitoring beam on the entrance slit of a monochromator (MONO). Both the monitoring and reference beams are detected by a vidicon or TV camera attached to the monochromator.

The echelon method was invented by Topp *et al.*[22] to obtain time resolution of emission intensity in picosecond region. It was also utilized to get time resolution of absorbance change on a picosecond time scale using single-shot lasers. The echelon used in picosecond experiments is simply a stack of glass plates or blocks. There are two kinds of echelons, transmissive and reflective, as shown in Fig. 3. These echelons are used to convert a monitoring pulse to a pulse train with appropriate interpulse separation in time and space. The light that travels through the different steps of the transmissive echelon experiences varying retardations. If the light pulses passing through the steps of an echelon can be resolved in space, they can give time-resolved information about the sample. The situation is similar when a reflective echelon is used.

When the step thickness of an echelon is d, the normal angle of incidence of the monitoring light passing through the echelon is θ, and the refractive index of the glass is n, then the interpulse separation is given by $\Delta t = 2d \cos \theta/c$ for the reflective echelon and $\Delta t = d[(n^2 - \sin^2 \theta)^{1/2} - \cos \theta]/c$ for the transmissive echelon. For example, in the case of an echelon with steps of 3 mm

(a) TRANSMISSIVE ECHELON   (b) REFLECTIVE ECHELON

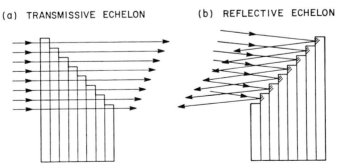

FIG. 3. (a) Transmissive and (b) reflective echelons used to get a pulse train of a probe or a reference beam to resolve in a picosecond region. A single pulse is converted into a pulse train passing through or being reflected by echelons.

thickness, the interpulse separation of the incident light pulse train is 20 psec for the reflective echelon and about 5 psec for the transmissive echelon at around 500 nm. Therefore, transmissive echelons are appropriate for experiments requiring high time resolution, while reflective echelons are suitable for experiments with relatively slow kinetics, up to about 1 nsec. With the transmissive echelon, care must be taken that the time separation is different from one wavelength to another since Δt is a function of the refractive index n of the echelon material. To get a high-quality transmitted and reflected image, each step of the stack must be in optical contact and each plate surface must be parallel to all the others.

If, for example, the absorbance changes with time on a picosecond time scale after an excitation pulse, the intensity distribution over the echelon steps after passing through the sample cell is different from that before passing through the cell. The intensity of each pulse in the monitoring pulse train is decreased or increased by transient absorption or bleaching of the sample.

A pulse train separated both spatially and temporally, formed by reflecting or passing through echelons from a monitoring pulse, can be observed by a detector which has spatial resolution, e.g., a silicon vidicon, TV camera, or photographic plate.

Sometimes the intensity distribution over echelon steps changes from one shot to the other because of fluctuation in the spatial distribution of the continuum intensity. This fluctuation is caused by the amplitude instability of laser pulses which induces fluctuation of the angular distribution of the picosecond continuum intensity in the optical medium where the incident laser is focused. The fluctuation of the spatial distribution of the monitoring light pulse may give artificial data if a single-beam (monitoring beam) method is applied. To remove the artifact caused by the changes in the intensity distribution across the echelon steps, a dual-beam (monitoring and reference beams) method was devised.[23a] Figure 2b shows the equipment for dual-beam picosecond absorption spectroscopy. If the light intensity corresponding to the ith step of the echelon is represented by $I^0(i)$ for the reference beam and $I(i)$ for the monitoring beam, and if subscripts ex and ne correspond to the case when the excitation light is exposed to the sample and the case when the light is blocked, respectively, then $\log[I_{ne}(i)I^0_{ex}(i)/I_{ex}(i)I^0_{ne}(i)]$ gives the absorbance change at delay time $\Delta t(i)$ corresponding to the ith echelon.

4.2.3.3. Spectroscopy for Time-Resolved Absorption Spectra.

A block diagram for the measurement of picosecond time-resolved absorption spectrum is shown in Fig. 4. A Nd/glass laser is used as the light source. The second harmonic generated by SHG1 is reflected by BS, VD, and M1, and then focused by L1 to excite a sample cell. The second harmonic generated

[23a] T. L. Netzel and P. M. Rentzepis, *Chem. Phys. Lett.* **29**, 337 (1974); T. Kobayashi and S. Nagakura, *Chem. Phys. Lett.* **43**, 429 (1976).

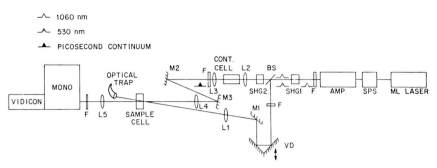

FIG. 4. Block diagram of an experimental system for measuring the time-resolved absorption spectrum at a fixed delay time with a single-shot laser.

by SHG2 is focused into a continuum generation cell by L2. The continuum for the use of monitoring light is reflected by M2 and M3 and then focused by L4 on the sample cell and refocused by L5 on the entrance slit of a monochromator (MONO). Either a photographic plate or a vidicon with an optical multichannel analyzer (OMA) can be used as a detector. To minimize fluctuations, the system shown in Fig. 4 can be modified to a dual-beam spectroscope similar to that of Fig. 2b. Two-dimensional-mode OMA is useful for dual-beam spectroscopy for measuring time-resolved absorption spectra.[23b] In the two-dimensional mode, the plane of the photocathode can be divided into two regions A and B. The A region covers the top half of each channel of the photocathode of the vidicon, and the B region covers the bottom half. One light beam passes through a pinhole which is exposed to excitation light and focused on the upper part of the entrace slit of the monochromator and then refocused onto the A region of the OMA. The other light beam, passing through a pinhole which is not exposed to excitation light, is focused on the lower part of the entrance slit of the monochromator and refocused on the B region of the OMA. If the counts of a channel corresponding to wavelength λ are $I_A(\lambda)$ and $I_B(\lambda)$ for the A and B regions, absorbance is given by $\log[I_A^{ne}(\lambda)I_B^{ex}(\lambda)/I_B^{ne}(\lambda)I_A^{ex}(\lambda)]$ as a function of λ. Changing the arrival time of the excitation light pulse at the sample by means of a variable delay (VD) will yield time-resolved absorption spectra at different delay times.

4.2.3.4. Multishot Emission Spectroscopy for the Measurement of the Time Dependence of Emission Intensity. A block diagram showing the experimental system for multishot emission kinetics is shown in Fig. 5a. A mode-locked Nd/glass laser is used as a light source. This method utilizes a Kerr shutter, first applied to the observation of picosecond phenomena by Duguay and Mattick.[24] The Kerr shutter used in picosecond emission

[23b] K. Suzuki, T. Kobayashi, H. Ohtani, H. Yesaka, S. Nagakura, Y. Shichida, and T. Yoshizawa, *Photochem. Photobiol.* **32**, 809 (1980).
[24] M. A. Duguay and A. T. Mattick, *Appl. Opt.* **10**, 2162 (1971).

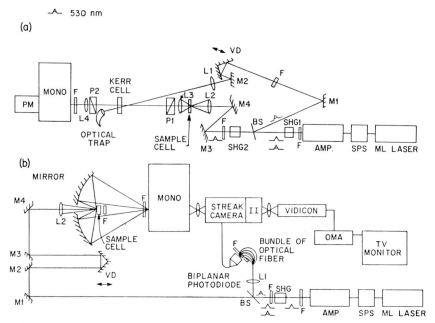

FIG. 5. Block diagram of a system of picosecond emission spectroscopy for the measurement of the time dependence of emission intensity in (a) multishot and (b) single-shot experiments. A mode-locked/Nd glass laser is used in both (a) and (b).

spectroscopy is opened at several delay times after excitation of the sample. The second harmonic generated by SHG1 is reflected by BS, M1, VD, and M2 and then focused by L1 on a Kerr cell to open a shutter composed of two parallel polarizers P1 and P2 and the Kerr cell. The second harmonic generated by SHG2 is reflected by M3 and M4 and then focused by L2 on a sample cell. Emission from the sample is collected by L3 and focused by L4 on the entrance slit of a monochromator (MONO) after passing through the Kerr shutter. The fluorescence is detected by an ordinary photodetector, such as a photomultiplier (PM), which does not have picosecond time resolution. The time resolution of the system is determined by the width of the pulse exciting the sample, the width of the shutter pumping pulse width, and the relaxation time of the Kerr medium.

Using subpicosecond pulses from a mode-locked CW dye laser, Ippen and Shank[25] have succeeded in operating a CS_2 gate at the Kerr resolution limit. In the process they made a direct measurement of the Kerr lifetime. The measured Kerr response time of CS_2 was determined to be 2.0 ± 0.3 psec.

[25] E. P. Ippen and C. V. Shank, *Appl. Phys. Lett.* **26**, 92 (1975).

The signal $S(\Delta t)$ at delay time Δt is proportional to the convolution of the time dependence of the fluorescence intensity $F(t)$ and gate function $g(t - \Delta t)$ of the Kerr shutter. $S(\Delta t)$ is given by the equation

$$S(\Delta t) = C \int_{-\infty}^{\infty} F(t)g(t - \Delta t)\, dt,$$

where C is the constant of proportionality.

If the width of the gate function is narrow enough compared with the time constant of the fluorescence decay, then $S(\Delta t) = F(\Delta t)$. When the time constants are not sufficiently long for the effect of convolution to be neglected, $F(t)$ must be obtained by deconvolution or convolution/simulation. Deconvolution can be done by applying a Laplace transformation to $S(\Delta t)$ and $g(t - \Delta t)$. $F(t)$ is the convolution of the excitation pulse shape and a function of the intrinsic fluorescence decay.

Another gate method which can be used instead of the Kerr shutter is the optical pulse gain amplification method.[26, 27]

4.2.3.5. Streak Camera Method to Measure the Time Dependence of Emission Intensity Using Single-Shot Laser.

A block diagram of the experimental system for emission kinetics measurement in a picosecond range by a single-shot laser and streak camera is given in Fig. 5b. A mode-locked Nd/glass laser is used as a picosecond light source. A tiny portion of the second harmonic is reflected by BS and focused by L1 on one end of a bundle of optical fibers which lead light to a biplanar photodiode used for triggering a streak camera.[28] The main portion of the second harmonic generated by SHG is reflected by M1, M2, VD, M3, and M4, and then focused on a sample by L2. The sample and the entrance slit of a monochromator are located at the focal points of an ellipsoidal mirror. The emission from the sample is focused by an ellipsoidal mirror on the entrance slit of a mono-chromator. The time-dependent spectrum is observed by the combination of the streak camera, an image intensifier, and a vidicon; two-dimensional emission information from the sample is monitored by the vidicon. This method is much easier than the multishot method explained in Section 4.2.3.4. However, there are two disadvantages to the method. One is that a streak camera is much more expensive than the Kerr shutter necessary for multishot experiments for emission kinetics. The other is that the camera has a jitter that makes it difficult to average the data. Therefore, in order to accumulate data without the effect of the jitter to improve the signal-to-noise ratio, it is necessary to have a standard time.[28] To get a standard time, or a time

[26] G. E. Busch, K. S. Greve, G. L. Olson, R. P. Jones, and P. M. Rentzepis, *Appl. Phys. Lett.* **27**, 450 (1975).

[27] T. Kobayashi, E. O. Degenkolb, and P. M. Rentzepis, *J. Appl. Phys.* **50**, 3118 (1979).

[28] T. Kobayashi, *J. Chem. Phys.* **69**, 3570 (1978).

reference, a tiny fraction of the excitation pulse is split by a glass plate and guided by an optical fiber to the inside position of the exit slit of the monochromator.[28] The length of the optical fiber is adjusted so that scattered light at the position of the sample arrives at the exit slit with some delay relative to the light after passing through the optical fiber (for example, 100 psec). In this way the light pulse through the optical fiber can be used as a prepulse which is a standard for the accumulation of fluorescence intensity obtained by each laser shot.

4.2.3.6. Single-Shot Experiment to Measurement Time-Resolved Emission Spectra. A block diagram of the equipment necessary to obtain time-resolved fluorescence spectra with a single-shot laser is shown in Fig. 6. In the diagram, a Nd/glass laser is used as a picosecond light source. The second harmonic generated by SHG1 is reflected by BS, M1, VD, and M2, and is then focused by L1 on a Kerr cell to open a shutter composed of two cross-polarized polarizers P1 and P2 and the Kerr cell. The second harmonic generated by SHG2 is reflected by M3 and M4 and then focused by L2 on a sample cell. Emission from the sample is collected by L3, passed through the shutter, and focused by L4 on the entrance slit of a monochromator (MONO). The time-resolved emission spectrum of a sample at different delay times can be obtained by moving the variable delay. The emission spectrum can be detected either by a photographic plate or by a vidicon attached to an OMA coupled to the monochromator.

4.2.3.7. Two-Dimensional Detection of Absorption Spectra as a Function of Delay Time After Excitation. We have discussed methods in which either the time dependence of absorbance or emission intensity is measured or the picosecond time-resolved absorption or emission spectra are measured. By modifying the methods, we can obtain both types of data at the same time. To acquire two-dimensional picosecond absorption spectroscopic data, the equipment must be slightly modified from that for the time

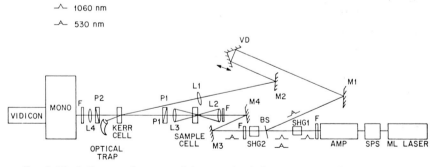

FIG. 6. Block diagram of a system of picosecond emission spectroscopy for the measurement of time-resolved emission spectrum at a fixed delay time by the use of a single-shot laser.

dependence measurements of absorbance described in Section 4.2.3.2. Instead of using a vidicon in a one-dimensional mode, it can be used in a two-dimensional mode. The new OMA from Princeton Applied Research EG & G, OMA II, can be easily operated in a two-dimensional mode. In this mode, the intensity distribution across the direction parallel to the entrance slit gives the data for time dependence, and the intensity distribution perpendicular to the slit gives the spectrum. This is also done using a photographic plate, but data processing is more difficult and the dynamic range of the photographic plate method is much narrower than that of the method utilizing OMA.

4.2.3.8. Two-Dimensional Detection of Emission Spectra as Functions of Delay Time after Excitation. It is possible to obtain data on the time dependence of emission intensity, as well as the spectrum, at several delay times after excitation. This can be done by modifying the experimental equipment for time dependence measurement of emission intensity mentioned in Section 4.2.3.5. By rotating the monochromator by 90° and setting a slit at the position of the ordinary exit slit of the monochromator perpendicular to the entrance slit, whole emission spectra can be measured at a delay time using one laser shot. The perpendicular slit is not for monochromatizing but for getting the total spectrum and for limiting the height of the spectrum. The image of the entrance slit on the screen of the streak camera then moves with time in the direction perpendicular to the exit slit. The two-dimensional image on the screen gives both spectra and time dependence. The exit slit width is one of the factors that determine time resolution of the monochromator–streak camera system. A two-dimensional streak camera is now available from Hamamatsu.

4.3. Applications to Photosynthesis and Vision

4.3.1. Photosynthesis

4.3.1.1. Introduction. An understanding of the process of the application of light energy to synthesize carbohydrates by plants through the photosynthetic process may someday lead to the development of biomimetic systems of solar energy conversion to chemical energy or to increased agricultural output through the use of additives or chemicals for higher photosynthetic efficiency. Photosynthesis takes place in chloroplasts, where the energy of light absorbed by antenna pigments is transferred to a reaction center. In the reaction center, reduction of CO_2 to form carbohydrates (CH_2O) takes place following the equation

$$H_2O + CO_2 \longrightarrow (CH_2O) + O_2.$$

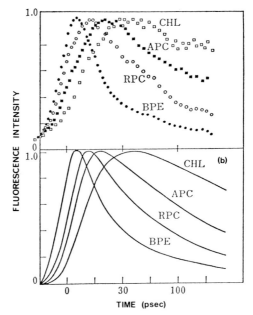

FIG. 7. (a) Observed and (b) calculated time dependence of the fluorescence intensity of B-phycoerythrin (BPE) at 576 nm, R-phycoerythrin (RPE) at 640 nm, allophycocyanin (APC) at 660 nm, and chlorophyll a (CHL) at 685 nm. From Porter *et al.*[29]

The preceding part of the primary process of photosynthesis, i.e., the transfer of energy from antenna pigments to a reaction center, was studied with the use of the emission spectroscopic technique for red algae by Porter *et al.*[29] The later part of the process, primary electron transfer, was studied by Windsor and Rockley[30] and Rentzepis *et al.*[31] for bacteriochlorophyll. Details of the study are given in the following subsections.

4.3.1.2. Picosecond Study of the Energy Transfer Process between Antenna Pigments by Emission Spectroscopy.

Porter *et al.*[29] utilized the second-harmonic pulse (530 nm, 6 psec) of a mode-locked Nd/glass laser to excite the accessory pigments in the intact alga of porphyridium cruentum. The rise and decay kinetics of emission from B-phycoerythrin (BPE) at 576 nm (which is excited by the picosecond pulse), R-phycoerythrin (PRE)

[29] G. Porter, C. J. Tredwell, G. F. W. Searle, and J. Barber, *Biochim. Biophys. Acta* **501**, 232 (1978).

[30] M. W. Windsor and M. G. Rockley, *in* "Lasers in Physical Chemistry and Biophysics" (J. Joussot-Dubien, ed.), p. 369. Elsevier, Amsterdam, 1975.

[31] P. M. Rentzepis, K. J. Kaufmann, P. Avouris, T. Kobayashi, and E. O. Degenkolb, *in* "Advances in Laser Chemistry" (A. H. Zewail, ed.), p. 126. Springer-Verlag, Berlin and New York, 1978.

at 640 nm, allophycocyanin (APC) at 660 nm, and chlorophyll a (Chl-a, CHL in Fig. 7) at 685 nm were monitored by a streak camera. All of the fluorescences except BPE fluorescence appeared to follow an exponential decay law. The mean fluorescence lifetimes of BPE, RPC, APC, and Chl-a were 70, 90, 118, and 175 psec, respectively. The fluorescence from the latter three pigments all showed finite rise times, to the maximum emission intensity, of 12 psec for RPE, 24 psec for APC, and 50 psec for Chl-a. The experimental results of the time dependences of the fluorescence intensities are shown in Fig. 7a.

Assuming that the energy transfer between chromophores takes place by long-range interaction in the order BPE to RPC to APC to Chl-a, Porter et al.[29] calculated the time dependences of the fluorescence intensities. The results, shown in Fig. 7b, are in good agreement with the experimental observations. This study is the first direct observation of energy transfer between chromophores in antenna pigments.

4.3.1.3. Picosecond Absorption Spectroscopic Study of Reaction-Center Oxidation in Bacteriochlorophyll Photosynthesis.

Photosynthetic bacteria are often used in studies of photosynthesis because they have relatively simple structure and the reaction-center chromophores are easily removed with detergents. A mutant which has no carotenoid, R26, is very often used. With the use of a picosecond absorption spectroscopic system with echelon, Netzel et al.[32] observed that the 864-nm band due to bacterio-chlorophyll bleaches within 10 psec of excitation with a 530-nm pulse. After this very preliminary experiment, Kaufmann et al.[33] and Rockley et al.[34] did independent picosecond experiments over a wide spectral range. The experimental results obtained by these two groups are quite similar to each other. The spectral change observed immediately (within 10 psec) after excitation with a 530 nm picosecond pulse is due to an intermediate state P^F. The intermediate state P^F disappears after several hundred picoseconds, and the spectrum of oxidized species P^+ becomes evident. The induced spectral change measured 240 psec after excitation has all the characteristics expected of the difference spectrum between P^+X^- and PX. The time dependences of absorbances at 680 and 610 nm are shown in Fig. 8.[34]. A positive absorbance change at 680 nm is due to the immediate appearance of the intermediate P^F after picosecond excitation; it decays in 246 ± 16 psec, while a negative absorbance change at 610 nm is due to the appearance of P^+ and it appears in 220 ± 14 psec.[34] Kinetics indicate, therefore, that P^+

[32] T. L. Netzel, P. M. Rentzepis, and J. Leigh, Science (Washington, D.C.) **182**, 238 (1973).

[33] K. J. Kaufmann, P. L. Dutton, T. L. Netzel, J. S. Leigh, and P. M. Rentzepis, Science (Washington, D.C.) **188**,1301 (1975).

[34] M. G. Rockley, M. W. Windsor, R. J. Cogdell, and W. W. Parson, Proc. Natl. Acad. Sci. U. S. A. **72**, 2251 (1975).

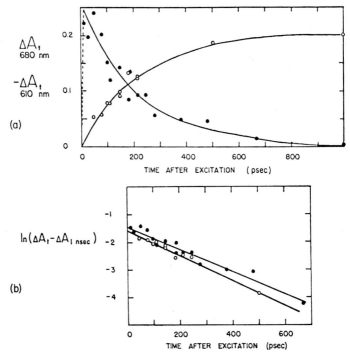

FIG. 8. (a) Rate of formation of P^+ (open circles) measured from the absorbance decrease at 610 nm, and the rate of disappearance of P^F (filled circles) measured from the absorbance increase at 680 nm after picosecond pulse excitation. (b) Semilogarithmic plot of Fig. 8a. P^F disappears in 246 ± 16 psec and P^+ forms in about 220 ± 14 psec, indicating that P^+ is formed directly from P^F. From Rockley et al.[34]

forms directly from P^F.[34] The time-resolved absorption spectrum 20 psec after excitation is very similar to that observed for chemically reduced PX^- after 20-nsec pulse excitation. From the result, it can be concluded that the intermediate P^F has a lifetime of about 200 psec and that the lifetime can be extended to 10 nsec if the electron transfer from P^F to X is blocked by the chemical reduction of X.

In order to clarify the electronic structure of P^+X, Dutton et al.[35] investigated the 1250-nm absorption band of the reaction center of R26 mutant. The oxidized dimer of BChl ($BChl^{\pm}BChl$) formed by the treatment with ferricyanide has absorption maxima at 1250 nm, whereas BChl monomer has no absorbance in the region. The absorbance at 1250 nm increased immediately after the excitation of neutral reaction center with 10 psec laser

[35] P. L. Dutton, K. J. Kaufmann, B. Chance, and P. M. Rentzepis, FEBS Lett. 60, 275 (1975).

pulse at 530 nm. As mentioned above, P^+X^- appears several hundred pico-seconds after excitation. From these experimental results, it was concluded[35] that the oxidized cation radical of the BChl dimer ([BChl$\pm$BChl]) is a part of the intermediate state P^F which is the precursor of P^+X^-. It has been suggested that the P^F state could be symbolized as [BChl$\pm$BChl I] where I represents another reduced component. The expected difference spectrum between [BChl$\pm$BChl BPh] and [BChl$\pm$BChl BPh$^-$] resembles the observed time-resolved spectrum of R26 at 20 psec after excitation. Therefore I is concluded to be BPh.

The concluding photochemical steps of bacteriochlorophyll obtained by the above experimental results are[31]

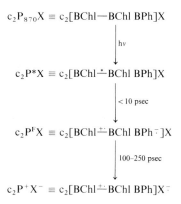

$$c_2P_{870}X \equiv c_2[BChl{-}BChl\ BPh]X$$

$$\downarrow h\nu$$

$$c_2P^*X \equiv c_2[BChl\stackrel{+}{-}BChl\ BPh]X$$

$$\downarrow <10\ psec$$

$$c_2P^FX \equiv c_2[BChl\stackrel{+}{-}BChl\ BPh^-]X$$

$$\downarrow 100{-}250\ psec$$

$$c_2P^+X^- \equiv c_2[BChl\stackrel{+}{-}BChl\ BPh]X^-$$

As can be seen from the above discussion, picosecond spectroscopy was very usefully applied to the clarification of the mechanism of the primary process of photosynthesis.

4.3.2. Vision

4.3.2.1. Introduction. Some of the photoreceptor molecules used in the very fast first steps of visual and photosynthetic processes are similar to each other. A class of molecules, the carotenoids, are present in most normal plants, and, as Wald[36,37] first showed, retinal, whose structure is approximately a half of a carotene is present in the eye. However, the physiological purposes of the photosynthetic and visual processes for light energy utilization are quite different from each other. In photosynthesis, light energy absorbed by antenna chlorophyl is transferred to reaction center and utilized

[36] G. Wald, *Nature (London)* **132**, 316 (1933).
[37] G. Wald, *J. Gen. Physiol.* **18**, 905 (1934–1935).

for driving the entire photosynthetic process and then finally converted to chemical energy. In visual process, on the other hand, an absorbed photon finally results in triggering nerve impulses whose energy comes from electro-chemical gradients across the cell membrane which are induced by light energy absorbed.

4.3.2.2. Picosecond Absorption Spectroscopic Study of the Primary Process in Vision. The absorption of the photons by the photoreceptor, rhodopsin, initiates the primary process in visual excitation. Several inter-mediates are formed successively after the photophysical event. Lifetimes of these intermediates are very short. Their spectra were at first observed by freezing samples in glass-forming solvents at low temperatures where the lifetimes of the intermediates are long enough to measure their absorption spectra with ordinary absorption spectroscopy equipments.

Two of the intermediates observed at low temperatures are batho-rhodopsin, observed by irradiating cattle rhodopsin at 77 K,[38] and hypso-rhodopsin, seen at 4K.[39] After the observation of hypsorhodopsin, the question arose whether hypsorhodopsin or bathorhodopsin is the first intermediate after the absorption of the photon. Picosecond spectroscopy equipment was used to study the very fast room-temperature kinetics of appearance and disappearance of intermediate species absorbing at 561 nm.[40] The second harmonic (530 nm, 6 psec) of a mode-locked Nd/glass laser was used as a picosecond light source to excite bovine rhodopsin solubilized in Ammonyx LO detergent. From the observed instantaneous rise of absor-bance at 561 nm, it was concluded that bathorhodopsin is formed within 6 psec after picosecond pulse excitation.[40] Furthermore, hypsorhodopsin could not be observed after 530 nm excitation of bovine rhodopsin in LDAO (i.e., Ammonyx LO).[41,42] Proton translocation was claimed to take place in the formation process of bathorhodopsin from the experiment of temperature dependence and deuterium substitution effect of the formation rate.[43] However, it was observed that hypsorhodopsin is formed from squid rhodopsin solubilized with digitonin with use of the second harmonic of the

[38] T. Yoshizawa and Y. Kito, *Nature* (*London*) **182**, 1604 (1958).

[39] T. Yoshizawa, *in* "Handbook of Sensory Physiology" (H. J. A. Dartnall, ed.), Vol. VII, Part 1, p. 146. Springer-Verlag, Berlin and New York, 1972.

[40] G. E. Busch, M. L. Applebury, A. A. Lamola, and P. M. Rentzepis, *Proc. Natl. Acad. Sci. U. S. A.* **69**, 2802 (1972).

[41] V. Sundstrom, P. M. Rentzepis, K. Peters, and M. L. Applebury, *Nature* (*London*) **267**, 645 (1977).

[42] B. H. Green, T. G. Monger, R. R. Alfano, B. Aton, and R. H. Callender, *Nature* (*London*) **269**, 179 (1977).

[43] K. Peters, M. L. Applebury, and P. M. Rentzepis, *Proc. Natl. Acad. Sci. U. S. A.* **74**, 3119 (1977).

mode-locked ruby laser as an excitation light source.[44,45] The time constant is < 20 psec at room temperature.[44,45] The hypso intermediate was found to convert into bathorhodopsin with a time constant of 50 ± 10 psec at room temperature.[44,45] The discrepancy between the formation time constant of bathorhodopsin for cattle (< 6 psec) and that for squid (50 psec) could possibly have been due to the difference in excitation wavelength (530 and 347 nm), in the detergent (LDAO and digitonin), or in the protein (cattle and squid).

After the picosecond spectroscopy of squid rhodopsin experiment the cattle rhodopsin experiment was repeated with the use of the second harmonic of a Nd/glass laser as an excitation light pulse.[46,47] It was found that the primary process of cattle rhodopsin is almost the same as that of squid rhodopsin.

The time dependence of absorbance at seven different wavelengths (430, 440, 480, 550, 570, 600, and 630 nm) was observed for bovine rhodopsin in LDAO- and octylglucoside-solubilized solutions after excitation with a picosecond pulse at 530 nm. The data obtained for 430, 440, 570, and 630 nm are shown in Fig. 9. The kinetics is observed to be triphasic at these wavelengths, i.e., a first phase appears at 0–20 psec and a final one at 80–400 psec.

The species corresponding to the kinetic stages are assigned to the excited singlet state of rhodopsin, hypsorhodopsin, and bathorhodopsin, respectively. The lifetime of the excited singlet state and the rise time of hypsorhodopsin were both found to be 15 ± 5 psec, and both the rise time of bathorhodopsin and decay time of hypsorhodopsin were 50 ± 20 psec, agreeing with the observed rise time of squid bathorhodopsin and decay time of squid hypsorhodopsin.[44,45]

From the analysis of the depth of the dip observed in the time dependence curve of absorbance change at 570 nm, it is concluded that $93^{+7}_{-18}\%$ $(100^{+0}_{-10}\%)$ of bathorhodopsin is formed through hypsorhodopsin in octylglucoside (LDAO)-solubilized samples and the rest is formed directly from the excited singlet state of rhodopsin; i.e., most cattle bathorhodopsins are formed by the following process at room temperature:

$$\text{rhodopsin }(S_0) \xrightarrow{\quad h\nu \quad} S_1 \xrightarrow{\quad 15 \pm 5 \text{ psec} \quad}$$

$$\text{hypsorhodopsin} \xrightarrow{\quad 50 \pm 20 \text{ psec} \quad} \text{bathorhodopsin.}$$

[44] Y. Shichida, T. Yoshizawa, T. Kobayashi, H. Ohtani, and S. Nagakura, *FEBS Lett.* **80**, 214 (1977).

[45] Y. Shichida, T. Kobayashi, H. Ohtani, T. Yoshizawa, and S. Nagakura, *Photochem. Photobiol.* **27**, 335 (1978).

[46] T. Kobayashi, *FEBS Lett.* **106**, 313 (1979).

[47] T. Kobayashi, *Photochem. Photobiol.* **32**, 207 (1980).

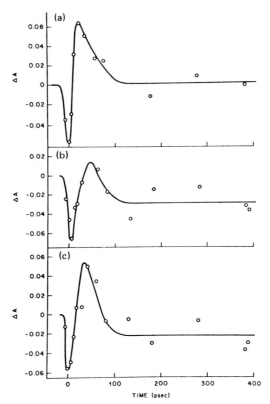

FIG. 9a. Time dependence of the absorbance change at 430 nm induced by 530-nm pico-second pulse excitation for (a) octylglucoside and (b) LDAO solubilized rhodopsin. Absorbance change at 440 nm for octylglucoside sample is shown by (c). Reprinted with permission from Kobayashi.[47] Copyright 1980, Pergamon Press, Ltd.

There may be a small amount of bathorhodopsin formed by the following process:

$$\text{rhodopsin } (S_0) \xrightarrow{\ hv\ } S_1 \xrightarrow{\ 15 \pm 5 \text{ psec}\ } \text{bathorhodopsin.}$$

There is no significant difference in the kinetic data and intermediates between 347-nm excitation of squid rhodopsin solubilized in digitonin and 530-nm excitation of bovine rhodopsin solubilized in LDAO or octylglucoside. From this fact it is concluded that, irrespective of the excitation wavelength, of the nature of the detergent, and of the protein, hypsorhodopsin is the main first intermediate of the visual process in bovine and squid rhodopsin. The formation and decay times of cattle hypsorhodopsin are 15 ± 5 and 50 ± 20 psec, respectively, while those of squid rhodopsin are < 20 and 50 ± 10 psec, respectively. Squid and bovine bathorhodopsin are both formed primarily via their respective hypsorhodopsins.

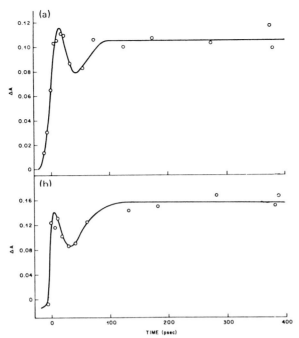

FIG. 9b. Time dependence of the absorbance change at 570 nm for (a) octylglucoside and (b) LDAO sample. Reprinted with permission from Kobayashi.[47] Copyright 1980, Pergamon Press, Ltd.

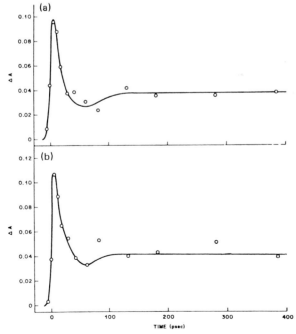

FIG. 9c. Time dependence of the absorbance change at 630 nm for (a) octylglucoside and (b) LDAO sample. Reprinted with permission from Kobayashi.[47] Copyright 1980, Pergamon Press, Ltd.

APPENDIX. Nonlinear Optical Phenomena, Optical Elements, and Detectors Related to Techniques of Picosecond Spectroscopy

A.1. Nonlinear Optical Phenomena[2, 3]

A.1.1. Multiphoton Absorption. At a high laser intensity the absorption cross section is not constant, but depends on the light intensity. Two or more photons can be simultaneously absorbed by a medium at sufficiently high light intensities. This phenomenon is called a two-photon or a multiphoton absorption process. In order for two-photon absorption to take place efficiently in organic dye molecules, the energy separation ΔE between the ground state and the lowest excited singlet state should be greater than the photon energy of the laser used, and it should be smaller than the twice the photon energy. Simultaneous two-photon absorption is different from a two-step absorption. In the latter process, an intermediate electronic excited state is actually excited by one photon, and then the transition from the state to the final state takes place. On the other hand, simultaneous two-photon absorption takes place via a virtual intermediate state which is not actually excited and works as a perturbing state to generate the two-photon transition probability. Two-photon absorption has been used for pulse width measurements of picosecond lasers.[48] Using the simple optical system shown in Fig. 10, the width of a picosecond pulse can be measured. The laser beam is split into two beams of equal intensity which are then reflected by mirrors M1 and M2. The mirrors are aligned to make the beams arrive at the same time at the center of a cell containing a highly fluorescent dye solution. Dye molecules used for pulse width measurement by two-photon fluorescence techniques should be highly fluorescent, and should have no single-photon absorption cross section and a high two-photon absorption cross section at the wavelength of the laser. For example, for the pulse width measurement of the fundamental of Nd/glass lasers or Nd/YAG lasers, rhodamine B or rhodamine 6G is used. Rhodamine 6G, rhodamine B, or fluorescein is used for the pulse width measurement of ruby lasers. The laser pulse optically pumps the dyes into an excited state with energy twice the photon energy. The dye molecules then relax to the lowest excited state and emit fluorescence. Weak fluorescence can be observed as background in the tracks of two oppositely directed beams outside the region where the two optical wave packets overlap. When the laser beam has noisy peaks with picosecond pulse widths and the total pulse width is much longer than the picosecond pulse,

[48] J. A. Giordmaine, P. M. Rentzepis, S. L. Shapiro, and K. W. Wecht, *Appl. Phys. Lett.* **11**, 216 (1967).

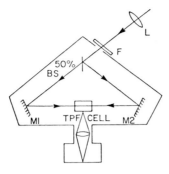

CAMERA or VIDICON

FIG. 10. Block diagram of apparatus for measurement of pulse width by two-photon fluorescence. The incident laser beam is split into two beams with equal intensity. Two mirrors are adjusted such that these two beams are coincident at the center of TPF cell. The two-photon fluorescence intensity is monitored by a camera or vidicon.

two-photon fluorescence patterns obtained by this method can have a sharp peak with a picosecond pulse equal to the pulse width of the noise spike.[49] In this case, the peak-to-noise ratio is 2:1, while if the pulse has no substructure and it is a single picosecond pulse, the ratio should be 3:1. Higher ratios can be obtained by applying the three-photon fluorescence method.[50]

A.1.2. Second-, Third-, and Fourth-Harmonic Generation. The oscillating wavelengths for various picosecond lasers are approximately 1060 nm for a Nd/glass laser, 1064 nm for a Nd/YAG laser, 694 nm for a ruby laser, and 610–630 nm for an Ar-laser-pumped rhodamine 6G laser. In many cases, it is necessary to convert the wavelength to the visible or near-UV regions to excite biological macromolecules. Nonlinear processes such as second-, third-, and fourth-harmonic generation are effective means of generating short-wavelength radiation. Higher-harmonic generations can be used to measure the pulse width of mode-locked lasers. The experiment can be performed in the following way. The laser beam is split into two beams of equal intensity. One beam passes through a fixed delay and the other through a variable delay. They are focused on the same spot on a nonlinear crystal to generate the second harmonic. The delay time is changed shot by shot when a single-shot laser is used. It is continuously changed by sliding a stage with a stepping motor for a repetitively pulsed laser. The intensity curve plotted against the position of the variable delay gives the autocorrelation function of the laser pulse from which a pulse width can be estimated.

[49] E. P. Ippen and C. V. Shank, in "Ultrashort Light Pulses" (S. L. Shapiro, ed.), p. 83. Springer-Verlag, Berlin and New York, 1977.
[50] P. M. Rentzepis, C. J. Mitschele, and A. C. Saxman, *Appl. Phys. Lett.* **17**, 122 (1970).

TABLE VI.   Nonlinear Crystals for Second-Harmonic Generation[a]

Abbreviated name	Crystal name	Wavelength (μm)	Phase matching angle
ADP	ammonium dihydrogen phosphate	1.06	42°
		0.6943	52°
KDP	potassium dihydrogen phosphate	1.06	40°
		0.6943	50°
CDA	cesium dihydrogen arsenate	1.06	90°
RDA	rubidium dihydrogen arsenate	0.6943	80°
RDP	rubidium dihydrogen phosphate	0.6943	66°

[a] Reprinted with permission from "Handbook of Lasers with Selected Data on Optical Technology" (R. J. Pressley, ed.), p. 497. CRC Press, Cleveland, Ohio, 1977. Copyright The Chemical Rubber Co., CRC Press, Inc.

In Table VI, nonlinear crystals, often used for higher-order harmonic generation are listed.

A.1.3. Picosecond Continuum Generation.[11] When intense laser light is focused into a cell containing various organic or inorganic liquids or solids, extreme spectral broadening can be observed. It sometimes extends several thousand wave numbers to both the Stokes and anti-Stokes sides of the oscillating laser wavelength. The pulse shape of the continuum is slightly modified from that of the pumping laser pulse because of the difference in the degree of modulation in the spectral region of the laser pulse. However, the half-width of the continuum pulse is close to that of the pumping pulse. Picosecond continuum light can be utilized as a monitoring light source for picosecond absorption spectroscopy. In order to generate continuum light with picosecond width, liquids and glasses or solids are usually used for self-phase modulation or four-wave mixing media.

Usually water, heavy water, ethanol, carbon tetrachloride, cyclohexane, or phosphoric acid are used for continuum generation. To obtain the monitoring continuum light in the visible region, the second harmonic of a Nd/glass laser or Nd/YAG laser is usually used. An intense continuum with very wide spectral range is obtained by focusing the fundamental of ruby laser into phosphoric acid.[51,52] The continuum covers the near UV, all of the visible region, and near IR. When a mode-locked ruby laser is used, polyphosphoric acid gives a more intense continuum than other liquids, and the spectral region of the continuum is broader than that obtained with normal phosphoric acid or isophosphoric acid.

[51] N. Nakashima and N. Mataga, Chem. Phys. Lett. 35, 487 (1975).
[52] T. Kobayashi, Opt. Commun. 28, 147 (1979).

A.1.4. Stimulated Raman Scattering Light Pulse.[11] The stimulated Raman scattering process, which is related to the third-order nonlinear susceptibility, can be used to obtain an excitation light source at different wavelengths from the fundamental or higher harmonics of lasers. In Table VII, the shifts of the stimulated Raman scattering from the pumping laser are given for various kinds of materials. From an experimental point of view, it is important to keep in mind that stimulated Raman scattering with picosecond pulses is accompanied by other nonlinear optical phenomena. One of them is self-focusing which may enhance the conversion efficiency but tends to break the beam up into several filaments, making accurate control of the geometry of the interaction difficult. Self-phase-modulation can also occur, producing considerable spectral broadening. If the above-mentioned effects occur, then the energy is dissipated and the light may not be as monochromatic as the fundamental. In order to get highly intense Raman light,

TABLE VII. Stimulated Raman
Scattering from Various Liquids[a]

Liquid	Frequency shift (cm^{-1})
bromoform	222
carbon tetrachloride	460
bromoform	539
carbon disulfide	656
chloroform	667
benzene	992
pyridine	992
toluene	1004
styrene	1315
nitrobenzene	1344
styrene	1629
1,2-dichlorobenzene	2202
methylcyclohexane	2817
cyclohexane	2852
cyclohexane	2863
dioxane	2967
styrene	3056
benzene	3064
water	3651

[a] Reprinted with permission from "Handbook of Lasers with Selected Data on Optical Technology" (R. J. Pressley, ed.), p. 527. CRC Press, Cleveland, Ohio, 1977. Copyright The Chemical Rubber Co., CRC Press, Inc.

the other effects mentioned above must be reduced. To reduce the probability of self-focusing, it is necessary to use a short cell containing a highly Raman-active medium. In order to get intense Raman scattering, it is necessary to use a lens with a somewhat longer focal length than that used to obtain picosecond continuum.

A.1.5. Parametric Interaction. There are two kinds of parametric emission processes that can be attained by ordinary-power ruby, Nd/YAG, or Nd/glass lasers. One is a three-photon parametric emission process and the other is a four-photon parametric emission, or mixing process. The former process is related to the second-order nonlinear susceptibility, while the latter is related to the third-order nonlinear suceptibility.[11] The two processes are written as

$$\omega_L \rightarrow \omega_s + \omega_i, \tag{A.1}$$

$$2\omega_L \rightarrow \omega_s + \omega_i. \tag{A.2}$$

where ω_L, ω_s, and ω_i are the frequencies of the laser, signal, and idler, respectively. Four-photon parametric interaction can also be obtained by mixing two waves at different frequencies. This process is represented as

$$\omega_1 + \omega_2 \rightarrow \omega_s + \omega_i. \tag{A.3}$$

A.1.6. Optical Kerr Effect. The optical Kerr effect has several causes. The three main causes are anisotropic polarizability of anisotropic molecules, electronic hyperpolarizability of isotropic molecules, and electrostriction. Depending on the causes of the Kerr effect, the relaxation time of an aniso-tropic refractive index is different. The reason why CS_2 is most often used for the Kerr shutter is that the relaxation time of the anisotropic refractive index is 2.1 psec, which is much shorter than that of nitrobenzene (30 psec), m-nitrotoluene (50 psec) or chlorobenzene (6–8 psec), while the non-linear refractive index of CS_2 is $(1-2) \times 10^{-11}$ esu, which is of the same order as that of nitrobenzene (1.4–1.6×10^{-11} esu) or chlorobenzene (1.6×10^{-11} esu).

When a Kerr shutter is used in high-time-resolution experiments, great care must be paid to the delay or time spreading of the white picosecond light with broad spectrum due to dispersion in the Kerr shutter.

A.2. Optical Elements and Detectors

A.2.1. Saturable Dye Solution and Cell. Kodak A9860 or A9740 dye in ethylene chloride solution is generally used as a saturable dye for Nd/glass and Nd/YAG lasers. Before using ethylene chloride as a solvent of the dye it is necessary to pass it through a silica gel column to remove impurities and to store it in the dark at low temperature to prevent degradation.

The light transmission at 1060 nm of the solution is adjusted to 62–70%. 1,1'-Diethyl-2,2'-quinodicarbocyanine iodide (DDI) or 1,1'-diethyl-4,4'-quinocarbocyanine iodide (cryptocyanine) in methanol solution is used for mode locking in ruby lasers. To avoid degradation the dye solution should be circulated. A contact mirror which forms one wall of the saturable dye cell is used to avoid the satellite pulses that may be generated by reflection from the walls of the cell when a separated saturable dye cell is used.

A.2.2. Cavity Mirrors. The back mirror of the cavity of a mode-locked solid-state laser should be highly reflective at the oscillation wavelength to increase the Q value of the resonator. The output mirror of a cavity usually has a reflectivity of 40–60%. One of the cavity mirrors is flat and the other is usually concave. This is to reduce both the loss of Q value by light diffraction and the thermal lensing effect of the rod. Both can be flat mirrors if two lenses are put inside the cavity to compose an inverted Galileian telescope, slightly divergent from one side to the other, to reduce the thermal lensing effect. They are also useful for protection against damage of the highly reflective mirrors of the cavity by multiplying the diameter of the beam. The lenses should be AR coated to eliminate the possibility of satellite pulse generation.

A.2.3. Single-Pulse Selector. A single-pulse selector is usually composed of two cross-polarized Glan–Thomson prisms (entrance and exit polarizers) and a Pockels cell between them. High voltage is applied either by an optically triggered spark gap or high-voltage pulse driver. The high-voltage driver is adjusted to generate a half-wave voltage to rotate the polarization plane of incident laser pulse by 90° for the beam to pass through the exit polarizer, which is cross polarized to the entrance one.

A.2.4. Detectors. Three groups of detectors are used for picosecond spectroscopy. The first group includes various photomultipliers and photodiodes that have neither spatial resolution nor picosecond time resolution. Detectors, such as vidicons and TV cameras, with spatial resolution form the second group. Streak cameras, which have picosecond time resolution, constitute the third group. When the first two groups of detector are used, time resolution is obtained by the optical system which makes the correlation between time and space. Multishot experiments must be performed when a detector of the first group is used to obtain a single kinetics curve. Detectors of the second group are free from jitter. Therefore, repetitive-pulse experiments can be easily carried out using group 2 detectors. Recently synchroscan streak cameras have become commercially available as new group 3 detectors. These cameras can be operated synchronously with a repetitive pulse. The camera has a time resolution of a few picoseconds when it is used in a single-scan mode. If it is operated in a synchronous-scan mode the resolution time becomes about 10 psec because of a jitter in each scan.

5. FLUORESCENCE METHODS FOR STUDYING MEMBRANE DYNAMICS

By Joseph Schlessinger and Elliot L. Elson

5.1. Introduction

Fluorescence measurements can reveal structural and dynamic properties of animal cell plasma membranes which are crucial to their physiological function. The plasma membrane not only separates the cell interior from the surroundings but also participates in many active processes. These include the reception and transmission of biochemical "signals" (from other cells and the environment) which control metabolism, differentiation, and other cellular functions; transport of molecules along or through the membrane; and the control of locomotion and morphology. These processes often require that cell surface components move relative to each other to permit interactions, for example, between membrane enzymes and substrates, to form specialized regions on the cell surface (e.g., "coated regions"[1]), or to share membrane components between progeny cells during cell division. This part describes fluorescence methods for measuring cell surface motions.

The fluid lipid bilayer that forms the basic matrix of biological membranes is expected to permit rapid translation and rotation of lipid molecules and embedded proteins. In fact the situation is more complex. The mobility of some membrane components is constrained by interactions with structures either within or outside the membrane. At least some of these interactions are likely to have functional importance. Therefore measurements of the mobility of specific cell surface components should both indicate the progress of particular physiological processes and provide information about their mechanisms.

Motions of membrane components can be classified as microscopic or macroscopic according to the distances over which they extend. Microscopic motions take place over distances corresponding to molecular dimensions and consist of molecular rotations and deformations. For example, the flexibility of the hydrocarbon chains of phospholipids has been mapped by

---

[1] R. G. W. Anderson, J. L. Goldstein, and M. S. Brown, *Nature (London)*, **270**, 695 (1977).

METHODS OF EXPERIMENTAL PHYSICS, VOL. 20

Kornberg and McConnell[2,3] and others.[4] Using spin label and nuclear magnetic resonance (NMR) methods, respectively, they showed that the flexibility of the chains increases from the head groups to the center of the bilayer. Indications of segmental flexibility in membrane proteins have appeared in measurements of fluorescence.[5] Magnetic resonance and fluorescence methods have also frequently been used to measure rotational motion. Rotational mobility is often taken as an indicator of the "microviscosity" of the membrane lipid bilayer. As is discussed below, however, this analysis in terms of a single viscosity parameter may be inadequate to describe the constrained rotations of asymmetric fluorophores in complex anisotropic membranes.

Macroscopic motions extend over distances large compared to molecular dimensions. In contrast to measurements of microscopic motions, which were adapted from methods previously used in bulk solutions, new methods were required for measurements of motions over a range of microns on cell surfaces. Macroscopic lateral motion of membrane proteins was first observed as the intermixing of histocompability antigens in mouse–human heterokaryons.[6,7] A photobleaching method was first used to measure the lateral mobility of rhodopsin in amphibian retinal disk membranes.[8,9] This approach has now been generalized to fluorescent labeled molecules and provides a versatile procedure for quantitative measurements of macroscopic lateral mobility.[10–17]

Translational motion over tens of angstroms can be measured using

---

[2] R. D. Kornberg and H. M. McConnell, *Proc. Natl. Acad. Sci. U. S. A* **68**, 2564 (1971).

[3] R. D. Kornberg and H. M. McConnell, *Biochemistry* **10**, 1111 (1971).

[4] D. Chapman, *in* "Biological Membranes" (D. Chapman and D. F. H. Wallach, eds.), Vol. 2, p. 91. Academic Press, New York, 1973.

[5] P. Wahl, M. Kasai, J. P. Changeux, and J. C. Auchet, *Eur. J. Biochem.* **18**, 332 (1971).

[6] L. D. Frye and M. Edidin, *J. Cell Sci.* **7**, 319 (1970).

[7] V. A. Petit and M. Edidin, *Science (Washington, D.C.)* **184**, 1183 (1974).

[8] M. Poo and R. A. Cone, *Nature (London)* **247**, 438 (1974).

[9] P. A. Liebmann and G. Entine, *Science (Washington, D.C.)* **185**, 457 (1974).

[10] R. Peters, J. Peters, K. H. Tews, and W. Bahr, *Biochim. Biophys. Acta* **367**, 282 (1974).

[11] M. Edidin, Y. Zagyansky, and T. J. Lardner, *Science (Washington, D.C.)* **196**, 466 (1976).

[12] Y. Zagyansky and M. Edidin, *Biochim. Biophys. Acta* **433**, 209 (1976).

[13] K. Jacobson, G. Wu, and G. Poste, *Biochim. Biophys. Acta* **433**, 215 (1976).

[14] J. Schlessinger, D. E. Koppel, D. Axelrod, K. Jacobson, W. W. Webb, and E. L. Elson, *Proc. Natl. Acad. Acad. Sci. U. S. A.* **73**, 2409 (1976).

[15] D. Axelrod, D. E. Koppel, J. Schlessinger, E. L. Elson, and W. W. Webb, *Biophys. J.* **16**, 1055 (1976).

[16] D. E. Koppel, D. Axelrod, J. Schlessinger, E. L. Elson, and W. W. Webb, *Biophys. J.* **16**, 1315 (1976).

[17] J. Schlessinger, D. Axelrod, D. E. Koppel, W. W. Webb, and E. L. Elson, *Science (Washington, D.C.)* **195**, 307 (1977).

fluorescence energy transfer.[18,19] Up to now, however, this method has found few applications.

The factors that control microscopic and macroscopic mobility in membranes are not yet well understood. The viscosity of the lipid bilayer provides a lower bound to the forces resisting the motion of molecules embedded in it. Viscous resistance should vary linearly with the viscosity of the bilayer for both macroscopic and microscopic motions. A dependence of rotational diffusion rate on membrane viscosity has been demonstrated experimentally.[20] These studies have shown that the viscosity of the bilayer is very sensitive to lipid composition. For example, cholesterol increases the apparent viscosity and suppresses the gel–liquid crystal phase transition of the bilayer. Interactions with membrane proteins and larger molecular aggregates may further impede all types of motions. These interactions are likely to be more specific than viscosity effects. They could influence differently microscopic and macroscopic motions and could vary among different types of membrane components.

The scope of this paper is limited to a description of fluorescence methods that are useful for studying dynamic behavior on cell membranes. Rotational motions are indicated by measurements of steady-state fluorescence polarization and of decay of fluorescence polarization and transient dichroism. The most versatile and convenient method for measuring macroscopic lateral motion is based on fluorescence photobleaching. A fluorescence fluctuation method, which may be useful in some limited conditions, is also briefly described.

Some specific fluorescent probes are considered as examples, but a detailed description of probes is not considered in this paper. A useful compilation of fluorescent probes can be found in Azzi.[20a]

5.2. Molecular Rotations in Membranes

The extent to which plane-polarized light is absorbed by a molecule depends on the extent to which one of its resonant transition dipoles is parallel to the plane of polarization. Similarly, light emitted as fluorescence is parallel to the emission transition dipole. The structure of a molecule fixes the relative orientation of the transition moments. Consequently, an array of stationary molecules should display a defined relationship between the polarizations of absorbed and emitted light. If the molecules rotate randomly during the interval between excitation and emission, this relationship is

[18] S. M. Fernandez and R. D. Berlin, *Nature (London)* **264**, 411 (1976).

[19] L. Stryer, *Annu. Rev. Biochem.* **47**, 819 (1978).

[20] M. Edidin, *Annu. Rev. Biophys. Bioeng.* **3**, 179 (1974).

[20a] A. Azzi, *Q. Rev. Biophys.* **8**, 237 (1975).

attenuated to a degree which depends on the extent of rotation. Therefore, if the mean fluorescence lifetime is comparable to the time required for molecular rotation, the rotation rate can be deduced from the extent of depolarization of fluorescence excited by plane-polarized light.

The interpretation of fluorescence polarization measurements in molecular terms requires that transition moments, fixed relative to the three-dimensional coordinates of the molecule, be related to the plane of polarization of the incident excitation light, which is fixed in a laboratory frame of reference. This arrangement is shown in Fig. 1. The (monochromatic) excitation light beam propagates in the y direction with vertical (z axis) polarization. Fluorescence is detected perpendicular to the excitation (x axis) through an analyzer which is oriented either parallel ($I_\parallel$) or perpendicular ($I_\perp$) to the polarization of the excitation light.

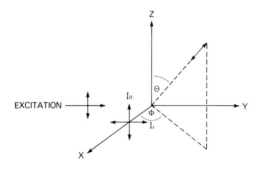

DETECTION

FIG. 1. Geometrical arrangement for measurement of fluorescence polarization. The transition dipole is described in polar coordinates. The light beam is propagated in the y direction and the emission intensity is detected through an analyzer which is oriented parallel ($I_\parallel$) or perpendicula ($I_\perp$) to the direction of polarization of the excitation light.

The fluorescence polarization is usually characterized by the polarization parameter P or by the fluorescence anisotropy r, which are defined as

$$P = (I_\parallel - I_\perp)/(I_\parallel + I_\perp), \qquad r = (I_\parallel - I_\perp)/(I_\parallel + 2I_\perp), \qquad (5.2.1)$$

so that

$$P = 3r/(2 + r). \qquad (5.2.2)$$

The steady-state level of fluorescence depolarization may be measured in a time-independent experiment. Using more complex apparatus one can observe the time-dependent decay of polarization or anisotropy, typically on the scale of nanoseconds.

5.2.1. Steady-State Fluorescence Polarization

The steady-state anisotropy of the fluorescence of spherical particles is determined by the relative magnitudes of two characteristic times, the fluorescence lifetime τ_0 and the time required for molecular rotation, specified by ρ, the rotational relaxation time. More precisely, the latter quantity measures the time required for the average projection of a molecular axis on its initial orientation to decay by e^{-1}. Hence if ϕ_0 is the angle between the axis and some reference orientation at time t_0 and ϕ is the value at time t, then

$$\overline{\cos \phi} = \overline{\cos \phi_0} \, e^{-t/\rho},$$

where the overbars indicate average values. In general, the rotational relaxation times are related to the rotational frictional coefficients, which depend on the size and shape of the molecule. A molecule of arbitrary shape can be described in terms of three rotational frictional coefficients. Then, for example, the relaxation time for rotation about one axis, say ρ_1, is related to the frictional coefficients for rotation about the other axes, f_2 and f_3, as

$$\rho_1 = (kT)^{-1}(1/f_2 + 1/f_3)^{-1},$$

where k is Boltzmann's constant. For a sphere of radius R, $f_2 = f_3 = 8\pi\eta R^3$ where η is the viscosity of the solvent. Hence for a sphere[21]

$$\rho = 4\pi\eta R^3/kT.$$

The measured anisotropy may be related to the rotational relaxation time with an equation originally derived by Perrin[22,23] in a different form:

$$r_0/r = 1 + 3\tau_0/\rho. \tag{5.2.3}$$

Here r_0 is the limiting fluorescence anisotropy in the absence of molecular rotation (e.g., in a solvent of high viscosity).

A slightly more complex relationship must be used for nonspherical particles which require more than a single relaxation time to describe their rotational behavior. More generally, Weber[21] showed that

$$r_0/r = 1 + c(t)T\tau_0/\eta, \tag{5.2.4}$$

where $c(r)$ is a function of molecular shape and the location of the transition dipoles in the molecular framework, T is the absolute temperature, and η is the viscosity of the solvent. For nonspherical particles $c(r)$ may vary with r.

[21] G. Weber, *Adv. Protein Chem.* **8**, 415 (1953).
[22] F. Perrin, *J. Phys. Radium Ser.* 7, **5**, 33 (1934).
[23] F. Perrin, *J. Phys. Radium Ser.* 7, **7**, 1 (1936).

For spherical particles, however, $c(r) = k/V$, a constant involving the Boltzmann constant k and the effective volume of the sphere V. Then

$$r_0/r = 1 + kT\tau_0/\eta V. \tag{5.2.5}$$

This equation has been extensively applied to the measurements of the microviscosity of membranes. Typically, a small, hydrophobic fluorescent molecule is dissolved in the lipid bilayer of a vesicle or cell membrane, and the steady-state polarization of a suspension of the cells or vesicles is measured. The fluorescent probes most commonly used are perylene or DPH (1,6-diphenyl-1,3,5-hexatriene). The fluorescence anisotropy of the embedded probe measured in the lipid bilayer system is then compared to the fluorescence anisotropy of the same probe when it is dissolved in a reference oil of known viscosity.[24–28] The membrane microviscosity can be calculated by applying Eq. (5.2.4) after determining the absolute temperature T and the fluorescence lifetime τ. Usually τ is measured at a single temperature and the relation between fluorescence intensity and fluorescence lifetime then used to evaluate τ at different temperatures[29]:

$$\tau_1/\tau_2 = I_1/I_2. \tag{5.2.6}$$

This simple approach has been used to study the "microviscosity" of artificial[30,31] and cell[25,32–36] membranes, phase transitions and degree of order in micelles, liposomes, and cell membranes.[30–32]

In a few cases the fluorescence polarization method has also been applied to measure the rotational relaxation time of fluorescein-conjugated Concanavalin A (Con A, a mannose binding protein) both in aqueous solution and bound to cell membranes.[37] In this experiment the rotational relaxation time of the cell-bound fluorescein-Con A was correlated with the "degree of rotational mobility" of the Con A receptor sites.[38]

[24] M. Shinitzky, A. C. Dianoux, C. Gitler, and S. Weber, *Biochemistry* **12**, 521 (1973).

[25] B. Rudy and C. Gitler, *Biochim. Biophys. Acta* **288**, 231 (1972).

[26] U. Cogan, M. Shinitzky, G. Weber, and T. Nishida, *Biochemistry* **12**, 521 (1973).

[27] K. Jacobson and D. Wobschall, *Chem. Phys. Lipids* **12**, 117 (1974).

[28] M. Shinitzky and Y. Barenholtz, *J. Biol. Chem.* **249**, 2652 (1974).

[29] F. Perrin, *Ann. Phys. (Leipzig)* [5] **12**, 169 (1929).

[30] B. R. Lentz, Y. Barenholtz, and T. E. Thompson, *Biochemistry* **15**, 4521 (1976).

[31] B. R. Lentz, Y. Barenholtz, and T. E. Thompson, *Biochemistry* **15**, 4529 (1976).

[32] M. Shinitzky and M. Inbar, *J. Mol. Biol.* **85**, 603 (1974).

[33] M. Inbar and M. Shinitzky, *Eur. J. Immunol.* **5**, 166 (1975).

[34] S. Toyoshima and T. Osawa, *J. Biol. Chem.* **250**, 1655 (1974).

[35] P. Fuchs, A. Parola, P. W. Robbins, and E. R. Blout, *Proc. Natl. Acad. Sci. U. S. A.* **72**, 3351 (1975).

[36] G. W. Stubbs, B. J. Litman, and Y. Barenholtz, *Biochemistry* **15**, 2766 (1976).

[37] M. Shinitzky, M. Inbar, and L. Sachs, *FEBS Lett.* **34**, 247 (1973).

[38] M. Inbar, M. Shinitzky, and L. Sachs, *J. Mol. Biol.* **81**, 245 (1973).

5.2.2. Nanosecond Time-Dependent Fluorescence Polarization

When a fluorescent molecule can be described as a spherical rotor in a three-dimensional liquid, then both the decay of the total emission and the decay of the emission anisotropy of the molecule can be described as mono-exponential functions. Consequently $c(r)$ of Eq. (5.2.4) is constant and the equation is linear in r. Frequently, however, the situation is more complex. The observed fluorophore may display several fluorescent processes with different lifetimes τ_i. Moreover, the fluorescent particle may undergo anisotropic rotation due either to an asymmetric shape or to restrictions imposed by the anisotropic fluid matrix on certain rotations. Then the behavior of the fluorescence anisotropy is correspondingly complicated. Multiexponential functions must be used to represent both the total emission:

$$I(t) = \sum_i a_i e^{-t/\tau_i} \qquad (5.2.7)$$

and the rotational correlation time:

$$r(t) = \sum_i \beta_i e^{-t/\rho_i}. \qquad (5.2.8)$$

This leads to a nonlinear Perrin plot.

For a system of this type the average fluorescence lifetime is defined as

$$\langle \tau \rangle = \int_0^\infty tI(t)\,dt \Big/ \int_0^\infty I(t)\,dt = \sum_i a_i \tau_i^2 \Big/ \sum_i a_i \tau_i, \qquad (5.2.9)$$

and the average rotational correlation time is defined as follows:

$$\langle \rho \rangle = \sum_i \beta_i \left(\sum_i \frac{\beta_i}{\rho_i} \right)^{-1}. \qquad (5.2.10)$$

While the steady-state depolarization measurement can give information about the average rotational relaxation time $\langle \rho \rangle$ using the average fluorescence lifetime $\langle \tau \rangle$, the nanosecond time-dependent fluorescence depolarization method can in principle resolve the different components of τ_i and ρ_i, detect possible asymmetric restrictions to rotation in the system, and make possible a more realistic analysis of the effects of bilayer "viscosity" on the motions of embedded particles.

The measuring system most commonly used is a simple modification of a single-photon-counting fluorometer.[39] Sample fluorescence is excited by pulses of polarized light from a thyratron-gated air flash lamp.[39] The decay of total emission and emission anisotropy are measured by collecting $I_\parallel(t)$

---

[39] J. H. Easter, R. P. DeToma, and L. Brand, *Biophys. J.* **16**, 571 (1976).

and $I_\perp(t)$ (see Fig. 1). Excitation and observation must be repeated many times and averaged to reach adequate precision. After appropriate deconvolution from the time profile of the excitation pulse, the time-dependent anisotropy $r(t)$ can be calculated.[40] Phase shift methods are also in use.[41]

Analysis of microviscosity determinations based on steady-state polarization data in terms of the component lifetimes and rotational relaxation times detected in time-dependent measurements has been carried out infrequently.[40,42,43] The time-dependent fluorescence polarization of DPH was measured in paraffin oil, in egg lecithin vesicles,[42] and in dipalmitoyl phosphatidyl choline vesicles.[40,43] A comparison of these data to the results of measurements of steady-state fluorescence polarization demonstrated that the former revealed a much more complicated picture of the behavior of DPH in oil or lipid vesicles than the latter did.

Kawato et al.[43] separated the anisotropy decay curves into two phases, a fast phase occurring in about one nanosecond and a much slower phase. Based on the temperature dependence of the anisotropy relaxation data, they proposed a model which allowed them to estimate the effective viscosity of the bilayer. Their values were about an order of magnitude smaller than those estimated from steady-state anisotropy data without considering the restrictions on rotation revealed in the relaxation experiments. A general theoretical discussion of the analysis of restricted rotation by fluorescence anisotropy decay has also appeared.[44]

5.2.3. Decay of Transient Dichroism

Steady-state and time-dependent fluorescence polarization measurements have suggested that the rotational relaxation time of membrane proteins is on the order of a microsecond or longer.[45] Measurement of rotational diffusion of a membrane protein was first accomplished by Cone, who made use of the natural photochemistry of rhodopsin in retinal disk membranes.[46] He exposed rod outer segments to 5-nsec pulses of polarized light. This exposure selectively bleached rhodopsin molecules whose absorption transition dipoles were parallel to the axis of polarization, thereby inducing a dichroism in the rhodopsin. As the rhodopsin molecules randomly reoriented because of rotational diffusion, the dichroism decayed. From the

[40] L. A. Chen, R. E. Dale, S. Roth, and L. Brand, J. Biol. Chem. **252**, 2163 (1977).
[41] G. Weber, J. Chem. Phys. **66**, 4081 (1977).
[42] R. E. Dale, L. A. Chen, and L. Brand, J. Biol. Chem. **252**, 7500 (1977).
[43] S. Kawato, K. Kinosita, and A. Ikegami, Biochemistry **16**, 2319 (1977).
[44] K. Kinosita, S. Kawato, and A. Ikegami, Biophys. J. **20**, 289 (1977).
[45] R. J. Cherry, FEBS Lett. **55**, 1 (1975).
[46] R. A. Cone, Nature (London), New Biol. **236**, 35 (1972).

rate of decay a value of 20 μsec was deduced for the rotational relaxation time of rhodopsin in the disk membrane. A very different result was obtained when a similar experiment was performed on bacteriorhodopsin in purple membrane fragments.[47] In this system the induced dichroism did not decay during the period of observation. From this experiment the rotational relaxation time of bacteriorhodopsin was calculated to be greater than 1 msec.

Cherry et al.[45,48] have developed a general method based on transient dichroism for studying the rotational relaxation of membrane proteins. Their approach was to label membrane proteins with probe molecules that have an excited-state lifetime in the millisecond range. Since triplet-state lifetimes are in this range, they used as a label eosin isothiocyanate, which is readily excited to a triplet level. Measurement of the transient dichroism of triplet–triplet absorption of labeled proteins yielded the rate of rotational motion. This method was used to study the rotational diffusion of band-3 proteins in human erythrocyte membranes. The initial dichroism was produced by polarized laser light pulses of 1–2 μsec duration. The anisotropy factor measured in this experiment was defined as

$$r(t) = [A_{\parallel}(t) - A_{\perp}(t)]/[A_{\parallel}(t) + 2A_{\perp}(t)]. \qquad (5.2.11)$$

$A_{\parallel}(t)$ and $A_{\perp}(t)$ are the time-dependent absorbance changes for light polarized parallel or perpendicular to the plane of polarization of the excitation light. These experiments yielded a rotational relaxation time of 0.5 msec for the band-3 protein, a major constituent of the erythrocyte membrane. Moreover, this time was not affected by removal of the erythrocyte spectrin network.

5.2.4. Special Features and Limitations

The interpretation of steady-state fluorescence polarization data of membranes in terms of microviscosity is based on a few crucial assumptions. It is possible to divide these assumptions into two categories, the conceptual and the experimental. In a few cases the limitations embodied in the assumptions have been ignored and data therefore misinterpreted.

First, the conceptual limitations. It is assumed that the fluorophore rotates in the bilayer (which is a two-dimensional fluid) as it does in isotropic fluids. It is clear that even a bilayer composed of a single lipid species is not an isotropic fluid. The fluorophore rotates in an anisotropic environment in which the rotations can be restricted to varying extents depending on the

[47] K. Razi-Naqvi, J. Gonzalez-Rodriguez, R. J. Cherry, and D. Chapman, Nature (London), New Biol. **245**, 249 (1973).
[48] R. J. Cherry, A. Bürkli, M. Busslinger, G. Schneider, and G. R. Parish, Nature (London) **263**, 389 (1976).

location of the fluorophore in the bilayer. The situation is even more complicated in a cell membrane which is composed of different types of lipids, proteins, glycolipids, and glycoproteins. The fluorescent probe may bind to different species, and therefore the average rotational relaxation time $\langle \rho \rangle$ which is obtained from the steady-state polarization data is determined by rotational behavior occurring in heterogeneous environments. It is possible that the hydrophobic fluorescent probe that is embedded in the lipid matrix could also bind to some hydrophobic domains of membrane proteins, thereby restricting the motion of the probe molecule.

A second problem originates in the typical complexity of the mechanism of fluorescence decay. In an isotropic medium without quenching processes the decay kinetics of a fluorophore can often be described by a simple exponential decay. Partial quenching or chemical reaction during the fluorescence lifetime would yield a complicated multiexponential decay. It is very common to use a single average lifetime $\langle \tau \rangle$ in the Perrin equation [Eq. (5.2.4)] instead of the individual components of the fluorescence decay processes, which are sensitive to the viscosity and anisotropy of the fluid. This assumption affects not only the rotational relaxation times calculated from the Perrin equation but also the estimated values of fluorescence lifetimes at different temperatures calculated from Eq. (5.2.6).

Experimental assumptions that can lead to misinterpretations can be summarized as follows:

(1) Localization of fluorescent probes. The measurements of fluorescence polarization are usually performed on suspensions of cells which have been labeled with a fluorescent probe. The location of the probe in the cell is often unknown and is difficult to ascertain. It is often assumed that the fluorescent hydrophobic probe is inserted in the lipid matrix of the plasma membrane. From our experience we conclude that most fluorescent hydrophobic molecules tend to enter very rapidly into the cell interior. This includes carbocyanine and merocyanine dyes and also fluorescent derivatives of lipids, NBD-phosphatidylethanolamine and rhodamine-phosphatidylethanolamine. Therefore the measured microviscosity does not necessarily reflect the fluidity of the plasma membrane but in fact is calculated from a complex average $\langle \rho \rangle$ of all the rotational relaxation times ρ_i of probes located at different sites weighted by their corresponding quantum efficiencies.

(2) Mixed population of cells. At least one million cells are necessary for a single fluorescence polarization experiment. Primary cultures are usually composed of heterogenous populations of cells at different stages of their cell cycles or differentiation. This can lead to serious problems of interpretation. Use of synchronized cell cultures can reduce this problem.[49]

[49] S. W. de Laat, P. T. van der Saag, and M. Shinitzky, *Proc. Natl. Acad. Sci. U. S. A.* **74**, 4458 (1977).

(3) Segmental flexibility in measurements of receptor rotational relaxation times.

In order to compare the rotational relaxation times of Con A binding sites on normal and transformed cells, cells of both types were labeled with fluorescein-Con A and the mean rotational relaxation times of the cell surface labeled lectins measured by steady-state fluorescence polarization.[37,38] The authors interpret the differences in the polarization of the fluorescein-Con A on the cell surface as an indication that the Con A receptors can rotate at different rates.[37,38] They found that the rotational relaxation time of fluorescein-Con A in solution is 58 nsec. When fluorescein-Con A is bound to the cells the polarization data yield relaxation times of 70–160 nsec. These values are at least 100-fold shorter than the values obtained for protein rotations on cell membrane by other methods.[45–48] Most values are on the order of microseconds or longer. It is very likely that the values deduced from steady-state polarization data for the rotation of fluorescein-Con A do not reflect the rotations of the Con A receptors but rather some kind of local rotation of the bound dye or a segment of the bound lectin. Is is noteworthy that at the high doses of Con A used in these experiments the cell surface receptors are "frozen" and cannot move laterally.[14] The fact that the rotational relaxation time for fluorescein-Con A in water is only two or three times shorter than the rotational relaxation time for fluorescein-Con A on the cell membrane is also consistent with the presumption of segmental flexibility.

Neither steady-state nor nanosecond time-dependent fluorescence depolarization measurements are adequate for studying the rotation of proteins in the cell membrane. The fluorescence lifetime, which is in the nanosecond range, is too short compared to the relaxation time of the proteins (microseconds or more). But this time domain is adequate for lipid probes which have rotational relaxation times in the nanosecond range. The major advantage of the steady-state fluorescence polarization method is its simplicity. Its major use is as a diagnostic tool for studying qualitative changes that are related to the fluidity and degree of order of membrane system.

Fluorescence polarization studies on artificial bilayers have precisely detected the gel–liquid crystal phase transition in phosphatidylcholine dispersions. The parameters of the phase transition are determined for both large, multilamellar liposomes and small, single-lamellar vesicles.[30,31] The results are qualitatively similar to those obtained by other methods.[20] Measurements of fluorescence polarization have also provided reasonable values of the phase transition temperature and degree of order in micelles and erythrocyte membranes.[20,24–28]

The interpretation of polarization data on intact cells is complicated by uncertaintites about the location of probes and about segmental motion of labeled proteins. Therefore the significance of measurements of apparent rotational motion of Con A receptors on normal and transformed cells[37, 38] and the effects of malignant transformation[37, 38] and cholesterol[50] on membrane fluidity cannot now be definitely known.

Generally, fluorescence polarization methods are suitable for studying the rotational relaxation of lipids but not of proteins in membranes. The method of decay of transient dichroism seems to be a promising general approach for studying the latter.[48]

5.3. Macroscopic Membrane Motions

5.3.1. Basic Concepts

Lateral motion may result either from random Brownian diffusion or from a systematic driven process, such as flow along the cell surface. We consider both diffusion, with diffusion coefficient D, and simple flow, uniform in speed and direction, with velocity V. The central experimental task in determining D or V is to measure the number of molecules of a specified kind in a defined open region of the membrane as a function of time. In these experiments the specified molecules are distinguishable and detectable by their fluorescence. Therefore their number is estimated from measurement of the fluorescence emitted from the observation region. In principle other molecular properties could be used, but fluorescence has the advantages of specificity, sensitivity, and simple, rapid measurement. The rate at which the number of fluorescent molecules changes is determined by the size of the observation region and the magnitudes of D and V. Once the size is known, D and V can be calculated from the measured time behavior of the fluorescence.

Two methods based on these ideas have been developed. In one, called fluorescence correlation spectroscopy (FCS), the concentration of fluorescent molecules on the membrane remains in overall equilibrium throughout the experiment. Therefore the mean number of fluorophores in the observation region is constant. Nevertheless, the number of fluorophores and consequently the fluorescence undergo transient *spontaneous* fluctuations about their mean values. The varying time course of the fluctuating fluorescence is analyzed statistically to yield D and V. The concentration of fluorescent molecules is minimally perturbed by measurement. The second method, fluorescence photobleaching recovery (FPR), generates an initial non-equilibrium gradient in the concentration of the fluorescent molecules by exposing them to an intense pulse of light. This irreversibly photobleaches a portion of the fluorophores in the observation region. Then the recovery of the fluorescence in the region due to redistribution of the bleached and

[50] M. Inbar and M. Shinitzky, *Proc. Natl. Acad. Sci. U. S. A.* **71**, 2128 (1974).

unbleached fluorophores is analyzed to yield the mobility parameters. We shall consider FPR in some detail since it has proved to be more convenient for measurement on cell surfaces. We shall then give a briefer description of FCS and compare the two approaches.

5.3.2. Fluorescence Photobleaching Recovery

An FPR experiment consists of two major phases: first, the irreversible photolysis of a portion of the fluorescently labeled molecules in a small region of the cell surface and, second, the monitoring of the fluorescence recovery due to redistribution of bleached and unbleached fluorophores. Although other experimental arrangements have been tried, it is convenient to use the same microscope optical system and laser excitation source for both phases. The observation region is defined as the area illuminated by the excitation light and is typically a small fraction of the cell surface ($<1\%$). The recovery time is determined by the size of the illuminated area and the rate of motion (D or V). The photobleaching pulse must be short compared to this recovery time. During the recovery phase the laser excitation is attenuated by several orders of magnitude relative to the bleaching pulse to minimize further bleaching. Fluorescence photobleaching experiments have been performed by Peters et al.,[10] Edidin and Zagyansky,[11,12] and Jacobson et al.[13] We call our approach fluorescence photobleaching recovery.[14-17] (See Fig. 2.)

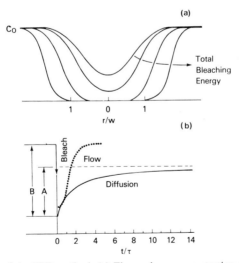

FIG. 2. Principle of the FPR method. (a) Fluorophore concentration profile induced by a brief pulse of an intense Gaussian laser beam for different bleaching energies. (b) Schematic recovery curves for diffusion and flow. The diffusion curve shows an immobile fraction $1 - A/B$. Subsequent bleach on the same spot would show increased fractional recovery.

The apparatus that we have developed for the photobleaching experiments is shown in Fig. 3. The beam from a krypton or argon ion laser is focused through a spatial filter, vertical illuminator, and the objective lens L2 of the microscope onto the specimen membrane. In a typical experiment the fluorescently labeled cells adhering to the dishes in which they were cultured are covered with complete growth medium or buffered saline solution.

The fluorescence radiation is collected by the same objective lens (L2), (usually water immersion 40 × NM 0.65), passes through the dichroic mirror (DM) of the vertical illuminator, through a field diaphragm (FD) and lens L3, and onto a sensitive photomultiplier tube (PMT) that is connected to photon counting electronics for measurement of the time-dependent fluorescence intensity.[16] The illuminated area on the fluorescence-labeled membrane is usually a spot of ~ 1-μ radius. The field diaphragm limits the volume "seen" by the photomultiplier. The appropriate laser line to excite the fluorophore is selected by a prism in the laser cavity. The beam is passed through a filter (PG) to remove plasma glow, a spatial filter (SF) to ensure a reasonably regular Gaussian intensity profile, a beam splitter that supplies a small portion of the radiation to a photodiode intensity monitor (MON), and a scatter aperture (A) and lens (L1) to control the size of the beam as it enters the vertical illuminator. In order to observe and photograph the cell, the lens (L1) may be removed to spread the laser illumination, or light can be supplied by the illuminator (IL). The mirrors (M) and the field diaphragm (FD) can be moved to present an image to viewing eyepieces or a camera (C).

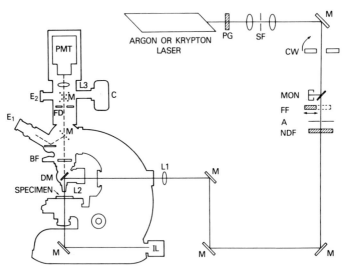

FIG. 3. Microscope geometry for FPR experiments. Dotted lines indicate removable components (see text for notation).

Eyepieces E_1 and E_2 also allow alignment of the field diaphragm with the center of the visual field and visual identification of the area selected for quantitative observation. The entire apparatus is mounted on a table with pneumatic supports that eliminate essentially all building vibrations.

To supply the photobleaching light pulse for FPR experiments, a solenoid-operated neutral density filter (FF, OD 4 or 3) is momentarily removed from the laser beam, and simultaneously the high-voltage bias potential is removed from the photomultiplier dynode chain to avoid damaging it. The duration of the bleaching pulse ranges from 5 msec to 3 sec. For the faster-diffusing phospholipids, bleaching times as short as 4 msec are preferred with brighter bleaching intensities. For slow diffusion, it is sometimes useful to reduce the average monitor beam intensity below that set by the filters FF and NDF by limiting the exposure to periodic pulses generated with a chopper wheel (CW). The bleaching-filter solenoid pulse is synchronized with the chopper wheel.

The fluorescence light intensity is measured by sensitive cold-operated photomultiplier to reduce dark current. The electronic system includes photon counting electronics, multichannel digital recording, recorders, etc.[16] The sensitivity for observations on membrane proteins with diffusion coefficients $\sim 10^{-10}$ cm^2/sec has been limited by autofluorescence of the cells to $\gtrsim 200$ fluorophores per square micron. Special efforts and/or multiple labeling may improve this sensitivity.

The theory for FPR[15] assumes that the photobleaching of the fluorophore to a nonfluorescent species is a simple irreversible first-order reaction with rate constant $\alpha I(r)$. Then the concentration $C(r, t)$ of unbleached fluorophores at position r and time t in the absence of transport can be calculated from

$$dC(r, t)/dt = -\alpha I(r)C(r, t), \tag{5.3.1}$$

where $I(r)$ is the bleaching intensity. For a bleaching pulse which lasts a time interval T_0 short compared to the recovery time, the fluorescence concentration profile at the beginning of the recovery phase ($t = 0$) is given by

$$C(r, 0) = C_0 e^{-\alpha T_0 I(r)}, \tag{5.3.2}$$

where C_0 is the initial uniform fluorophore concentration. The "amount" of bleaching induced in time T_0 is expressed by a parameter K:

$$K = \alpha T_0 I(0). \tag{5.3.3}$$

For a Gaussian intensity profile, $I(r)$ is given by

$$I(r) = (2P_0/\pi W^2)e^{-2r^2/W^2}, \tag{5.3.4}$$

where W is the half-width at e^{-2} height and P_0 is the total laser power.

Axelrod et al.[15] showed that for $K \ll 1$ for simple diffusion the fluorescence recovery curves can be represented in simple form:

$$F(t) = F(0)(1 + t/\tau_0)^{-1}, \tag{5.3.5}$$

where

$$\tau_0 = W^2/4D \tag{5.3.6}$$

and $F(t)$ is the fluorescence intensity at time t.

Theory was also developed for uniform flow at velocity V_0, or for simultaneous diffusion and flow. More complicated expressions are required for arbitrary K values.[15]

For pure diffusion, $F(t)$ has a finite slope at $t = 0$ but approaches its final value very slowly. For pure flow, $F(t)$ has a sigmoidal shape with zero slope at $t = 0$ and then a relatively rapid approach to the final value (see Fig. 2). Simultaneous diffusion and flow would yield recovery curves of intermediate properties.

It is often simplest to determine a transport rate from $\tau_{1/2}$, the time required for half of the observed fluorescence recovery to occur:

diffusion

$$D = (W^2/4\tau_{1/2})\gamma_D, \qquad \gamma_D = \tau_{1/2}/\tau_D, \tag{5.3.7}$$

flow

$$V_0 = (W^1\tau_{1/2})\gamma_F, \qquad \gamma_F = \tau_{1/2}/\tau_F, \tag{5.3.8}$$

where γ_D and γ_F are constants which depend on K. The values of γ_D and γ_F versus K, as well as expressions for $F(t)$ for diffusion, flow, and simultaneous diffusion and flow are presented in the paper by Axelrod et al.[15]

If the fluorophore is attached to a variety of molecules with a range of diffusion coefficients, the recovery curve reflects a weighted average D which depends on the distribution of individual diffusion coefficients. Only the existence of some independent information on the distribution or a special case such as clear bimodal recovery due to two widely separated values of D allow the deconvolution of the recovery curve to be carried out quantitatively. If some of the fluorophores are immobile the recovery of fluorescence intensity is incomplete and the fraction recovered at long times gives the mobile fraction A/B (see Fig. 2) of fluorophores. Values of D and A/B are determined by curve-fitting procedures described elsewhere.[15]

5.3.3. Fluorescence Correlation Spectroscopy

The experimental apparatus and theoretical analysis of FCS and FPR measurements are similar in many respects. The major differences between the two methods arise from the smallness of the typical spontaneous fluctuations of the number of observed fluorophores and the fact that the time course of an individual fluctuation does not *deterministically* define the diffusion coefficients of the labeled molecules. Therefore the fluctuating fluorescence signal in an FCS experiment must be analyzed statistically to yield precise values of transport coefficients. This is appropriately done by computing the autocorrelation function $g(t)$ of the measured fluorescence. A detailed discussion of the measurement and interpretation of the autocorrelation function has been presented.[51–54] Development of $g(t)$ requires that the fluorescence signal be observed over periods of time long compared to an individual fluctuation. This contrasts with FPR, in which a single recovery allows an estimate of D or $\mathbf{V}$.

The time dependence of $g(t)$ is identical to that of $F(t)$ [Eq. (5.2.6)] in an FPR experiment (for $K \ll 1$) for diffusion or flow. Unlike $F(t)$, however, the amplitude of $g(t)$ yields directly the average number of independently diffusing fluorescent units in the illuminated area.[52]

FCS was originally developed to measure coupled diffusion and chemical kinetics in chemical reaction systems in equilibrium.[53] FPR may also be used to measure chemical kinetics in suitable systems although this has not yet been demonstrated in practice.

5.3.4. FPR and FCS: Special Features and Limitations

Both FCS and FPR have features which are favorable for measuring the mobilities of surface components of living cells:

(1) The use of a fluorescence microscope enables a fairly precise localization of the observation region relative to visible cellular features. Furthermore, the narrow depth of focus of the microscope makes possible a determination that the fluorescent marker is on or near the cell surface.

(2) Since the observation region is small ($\sim 1~\mu$m radius), the mobilities of components in different regions on the same cell can be measured and compared.

(3) Lateral motion over micron distances is measured *directly*. It is not necessary to infer rates of macroscopic transport from measurements of microscopic motions as in magnetic resonance or fluorescence depolarization.

(4) Mobility of specific fluorescently labeled components is measured. Up to now these have included surface antigens, lectin binding sites, hormone receptors, ganglioside analogs, and lipid probes.

[51] D. Magde, E. L. Elson, and W. W. Webb, *Phys. Rev. Lett.* **29**, 705 (1972).
[52] E. L. Elson and D. Magde, *Biopolymers* **13**, 1 (1974).
[53] D. Magde, E. L. Elson, and W. W. Webb, *Biopolymers* **13**, 29 (1974).
[54] E. L. Elson and W. W. Webb, *Annu. Rev. Biophys. Bioeng.* **4**, 311 (1975).

(5)   A wide range of mobilities is accessible. In terms of diffusion co-efficients this extends from 10^{-12} to 10^{-6} cm^2/sec. (As indicated below, however, FCS is less useful than FPR in the lower part of this range.)

(6)   The methods are quantitatively precise and in principle capable of distinguishing among various mechanisms of transport (e.g., diffusion or flow).

(7)   Transport rates are measured in individual cells (under physiological conditions), thereby enabling a correlation of physiological events and morphology with dynamic properties.

At least in principle the FCS method has three significant advantages over FPR. First, it directly determines the average number of fluorescent diffusing units (differentially weighted by excitation and emission coefficients in heterogeneous mixtures[52]). This capability could be useful in studying aggregation equilibria.[53,54] Second, FCS requires no assumptions about the photochemistry of the observed fluorophores. In FPR, however, it is necessary to assume (and verify) a rate law for photobleaching. As previously indicated, the currently available and simplest interpretive theory assumes irreversible first-order photobleaching. It is straightforward, however, to develop interpretive equations for more complex mechanisms of the bleaching kinetics.[55] Third, FCS avoids observable perturbations of the measured system. This is in contrast with FPR, which requires the destruction of a substantial portion of the fluorophores in a small region of the cell surface. The requirement for a statistical analysis of spontaneous fluctuations, however, creates difficulties in applying FCS to living systems. Unlike a photobleaching experiment, in which a diffusion coefficient can be determined from a single recovery curve, in an FCS experiment data must be collected over at least several hundreds of correlation times τ_D to obtain a precise estimate of $g(t)$. An FPR measurement of the diffusion coefficient of a typical membrane protein ($\tau_D \sim 20$ sec) would last approximately three minutes, while FCS on the same system would consume more than two hours. During this prolonged period of data collection locomotion of the cell or local motion of the membrane could overwhelm the small fluorescence changes due to spontaneous fluctuations of fluorophore numbers and prevent the development of a precise correlation function. Unavoidable slow photo-bleaching of fluorophores over these long "integration" times could also lead to a significant slow decay in cell-surface fluorophore concentration which would appear as an additional time-dependent complexity in the measured autocorrelation function. In practice FCS is useful for measuring diffusion coefficients in the range $10^{-5} > D > 10^{-8}$ cm^2/sec. For slower processes it is usually necessary to use FPR. Indeed, on living cells, which

---

[55] E. L. Elson, in preparation.

typically show enough membrane activity to distort FCS measurement even of relatively fast processes, FPR is usually the preferred method over the entire range of mobilities.

Therefore it is essential to evaluate the possibility that the bleaching pulse in an FPR experiment damages the cell or the plasma membrane. The damage could arise either from heating or from chemical effects on membrane lipids or proteins due either to direct photochemistry or to free radicals generated by the bleaching pulse.

The laser power density during the bleaching pulse is approximately 10^{-3} W/μm^2, the concentration of fluorophores ~ 1000 molecules/μm^2, the duration of the bleaching pulse 1 sec, and the extinction coefficient of the fluorophore at the wavelength of the laser line $\sim 10^5$ M cm^{-1}. These parameters yield a heating rate of $10^{5\circ}$C/sec. This heat is rapidly dissipated owing to the high thermal conductance of water so that a thermal steady state is reached within 10^{-6} sec. Consequently, the maximum temperature rise is small, during a typical bleaching pulse $\sim 0.03^\circ$C and during the fluorescence recovery phase $\sim 10^{-5\circ}$C.[56] Although these increments are probably negligible, the temperature rise could become significant if the fluorophore concentration or beam intensity were increased by several orders of magnitude or if the cell were to contain other pigments which absorb the excitation light, as erythrocytes or plant cells do. An experimental demonstration of the small extent of heating during an FPR measurement has been carried out by Wu et al.[57] They used FPR to measure the temperature dependence of the lateral diffusion coefficients D of two fluorescent lipid analogs in phospholipid multilayers of varying composition. A greater than 100-fold change in D was detected in dimyristol phosphatidylcholine multilayers at $\sim 23^\circ$C, the temperature at which the gel–liquid crystal transition has been shown to occur by many other methods. This coincidence of the transition temperature determined by FPR and by other methods demonstrates that the temperature perturbation in an FPR measurement must be quite small. These results have been confirmed in similar experiments by Fahey and Webb.[58]

Direct detection of photochemical effects in the bleached region is very difficult because the number of photolabile molecules is so small (500–2000). Therefore information on this matter is mainly indirect but in its cumulative weight provides a persuasive argument against significant adverse photochemical effects.

(1) To minimize cell damage, we choose fluorophores excitable at wavelengths at which most cell components do not absorb, e.g., krypton laser lines 568.2, 520.8, and 482 nm.

[56] D. Axelrod, *Biophys. J.* **18**, 129 (1977).

[57] E.-S. Wu, K. Jacobson, and D. Papahadjopoulos, *Biochemistry* **16**, 3936 (1977).

[58] P. F. Fahey and W. W. Webb, *Biochemistry* **17**, 3046 (1978).

(2) In most cases the fluorophore is not directly attached to a cell structure, but rather is bound to a large protein ligand, such as an antibody, hormone, toxin, or lectin. The fluorophores and resultant free radicals are in the aqueous phase, not embedded in the membrane. They are therefore exposed to rapid quenching processes.

(3) Increasing the laser intensity tenfold [into the range 10^{-3}–10^{-2} W/$(\mu m)^2$], changing excitation wavelength from 568.2 to 520.8 nm, and changing the fluorescence probe from rhodamine to fluorescein did not alter the measured diffusion coefficients or the fraction of mobile fluorophores. Therefore there were no perceptible effects which depended on excitation wavelength or intensity or on the identity of the fluorophore, as would be expected if substantial photochemical processes were occurring.

(4) Similar values were obtained using both FPR and FCS to measure the diffusion coefficient of the lipid probe diI in myoblast membranes[14,17] and in lipid bilayers.[59] Similar values were also obtained by the two methods for the diffusion coefficients of fluorescent lipopolysaccharides in artificial bilayers.[60] FCS experiments do not require an exposure at high laser intensity as do FPR experiments.

Jacobson *et al.* have recently carried out experiments which seek to assess in detail the extent and kind of local photochemical damage that occurs in an FPR measurement. They have used scanning electron microscopy to examine the irradiated area after bleaching with different light intensities. They have also tested the effects of free-radical quenchers and light intensity on the incorporation of trypan blue, a standard test of cell viability. Up to now they have detected no photochemical damage in the bleached area.[61]

5.4. Applications

5.4.1. Translational and Rotational Diffusion of Molecules in Membranes and Membrane Viscosity

Published data summarized in Tables I and II clearly indicate the diversity in both the translational and rotational mobility of different cell membrane components. Measurement of the mobility of the lipid probe diI (3,3′-dioctadecylindocarbocyanine iodide) in different cell types confirms the

[59] P. F. Fahey, D. E. Koppel, L. S. Barak, D. E. Wolf, E. L. Elson, and W. W. Webb, *Science* (*Washington, D.C.*) **195**, 305 (1977).

[60] D. Wolf, J. Schlessinger, E. L. Elson, W. W. Webb, R. Blumenthal, and P. Henkart, *Biochemistry* **16**, 3476 (1977).

[61] K. Jacobson, Y. Hou, and J. Wojcieszyn, *Exp. Cell Res.* **116**, 179 (1978).

TABLE I.   Lateral Diffusion of Cell Membrane Components

Membrane component	Diffusion coefficient (cm^2/sec)	Mobile fraction	References
Lipid probe (diI)	$\sim 10^{-8}$	1	[b–f]
Analog of ganglioside GM_1	$\sim 5 \times 10^{-9}$	1	[e,g]
Lipid probe (diI) on unfertilized mouse eggs	$\sim 10^{-8}$	~ 0.25	[d]
Lipid probe (diI) on fertilized mouse eggs	$10^{-9}, < 10^{-10a}$	—	[d]
Unselected surface antigens on unfertilized eggs	$10^{-8}, 10^{-9a}$	0.24–0.60	[d]
Unselected surface antigens on fertilized eggs	$10^{-8}, < 10^{-11a}$	~ 0.14	[83d]
Rhodopsin in amphibian disks	$\sim 5 \times 10^{-9}$	1	[h,i]
Surface antigens on mouse–human heterokaryons	2×10^{-10}–10^{-9}	—	[j–l]
Unselected surface "proteins" (labeled chemically)	$\sim 2 \times 10^{-10}$	0.3–0.8	[b,c,m]
Unselected surface antigens (labeled by anti-P388)	$\sim 2 \times 10^{-10}$	0.3–0.8	[f,n]
Labeled proteins of erythrocyte membrane (mainly band 3)	$< 3 \times 10^{-12}$	0	[o]
Con A binding sites (S–Con A on Con A at low dose)	$\sim 4 \times 10^{-11}$	0.5–0.7	[b,f]
Con A binding sites (high dose)	10^{-11}–$< 3 \times 10^{-12}$	0–0.5	[b,e,n,p,q]

(*continued*)

[a] Shows two distinct components. See K. M. Yamada, S. S. Yamada, and I. Pastan, *Proc. Natl. Acad. Sci. U. S. A.* **73**, 1217 (1976).

[b] J. Schlessinger, D. E. Koppel, D. Axelrod, K. Jacobson, W. W. Webb, and E. L. Elson, *Proc. Natl. Acad. Sci. U. S. A.* **73**, 2409 (1976).

[c] J. Schlessinger, D. Axelrod, D. E. Koppel, W. W. Webb, and E. L. Elson, *Science (Washington, D.C.)* **195**, 307 (1977).

[d] M. Johnson and M. Edidin, *Nature (London)* **272**, 448 (1978).

[e] S. de Laat, P. T. van der Saag, E. L. Elson, and J. Schlessinger, *Proc. Natl. Acad. Sci. U. S. A.* **77**, 1526 (1980).

[f] E. L. Elson and J. Schlessinger, *in* "The Neurosciences: Fourth Study Program" (F. O. Schmitt and F. G. Worden, eds.), 691.. MIT Press, Cambridge, Massachusetts, 1978.

[g] J. Reidler, C. A. Eldridge, J. Schlessinger, E. L. Elson, and H. Wiegandt, in preparation.

[h] M. Poo and R. A. Cone, *Nature (London)* **247**, 438 (1974).

[i] P. A. Liebmann and G. Entine, *Science (Washington, D.C.)* **185**, 457 (1974).

[j] L. D. Frye and M. Edidin, *J. Cell Sci.* **7**, 319 (1970).

[k] V. A. Petit and M. Edidin, *Science (Washington, D.C.)* **184**, 1183 (1974).

[l] M. Edidin, *Annu. Rev. Biophys. Bioeng.* **3**, 179 (1974).

[m] M. Edidin, Y. Zagyansky, and T. J. Lardner, *Science (Washington, D.C.)* **196**, 466 (1976).

[n] J. Schlessinger, E. L. Elson, W. W. Webb, I. Yahara, U. Rutishauser, and G. M. Edelman, *Proc. Natl. Acad. Sci. U. S. A.* **74**, 1110 (1977).

[o] R. Peters, J. Peters, K. H. Tews, and W. Bahr, *Biochem. Biophys. Acta* **367**, 282 (1974).

[p] Y. Zagyansky and M. Edidin, *Biochim. Biophys. Acta* **433**, 209 (1976).

[q] K. Jacobson, G. Wu, and G. Poste, *Biochim. Biophys. Acta* **433**, 215 (1976).

TABLE I (*continued*)

Membrane component	Diffusion coefficient (cm^2/sec)	Mobile fraction	References
Acetylcholine receptor (diffuse)	$\sim 3 \times 10^{-11}$	0.8	[r]
Acetylcholine receptor (patch)	$< 10^{-12}$	0	[r]
IgE–Fc receptor	$\sim 3 \times 10^{-10}$	0.5–0.8	[f,s]
Fibronectin	$< 3 \times 10^{-12}$	0	[f,t]
Insulin receptor	$\sim 4 \times 10^{-10}$	0.8	[u]
EGF receptor	$\sim 3 \times 10^{-10}$	0.8	[u]

[r] D. Axelford, P. Ravdin, D. E. Koppel, J. Schlessinger, W. W. Webb, E. L. Elson, and T. R. Podleski, *Proc. Natl. Acad. Sci. U. S. A.* **73**, 4594 (1976).

[s] J. Schlessinger, W. W. Webb, E. L. Elson, and A. Metzger, *Nature* (*London*) **264**, 550 (1976).

[t] J. Schlessinger, L. S. Barak, G. G. Hammes, K. M. Yamada, I. Pastan, W. W. Webb, and E. L. Elson, *Proc. Natl. Acad. Sci. U. S. A.* **74**, 2909 (1977).

[u] J. Schlessinger, Y. Shechter, P. Cuatrecasas, M. Willingham, and I. Pastan, *Proc. Natl. Acad. Sci. U. S. A.* **75**, 5353 (1978).

TABLE II. Rotational Diffusion of Cell Membrane Components

Membrane component	Method	Rotational relaxation time	Viscosity (poise)	References
Perylene in erythrocyte ghosts	Fluorescence polarization	—	1.3	[a]
Retinol in erythrocyte ghosts	Fluorescence polarization	—	1–10	[b]
Rhodopsin in amphibian disks	Decay of transient dichroism	20 μsec	2	[c]
Bacteriorhodopsin in purple membrane	Decay of transient dichroism	< 20 msec	—	[d]
Eosin labeled band-3 protein of erythrocyte ghosts	Decay of transient dichroism	0.5 msec	—	[e]
Fluorescein-Con A on normal and transformed cells	Fluorescence polarization	70–160 nsec	—	[f,g]
ANS or DNS bound to electroplax membrane fragments	Nanosecond time-dependent fluorescence polarization	> 700 nsec	—	[h]

[a] B. Rudy and C. Gitler, *Biochem. Biophys. Acta* **288**, 231 (1972).

[b] G. K. Radda and D. S. Smith, *FEBS Lett.* **9**, 287 (1970).

[c] R. A. Cone, *Nature* (*London*), *New Biol.* **236**, 35 (1972).

[d] K. Razi-Naqvi, J. Gonzalez-Rodriguez, R. J. Cherry, and D. Chapman, *Nature* (*London*), *New Biol.* **245**, 249 (1973).

[e] R. J. Cherry, A. Bürkli, M. Busslinger, G. Schneider, and G. R. Parish, *Nature* (*London*) **263**, 389 (1976).

[f] M. Shinitzky, M. Inbar, and L. Sachs, *FEBS Lett.* **34**, 247 (1973).

[g] M. Inbar, M. Shinitzky, and L. Sachs, *J. Mol. Biol.* **81**, 245 (1973).

[h] P. Wahl, M. Kasai, J. P. Changeux, and J. C. Auchet, *Eur. J. Biochem.* **18**, 332 (1971).

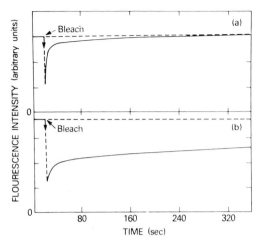

FIG. 4. FPR curves for (a) the lipid probe diI and (b) proteins on cell membrane labeled with TNBS and rhodamine anti-TNP antibodes. Note complete and fast recovery for diI and incomplete recovery with slower diffusion of the labeled proteins.

existence of a fluid membrane matrix in which membrane molecules are free to move over wide areas of the cell membrane (see Fig. 4a). The similarity of diI diffusion in different kinds of cells suggests that the viscosity does not depend strongly on cell type. The behavior of the fluorescent analog of the ganglioside GM_1[62] is similar to that of diI. Its diffusion coefficient is only a factor of 2 smaller. The consistency of direct measurements of macroscopic lateral diffusion of the lipid probe with estimates derived from microscopic motions of lipids suggests that the same forces limit both. Nevertheless, it may be wrong to assume that the major factor limiting the mobility is membrane viscosity. It is not certain that only fluid dynamic factors limit the lateral mobility even of so simple a molecule as diI. Interaction of real membrane lipids and of diI with hydrophobic membrane proteins could also limit their rate of motion. Evidence for the immobilization of membrane lipids by hydrophobic proteins has been obtained in studies using spin-labeled fatty acids.[63] The formation of a lipid boundary layer could influence the mobility of both the lipid and the protein. In order to study this phenomenon using fluorescence methods, it is necessary to develop fluorescently labeled lipids which retain their capacity to bind to membrane proteins.

[62] J. Reidler, C. A. Eldridge, J. Schlessinger, E. L. Elson, and H. Wiegandt, in preparation.

[63] P. C. Jost, O. H. Griffith, R. A. Capaldi, and G. Vanderkooi, *Proc. Natl. Acad. Sci. U. S. A.* **70**, 480 (1973).

The mobility of membrane proteins is more complex. The membrane proteins occur in both mobile and immobile classes (see Fig. 4b). The existence of immobile protein molecules suggests an interaction between them and some other molecules or supermolecular structures in the membrane or in the cytoplasm. The mobile membrane proteins fairly consistently diffuse about 50–100-fold slower than the membrane lipids. It is impossible at this time to give a quantitative interpretation of the difference in diffusion coefficients of diI and the mobile labeled proteins. One possible interpretation is hydrodynamic: the proteins are larger and therefore move more slowly because they experience greater viscous drag. An analysis of this possibility could be carried out using the theory of Saffman and Delbruck.[64] Let us assume that membrane proteins have reasonable size (radius of 50 Å) and that the diffusion coefficient is $\sim 2 \times 10^{-10}$ cm^2/sec. The calculated viscosity would be ~ 20 poise, which is approximately tenfold higher than the viscosity estimated for the lipid matrix by many different methods. Using the same approximation, a viscosity of 1–2 poise is calculated from the diffusion coefficients of diI and the fluorescent ganglioside analog in the cell membrane. The calculation suggests that the lipid viscosity does not control the lateral mobility of the labeled surface proteins. Rhodopsin in amphibian disk membranes seems to be the only system in which the lateral diffusion of a membrane protein is determined by the viscosity of the lipid matrix. The diffusion coefficient of rhodopsin is $\sim 5 \times 10^{-9}$ cm^2/sec[8], which is approximately 20 times faster than nonselected proteins and antigens on cell membranes and IgE receptors on rat mast cells. It is approximately ten times faster than the diffusion of insulin or epidermal growth factor receptor complexes on 3T3 cells. It is noteworthy that in retinal disk membranes rhodopsin is the major protein constituent and cannot have interactions with cytoskeletal elements. If we assume that retinal disks present an ideal fluid membrane in which the diffusion rate of the rhodopsin molecules is mainly limited by the viscosity of the lipid matrix, then the mobility of membrane proteins in other cell types seems to be limited by interactions in addition to viscous drag.† The additional interactions responsible for limiting the lateral mobility of membrane proteins could originate from many sources including (among others): interactions with cytoskeletal structures, nonspecific collisions of membrane proteins with each other, and

[64] P. G. Saffman and M. Delbruck, *Proc. Natl. Acad. Sci. U. S. A.* **72**, 3111 (1975).

[65] N. C. Nielsen, S. Fleischer, and D. G. McConnell, *Biochim. Biophys. Acta* **211**, 10 (1970).

† A quantittative comparison of mobility in disk and in plasma membranes would require that account be taken of the unusually high content of unsaturated lipid in the former.[65]. The data necessary to do this are not yet available. Nevertheless, we feel that the presence of these unsaturated lipids will not affect the qualitative validity of this discussion.

interactions of membrane proteins with the extracellular matrix composed of collagen and fibronectin.[66-70]

The rotational diffusion of cell membrane components is in many respects similar to their translational motion. Lipid probes rotate at a rate which seems to be controlled by the viscosity of the lipid matrix (in the range of 1–10 poise). Again rhodopsin in amphibian disks is an exceptional protein. Its rotational relaxation time equals 20 μsec, which corresponds to a viscosity of ~ 2 poise. A similar value was obtained when the viscosity of the amphibian disk membrane was calculated from the lateral diffusion coefficient. It seems that the retinal disk presents an ideal fluid membrane in which both the rates of rotation and translational of rhodopsin are limited by the viscosity of the lipid matrix.

Unlike visual rhodopsin, bacteriorhodopsin does not rotate rapidly in the purple membrane.[47] This difference is consistent with x-ray diffraction data which indicate that disk membrane is a planar fluid array of rhodopsin molecules,[71] while bacteriorhodopsin is packed in a crystal-like arrangement in the purple membrane.[72,73]

The rotational relaxation time of band-3 protein in erythrocyte ghosts is considerably longer than the values obtained for small hydrophobic molecules by steady-state fluorescence polarization. The viscosity of the lipid matrix of erythrocyte ghosts is in the range of 1–10 poise.[23,25] The viscosity calculated from the rotational relaxation time of band-3 proteins corresponds to 25–100 poise.[48] This result seems to indicate that the viscosity of the erythrocyte membrane is not the major limit for the rotational diffusion rate of band-3 proteins. In addition, it was found that depletion of spectrin did not affect the rotational relaxation of band-3 proteins in the erythrocyte ghosts.[48] It is not clear what interactions impede the motion of this protein. The fact that band-3 proteins can possess rotational motion without being able to move laterally over macroscopic distances[10] demonstrates an important distinction between local diffusion (around the axis of the molecule) and macroscopic lateral diffusion.

The fast relaxation times measured for fluorescein-Con A on cell membranes by steady-state fluorescence polarization[37,38] and for ANS and DNS

[66] K. M. Yamada and I. Pastan, *Trends Biochem. Sci.* **1**, 222 (1976).

[67] R. O. Hynes, *Biochim. Biophys. Acta* **456**, 73 (1976).

[68] K. M. Yamada and J. A. Weston, *Proc. Natl. Acad. Sci. U. S. A.* **71**, 3492 (1974).

[69] R. O. Hynes and J. M. Bye, *Cell* **3**, 113 (1974).

[70] J. Schlessinger, L. S. Barak, G. G. Hammes, K. M. Yamada, I. Pastan, W. W. Webb, and E. L. Elson, *Proc. Natl. Acad. Sci. U. S. A.* **74**, 2909 (1977).

[71] J. K. Blasie and C. R. Worthington, *J. Mol. Biol.* **39**, 417 (1969).

[72] A. E. Blaurock and W. Stoeckenius, *Nature (London), New Biol.* **233**, 152 (1971).

[73] R. Henderson, *Annu. Rev. Biophys. Bioeng.* **6**, 87 (1977).

bound to electroplax proteins by nanosecond fluorescence polarization[5] do not seem to reflect the rotational relaxation of the entire protein molecule but rather some kind of independent, restricted rotation of the probe molecule or the flexibility of a segment of the protein.

5.4.2. Possible Significance of Mobility and Immobility of Membrane Proteins

The scope of this paper does not include detailed analysis of the mobility of different membrane components. Nevertheless, to illustrate the biological significance of membrane mobility we consider two systems in which the mobility or immobility of membrane components seem to be consistent with their physiological function. The two systems are the receptors for the hormones insulin or epidermal growth factor (EGF) and the major cell-surface protein of fibroblasts fibronectin.[66-69]

Several reports have already proposed that the fluidity of the plasma membrane plays a role in the mechanism of hormone action.[74-77] Lateral motion of the hormone receptor in the plasma membrane would enable the hormone to interact with various effector molecules, some of which might be mobile, and others located at fixed sites. The existence of a mobile hormone receptor is consistent with the fact that different hormones that bind to different receptors can apparently activate the same adenylate cyclase molecule. Furthermore, β-adrenergic receptors from turkey erythrocytes can activate the adenylate cyclase of other cell types after the two are fused by Sendai virus.[78] In order to measure the mobility of the receptors for insulin and EGF by FPR, fluorescent derivatives of those two hormones were prepared.[79] The fluorescent analogs retained substantial binding activity[79,80] and biological potency.[79]

3T3 fibroblasts were labeled with the fluorescent derivatives of insulin or EGF and the FPR method was employed to study their mobility. The conclusion of this study was that the receptors for insulin and EGF are mobile on the cell membrane with similar diffusion coefficients $D \sim (3-5) \times 10^{-10}$ cm^2/sec at 23°C.[80] Increasing temperature to 37°C yields pronounced

---

[74] J. P. Perkins, *Adv. Cyclic Nucleotide Res.* **3**, 1 (1973).

[75] P. Cuatrecasas, *Annu. Rev. Biochem.* **43**, 169 (1974).

[76] P. DeMeyts, A. R. Bianco, and J. Roth, *J. Biol. Chem.* **251**, 1877 (1976).

[77] C. R. Kahn, *J. Cell Biol.* **70**, 261 (1976).

[78] J. Orly and M. Schramm, *Proc. Natl. Acad. Sci. U. S. A.* **73**, 4410 (1976).

[79] Y. Shechter, J. Schlessinger, S. Jacobs, K. W. Chang, and P. Cuatrecasas, *Proc. Natl. Acad. Sci. U. S. A.* **75**, 2135 (1978).

[80] J. Schlessinger, Y. Shechter, P. Cuatrecasas, M. Willingham, and I. Pastan, *Proc. Natl. Acad. Sci. U. S. A.* **15**, 5353 (1978).

receptor immobilization. In another study Schlessinger et al.[80,81] showed that at 37°C the fluorescent hormone receptor complexes rapidly aggregate into patches on the cell surface and soon thereafter appeared in endocytic vesicles within the cell. Therefore the immobilization of the hormone receptor complexes is presumably the consequence of receptor aggregation, internalization, or a combination of both. It seems likely that the behavior seen in this study is involved in the regulation of receptor levels on the cell surface and also, perhaps, in the mechanisms of hormonal activation of certain cellular functions.[80,81]

The large glycoprotein known as fibronectin appears in an immobile structure on cell surfaces.[69,70] It forms fibrils concentrated at the edges of cells and seems to connect cells with each other and with the substrate on which they are growing. This pattern is consistent with the apparent involvement of fibronectin in cell–cell and cell–substrate adhesion. The amount of this protein decreases substantially after transformation. Addition of exogenous fibronectin to transformed cells partially restores the morphology, adhesiveness, and parallel alignment typical of normal fibroblasts.[82] These observations prompted a study of how fibronectin binds to cell surfaces using fluorescence microscopy and photobleach methods.[70]

It was found that both endogenous fibronectin labeled with rhodamine antifibronectin antibodies and fluorescein-labeled exogenous fibronectin bound to the cell in a fibrillar pattern and were immobile on the experimental time scale of the FPR method ($D < 3 \times 10^{-12}$ cm^2/sec). Azide, vinblastine, and cytochalasin B did not alter the immobility and the cell surface distribution of the fibronectin molecules. Therefore oxidative phosphorylation and some cytoskeletal elements do not seem to be responsible for the properties of the bound fibronectin. The presence of immobile fibronectin fibrils did not impede the diffusion of a lipid probe, a fluorescent ganglioside analog, or various surface antigens. Therefore the fibrils do not form a barrier across the lipid phase of the plasma membrane. In contrast, Con A binds to fibronectin and is largely immobile in areas rich in fibronectin. These static properties of fibronectin are consistent with its apparent structural function as an adhesive or anchoring component.

These examples should provide some idea of the kinds of information that can be obtained about the dynamic properties of cell surface components from fluorescence measurements. The capability for detailed analysis of the mechanisms of physiologically important processes will further improve as we gain greater understanding of the physical forces and structures that limit

[81] J. Schlessinger, Y. Shechter, M. C. Willingham, and I. Pastan, *Proc. Natl. Acad. Sci. U. S. A.* **75**, 2659 (1978).

[82] K. M. Yamada, S. S. Yamada, and I. Pastan, *Proc. Natl. Acad. Sci. U. S. A.* **73**, 1217 (1976).

the mobility of cellular components, when we combine measurements of microscopic and macroscopic motions to provide a more complete picture of the dynamic state of various cellular structures, and as we further refine our capabilities both of making quantitative dynamic measurements and of obtaining two-dimensional images of the distribution of specific cellular components in and on individual cells.

Fluorescence measurements of rotation and macroscopic translation of macromolecules in solution have proved useful in a number of applications. In particular, both steady-state and time-resolved measurements of fluorescence depolarization have been used for some time in studies of proteins[23,83] and nucleic acids.[84] The newer techniques of FCS and FPR have rarely been applied to macromolecular solutions. The original demonstration of the FCS method concerned coupled translation and chemical reaction of ethidium bromide with DNA.[53] More recently, Borejdo[85] workers have used FCS to study interactions among muscle components.

5.5. Recent Developments

5.5.1. Technical Developments

Measurements of polarized emission from triplet states (anisotropy of phosphorescence or delayed fluorescence) are now being used to characterize rotational diffusion of biological macromolecules in proteins.[86] This provides a sensitive and versatile complement to the transient dichroism method described in Section 5.2.3.

A small but important change in the optical system for FPR measurements has been the replacement of the removable neutral density filter (FF in Fig. 3) by an arrangement of beam splitters and a shutter.[87] The incident laser beam is split into a weak monitoring component and an intense bleaching component. The latter is blocked by a fast shutter except during the bleaching interval. The bleaching and monitoring components are recombined into a single colinear beam after passing the shutter. This arrangement eliminates problems due to displacement of the axes of the bleaching and monitoring beams from optical wedge effects in the neutral density filter.[88]

[83] L. Stryer, *Science* (*Washington, D.C.*) **162**, 526 (1968).

[84] C. R. Cantor and T. Tao, *in* "Procedures in Nucleic Acid Research" (G. L. Cantoni and D. R. Davies, eds.), Vol. 2, p. 31. Harper, New York, 1971.

[85] J. Borejdo, *Biopolymers* **18**, 2807 (1979).

[86] T. M. Jovin, M. Bartholdi, and W. L. C. Vaz, *Ann. N. Y. Acad. Sci.* **366**, 176 (1981).

[87] D. E. Koppel, *Biophys. J.* **28**, 281 (1979).

[88] B. G. Barisas, *Biophys. J.* **29**, 545 (1980).

Several new bleaching geometries have been proposed. One is suitable for simple analysis of diffusion on a sphere.[89] Another is a pattern photobleaching method, which has the advantage of allowing measurement of slow or anisotropic diffusion on cell surfaces.[90]

A potential problem in the interpretation of FPR data is the possibility of photolysis-induced cross-linking of cell surface components. This type of cross-linking has been experimentally demonstrated under excitation conditions quite different from those of FPR measurements.[91,92] Fortunately, recent work has provided evidence that these effects have no detectable influence on measurements of lateral diffusion on cell surfaces.[93,94]

5.5.2. Origin of Constraints on Membrane Protein Mobility

Only recently have the structural bases of these constraints begun to emerge. The erythrocyte provides the most advanced example up to now. Elegant biochemical work has demonstrated that the band 3 protein, the erythrocyte anion transporter, is linked to the spectrin-based erythrocyte cytoskeleton through a protein with molecular weight 215,000 named ankyrin.[95] Using a cell fusion method, it has been shown that displacement of ankyrin from spectrin enhances the lateral mobility of erythrocyte integral membrane proteins.[96] More quantitative studies using FPR have confirmed and extended these studies.[97,98] Measurements of the rotational mobility of band 3 protein, however, suggest that it is free to rotate at a rate limited only by the viscosity of the lipid bilayer in which it is embedded.[99] The apparent contradiction between the observed rotational and lateral mobility can be resolved by supposing that this protein is confined to a small domain in which it is free to rotate.[99] This confinement could be due to steric interactions with the spectrin matrix. A model for constraint of the lateral mobility of membrane proteins by steric interactions with a labile cytoskeletal matrix has been

---

[89] D. E. Koppel, M. P. Sheetz, and M. Schindler, *Biophys. J.* **30**, 3114 (1979).

[90] B. A. Smith, W. R. Clark, and H. M. McConnell, *Proc. Natl. Acad. Sci. U. S. A.* **76**, 5641 (1979).

[91] J. R. Lepock, J. E. Thompson, J. Kruuv, and D. F. H. Wallach, *Biochem. Biophys. Res. Commun.* **85**, 344 (1978).

[92] M. P. Sheetz and D. E. Koppel, *Proc. Natl. Acad. Sci. U. S. A.* **76**, 3314 (1979).

[93] D. E. Wolf, M. Edidin, and P. R. Dragsten, *Proc. Natl. Acad. Sci. U. S. A.* **77**, 2043 (1980).

[94] C.-L. Wey, R. A. Cone, and M. A. Edidin, *Biophys. J.* **33**, 225 (1981).

[95] V. Bennett and P. Stenbuck, *J. Biol. Chem.* **255**, 2540 (1980).

[96] V. Fowler and V. Bennett, *J. Supramol. Struct.* **8**, 215 (1978).

[97] M. P. Sheetz, M. Schindler, and D. E. Koppel, *Nature (London)* **285**, 510 (1980).

[98] D. E. Golan and W. Veatch, *Proc. Natl. Acad. Sci. U. S. A.* **77**, 2537 (1980).

[99] R. J. Cherry, *Biochim. Biophys. Acta* **559**, 289 (1979).

proposed.[100] The relative importance of nonspecific steric interactions and of specific links via ankyrin in restricting mobility of erythrocyte membrane proteins is yet to be determined.

The origin of mobility restrictions in the plasma membrane of nucleated cells is less clear. Very recent work has demonstrated that the apparent detachment of the plasma membrane from the underlying cytoskeleton allows membrane proteins to diffuse much more rapidly than in the normal membranes and at a rate appropriate for the operation of bilayer viscosity as the major constraint of lateral motion.[101] It has also recently been demonstrated that cross-linking a critical number of concanavalin A receptors on B-lymphocytes causes a large reduction in the lateral mobility of surface immunoglobulin on these cells.[102] This reduction can be reversed by treating the cell simultaneously with both colchicine and cytochalasin B. Hence the integrity of both microtubules and microfilaments is required for the effect.[102]

5.5.3. Receptor Mobility

There appears to be a common mechanism for the internalization of cell surface polypeptide hormone–receptor complexes. This involves the binding of the hormone to initially diffusely distributed mobile membrane receptors; the aggregation and immobilization of these receptors, which then cluster mainly over coated pits; and, finally, the pinching off of vesicles containing the hormone–receptor complexes. This process has been seen for insulin,[103] epidermal growth factor (EGF),[104, 105] nerve growth factor (NGF),[106] thyrotropin (TSH),[107] and α_2-macroglobulin.[103] An early phase of the aggregation of EGF–receptor complexes, which may be related to the triggering of cellular responses, has been detected by measurements of phosphorescence anisotropy.[108]

[100] D. E. Koppel, M. P. Sheetz, and M. Schindler, *Proc. Natl. Acad. Sci. U. S. A.* **78**, 3576 (1981).

[101] W. W. Webb, personal communication.

[102] Y. I. Henis and E. L. Elson, *Proc. Natl. Acad. Sci. U. S. A.* **78**, 1072 (1981).

[103] F. R. Maxfield, J. Schlessinger, Y. Shechter, I. Pastan, and M. C. Willingham, *Cell* **14**, 805 (1978).

[104] J. Schlessinger, Y. Shechter, M. C. Willingham, and I. Pastan, *Proc. Natl. Acad. Sci. U. S. A.* **75**, 2135 (1978).

[105] P. Gorden, J.-L. Carpentier, S. Cohen, and L. Orci, *Proc. Natl. Acad. Sci. U.S.A.* **75**, 5025 (1978).

[106] A. Levi, Y. Shechter, E. J. Neufeld, and J. Schlessinger, *Proc. Natl. Acad. Sci. U. S. A.* **77**, 3469 (1980).

[107] D. Tramontano, A. Avivi, F. S. Ambesi-Impiombato, and J. Schlessinger, *Mol. Cell Endocrinol.*, submitted (1982).

[108] R. Zidovetski, Y. Yarden, J. Schlessinger, and T. M. Javin, *Proc. Natl. Acad. Sci. U. S. A.* **78**, 6981 (1981).

Two receptors which stimulate production of adenosine 3′,5′-cyclic monophosphate (cAMP) show quite different mobility properties. The β-adrenergic receptor on liver cells has been shown to be present in a patchy, largely immobile state but to be released and mobilized by the agonist (−)-isoproterenol.[109] The response to the agonist is, however, too slow to be involved in the stimulation of cAMP synthesis. Quite different behavior is seen for the receptor for TSH on thyroid cells. In contrast to β-receptors, TSH receptors are mobile and randomly distributed over the cell membrane.[110] Pretreatment of the thyroid cells with 8-bromo-cAMP induces the clustering of TSH receptors in the absence of TSH. Moreover, cAMP reduces the potency of TSH to induce the production of cAMP in thyroid cells. This suggests that cAMP acts both as a second messenger and as a regulator of the level of TSH on the surface of thyroid cells.[110]

Acknowledgments

We are especially grateful to our colleagues, D. Axelrod, L. Barak, C. Eldridge, P. Fahey, D. Koppel, J. Reidler, W. W. Webb, and D. Wolf, who have worked with us on the various studies from our laboratory discussed in this paper. This work was supported by NIH grants GM 21661 and CA 14454 and by NSF grant DMR 75-04509 (to W. W. Webb).

[109] Y. I. Henis, M. Hekman, E. L. Elson, and E. J. M. Helmreich, *Proc. Natl. Acad. Sci. U. S. A.*, in press (1982).

[110] A. Avivi, D. Tramontano, D. Ambesi, F. S. Ambesi-Impiombato, and J. Schlessinger, *Science* (*Washington, D.C.*), submitted (1982).

6. STRUCTURE DETERMINATION OF BIOLOGICAL MACROMOLECULES USING X-RAY DIFFRACTION ANALYSIS

By Eaton E. Lattman and L. Mario Amzel

6.1. Introduction

X-ray diffraction analysis (XRDA) of single crystals and of fibers has been the source of our knowledge of the atomic structures of over a hundred biological macromolecules, including such important ones as hemoglobin, immunoglobulins, and DNA. By atomic structures we mean models in which the positions of the individual atoms are known to within several tenths of an angstrom. Such models provide the basis for our understanding of the mechanism of enzyme catalysis, ligand binding, and many other macromolecular functions. In addition, they serve to direct and organize other types of studies of the molecules involved. Recently, crystals of virus particles weighing several million daltons have been analyzed in nearly atomic detail, while correspondingly complex fiber structures have also been studied. No clear limits to size and complexity are yet in view. In this part we present briefly the principles of single-crystal and fiber XRDA, describe data collection technology, and then present some case studies that illuminate the versatility and limitations of the methods. For more detailed material on this subject see Blundell and Johnson[1] and Holmes and Blow.[2] We touch only briefly upon the closely related technique of small (~ 100 atoms)-molecule crystallography, since it is better known and covered in many textbooks.[3,4] Small-angle x-ray diffraction analysis, which is concerned with the shapes and interactions of molecules rather than with their atomic structures, is the subject of Part 8 of this volume.[5]

[1] T. L. Blundell and L. N. Johnson, "Protein Crystallography." Academic Press, New York, 1976.

[2] K. C. Holmes and D. M. Blow, "The Use of X-Ray Diffraction in the Study of Protein and Nucleic Acid Structure." Wiley (Interscience), New York, 1966.

[3] G. H. Stout and L. H. Jensen, "X-Ray Structure Determination: A Practical Guide." Macmillan, New York, 1968.

[4] H. Lipson and W. Cochran, "The Crystalline State. Vol. III, The Determination of Crystal Structures," 3rd ed. Bell, London, 1966.

[5] P. B. Moore, this volume, Part 8.

METHODS OF EXPERIMENTAL PHYSICS, VOL. 20

In some ways XRDA might better be termed lensless x-ray microscopy. In a typical microscopy experiment a thin, usually aperiodic, specimen is illuminated by a collimated beam of radiation, which is diffracted by fluctuations in scattering density within the specimen. A lens placed behind the specimen focuses the diffracted radiation into an image. An XRD experiment differs in several ways from this prototype. High-quality x-ray lenses do not exist. In the absence of a lens the distribution of radiation scattered far from the specimen is termed the Fraunhofer diffraction pattern. This pattern becomes the only observable in an XRD experiment. The goal of the method (XRDA) is the conversion of this pattern into an image. The process is an intricate one, with the computer playing the role of the lens. The method also differs from conventional microscopy in the characteristics of the specimen. To preserve the analogy, an XRD specimen should perhaps consist of a single molecule held delicately in front of the beam. The contrast in such an experiment would be vanishingly small. The situation can be improved by using periodic specimens such as crystals or fibers. These increase contrast both by increasing the number of molecules in the beam and, as we discuss below, by concentrating the diffracted radiation into certain defined directions.

We show below that, given the phase and amplitude of the scattered radiation, reconstruction of the image is, in principle, a simple process. However, available x-ray detectors measure the amplitude but not the phase of the radiation. The loss of phase information, sometimes called "the phase problem," is the central difficulty in the practice of XRDA. Much of this paper is devoted to the discussion of techniques for dealing with this problem.

6.2. Diffraction by a General Object

We begin by defining some of the notation used in this paper. Position vectors in direct or diffraction space will be denoted by lower-case, boldface letters, such as $\mathbf{s}$ or $\mathbf{x}$. Matrices will be written with upper-case, boldface letters, such as $\mathbf{A}$. Diffracted radiation can be described by a two-dimensional vector in the complex plane: the modulus and azimuth of the vector represent the amplitude and phase of the radiation. We frequently write this vector, the structure factor, as $\mathbf{F}$. We can break $\mathbf{F}$ into components in two ways:

$$\mathbf{F} = |\mathbf{F}|\exp(i\alpha) = F \exp(i\alpha), \qquad \mathbf{F} = A + iB.$$

The use of F for $|\mathbf{F}|$ is conventional in the protein crystallographic literature.

The most conventional approach to the theory of XRD in crystals is through Bragg's law. It states that a beam of x rays incident upon a stack of parallel, equally spaced planes appears to be reflected from these planes, as a

mirror reflects light, when

$$n\lambda = 2d \sin \theta. \qquad (6.2.1)$$

Here n is any positive integer, d is the spacing between planes, and θ, the Bragg angle, is the angle between the incoming beam and the planes. This is illustrated in Fig. 1a. When the angle of incidence differs from θ no Bragg

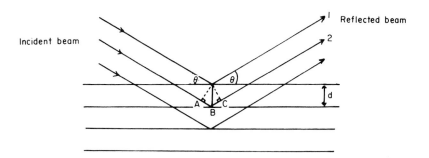

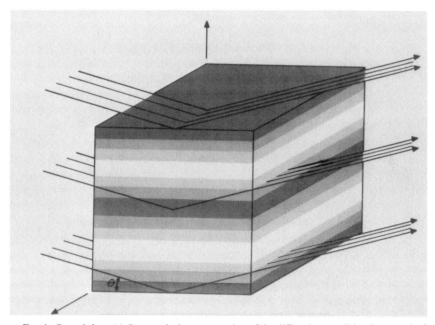

FIG. 1. Bragg's law. (a) Geometrical representation of the diffraction condition for a stack of equally spaced planes. Rays "reflected" from adjacent planes have a path difference $\overline{AB} + \overline{BC} = 2d \sin \theta$. For constructive interference this distance must be an integral number of wavelengths; thus $n\lambda = 2d \sin \theta$. (From Blundell and Johnson.[1]) (b) Diffraction by a sinusoidal density fluctuation of wavelength $1/s$. For this object only a path difference of λ produces constructive interference; thus $\lambda = (2/s) \sin \theta$. This artist has represented the continuous density fluctuation by four shades of grey.

reflection is observed. This approach can be extended to analyze a general object if one considers diffraction from a spatial, sinusoidal density fluctuation $\rho(\mathbf{x}) = \rho_0 \exp(-2\pi i \mathbf{s}_0 \cdot \mathbf{x})$. The situation is diagrammed in Fig. 1b. This density $\rho(\mathbf{x})$ has certain key features in common with the stack of Bragg planes: both have constant value on planes normal to a particular axis, and both are periodic along this axis. It is perhaps not surprising, then, that diffraction from the density fluctuation occurs when a slight variant of Bragg's law is satisfied:

$$\lambda = (2/|\mathbf{s}_0|) \sin \theta. \tag{6.2.2}$$

The factor $1/|\mathbf{s}_0|$ is the period of $\exp(-2\pi i \mathbf{s}_0 \cdot \mathbf{x})$ and is analogous to the spacing d in the conventional Braggs's law. Note that only one angle of incidence θ is permitted for this function. Equation (6.2.2) is not difficult to obtain, and we omit a derivation for lack of space.

It is also clear that $F(\mathbf{s}_0)$, the amplitude of the diffracted wave, is proportional to the amplitude of the spatial fluctuation. We can then write

$$F(\mathbf{s}_0) = \rho_0,$$

and therefore

$$\mathbf{F}(\mathbf{s}_0) = \rho_0 \exp(i\alpha).$$

If the origin of the spatial wave is shifted to $\mathbf{x}_0$, we have

$$\rho(x) = \rho_0 \exp[-2\pi i \mathbf{s}_0 \cdot (\mathbf{x} - \mathbf{x}_0)], \tag{6.2.3}$$

or

$$\rho(x) = \rho_0 \exp(2\pi i \mathbf{s}_0 \cdot \mathbf{x}_0) \exp(-2\pi i \mathbf{s}_0 \cdot \mathbf{x}),$$

and therefore

$$\mathbf{F}(\mathbf{s}_0) = \rho_0 \exp(2\pi i \mathbf{s}_0 \cdot \mathbf{x}_0) \exp(i\alpha). \tag{6.2.4}$$

Thus translation of the spatial fluctuation by $\mathbf{x}_0$ produces a phase shift of $2\pi \mathbf{x}_0 \mathbf{s}_0$ in the diffracted radiation. Experimentally, only differences of phase can be measured, and so α in the above equations is generally defined to be zero.

From the diffraction by an isolated spatial wave one can thus determine the period (through θ), the amplitude, and the origin (i.e., phase) of that wave.

It is well known that any object can be represented through a Fourier integral as a combination of such spatial waves (Fourier components), so that

$$\rho(\mathbf{x}) = \int\!\!\!\int\!\!\!\int_{-\infty}^{\infty} \mathbf{G}(\mathbf{s}) \exp(-2\pi i \mathbf{s} \cdot \mathbf{x}) \, dV. \tag{6.2.5}$$

The amplitude and phase of **G(s)** indicate the weight and position of the Fourier component. Clearly, **G** and **F** are identical, apart from a constant of proportionality. We can thus identify the diffraction pattern of an object with its Fourier transform and write

$$\rho(\mathbf{x}) = \int\!\!\!\int\!\!\!\int_{-\infty}^{\infty} \mathbf{F}(\mathbf{s})\exp(-2\pi i \mathbf{s} \cdot \mathbf{x})\, dV. \qquad (6.2.6)$$

The Fourier inversion theorem can be applied to Eq. (6.2.6) to give

$$\mathbf{F}(\mathbf{s}) = \int\!\!\!\int\!\!\!\int_{-\infty}^{\infty} \rho(\mathbf{x})\exp(2\pi i \mathbf{x} \cdot \mathbf{s})\, dV, \qquad (6.2.7)$$

a result which is used extensively below. The textbook by Goodman[6] provides an excellent introduction to Fourier theory and diffraction.

For any given orientation of the specimen only certain of the Fourier components satisfy the condition in Equation (6.2.2). To measure the full, three-dimensional diffraction pattern in ordered specimens it is necessary to rotate the specimen systematically so that diffraction from all Fourier components of interest occurs and is recorded. Strategies and instrumentation to achieve this goal are discussed in Section 6.4.3. We describe now a simple geometrical construction developed by Ewald which tells which Fourier components are in the diffracting condition for a given orientation of the object. Figure 2 shows the vectors **OA, OB**, and **s** drawn in a sphere of radius $1/\lambda$ (Ewald's sphere). **OA** denotes the direction of the incident beam, and is drawn from the center of the sphere to a point **A** which is taken as the origin of all vectors **s**, the wave vectors of the various Fourier components of the object. The vectors **s** are defined with respect to the object's coordinate system. If the object is rotated they are, perforce, rotated in the same way. Consider any **s** which terminates on the surface of the sphere, and draw from the center of the sphere to its terminus the vector **OB**. By comparison with Fig. 1 it is clear that the vector triangle **OA-s-OB** is equivalent to the one formed by the incoming and reflected rays, and that whenever the terminus of a vector **s** is on the surface of Ewald's sphere its corresponding Fourier component is in the diffracting condition. Diffraction is observed along the direction of **OB**. For any orientation of the specimen diffraction along all vectors **OB** can be recorded by placing a film (or any other x-ray detector) at some distance from the object. The signal recorded at any point represents the diffraction from one Fourier component, denoted **s**.

[6] J. W. Goodman, "Introduction to Fourier Optics." McGraw-Hill, New York, 1968.

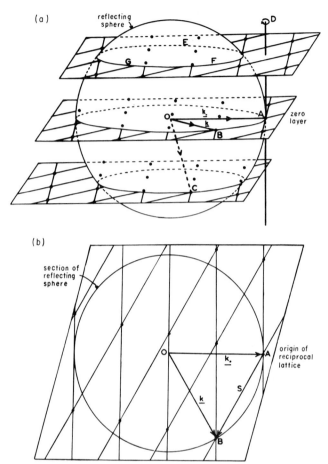

FIG. 2. The Ewald construction. (a) A sphere of radius $1/\lambda$ is drawn so as to pass through the origin of the reciprocal lattice A. If lattice point B lies on the sphere there will be a diffracted beam in the direction of the line joining the center of the sphere to this point. (b) A section through the same drawing in the plane of the zero-level section of reciprocal space (see also Fig. 19). (Holmes and Blow.[2] Copyright © 1966. John Wiley & Sons, Inc. Reprinted by permission of John Wiley & Sons, Inc.)

We have thus far dealt with diffraction from completely general specimens, and have kept the discussion general as well (for a more complete discussion of the Fourier approach to diffraction see Goodman[6]). We next discuss crystals and fibers, periodic specimens which produce simplified diffraction patterns (see James[7]). Before we do so, however, two additional mathematical

[7] R. W. James, "The Crystalline State. Vol. II, The Optical Principles of the Diffraction of X-Rays." Bell, London, 1957.

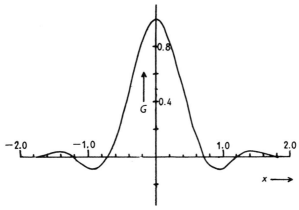

FIG 3. Image of a point after Fourier reconstruction with wavelength cutoff at $1/s = 1$ Å; x is the distance in angstroms from the point. (Reprinted from Rossmann and Blow[38] with permission.)

devices must be introduced: the application of the convolution theorem to diffraction[6] and the projection theorem.

If F and f are a Fourier transform pair and so are G and g, the convolution theorem states that

$$F * G = \mathscr{F}(fg) \tag{6.2.8a}$$

and that

$$FG = \mathscr{F}(f * g), \tag{6.2.8b}$$

where $\mathscr{F}$ represents the Fourier transform and the asterisk represents convolution, which is defined by

$$f(x) * g(x) = \int f(u)g(x - u)\, du. \tag{6.2.9}$$

This theorem has wide application. As an illustration we consider the effects of the nonzero wavelength of x rays on our ability to form an accurate image. In Fig. 2 it is clear that no vector **s** of magnitude greater than $2/\lambda$ can touch Ewald's sphere; thus all diffraction from such components must remain unobserved. Formally this means that the diffraction pattern of the specimen is multiplied by a function with value one within a sphere of radius $2/\lambda$ and zero outside this sphere. The observed image is therefore the convolution of the true image with the transform of the solid sphere. This transform is shown in Fig. 3[8] and represents the image of a point object after wavelength cutoff.

[8] M. G. Rossmann, ed., "The Molecular Replacement Method." Gordon & Breach, New York, 1972.

In protein crystallography wavelength cutoff is not usually the factor limiting the amount of diffraction data available; sometimes it is expedient for data collection to be limited by the investigator. On other occasions the intensity of the diffraction pattern falls effectively to zero at values of s —or simply s—which are smaller than $2/\lambda$. This gives rise to the crystallographic definition of resolution in an image. Resolution is given by $1/s_{max}$, where s_{max} is the radius in reciprocal space beyond which no data are observed. Thus $1/s_{max}$ is the wavelength of the highest-frequency component in the synthesized electron density function.

We now discuss the projection theorem, which states that the image synthesized using only a central section from the full diffraction pattern is the projection of the full, three-dimensional image down the normal to the central section. This theorem is of great practical importance, since it is frequently very simple to record diffraction from such central sections, especially for crystals, where the precession camera, (Section 6.4.3) automatically provides photographs of particular planes in the diffraction pattern. To derive this, we rewrite Eq. (6.2.7) using the components of $\mathbf{x}$ and $\mathbf{s}$:

$$F(s, t, u) = \iiint \rho(x, y, z)\exp[2\pi i(sx + ty + uz)]\,dx\,dy\,dz.$$

Setting u equal to zero and rearranging, we have

$$F(s, t, 0) = \iint \exp[2\pi i(sx + ty)] \int \rho(x, y, z)\,dx\,dy\,dz.$$

If $\sigma(x, y)$ represents the integral over z (that is, the projection down z), then

$$F(s, t, 0) = \iint \sigma(x, y)\exp[2\pi i(sx + ty)]\,dx\,dy. \tag{6.2.10}$$

Thus a two-dimensional Fourier transform of the z-axis projection of the electron density function is the $u = 0$ section through the three-dimensional diffraction pattern. This result is independent of the projection axis chosen, and has a closely corresponding form for crystals. Many crystals have projections which are centrosymmetric—i.e., $\sigma(x, y) = \sigma(-x, -y)$. It is easy to show that such projections have structure factors of phase either $0°$ or $180°$, instead of any value between $0°$ and $360°$. They are thus signed, real numbers. Phases for such reflections are particularly easy to determine, so that centrosymmetric projections are of great interest experimentally.

Application of the Fourier inversion theorem to Eq. (6.2.10) demonstrates that that projection $\sigma(x, y)$ is the Fourier transform of the central section $F(s, t, 0)$, which is the form most commonly used.

6.3. Crystallography

6.3.1. Diffraction by Crystals

A lattice is a regular grid of points in three dimensions. Formally, it is the set of points at the termini of the vectors

$$\mathbf{H} = m\mathbf{a} + n\mathbf{b} + p\mathbf{c},$$

where m, n, and p are any integers and $\mathbf{a}$, $\mathbf{b}$, and $\mathbf{c}$ are any noncoplanar vectors. Each lattice point is in the same environment as all others and can be imagined to lie at a vertex of a parallelepiped called the unit cell, which is defined by the vectors $\mathbf{a}$, $\mathbf{b}$, and $\mathbf{c}$, the unit cell vectors. Lattices are divided into classes depending upon special relations among the unit cell vectors, but we shall only consider specific examples as they arise. In a crystal each unit cell of the lattice is occupied by a number (usually small) of molecules which are in most cases related to one another by simple operations such as rational rotations, reflections, and translations. The crystal can be regarded as the convolution of a single unit cell's worth of molecules, ρ_M, with a set of δ functions at the lattice points. It is clear that the convolution operation here amounts to placing a copy of ρ_M at each lattice point in the crystal. Using Eq. (6.2.8b) one can show that the Fourier transform or diffraction pattern of a crystal is the product of $\mathbf{F}_M$, the Fourier transform of ρ_M, and $\mathbf{h}$, the Fourier transform of $\mathbf{H}$. It is easy to show that $\mathbf{h}$ is itself a lattice (called the reciprocal lattice) with unit cell vectors

$$\mathbf{a}^* = \mathbf{b} \times \mathbf{c}/V, \qquad \mathbf{b}^* = \mathbf{c} \times \mathbf{a}/V, \qquad \mathbf{c}^* = \mathbf{a} \times \mathbf{b}/V, \qquad (6.3.1)$$

where V is the volume of the unit cell and $\times$ denotes the cross (vector) product.

It is clear from Eqs. (6.3.1) that $\mathbf{a} \cdot \mathbf{a}^* = \mathbf{b} \cdot \mathbf{b}^* = \mathbf{c} \cdot \mathbf{c}^* = 1$. These relations are the source of the name reciprocal lattice. The reciprocal lattice is an important tool in dealing with the geometry of Bragg diffraction. Inspection of Fig. 7 (which is discussed more fully below) reveals that the Bragg reflections lie neatly arranged at the points of a two-dimensional net. This net, in fact, is a section from the three-dimensional lattice defined by $\mathbf{h}$. The reciprocal lattice is thus intimately related to the geometry of diffraction, and some discussion of its characteristics is essential for understanding x-ray photographs and data collection in general. Also, it is a unification that a given $\mathbf{h}$ will serve to label a reciprocal lattice point, the corresponding reflection, and the Fourier component giving rise to the reflection.

As we stated above, the diffraction pattern of a crystal is given by $\mathbf{F}_M(\mathbf{h})$, which is zero except at points of the reciprocal lattice.

The vector $\mathbf{h}$ can be written as

$$\mathbf{h} = h'\hat{a}^* + k'\hat{b}^* + l'\hat{c}^* \tag{6.3.2a}$$

or

$$\mathbf{h} = h\mathbf{a}^* + k\mathbf{b}^* + l\mathbf{c}^*, \tag{6.3.2b}$$

where a caret denotes a vector of unit length. In the first case h', k', and l' are measured in reciprocal angstroms; in the second h, k, and l are in units of the reciprocal lattice translations and are always integers. The latter notation is universal in crystallography, where the triplet hkl is often called the Miller indices.

Similarly, the position vector $\mathbf{x}$ can be written as

$$\mathbf{x} = x'\hat{a} + y'\hat{b} + z'\hat{c} \tag{6.3.3a}$$

or

$$x = x\mathbf{a} + y\mathbf{b} + z\mathbf{c}. \tag{6.3.3b}$$

Of course, the inner product $\mathbf{h} \cdot \mathbf{x}$ is independent of the choice of formalism. Using Miller indices, Eq. (6.3.2b), and fractional coordinates, Eq. (6.3.3b), we have

$$\mathbf{h} \cdot \mathbf{x} = hx\mathbf{a} \cdot \mathbf{a}^* + ky\mathbf{b} \cdot \mathbf{b}^* + lz\mathbf{c} \cdot \mathbf{c}^* + \text{terms in } \mathbf{a} \cdot \mathbf{b}^*, \text{ etc.}$$

From the cross-product relations in Eq. (6.3.1) all terms containing $\mathbf{a} \cdot \mathbf{b}^*$, etc., must vanish. Recalling that

$$\mathbf{a} \cdot \mathbf{a}^* = \mathbf{b} \cdot \mathbf{b}^* = \mathbf{c} \cdot \mathbf{c}^* = 1,$$

we have

$$\mathbf{h} \cdot \mathbf{x} = hx + ky + lz, \tag{6.3.4}$$

which is the usual inner product form.

This rather abstract derivation has shown us that only certain well-defined Fourier components exist in a crystal, and so diffraction can only be observed when one of these satisfies Eq. (6.3.2). It is important to realize that the same Fourier components would have emerged if we had treated the crystal conventionally as an object periodic in three dimensions, and Fourier-analyzed it accordingly. Thus the hkl component repeats h times along the a axis, k times along the b axis, and l times along c. If one rewrites (6.2.5) to reflect the periodic nature of the crystal, one finds

$$\rho(\mathbf{x}) = \frac{1}{V} \sum_{\mathbf{h}} \mathbf{F}(\mathbf{h}) \exp(-2\pi i \mathbf{h} \cdot \mathbf{x}). \tag{6.3.5}$$

This is the equation used to calculate electron density function images. Again, **h** and **x** are expressed in integer and fractional coordinates by convention. One can also apply the Fourier inversion theorem to (6.3.5):

$$\mathbf{F(h)} = \iiint\limits_{\text{unit cell}} \rho(\mathbf{x}) \exp(2\pi i \mathbf{h} \cdot \mathbf{x}) \, dV, \qquad (6.3.6)$$

giving the structure factors as a function of the electron density.

Although symmetry is one of the most important fields of study in crystallography, (refs. 9, Vol. I, and 10), we pass over this entire topic by noting that symmetry operations which superimpose the whole crystal on itself do exist, and that for any crystal the set of such operations forms a group when the product of two operators is defined as the result of their successive application. These groups are termed space groups, and the symmetry operators are the composition of the translational symmetry of the lattice with operations that relate the molecules within one unit cell. Again, we do not attempt a general discussion, but deal with particular examples as they arise. It is worth noting that almost all biological molecules are handed (i.e., are non-superimposable upon their mirror image.) This precludes the existence of mirror planes or other inverting operations in their crystals.

Equation (6.3.6) relates the diffraction pattern to the continuous electron density in the crystal. Crystals are made of atoms, and we interpret their electron density functions in terms of atoms. Formally, one can write that

$$\rho_M(\mathbf{x}) = \sum_j \rho_j(\mathbf{x} - \mathbf{x}_j), \qquad (6.3.7)$$

where ρ_j is the electron density function of the jth atom, and x_j is the coordinate of its center. It is therefore convenient to have a formula, called the structure factor equation, which expresses the diffraction pattern in terms of atomic positions and types. The diffraction pattern of one unit cell is the sum of the diffraction patterns of all the atoms in the cell. Diffraction patterns for each atom type, often called atomic scattering factors, have been tabulated (Ref. 9, Vol. IV), and are usually denoted $f(s)$. Since atoms in this model are spherically symmetric, $f(s)$ is also. Now $f(s)$ represents diffraction by an atom situated at the origin. If the atom is shifted to point $\mathbf{x}_0$, its diffraction pattern is phase shifted by the factor $\exp(2\pi i \mathbf{h} \cdot \mathbf{x}_0)$. The sum of the diffraction by all atoms is therefore

$$\mathbf{F(h)} = \sum_j f_j(\mathbf{h}) \exp(2\pi i \mathbf{h} \cdot \mathbf{x}_j), \qquad (6.3.8a)$$

[9] "International Tables for X-Ray Crystallography." Kynoch Press, Birmingham, England, 1952.

[10] M. J. Buerger, "X-Ray Crystallography." Wiley, New York, 1942.

where x_j is the position of atom j. Equation (6.3.8a) is based on diffraction by atoms at rest. It is well known that the atoms in a crystal are undergoing thermal motion. The contribution to the structure factors is different from resting and from vibrating atoms. To take into account the effects of atomic thermal motion (James,[7] Chapter 5) each $f_j(s)$ must be multiplied by the factor $\exp(-B_j s^2/4)$, where the value of B_j, usually determined experimentally, can be related to the root-mean-square excursion of the jth atom. The quantity B is usually called the temperature or Debye–Waller factor. The more general form for this factor, which allows for anisotropic atomic motion, is rarely used in protein work.

The final form of the structure factor equation (6.3.8a) is

$$\mathbf{F(h)} = \sum_j f_j \exp \frac{-B_j s^2}{4} \exp(2\pi i \mathbf{h} \cdot \mathbf{x}_j). \tag{6.3.8b}$$

This equation enables us to calculate the diffraction pattern from a preliminary atomic model, and forms the basis of the process for refining such models in order to minimize differences between observed and calculated structure factor magnitudes.

6.3.2. The Patterson Function

Electron density maps can be calculated from the structure factors $\mathbf{F(h)}$ using the Fourier synthesis described by Eq. (6.3.5) when the phase of $\mathbf{F(h)}$ is known.

Normally the phase of $\mathbf{F(h)}$ is not known. However, a synthesis using the observed $F^2(\mathbf{h})$—which have a phase of zero—as coefficients can always be calculated. This synthesis and its significance were first discussed by A. L. Patterson[11] in 1934, and it is commonly named after him. Properly interpreted, it can provide clues leading to a structure determination.

We show below that the Patterson function is the autocorrelation of the electron density function and represents a complete, alternative description of the observations. (Some texts say, loosely, that the Patterson function is the self-convolution of the electron density function, but in fact it is the convolution of $\rho(\mathbf{x})$ with $\rho(-\mathbf{x})$. In this nomenclature one deconvolutes the Patterson function to recover the electron density.) Until about 1960 the Patterson function was the dominant tool for solving the phase problem in crystallography. As we shall see, the structures solved typically had a limited number of atoms. In protein crystallography we use the Patterson function for a problem of similar complexity, the location of heavy atoms that have been bound to the protein molecules. These locations are essential data for the procedure that provides phases for the protein diffraction pattern.

---

[11] A. L. Patterson, *Phys. Rev.* **46**, 372 (1934).

Use of the Patterson function depends upon interpreting it in terms of the autocorrelation of a function made of atoms or, ideally, points. There is an extensive literature discussing this interpretation, which depends heavily upon symmetry, and upon the concept of vectors running between atoms. To be frank, the topic can be tedious, and we cover it as briefly as possible. In crystallographic nomenclature the Patterson function $P(\mathbf{u})$ is defined as

$$P(\mathbf{u}) = \frac{1}{V} \sum_{\mathbf{h}} F^2(\mathbf{h}) \cos(2\pi \mathbf{h} \cdot \mathbf{u}). \tag{6.3.9}$$

The function $P(\mathbf{u})$ is centrosymmetric $[P(\mathbf{u}) = P(-\mathbf{u})]$. The interpretation of this synthesis requires some further derivations. Substituting

$$\mathbf{F}(\mathbf{h})\mathbf{F}^*(\mathbf{h}) = F^2(\mathbf{h})$$

in expression (6.3.9), where $\mathbf{F}^*(\mathbf{h})$ is the complex conjugate of $\mathbf{F}(\mathbf{h})$, we obtain

$$P(\mathbf{u}) = \frac{1}{V} \sum_{\mathbf{h}} [\mathbf{F}(\mathbf{h})\mathbf{F}^*(\mathbf{h})] \cos(2\pi \mathbf{h} \cdot \mathbf{u}). \tag{6.3.10}$$

With expression (6.3.5) for the electron density we can use the convolution theorem [Eqs. (6.2.8b) and (6.2.9)] and the fact that $\rho(\mathbf{x})$ is real to obtain

$$P(\mathbf{u}) = V \iiint \rho(\mathbf{x})\rho(\mathbf{x} + \mathbf{u}) \, dV. \tag{6.3.11}$$

This expression indicates that the Patterson function is the autocorrelation of the electron density of the crystal. The expression for the electron density based on atomic electron densities can be used to further analyze Eq. (6.3.11). As shown before [Eq. (6.3.7)], $\rho(\mathbf{x})$ can be expressed as

$$\rho(\mathbf{x}) = \sum_j \rho_j(\mathbf{x} - \mathbf{x}_j),$$

where $\mathbf{x}_j$ is the coordinate of the jth atom and N is the total number of atoms in the cell. Substituting in (6.3.11) we obtain

$$P(\mathbf{u}) = V \iiint \left(\sum_j \rho_j(\mathbf{x} - \mathbf{x}_j) \right) \left(\sum_k \rho_k(\mathbf{x} - \mathbf{x}_k + \mathbf{u}) \right) dV,$$

which can be rearranged to yield

$$P(\mathbf{u}) = \sum_{j=1}^{N} \sum_{k=1}^{N} V \iiint \rho_j(\mathbf{x} - \mathbf{x}_j)\rho_k(\mathbf{x} - \mathbf{x}_k + \mathbf{u}) \, dV. \tag{6.3.12}$$

The change of variables $\mathbf{x}' = \mathbf{x} - \mathbf{x}_j$ simplifies the above to

$$P(\mathbf{u}) = \sum_{j=1}^{N} \sum_{k=1}^{N} V \iiint \rho_j(\mathbf{x}')\rho_k(\mathbf{x}' + \mathbf{u} - \mathbf{u}_{jk})\, dV, \qquad (6.3.13)$$

where $\mathbf{u}_{jk} = \mathbf{x}_j - \mathbf{x}_k$.
Define the integral in the above expression as

$$P_{jk}(\mathbf{u}') = V \iiint \rho_j(\mathbf{x}')\rho_k(\mathbf{x}' + \mathbf{u}')\, dV, \qquad (6.3.14)$$

where $\mathbf{u}' = \mathbf{u} - \mathbf{u}_{jk}$. P_{jk} is the convolution of the electron density profiles of the ith and jth atoms. Substituting (6.3.14) in (6.3.13) we obtain

$$P(\mathbf{u}) = \sum_{j=1}^{N} \sum_{k=1}^{N} P_{jk}(\mathbf{u} - \mathbf{u}_{jk}). \qquad (6.3.15)$$

This expression describes the most useful feature of the Patterson function: its maxima occur at points $\mathbf{u}_{jk}$ which correspond to the interatomic vectors of the structure, $\mathbf{x}_j - \mathbf{x}_k$, translated to the Patterson origin. The peaks P_{jk} are the convolution of the individual atomic profiles. There are $N^2 - N$ peaks for the vectors $\mathbf{u}_{jk}$ with $k \neq j$. In addition, this synthesis has a large peak at the origin corresponding to the interatomic vectors $\mathbf{u}_{jj}$ that contains the sum of the self-convolutions of the electron density profiles of individual atoms. If all atoms in the structure have similar radial electron density profiles, the height of the peaks is approximately proportional to the product of the atomic numbers of the relevant atoms. If, for example, the atomic electron density profiles of the kth and jth atoms are Gaussians of variances σ_j^2 and σ_k^2, the peak centered at $\mathbf{u}_{jk}$ will also be a Gaussian of variance $\sigma_{jk}^2 = \sigma_j^2 + \sigma_k^2$.

The peaks in the Patterson function can be "sharpened" if instead of the observed $F^2(\mathbf{h})$ modified values are used which correspond to the same structure but composed of point atoms. This can be done by dividing the $F^2(\mathbf{h})$ by the square of the Fourier transform of the average atomic electron density profile. For a crystal structure containing a few atoms in the unit cell the Patterson function can be "solved" (deconvoluted). Sophisticated trial and error methods are used to obtain the coordinates of the atoms in the unit cell from the set of interatomic vectors. In addition, the function can be partially deconvoluted in structures which contain a few heavy atoms that dominate the diffraction pattern. A very detailed analysis of the Patterson function and of different deconvolution methods can be found in Buerger[12] (and references therein).

[12] M. J. Buerger, "Vector Space and Its Application in Crystal-structure Investigation." Wiley, New York, 1959.

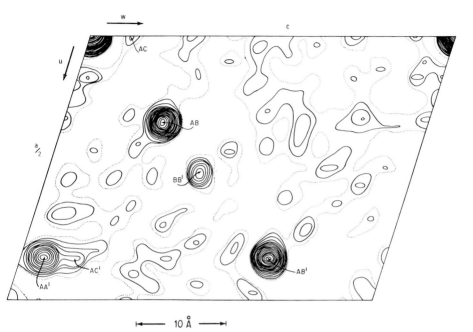

FIG. 4. Projection along the b axis of the difference Patterson function of a gold heavy-atom derivative (AuCl$_4^-$) of glycera hemoglobin crystals. The contour lines are drawn at equal but arbitrary intervals. The zero contour is dotted, and the negative contours are omitted. The interpretation of this Patterson function involves two major (A and B) and one minor (C) heavy-atom sites. The crystal in this projection contains twofold axes of symmetry perpendicular to the plane of the drawing. For the major heavy-atom sites the map shows the peaks corresponding to the interatomic vectors between a heavy-atom site and its symmetry mate (AA' and BB') and between the two heavy-atom sites (AB and AB'). The vectors between site A and the minor site C (AC and AC') are also present. This assignment can be used to obtain the coordinates of the heavy-atom sites. (Courtesy of Dr. Eduardo Padlan.)

The main use of the Patterson function in protein crystallography is for the determination of the positions of heavy atoms (Section 6.4.4). For this purpose, in many cases, projections of the Patterson function are used [Eq. (6.2.10)]. These functions can be calculated using data from one section of the reciprocal lattice, for example,

$$P(u, v) = \frac{1}{A} \sum_h \sum_k F^2(h, k, 0) \cos[2\pi(hx + ky)]. \qquad (6.3.16)$$

Data for such a synthesis can be collected very easily. A single precession photograph or a short diffractometer run (Section 6.4.3) can be used for this purpose. Heavy-atom derivatives are usually screened and the heavy atoms located using these projections (Section 6.4.4). The analysis is systematic trial and error, making careful use of the symmetry of the crystal. In one version, for example, a computer search assumes in turn that each grid point in the unit cell is a possible heavy-atom location. The symmetry-equivalent points are calculated, and the putative interatomic vectors generated. The computer then checks to see if these vectors correspond to peaks in the Patterson function. If so, the grid point is flagged as a possible site. Figure 4 shows an example of how the positions of a heavy-atom derivative can be located using a projection of the Patterson function.

6.4. Protein Crystallography

6.4.1. Introduction

The oxygen transport protein hemoglobin was crystallized in the latter part of the nineteenth century,[13] and the enzyme (protein with catalytic activity) urease was crystallized by Sumner in 1926.[14] Since then a great variety of proteins have been crystallized. The crystals often have a very beautiful external morphology, with well-developed faces and edges. An example is shown in Fig. 5. In 1934 Bernal and Crowfoot[15] showed that these crystals produce a rich and complex x-ray diffraction pattern. They realized that protein crystals contain a large percentage of mother liquor filling the space between molecules, and that crystals become disordered as they dry out. Their photographs were obtained by enclosing the crystals in glass capillaries. For the first time there was a tantalizing possibility that the molecular structure of the proteins within the crystal might be determined. The problems to be overcome were formidable. Most challenging of all was the difficulty in measuring the phase of all the protein structure factors. The synthesis of the electron density function [Eq. (6.3.5)] requires the amplitude and phase of each structure factor. The intensity of the individual Bragg reflections is proportional to the squared modulus of the structure factors, but all phase information is lost. In the case of small (nowadays about 100 atoms) molecules this information is recovered by a number of procedures based on the Patterson function (Section 6.3.2) or upon various algebraic relations between the structure factors. The latter are collectively called direct methods (a selection of papers on this topic appears in Ahmed[16]). All of these procedures fail when extrapolated to protein crystals, which contain thousands of atoms per unit cell. It was not until the middle of the

[13] K. E. Reichert, *Muellers Arch. Anat., Physiol. Wiss. Med.* p. 198 (1849).

[14] J. B. Summer, *J. Biol. Chem.* **69**, 435 (1926).

[15] J. D. Bernal and D. C. Crawfoot, *Nature (London)* **133**, 794 (1934).

[16] F. R. Ahmed, ed., "Crystallographic Computing." Munksgaard, Copenhagen, 1970.

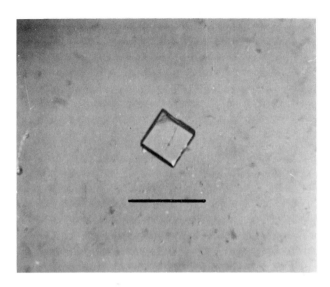

FIG. 5. Photograph of a crystal of the F1 sector of mitochondrial ATPase. The bar represents 1 mm in the scale of the micrograph.

1950s that the laboratory of Max Perutz[17] introduced the method of multiple isomorphous replacement (Section 6.4.4), which has proved astonishingly successful in the intervening years.

Many other difficulties also obtruded. The number of Bragg reflections within a reciprocal space sphere of fixed radius is proportional to the volume of the unit cell. Thus protein crystals give rise to hundreds or thousands as many Bragg reflections as small-molecule crystals. Automated procedures to measure these reflections awaited development (Section 6.4.3). Further, the diffraction patterns from even the best protein crystals fade out at much lower resolution than the patterns from small molecules (Chapter 6.2)—typically in the range 1.5–3.0 Å. The images synthesized from such patterns are deficient in detail, and individual atoms are not visible as they are in the small-molecule case. To provide accurate models of protein molecules these images have to be interpreted, usually in light of a known sequence of amino acids within the protein (Section 6.4.5). Difficulties associated with the interpretation of electron density maps have, to a large extent, been overcome by a combination of manual and computer-based model fitting and refinement techniques. These are also discussed below (Section 6.4.5).

[17] D. W. Green, V. M. Ingram, and M. F. Perutz, *Proc. R. Soc. London, Ser. A* **225**, 287 (1954).

TABLE I.   Steps in a Protein Crystal Structure Determination

1. Crystallize protein.
2. Make heavy-atom derivatives.
3. Measure structure factor amplitudes for native and heavy-atom derivatives.
4. Determine position of heavy atoms.
5. Using above information calculate phases for structure factors.
6. Calculate an electron density synthesis of the crystal.
7. Fit a molecular model to this synthesis.[a]
8. Calculate structure factors F_c and ϕ_c from the model.
9. Make a new synthesis using calculated phases ϕ_c but observed amplitudes F_o—i.e., combine model and data.

[a] The repetition of steps 7, 8, and 9 illustrates Fourier refinement, one of several methods of improving the initial model. The cycle is abandoned when improvement ceases.

Table I is a flowchart of the major steps in the determination of a protein crystal structure. In what follows we discuss each of these steps in more detail. We then deal with examples of protein structure determination.

6.4.2. Crystallization

The crystallization of proteins retains large elements of black magic, and the literature is filled with examples of apparently random procedures that allow the crystallization of previously resistant molecules. Certain conditions, however, are almost universally met. One begins with a concentrated protein solution often in the range 1 % by weight of protein. To this is added either gradually or suddenly a precipitant, such as salt or organic solvent, which causes the protein to come out of solution. In most cases such precipitates are amorphous or microcrystalline, but occasionally single crystals result. Variables explored are precipitant type and concentration, pH, temperature, and rate of precipitant addition. In addition, certain ions such as Cl^- or Ca^{2+} could be required in low concentration for crystals to form. A variety of microtechniques have been developed to allow the systematic investigation of many conditions using a minimum of material. The crystallization of proteins has been reviewed by McPherson,[18] who appears to have a green thumb in this regard.

Protein crystals suitable for x-ray diffraction studies are usually larger than 0.2 mm in all dimensions. The final criterion of adequacy is their ability to diffract to the desired resolution. Once obtained protein crystals must be kept in the presence of mother liquor. Usually about 50% of the crystal volume is occupied by solvent,[19] and therefore the crystals are destroyed

[18] A. McPherson, *Methods Biochem. Anal.* **23**, 249 (1976).
[19] B. W. Matthews, *J. Mol. Biol.* **33**, 491 (1968).

when allowed to dry. For diffraction experiments the crystals are sealed in glass or quartz capillaries in the presence of mother liquor. Under these conditions they are in general very stable and can be used for data collection.

6.4.3. Collection of X-Ray Diffraction Data

Two types of information can be obtained from x-ray diffraction experiments using single crystals. The first type involves the determination of the unit cell parameters and the symmetry of the crystal.[3] The second consists of the measurement of the diffracted intensity of all observable reflections.

Experiments for the determination of the crystal lattice and its symmetry are simple to perform and easy to interpret. They consist essentially in recording (usually photographically) several selected regions of the reciprocal lattice. Using Bragg's law [Eq. (6.2.1)] one can obtain the unit cell parameters from measurements of the diffraction angle θ. The symmetry of the crystal can be determined by careful visual inspection of the recorded regions of the reciprocal lattice.

The measurement of the intensities that provide the desired structure factor amplitudes is often a long and tedious process. The number of reflections which must be measured for the determination of the structure of a protein to high resolution is very large (from approximately five thousand to several millions). Therefore the strategy of data collection is of great concern to protein crystallographers. The method of choice depends on the particular problem under consideration. One should take into account the cell dimensions of the crystal, how well it diffracts, the sensitivity to damage by radiation, the size of the crystals, etc., when deciding how to collect the data. Some of the advantages of the individual methods are discussed in more detail this section.

X rays for protein crystallography are usually obtained from copper anodes. Radiation from storage rings of high-energy accelerators is also used,[20] but its use is obviously restricted by the access to such sources. When Cu x-ray tubes are employed the K_α line (8.3 keV, 1.54 Å) is selected by means of one of three techniques.[21] In the first, a filter of Ni, which has a sharp absorption increase for wavelengths just shorter than 1.54 Å, is used to reduce the amount of contaminating radiation. The second method involves the use of crystal monochromators.[21] In this case the x-ray beam is reflected by means of an intense Bragg reflection of the monochromator crystal. The condition in

---

[20] H. Winick and G. Brown, eds., "Workshop on X-Ray Instrumentation for Synchrotron Radiation Research," SSRL Rep. No. 78/04. Stanford Synchrotron Radiation Laboratory, Stanford, Connecticut, 1978.

[21] A. Guinier, "X-Ray Crystallographic Technology." Hilger & Watts, London, 1952.

Equation. (6.2.1) will permit only a single wavelength to be reflected, thus providing a high degree of monochromatization. Crystals of quartz, LiF, and graphite are the most widely used for this purpose. The third method consists of using grazing-incidence reflection from metallized glass flats (mirrors).[22] For a given grazing angle there is a strong dependence of the reflected intensity on wavelength. Reflection angles can be chosen such that the shorter wavelengths (the most common contamination) can be eliminated. The mirrors can be slightly bent to focus the x-ray beam for higher luminosity.[22] In the other systems collimation is used to provide an x-ray beam of the appropriate cross section (usually from 100 microns to a few millimeters).

Two general methods exist to record diffracted intensities, namely photographic[23,24] and counter techniques.[25] We discuss briefly below the most common implementations of these methods. The choice of a particular configuration for data collection involves a variety of considerations. These will emerge as we discuss particular devices.

6.4.3.1 The Precession Camera. Since its introduction by Martin Buerger in the 1940s, the precession camera has become a favorite among protein crystallographers.[24] All preliminary photographs and initial determinations of cell dimensions and symmetry are done with this camera. The precession camera permits the undistorted recording of a section of the reciprocal lattice of a crystal on a photograph. This is accomplished by orienting the crystal so that the chosen reciprocal lattice plane is perpendicular to the x-ray beam and parallel to the film (Fig. 6). A precession angle $\bar{\mu}$ is introduced and the normal to the plane precessed around the x-ray beam with semiangle $\bar{\mu}$. The film is always maintained parallel to the plane being recorded to assure that the image is undistorted. A layer line screen of the appropriate radius (Fig. 6) is located at a predetermined distance from the crystal and moved so as to isolate reflections from a single plane of reciprocal space. An example of a precession photograph is shown in Fig. 7. These photographs are ideally suited for the measurements of unit cell parameters. Intensity data can also be obtained from precession photographs. Films can be digitized using automatic microdensitometers and the data further processed to give structure factor amplitudes.[26] Precession photographs are particularly easy to deal with. All reflections have the same size and shape and fall in positions on the film that can be simply and accurately predicted.

[22] S. Harrison, *J. Appl. Crystallogr.* **1**, 84 (1968).

[23] U. W. Arndt and A. J. Wonocott, eds., "The Rotation Method in Crystallography." North-Holland Publ., Amsterdam, 1977.

[24] M. J. Buerger, "The Precession Method in X-Ray Crystallography." Wiley, New York, 1964.

[25] U. W. Arndt and B. T. M. Willis, "Single Crystal Diffractometry." Cambridge Univ. Press, London and New York, 1966.

[26] C. E. Nockolds and R. H. Kretsinger, *J. Sci. Instrum.* **3**, 842 (1970).

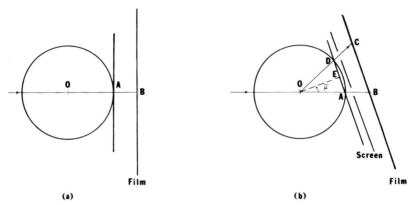

FIG. 6. Ewald's sphere construction for the precession camera: (a) the incident x-ray beam (arrow) intersects a reciprocal plane at a point A. The plane is oriented perpendicular to the beam. A film is placed at point B and is oriented parallel to the reciprocal plane. (b) A precession angle $\bar{\mu}$ is introduced such that $\overline{OE}$, the normal to the reciprocal plane (and to the film), forms an angle μ with the x-ray beam. A layer line screen is introduced to isolate a cone of semiangle $\bar{\mu}$. To record the photograph, the normal $\overline{OE}$ is precessed around the x-ray beam. This allows all reflections of the reciprocal plane within a circle of radius $\overline{AD}$ to pass through the diffraction condition. An undistorted image of the reciprocal plane is recorded in the film within a circle of radius $\overline{CB}$ (see Fig. 7).

However, for crystals with large cell dimensions ($> 50\,\text{Å}$) and for high-resolution data $1/s_{max} \approx 2\,\text{Å}$, the number of long-exposure photographs needed to collect the data is unreasonably large. The reason for this disadvantage can be traced in part to the use of layer line screens. The presence of the screen does not permit all diffracted beams to reach the film (this is the reason for its inclusion to start with). Not all the intensity diffracted at a given time is recorded, and therefore extended exposure times are needed to collect a complete data set. Screenless precession photographs can be used to avoid this problem. Small precession angles have to be used to avoid spot overlapping. Screenless precession photographs have been used successfully in protein structure determinations.[27] However, other problems intrinsic to this method of data collection make it less convenient than those using other geometries.

Screened precession photographs are ideal for recording one section of the reciprocal lattice to use for the calculation of projection Pattersons [Eq. (6.3.16)].

6.4.3.2 The Oscillation Camera. The simplest geometry for recording screenless photographs consists of rotating the crystal in the x-ray beam while recording the diffracted beams on a stationary film. This is the general

[27] N.H. Xuong and S. Freer, *Acta Crystallogr.*, *Sect. B* **B27**, 2380 (1971).

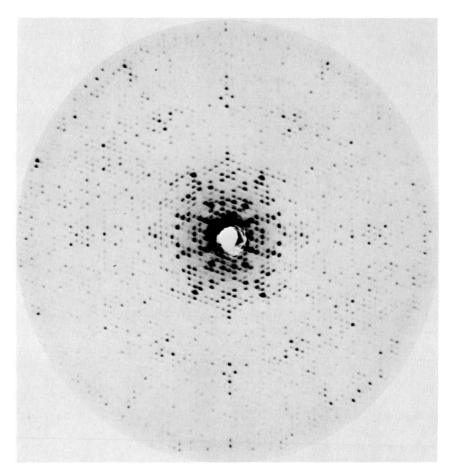

FIG. 7. Precession photograph of a uranyl derivative of crystalline southern bean mosaic virus. The photograph was taken using a precession angle of 6°, corresponding to a resolution cutoff of approximately 7.5 Å. Symmetry is easily recognized in the photograph. Cell dimensions simply related to the distance between spots. Note also the very large number of reflection present. (Courtesy of Michael Rossmann and John Johnson.)

principle of the oscillation camera.[23] Usually, to facilitate the identification of the reflections, the crystal is oriented with a unit cell axis parallel to the rotation axis and perpendicular to the x-ray beam. Also, to avoid overlap of reflections, each photograph is obtained while the crystal rotates over a very small angle (from a few minutes to a few degrees). For long exposures, the crystals are oscillated back and forth through this angle. Speical precautions are taken to avoid backlash and overexposure of the reflections that are in the diffraction condition when the movement is reversed.

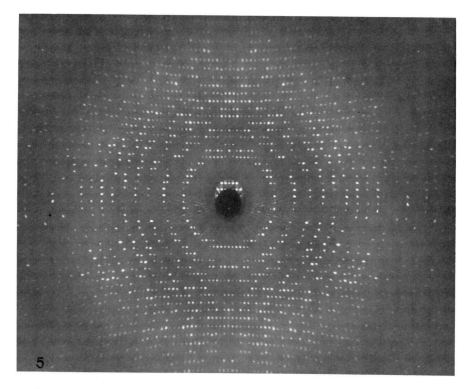

FIG. 8. Oscillation photograph of crystal porcine pancreatic α-amylase. The oscillation angle is 1.5°. The edge of the photograph shows reflections at 2-Å resolution. The curved rows of spots arise (see Fig. 2) from the intersection of Ewald's sphere with reciprocal lattice planes. (Courtesy of Paula Fitzgerald.)

A complete data set is obtained by recording small-angle oscillation photographs for all the necessary different orientations of the crystal around the rotation axis. The problems involved in obtaining a complete data set using the oscillation camera and in measuring the films are outside the scope of this chapter and will not be discussed. (For a complete discussion see Arndt and Wonocott[23]). It is sufficient to note that there are several good computer programs available for processing oscillation photographs, such as the one shown in Fig. 8.

6.4.3.3. The Four-Circle Diffractometer. To measure diffracted intensities with conventional proportional or scintillation counters it is most convenient to be able to orient the crystal and the counter in any arbitrary orientation with respect to the x-ray beam. However, several simplified geometries have been used successfully for this purpose. Since their introduction, computer-controlled four-circle diffractometers have become the method of choice.[25] In this geometry the counter can be rotated in the equatorial plane that contains the x-ray beam. The crystal is mounted at the center of rotation of an Eulerian cradle (with angles ω, χ, ϕ) which allows the crystal to be set at any desired orientation. With this arrangement any reflection can be made to form in the equatorial plane. The counter can then be set at an angle 2θ from the x-ray beam to record the intensity of the diffracted beam. Computer control of the circles permits the collection of a large number of reflections with very limited operator intervention. Four-circle diffractometers are very convenient to use and are usually the method of choice for proteins with cell dimensions smaller than approximately 100 Å. The major drawback of diffractometers is that they measure one reflection at a time and miss all the others that are formed simultaneously. This problem is especially serious for crystals that are very sensitive to radiation. The development of large, bidimensional, position-sensitive detectors has resolved this difficulty.[28] Unfortunately, equipment incorporating these detectors is still in the developmental stage. At least one such machine has been functional for several years.[28] In any case, the present cost of these instruments does not put them in the category of equipment that can be acquired by a single laboratory.

6.4.4. Phase Determination Using Isomorphous Replacement

As stated before (Section 6.4.1) neither the Patterson function nor direct methods can be used to obtain the structure of a protein crystal from the observed structure factor amplitudes $F(\mathbf{h})$. The method most commonly used for solving the phase problem for crystalline proteins is called multiple isomorphous replacement or the heavy-atom method.[29] In this method protein crystals are modified by the introduction of atoms of high atomic number. The process whereby these modified crystals can lead to the determination of phase is quite simple. Using the diffraction patterns from the native and modified crystals one determines the positions of heavy atoms bound to the protein. Because there are few of these, and because their high electron density makes them stand out above the protein, the heavy atoms can be located using the Patterson function or other methods from small-molecule crystallography. Once the positions of the heavy atoms are known, the phase of the scattering from them can be *calculated* [Eq. (6.3.8b)] and can serve as a reference to measure the phase of the protein structure factors. Introduction of heavy atoms is usually accomplished by soaking preformed

[28] N.-H. Xuong, S. T. Freer, R. Hamlin, C. Nielsen, and W. Vernon, *Acta Crystallogr.*, *Sect. A* **A34**, 289 (1978).

[29] R. E. Dickerson, J. C. Kendrew, and B. E. Strandberg, *Acta Crystallogr.* **14**, 1188 (1961).

crystals in solutions containing, in addition to mother liquor, varying concentrations of salts of high-atomic-number elements (e.g., Hg, Au, Pt, or Sm). Most heavy-atom derivatives are found by trial and error. However, specific modification of the protein using heavy-atom-containing groups has also been used.

For a heavy-atom derivative to be useful it must be isomorphous to the native crystal. This means, ideally, that the heavy atoms have replaced an equivalent volume of solvent in the crystal, while causing no other changes. Heavy atoms normally bind to the surface of the molecule, which is exposed to solvent. For the most part they make interactions with amino acid side chains in ways that cannot be predicted before the structure is known. However, certain reactions, such as the binding of Hg to –SH groups are generally useful. Heavy-atom substitution and isomorphism can be monitored by comparing intensities and lattice constants and derivative crystals in precession photographs.

Once a suitable heavy-atom derivative has been prepared and its intensity data collected, it can be used to determine the phases of the native crystals. The structure determination of cytochrome C^{30} and rubredoxin (Section 6.4.9) illustrate many of the facets of this method which we now describe.

The structure factors of the native and the heavy-atom derivative $F_P(h)$ and $F_{PH}(h)$ are

$$F_P(h) = \sum_{j=1}^{N} f_j \exp(2\pi i h \cdot x_j), \qquad (6.4.1)$$

where x_j are the coordinates of the N protein atoms.

$$F_{PH}(h) = \sum_{j=1}^{N} f_j \exp(2\pi i h \cdot x_j) + \sum_{r=1}^{N_H} f_r \exp(2\pi i h \cdot x_r), \qquad (6.4.2)$$

here N_H is the number of heavy atoms in the unit cell and x_r are their fractional coordinates. It is convenient to define the heavy-atom structure factor

$$f_H(h) = \sum_r f_r \exp(2\pi i h \cdot x_r), \qquad (6.4.3)$$

which represents the diffraction that would arise from the heavy atoms alone. Substituting (6.4.1) and (6.4.3) into (6.4.2) yields

$$F_{PH}(h) = F_P(h) + f_H(h). \qquad (6.4.4)$$

To use the heavy-atom derivative for the determination of the phases of the native crystals it is necessary to find the position of the heavy atoms in

[30] R. E. Dickerson, M. L. Kopka, C. L. Borders, Jr., J. Varnum, J. E. Weinzierl, and E. Margoliash, *J. Mol. Biol.* **29**, 77 (1967).

the cell. However, the heavy-atom structure factors f_H cannot be directly obtained from the experimentally accessible magnitudes $F_P(h)$ and $F_{PH}(h)$.

Most space groups have at least one centrosymmetric projection. The structure factors that contribute to this projection are real, and therefore

$$F_{PH}(\mathbf{h}) = F_P(\mathbf{h}) \pm f_H(\mathbf{h}). \tag{6.4.5}$$

For most reflections $F_P(\mathbf{h}) \geq f_H(\mathbf{h})$ and thus $f_H(\mathbf{h})$ can be calculated from the structure factor amplitudes using

$$\Delta F(\mathbf{h}) \equiv F_{PH}(\mathbf{h}) - F_P(\mathbf{h}).$$

$\Delta F(\mathbf{h})$ is normally used as the estimate of the heavy-atom structure factor amplitude for *all* reflections contributing to a centrosymmetric projections. When $F_P(\mathbf{h}) < f_H(\mathbf{h})$ and F_P and f_H have opposite sign (the so-called crossover condition), ΔF is an underestimate of f_H, since in this case

$$f_H(\mathbf{h}) = F_{PH}(\mathbf{h}) + F_P(\mathbf{h}).$$

This approximation does not introduce serious difficulties and can be fully corrected once the coordinates of the heavy atoms are known.

The preferred method for locating the heavy atoms involves the use of a Patterson function in suitable centrosymmetric projections with coefficients $\Delta F(h)^2$. This synthesis is identical to the projection of the Patterson function of a crystal containing only the heavy atoms if the problem of "crossovers" is ignored. A "good" heavy-atom derivative should contain only a few major heavy-atom sites, and its Patterson function can in general be unscrambled. In crystals without suitable centrosymmetric projections a three-dimensional Patterson synthesis with coefficients $\Delta F(h)^2$ is usually calculated. This function contains as its major peaks the heavy atom–heavy atom interatomic vectors and can in principle be solved in the usual manner (see Fig. 4). The positional parameters of the heavy atoms, their percent substitution (occupancy), and their temperature factors can be refined using least-squares procedures (Section 6.4.8, see also Dickerson et al.[31] and Sygusch[32]). These parameters can be used to calculate the heavy-atom structure factors f_H. These structure factors can be used to determine native phases through Eq. (6.4.4). If moduli are taken on both sides of this equation we obtain

$$|\mathbf{F}_{PH}|^2 = |\mathbf{F}_P + \mathbf{f}_H|^2. \tag{6.4.6}$$

This can be shown to yield

$$F_{PH}^2 = (F_P \cos \alpha + a_H)^2 + (F_P \sin \alpha + b_H)^2. \tag{6.4.7}$$

[31] R. E. Dickerson, J. E. Weinzierl, and R. A. Palmer, *Acta Crystallogr.*, Sect. B **B24**, 997 (1968).

[32] J. Sygusch, *Acta Crystallogr.*, Sect. A **A33**, 512 (1977).

In this expression α is the phase of the native structure factor and a_H and b_H are the real and imaginary components of $\mathbf{f}_H$. This equation can be solved to obtain the value of α. There will be in general two values of α that satisfy Eq. (6.4.7). This ambiguity can be eliminated by using several heavy-atom derivatives. The correct value of the phase will then correspond to the common roots of Eq. (6.4.7) for all heavy-atom derivatives.

In an actual structure determination the situation is more complicated. All magnitudes in Eq. (6.4.7) are subject to errors. F_P and F_{PH} are experimentally determined magnitudes and a_H and b_H are affected by errors in the heavy-atom model (i.e., errors in the atomic parameters, addition or deletion of sites, etc.). Therefore, even though Eq. (6.4.7) gives two possible values of α for each individual heavy-atom derivative, all the obtained pairs will not have a clear common value. The correct approach is then to recognize that this is an observational equation which depends on the phase α and on the heavy-atom parameters. For each reflection all heavy-atom derivatives should give a common value of α; it is therefore convenient to incorporate all heavy-atom derivatives in a single observational equation. The resulting set of nonlinear equations can be solved for the protein phase and for corrected heavy-atom parameters that will minimize a conveniently chosen residual. Several formulations have been used which differ mainly in the residual chosen and in the detailed form of the observational equations.[31,32] The "quality" of the phases can be estimated from the statistics of the minimization procedure.

The estimated phases and the values of F_P are then combined to calculate an electron density map using the synthesis

$$\rho(\mathbf{x}) = \frac{1}{V} \sum F_P(\mathbf{h}) \exp(i\alpha) \exp(-2\pi i \mathbf{h} \cdot \mathbf{x}). \tag{6.4.8}$$

This expression is identical to Eq. (6.3.5) with $\mathbf{F}(\mathbf{h}) = F(\mathbf{h})\exp(i\alpha)$. Usually, each reflection in this synthesis is weighted according to the estimated accuracy of its phase. The most commonly used weight is called the "figure of merit," $m(\mathbf{h})$. Use of $m(\mathbf{h})$ weighting minimizes the root-mean-square error in the electron density when certain error models which are outside the scope of this article are assumed. It is roughly true that $m(\mathbf{h}) = \cos[\text{error in } \alpha(\mathbf{h})]$. The synthesis using $m(\mathbf{h})$ weighting is often called the "best Fourier"[33] and is given by

$$\rho(x) = \frac{1}{V} \sum m(\mathbf{h})F_P \exp(i\alpha) \exp(-2\pi i \mathbf{h} \cdot \mathbf{x}). \tag{6.4.9}$$

The average value of the figure of merit over all reflections, $\langle m(\mathbf{h}) \rangle$, provides a useful indication of the overall quality of the phase determination.

[33] D. M. Blow and F. H. C. Crick, *Acta Crystallogr.* **12**, 794 (1959).

6.4.5. Interpretation of Electron Density Maps

The calculation of $\rho(\mathbf{x})$ provides an estimate of the electron density in the unit cell of the crystal. This map represents the distribution of diffracting matter in the crystal. However, in protein crystallography, we want to know the structure of the protein molecules forming the crystal. Because the crystals are periodic, the electron density belonging to an isolated molecule cannot always be identified immediately. The ease of this procedure and the detail with which the structure of the protein is determined depend mainly on the resolution of the data and on the quality of the phases used. Low-resolution maps ($1/s_{max} = 6$ Å), when interpreted, can indicate the general shape of the molecule, subunit organization, presence of helices, and other gross structural features.

The maps are usually contoured on lines of constant electron density and plotted on transparent sheets, each containing the electron density corresponding to one section through the unit cell of the crystal. The transparent sections are stacked with the proper spacings and illuminated for visual inspection (Fig. 9). Computer displays can also be used, but they are in general not very helpful for the interpretation of low-resolution maps. In many cases molecular boundaries can be located as regions of low electron density separating continuous globular regions of high density. When this separation is unambiguous, a model of the protein molecule can be built. Balsa wood or styrofoam sheets of appropriate thickness are usually cut to follow some chosen contour with the shape of electron density function and

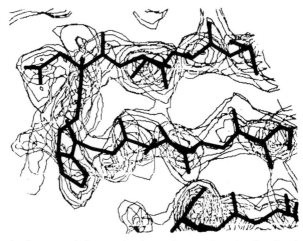

FIG. 9. Stack of contoured electron density sections used to visualize the electron density function of the immunoglobulin fragment Fab New. The atomic model derived from this density is shown superimposed.

FIG. 10. Balsa wood model of the hemoglobin molecule from the bloodworm, *Glycera dibranchiata* at 6-Å resolution. Individual pieces fill the regions within contour lines drawn at some constant level of electron density. The rodlike features of the model are α helices. (Courtesy of E. A. Padlan and W. E. Love.)

assembled to represent the molecule (Fig. 10). With higher-resolution maps ($1/s_{max} \cong 2$ Å) the polypeptide chain of the protein can be traced. Knowledge of the amino acid sequence of the protein (primary structure) greatly simplifies this task. The high-resolution maps can also be plotted on transparent sheets, and the electron density features interpreted using this information. To aid in the interpretation of the maps they are displayed in optical comparators like those shown in Fig. 11.[34] The comparators allow one to fit skeletal models of the amino acids to the electron density. For this procedure a few (5–10) sections of electron density are displayed at a time and viewed through a half-silvered mirror. One then builds a brass model while looking at the reflected image, which can be superimposed over the density being interpreted. The illusion that the brass model is floating in the contour density is often quite powerful, and this makes it possible to adjust the model to fit the density. The procedure is continued until all the density in the crystal is used and the complete amino acid sequence built. An equivalent procedure can be followed using computer displays.[35] In both cases the final product is a three-dimensional atomic model of the protein molecule. A model of this

[34] F. M. Richards, *J. Mol. Biol.* **37**, 225 (1968).

[35] J. Hermans, Jr., J. E. McQueen, G. A. Petsko, and D. Tsernoglou, *Science (Washington, D.C.)* **197**, 1378 (1977).

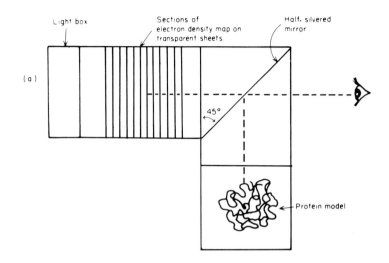

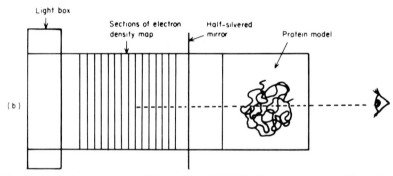

FIG. 11. Optical comparator or 'Richard's box' which facilitates comparison of the molecular model with the electron density map. (a) Mirror tilted 45° toward sections of map. (b) Mirror parallel to sections of map. (From Blundell and Johnson.[1])

kind contains a remarkable amount of information. Physical, chemical, and physiological properties of the protein can be correlated with structural characteristics. The geometrical and chemical properties of the active site(s) and other functional regions can be analyzed. Important insight can be gained into the mechanism of action of the molecule. Solution and crystallographic experiments can be designed to test or exclude proposed modes of action.

The procedure just described for the determination of the three-dimensional structure of a protein molecule using single-crystal x-ray diffraction is long and involved, but experience has shown that the wealth of information that can be obtained from such a determination is always worth the effort.

6.4.6. Difference Fourier Syntheses

One of the goals of the determination of a protein structure is the characterization of its functional and physiological properties. When enough biochemical and physical chemical information is available, structural correlation of functional properties can be made using the three-dimensional structure of the native protein. For many proteins, however, most functions are dependent on interactions with other molecules. In the cases of enzymes these molecules are the substrates, products, and effectors of the catalyzed reactions. Carrier proteins (i.e., hemoglobin), antibodies, and binding proteins (i.e., repressors, receptors, etc.) represent other examples of proteins for which the interaction with other molecules (which we shall call ligands) can be studied. In principle these studies involve a completely new structure determination. The complex of the protein and its ligand must be crystallized and the structure determined using the usual techniques. This problem can be simplified enormously if a suitably small ligand such as a substrate analog, can be used for these experiments. Sometimes these small molecules can diffuse into preformed protein crystals (as happens in making heavy-atom derivatives) and bind isomorphously to their specific binding site, just as if the proteins were in solution. Not surprisingly, the binding constants are generally lower in the crystallized protein. It must be realized that in most cases enzymes also retain their activity in the crystals. Therefore in the case of hydrolytic enzymes if substrates are diffused into the crystal they will be hydrolyzed before x-ray data can be collected. The usual way to overcome this problem is to use nonhydrolyzable substrate analogs. Low-temperature techniques have also been able to "freeze" substrates so they cannot react.

When protein–ligand crystals are isomorphous to the parent crystals, most of the structure of the protein–ligand complex is known (i.e., the protein part.) A simple technique termed the difference-Fourier synthesis can then be used to localize the ligand.

The electron density functions of the protein and protein–ligand crystals are given by

$$\rho_P(\mathbf{x}) = \frac{1}{V} \sum F_P(\mathbf{h}) \exp(i\alpha_P) \exp(-2\pi i \mathbf{h} \cdot \mathbf{x}), \qquad (6.4.10a)$$

$$\rho_{PL}(\mathbf{x}) = \frac{1}{V} \sum F_{PL}(\mathbf{h}) \exp(i\alpha_{PL}) \exp(-2\pi i \mathbf{h} \cdot \mathbf{x}). \qquad (6.4.10b)$$

If the substrate has a small number of electrons the phases of the protein–ligand crystals will be very close to those of the protein crystals. If we use the approximation $\alpha_{PL} = \alpha_P$ we can calculate the difference $\rho_{PL} - \rho_P$ (difference map), which is given by

$$\rho_{PL} - \rho_P = \frac{1}{V} \sum (F_{PL} - F_P) \exp(i\alpha_P) \exp(-2\pi i \mathbf{h} \cdot \mathbf{x}). \qquad (6.4.11)$$

This map can be calculated from the structure factor amplitudes of the native and protein–ligand crystals using the native phases and provides, in principle, the electron density corresponding to the ligand. These maps are subject to large errors due to the approximation $\alpha_{PL} = \alpha_P$[36] and the need to measure small amplitude differences between F_{PL} and F_P. In practice, however, difference maps of ligand are easy to interpret, especially if the general location of the binding site is known from sequence and chemical data.

Usually, ligand difference maps are interpreted by copying the contoured density into transparent sheets which are then displayed in an optical comparator (Section 6.4.5). The protein model (already built) is kept in its correct position and the ligand built into the density while being checked for favorable interactions with the protein. The protein–ligand complexes obtained in this manner can be studied to characterize the interactions between residues of the protein and ligand, to locate the residues involved in the catalytic steps of the enzymatic reaction,[37] etc.

Movements of side chains of the protein that sometimes occur upon binding can also be detected in difference-Fourier maps. If a group of protein atoms moves upon ligand binding its original position will appear as negative density in the difference map. The new position of the group will appear as positive electron density.

Experiments using difference-Fourier maps provide invaluable information about the biochemical and physiological properties of protein molecules. The simplicity of the experiments (once the structure of the protein is solved) make difference Fouriers a powerful tool for the study of the function of proteins.

6.4.7. Molecular Replacement

The method of difference-Fourier synthesis just described is invaluable in simplifying the crystal structure determination of a protein–small-molecule complex whenever the crystal is isomorphous to a native protein crystal whose structure has already been determined. The perversity of nature

[36] R. Henderson and J. K. Moffat, *Acta Crystallogr.*, *Sect. B* **B27**, 1414 (1971).

[37] G. C. F. Blake, L. N. Johnson, G. A. Mair, A. C. T. North, D. C. Phillips, and V. R. Sarma, *Proc. R. Soc. London, Ser. B* **167**, 378 (1967).

ensures, however, that such crystals are not always isomorphous to native ones. Further, many other circumstances arise when it is of interest to determine a crystal structure when the structure of the constituent molecules is already approximately known. The study of the evolutionary relationship of molecules performing identical functions in different species is one example. Another is the study of the same molecule crystallized under different conditions—from organic solvent and from salt perhaps—to check the constancy of conformation. A third is the study of functional conformational changes which occur upon ligand or substrate binding. The molecular replacement method, inspired by Rossmann and Blow,[38] allows known molecular structures (search molecules) to be used as building blocks to construct models of crystals of similar or related molecules. These models can then be refined (Section 6.4.8) to give accurate representations of the new molecules, unbiased by the original search molecule structure. This method is neither as straightforward nor as foolproof as the difference-Fourier method, but it is very efficient compared with isomorphous replacement. Structures of hemoglobin, immunoglobulin, protease, and other molecules have been successfully determined in this way.[39–41] A collection of papers about molecular replacement appears in Rossmann.[8]

To construct a model of a crystal using a search molecule requires, in the usual case, only six parameters, three to specify the position of the search molecule in the unit cell and three its orientation. The known symmetry operations of the crystal allow the other molecules in the unit cell to be generated. To find the correct model one could, in principle, perform a six-dimensional search in which the diffraction pattern or Patterson function corresponding to every possible value of the parameter sextet is compared with corresponding experimental values from the crystal. One must use the diffraction pattern or Patterson function in the test because there is no experimental electron density function with which to compare the model. A particular parameter set may provide a clear best fit with experiment, thus generating the desired preliminary model for the crystal structure. Such six-dimensional searches, however, are far beyond the scope of even the fastest computers, and some simplification must be found. Rossmann and Blow[38] realized that it was possible to separate the rotational and translational parts of the search, reducing a six-dimensional problem to a sequence of two three-dimensional ones. The rotational search or rotation function is based upon certain simple properties of the Patterson function.

[38] M. G. Rossmann and D. M. Blow, *Acta Crystallogr.* **15**, 24 (1962).

[39] B. C. Wishner, K. B. Ward, E. E. Lattman, and W. E. Love, *J. Mol. Biol.* **98**, 179 (1975).

[40] H. Fehlhammer, M. Schiffer, O. Epp, P. M. Colman, E. E. Lattman, P. Schwager, and W. Steigemann, *Biophys. Struct. Mech.* **1**, 139 (1975).

[41] M. Schmid, E. E. Lattman, and J. R. Herriott, *J. Mol. Biol.* **84**, 97 (1974).

In the Patterson function of a crystal one can mentally divide up the interatomic vectors into two classes: vectors running within one molecule (self-vectors), and vectors running between molecules (cross vectors). All self-vectors are contained within a sphere of radius D about the origin of the Patterson function, where D is the maximum dimension of the molecule. This sphere of radius D will contain one family of self-vectors for each molecule in the unit cell, plus whatever cross vectors happen to be of length less than D. The families of self-vectors are identical, but appear in various orientations about the origin, corresponding to the orientations of the molecules within the crystal.

If one knows the structure of the search molecule, one can readily calculate its Patterson function $P_M(\mathbf{u})$. P_M is the Patterson function of a single, isolated molecule and is therefore not periodic. Indeed, it is best thought of as the set of interatomic vectors within one test molecule. As such it is formally identical (and numerically similar) to the self-vector sets clustered about the origin of the experimental crystal Patterson function $P(\mathbf{u})$. To find the orientation of the search molecule in the crystal, one need only seek the orientation for which P_M superimposes upon one of these. The correlation function most frequently used for this is

$$R(\mathbf{C}) = \iiint P_M(\mathbf{Cu})P(\mathbf{u})\, dV. \qquad (6.4.12)$$

The matrix $\mathbf{C}$ is a rotation matrix, often based upon Eulerian angles.

The integral above is evaluated numerically for each value of $\mathbf{C}$, and a maximum in R is sought. $R(\mathbf{C})$ is large when peaks in P_M coincide with peaks in P, and valleys in P_M coincide with valleys in P. It is small when peaks do not coincide. In principle the absolute maximum of R occurs for a rotation $\mathbf{C}$ which superimposes the whole set of vector peaks in P_M upon a self-vector family in P, thereby providing as estimate of the molecular orientation in the crystal. The volume of integration is a sphere of radius D, but may be written as an integral over all space, since P_M vanishes outside the sphere in any case. The latter bit of formalism allows one to apply Parseval's theorem[6] to (6.4.12), showing that

$$R(\mathbf{C}) = \sum F_M^2(\mathbf{Ch})F^2(\mathbf{h}), \qquad (6.4.13)$$

where the Fs are the structure factor amplitudes of the model and crystal. Both the direct and reciprocal space formulations have been used, depending upon convenience. In practice the success of the method varies widely. In some cases isolated, highly significant peaks in $R(\mathbf{C})$ are seen; in others several peaks of nearly equal height are found, and considerable difficulty may arise in attempting to pick the correct one.

The above formulation of the rotation function[42] is a simplified version of the original function of Rossmann and Blow, who put it forward for slightly different purposes. In an important advance Crowther[43] has expanded the functions of the right-hand side of Eq. (6.4.12) in spherical harmonics and has shown that two out of the three integrals can thereby be reduced to fast Fourier transforms, thus speeding up the calculation enormously.

Once the orientation of the molecule is known, only its position remains to be found. This is usually accomplished by the use of a 'translation function.' Full details of the translation function are outside the scope of this chapter. It is sufficient to note that by assigning a trial position to a properly oriented molecule one obtains a complete trial model of the crystal. The Patterson function of this model (which now represents one unit cell of a crystal) can be compared with the experimental Patterson function using an equation like (6.4.14):

$$T(\mathbf{x}) = \iiint_{\text{unit cell}} P_{\text{Mx}}(\mathbf{u}, \mathbf{x}) P(\mathbf{u}) \, dV. \tag{6.4.14}$$

Here P_{Mx} is the Patterson function of the model with the properly oriented search molecule at position $\mathbf{x}$. An extensive literature exists describing variations on $T(\mathbf{x})$ and discussing the most efficient ways of evaluating it.[8]

Again, the success of $T(\mathbf{x})$ is quite variable, but in favorable cases it gives a clear and unambiguous indication of the molecular position, thus completing description of the model structure. Using a structure factor equation this model can be a source of phases which can be combined with the experimental amplitudes to produce a Fourier synthesis of the experimental crystal. Such a synthesis is, of course, a hybrid of the search and correct structures, and it is the function of the refinement process to convert it into an accurate image.

6.4.8. Refinement of Protein Structures

Protein models obtained from the interpretation electron density maps (Section 6.4.5) provide coordinates for all atoms in the structure. However, these coordinates are subject to considerable error. At the resolution usually attainable in protein crystals ($1/s_{\text{max}} \approx 2$ Å) the maps do not have enough detail to visualize individual atoms. Also, errors in the estimation of the phases introduce errors and distortions into the electron density and into

[42] E. E. Lattman and W. E. Love, *Acta Crystallogr., Sect. B* **B26**, 1854 (1970).

[43] R. A. Crowther, *in* "The Molecular Replacement Method" (M. G. Rossmann, ed.), Gordon and Breach, New York, 1972.

the model. Hence the models obtained in this manner represent only the best efforts of the builders. In addition, if the electron density maps are interpreted in an optical comparator the atomic coordinates do not conform to the standard geometry of the amino acids. The correct stereochemistry is usually imposed upon the measured coordinates using "model building" programs.[44,45]

These models are in general good enough to study most of the relevant biochemical properties of the protein. However, from a physical point of view the models are highly inaccurate.

There is in principle one set of observed magnitudes that can be used to evaluate the accuracy of the parameters, namely the structure factor amplitudes of the native crystals. An estimation of the values of this set of observed magnitudes can be calculated from the parameters of the model using Eqs. (6.3.18). The most commonly used index of agreement between observed and calculated magnitudes $F_0(\mathbf{h})$ and $F_c(\mathbf{h})$ is called the "R factor" and is defined as

$$R = \sum |\Delta F(\mathbf{h})| / \sum F_o(\mathbf{h}), \qquad (6.4.15)$$

with

$$\Delta F(\mathbf{h}) = F_o(\mathbf{h}) - F_c(\mathbf{h}). \qquad (6.4.16)$$

Equation (6.4.16) is an observational relation which can be used to refine the atomic parameters. Usually least-squares procedures are used to minimize $\sum W(\mathbf{h})\Delta F^2(\mathbf{h})$, where $W(\mathbf{h})$ is a properly chosen weight. It is important to realize that in view of Parseval's theorem[6]

$$\frac{1}{V} \sum \Delta F^2(\mathbf{h})$$

is in principle equal to

$$\int_{\text{unit cell}} (\rho_o - \rho_c)^2 \, dV$$

if phases α_c are used to calculate ρ_o. This clearly indicates that the minimization procedure can be performed both in direct (real) or reciprocal space. Both kinds of methods have been used for the refinement of proteins. (For a complete evaluation of some of the methods, see Blundell and Johnson[1] and Diamond.[46])

[44] R. Diamond, *Acta Crystallogr.* **21**, 253 (1966).

[45] J. Hermans and J. E. McQueen, *Acta Crystallogr., Sect. A* **A30**, 730 (1974).

[46] R. Diamond, *in* "Crystallographic Computing" (F. R. Ahmed, ed.), p. 292. Munksgaard, Copenhagen, 1975.

Independent of the particular residual minimized, the resulting system of equations is nonlinear. Therefore all methods of solution necessarily involve iterative procedures. Also, the problem has to be overdetermined to provide for a stable refinement. For a typical protein this condition is not easily met, since the number of observations per atomic parameter is usually quite low (from less than 1 to 2 or 3). The known stereochemistry of the amino acids is normally used to overcome this problem. This information can be used in two different ways: (i) to reduce the number of parameters, and (ii) to increase the number of observations. For convenience we summarize some of the possible approaches in Table II.

The traditional real-space methods use either electron density or difference electron density maps. The atomic positions are refined by shifting atoms up gradients to the summit of electron density peaks. These procedures are usually performed without stereochemical constraints, which are then introduced after each cycle.[47,48]

Recently Diamond developed a method[49] in which the residual

$$\int [\rho_o(\mathbf{x}) - \rho_{model}(\mathbf{x})]^2 \, dV$$

is minimized. The electron density ρ_o is ρ_{best} obtained from multiple iso-morphous replacement (Section 6.4.4) or from a map calculated using $F_o(\mathbf{h})\exp(i\alpha_c)$. After shifts have been imposed a new set of calculated phases α_c can be obtained through a structure factor calculation [Eqs. (6.3.8)]. These α_c can then be combined with the observed amplitudes F_o to produce coefficients for an improved electron density map, which serves to begin a new cycle of refinement. The electron density ρ_{model} is calculated based on the atomic coordinates of the model using a spherical Gaussian to represent the individual atoms. The shifts calculated in each minimization cycle are not used directly. Instead they are modified by a least-squares fit so as to preserve the stereochemistry of the model. The independent variables of the refinement are then rotations around single bonds in the structure, thus reducing the number of model parameters by a factor of approximately three.

Reciprocal space methods refine directly the residual $\sum W(\mathbf{h})\Delta F^2(\mathbf{h})$. For a problem with M observed structure factor amplitudes and N parameters the linearized observational equations can be written in matrix representation

$$\mathbf{A}_{N,M}\delta\mathbf{x}_{1,N} = \mathbf{\Delta}_{1,M}, \qquad (6.4.17)$$

where the subindices represent the number of columns and rows of the matrices. The matrix elements are derived from minimizing $\sum W(\mathbf{h})\Delta F^2(\mathbf{h})$

[47] K. D. Watenpaugh, L. C. Sieker, and L. H. Jensen, *J. Mol. Biol.* **131**, 509 (1979).

[48] S. T. Freer, R. A. Alden, C. W. Carter, Jr., and J. Kraut, *J. Biol. Chem.* **250**, 46 (1975).

[49] R. Diamond, *Acta Crystallogr., Sect. A* **A27**, 436 (1971).

TABLE II. Summary of Refinement Options

Important quantities in refinement programs	Options used in different refinement programs
Residual minimized (explicitly)	$\int (\rho_{obs} - \rho_{model})^2 \, dV$, $\quad \sum \sigma r^2$, $\quad \sum \omega \Delta^2 F(h)$, $\quad \sum \omega \Delta F^2(h)$, $\quad \sum \omega \Delta F^2(h) + \sum \omega(p_m - p_{ideal})^{2a}$; $\quad \sum \omega(p_m - p_{ideal})^2$
Weights of F (explicit or implied)[b]	$1, f^{-1}, \sigma^{-2}$ (observational or approximate)
Atom shape[b]	Gaussian, point, transform of atomic scattering factor
Parameters refined (explicitly)	(x,y,z), B, occupancy; conformational angles, atomic radii
Stereochemistry of final model	ideal, optimized
Number of partial derivatives calculated for N atoms	nN, $\quad (mN)^{2c}$

[a] p_{ideal} and p_m are the values of a stereochemical parameter (i.e., bond length, bond angle, planarity condition, etc.) obtained as an average of a series of very accurate small-molecule observations (p_{ideal}) and as calculated from the atomic coordinates of the protein being refined (p_m).

[b] This is the implicit or explicit atom shape at the minimization step. All structure factor calculations are done with atomic scattering factors.

[c] The values of n and m vary for different implementation of these methods, but they range from 3 to 50 or more.

with $\Delta F(\mathbf{h})$ made linear on the shifts by using the zeroth- and first-order terms in a Taylor expansion. They can be shown to be

$$A_{i,h} = \sqrt{w(\mathbf{h})}\,\frac{\partial F_c(\mathbf{h})}{\partial x_i} \quad \text{and} \quad \Delta_h = \sqrt{w(\mathbf{h})}\Delta F(\mathbf{h}),$$

where $F_c(\mathbf{h})$ and $\Delta F(\mathbf{h})$ are calculated with the present atomic parameters. This expression can be solved for the shifts $\delta\mathbf{x}$ in the parameters by transforming to the normal equations. Multiplying on the left by $\mathbf{A}^T$ (transpose of $\mathbf{A}$) we obtain

$$\mathbf{A}^T\mathbf{A}\delta\mathbf{x} = \mathbf{A}^T\Delta.$$

Defining

$$\mathbf{B}_{N,N} = \mathbf{A}^T_{M,N}\mathbf{A}_{N,M} \quad \text{and} \quad \mathbf{C}_{1,N} = \mathbf{A}^T_{M,N}\Delta_{1,M},$$

we obtain

$$\mathbf{B}\delta\mathbf{x} = \mathbf{C}. \tag{6.4.18}$$

The elements of $\mathbf{B}$ and $\mathbf{C}$ are

$$B_{ij} = \sum_{\mathbf{h}} W(\mathbf{h})\,\frac{\partial F_c(\mathbf{h})}{\partial x_i}\,\frac{\partial F_c(\mathbf{h})}{\partial x_j}$$

and

$$C_j = \sum_{\mathbf{h}} W(\mathbf{h})\,\frac{\partial F_c(\mathbf{h})}{\partial x_j}\,\Delta F(\mathbf{h}).$$

These normal equations can be solved for the shifts $\delta\mathbf{x}$ by multiplying (6.4.18) by $\mathbf{B}^{-1}$ (inverse of $\mathbf{B}$), so that

$$\delta\mathbf{x} = \mathbf{B}^{-1}\mathbf{C}. \tag{6.4.19}$$

As mentioned above, there are two problems in applying this method directly to the refinement of proteins. First, M is usually not much larger than N. Second, the number of parameters is in general very large. In a typical case there will be more than 3000 parameters, and the normal equations will thus require the generation and inversion of a 3000×3000 matrix for a large number of cycles. This is, in principle, a very expensive task. The method was, however, used for a small protein (rubredoxin, Section 6.4.9). In this case, large blocks of 200 parameters were refined to avoid working with the full matrix.

Another method used to diminish the number of parameters consists of refining the positions and orientation of rigid groups. This method has the additional advantage that the stereochemistry of the molecule is automatically preserved.[50]

[50] J. L. Sussman, S. R. Holbrook, G. M. Church, and S. H. Kim, *Acta Crystallogr., Sect A* **A33**, 800 (1977).

In a recently developed method Hendrickson and Konnert[51] proposed refining the residual

$$\sum_{\mathbf{h}} W(\mathbf{h}) \Delta F^2(\mathbf{h}) + \sum_k W_k(p_m - p_{\text{ideal}})^2$$

(see Table II for definitions), thus using the stereochemical information as additional observations. This implies that the matrix elements of **B** and **C** will now include terms from the stereochemical equations. These terms can be very large if parameters are related by a stereochemical restraint (for example, if they are coordinates of two atoms that are bonded or belong to a rigid group). Otherwise they are very small. Konnert and Hendrickson found that the system of normal equations is very stable even if the matrix elements not containing stereochemical constraints are omitted. With this approximation, in a typical case only 1% of the matrix elements have to be calculated and used. In addition, for this 1% of the matrix elements $\sum_{\mathbf{h}} W(\partial F_c/\partial x_i)(\partial F_c/\partial x_j)$ is small compared to the terms arising from stereo-chemical constraints when x_i and x_j are not coordinates of the same atom. These approximations lead to very substantial savings in computing time for the generation of the sparse matrix. In addition, the method of conjugate gradients[52] used for solving the system of equations takes full advantage of the sparseness of the matrix, providing further savings in computing time.

Refinement has led to greatly increased rigor and quality in the protein models obtained by XRDA. Such models have become of interest to a wide range of investigators, leading to the establishment of a Protein Data Bank (Brookhaven National Laboratory, Upton, New York 11973) from which atomic coordinates from a rapidly expanding number of molecules can be obtained.

6.4.9. A Case History: Rubredoxin

To illustrate the different aspects of protein structure determination we next analyze the process as applied to a specific case. There are probably about 100 proteins whose structures have been determined (to some resolution) using single-crystal x-ray diffraction techniques. They differ in function, biological relevance, size, complexity, etc. They were solved using different methods for data collection, phase determination, and refinement. It is therefore very difficult to choose a "typical" case; we have settled on a

[51] W. A. Hendrickson and J. H. Konnert, *in* "Computing in Crystallography" (R. Diamond, S. Ramaseshan, and K. Venkatsan, eds.), p. 13.01. Indian Academy of Sciences, Bangalore, 1980.

[52] M. R. Hestenes and E. Stiefel, *J. Res. Natl. Bur. Stand. (U.S.)* **49**, 409 (1952); F. S. Beckman, *in* "Mathematical Methods for Digital Computers" (A. Ralston and H. S. Wilf, eds.), p. 62, Wiley, New York, 1960.

protein that is small, easy to describe, has been solved to high resolution, and for which more than one refinement method has been used; namely, rubredoxin. At least 25 investigator years have been invested in this project.

Rubredoxin is a nonheme, iron–sulfur protein that has been found in a number of anaerobic bacteria. Most rubredoxins have an iron atom and four cysteines per molecule of approximately 6000 daltons. The structure of the rubredoxin of *Clostridium pasterianum* was determined to 1.5 Å resolution by Watenpaugh et al.[47,53,54] The protein was crystallized by precipitation with near-saturated $(NH_4)_2SO_4$ at pH 4. The crystals are flattened rhombs and can be grown to more than 1 mm per side in a few weeks. X-ray diffraction experiments indicated that the only symmetry of the crystal is the presence of a threefold axis at every lattice point. The crystal is then said to be rhombohedral and to belong to space group R3.[9] The unit cell of the crystal can then be represented in terms of three identical axis of length a with the same angle α between them. The cell parameters of these crystals are $a = 38.77$ Å and $\alpha = 112.40°$. The rubredoxin literature uses a hexagonal unit cell (as is usually done), but the rhombohedral cell is more appropriate for the scope of this book.

The structure was solved by the multiple isomorphous replacement method. Heavy atoms were prepared by the usual soaking technique. A single-site HgI_4^{2-} derivative and a multiple-site UO_2^{2+} were used for phase determination.

Data were collected using a four-circle diffractometer.

The positions of the heavy atoms were determined using $\Delta F^2(\mathbf{h})$ Patterson syntheses. Heavy-atom positions were refined and phases were calculated using a program written by Dickerson et al.[31] For phase determination the authors used additional diffraction information (anomalous scattering),[1,7] but its discussion is outside the scope of this book.

Several electron density maps were calculated and interpreted. The final map calculated with phases obtained from multiple isomorphous replacement included data to a resolution of 2.0 Å. Examples of the electron density corresponding to several side chains are shown in Fig. 12.

The tracing of the polypeptide chain was unambiguous and the iron atom and the four coordinated cysteine–sulfurs were located. The sequence of the rubredoxin from *C. pasterianum* was not known at this time. Using sequence data from other rubredoxins, the authors built 54 amino acids and were able to identify about 40 of them with "reasonable certainty." The four cysteines were assigned as amino acids 6, 9, 39, and 42 in agreement

[53] K. D. Watenpaugh, L. C. Sieker, J. R. Herriott, and L. H. Jensen, *Cold Spring Harbor Symp. Quant. Biol.* **36**, 359 (1971).

[54] K. D. Watenpaugh, L. C. Sieker, J. R. Herriott, and L. H. Jensen, *Acta Crystallogr., Sect. B* **B29**, 943 (1973).

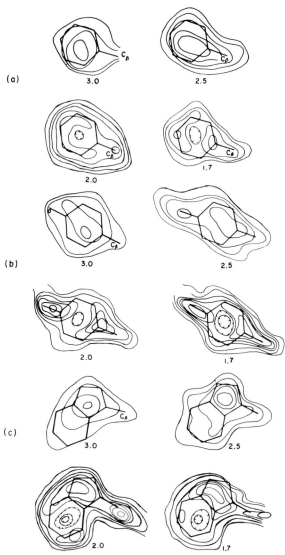

FIG. 12. Electron density through aromatic side chains in rubredoxin, at resolutions of 3.0, 2.5, 2.0, and 1.7 Å. The side chains represented are (a) phenylalanine, (b) tyrosine, and (c) tryptophan. Note how the rendition of the groups improves with resolution. (Reproduced from Watenpaugh et al.[53] with permission. Copyright 1971 Cold Spring Harbor Laboratory.)

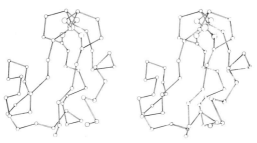

FIG. 13. Polypeptide chain backbone of rubredoxin, represented by line segments connecting α-carbon centers. The iron–sulphur complex involved in the redox activity of the protein is also shown. The large circle represents iron, the intermediate circles sulphur, and the small ones carbon. (Reproduced from Watenpaugh et al.[54] with permission.)

with the sequence data of the rubredoxin from *Micrococcus aerogenese*. Figure 13 is a stereo pair drawing of the Fe–S complex and the path of the polypeptide chain.

Intensity data for reflections with $1/s \geq 1.5$ Å were then collected in order to improve the resolution of the model. It was found that the average intensity of these reflections decreased rapidly with decreasing $1/s$ (Table III). For example, only 45 % of the reflections between 2.0 and 1.5 Å as measured by the authors had intensities larger than two standard deviations above background. An electron density map including data to 1.7-Å resolution was calculated and showed the expected improved resolution (Fig. 12).

Although a chemical sequence was still not available, about 95 % of the nonhydrogen atoms in the structure were assigned atomic coordinates. Also, 23 peaks in the electron density map were interpreted as water molecules and their oxygen atoms assigned coordinates. This gave a grand total of 424 atoms. A temperature factor B (Section 6.3.1) of 15 Å^2 was assigned to all atoms in the structure and an R factor [Eq. (6.4.15)] was calculated. The value obtained (0.389) indicated that the quality of the atomic parameters

TABLE III. Summary of Intensity Data for Rubredoxin[a,b]

Range of d (Å)	Possible number of reflections	Number of reflections exceeding 2σ	Reflections exceeding 2σ (%)
$5.0 < d$	224	218	97.3
$3.0 < d < 5.0$	816	765	93.7
$2.0 < d < 3.0$	2478	2151	86.9
$1.5 < d < 2.0$	4254	1899	44.6

[a] From K. D. Watenpaugh, L. C. Sieker, J. R. Herriott, and L. H. Jensen, *Cold Spring Harbor Symp. Quant. Biol.* **36**, 359 (1971). Copyright 1971 Cold Spring Harbor Laboratory.

was reasonably high for an initial model. Six reflections with $s \leq 0.06$ were found to have very large errors and were eliminated from the calculation. The temperature factor was then adjusted in two steps to a value of 12 Å^2, giving an R factor of 0.372.

The refinement was then carried out using difference-Fourier maps. The course of the refinement through this and the subsequent steps is shown in Table IV. As the refinement proceeded the authors introduced additional water molecules to explain properly located positive peaks in their difference maps. Furthermore, detailed analysis of the maps indicated that the temperature factor used (12 Å^2) was not suitable for all atoms in the structure. Atoms at the inside of the molecule seemed to require lower B values, while the atoms at the outside required larger values. In order to restrict the number of parameters only seven B values ranging from 9 to 25 Å^2 were included. The assignment was made depending roughly on the distance of the atoms to the center of the molecule. Also, occupancy factors proportional to the peak height were introduced for all atoms. The value of these factors was restricted to the range 0.3–1.0.

At this stage a least-squares refinement minimizing $\sum w(\mathbf{h}) \Delta F^2(\mathbf{h})$ (Section 6.4.8) was initiated. In this refinement individual temperature factors were used. With this addition, only two observations per parameter were available. Problems intrinsic to this low overdeterminancy were overcome, probably because of the high quality of the diffraction data.

Even for this extremely small protein, the refinement involves a 2000 × 2000 normal matrix. The authors chose the simplest possible approach. They partitioned the matrix in 10–12 blocks involving 200 parameters each. The apprehensions derived from the imposed limitations (i.e., low overdeterminancy and use of diagonal blocks) were soon dissipated. The least-squares refinement behaved surprisingly well. The authors found no general tendency for the parameters to diverge or oscillate. The few cases where it did occur were corrected using difference-Fourier maps between cycles. In most of these cases the atomic parameters stabilized with subsequent refinement, leaving only a few atoms in the structure with uncertain atomic parameters. The final R factor of 0.12 (Table IV) is remarkably low.

The standard deviations of the interatomic distances for C, N, and O atoms were found to be in the range 0.1–0.2 Å.

The course of the refinement was also analyzed by computing the R factor for reflections grouped in ranges of s. It is clear from Fig. 14 that the reflections in the intermediate ranges have the best agreement between observed and calculated structure factor amplitudes. This distribution reflects several of the intrinsic problems of calculating structure factors for protein crystals. First, only the protein molecule and the tightly bound water molecules are included in the calculation. The rest of the unit cell is filled with randomly

TABLE IV. Course of Refinement for Rubredoxin[a]

Coordinates	Overall R	R and number of reflections as function of $\sin \theta/\lambda$				Number of reflections	B	Number of H_2O and occupancy	
		0–0.10	0.10–0.167	0.167–0.25	0.25–0.333				
From 2 Å res. F_0 map	0.389	0.663 218	0.312 765	0.385 2151	0.397 1899	5033	15	23	1
From 2 Å res. F_0 map	0.376[a]	0.504 212[a]	0.307 765	0.390 2151	0.392 1899	5027	13.5	23	1
From 2 Å res. F_0 map	0.372	0.461 212	0.301 765	0.392 2151	0.394 1899	5027	12	23	1
From 1st ΔF map	0.321	0.389 212	0.271 765	0.328 2151	0.346 1899	5027	12	22	1
From 2nd ΔF map	0.289	0.389 212	0.239 765	0.292 2151	0.312 1899	5027	9–20	22	1
From 3rd ΔF map	0.262	0.389 212	0.209 765	0.261 2151	0.288 1899	5027	7–30	22	1
From 4th ΔF map	0.242	0.373 212	0.192 765	0.239 2151	0.263 1899	5027	6–40	23	1
From 4th ΔF map	0.224	0.306 212	0.170 765	0.225 2151	0.260 1899	5027	6–40	100	0.3–1
1st L. S. cycle	0.179	0.192 212	0.123 765	0.181 2151	0.241 1899	5027	–5–50	106	0.3–1
2nd L. S. cycle	0.148	0.159 212	0.092 765	0.154 2151	0.222 1899	5027	3–50	127	0.3–1
3rd L. S. cycle	0.132[b]	0.118 190[b]	0.080 765	0.139 2151	0.188 1899	5005	2–107	130	0.3–1
F_c including 256 H atoms	0.141	0.153 190	0.092 765	0.145 2151	0.192 1899	5005	2–107	130	0.3–1
4th L. S. including 256 H	0.126	0.112 190	0.078 765	0.131 2151	0.183 1899	5005	2–86	130	0.3–1

[a] From K. D. Watenpaugh, L. C. Sieker, J. R. Herriott, and L. H. Jensen, *Cold Spring Harbor Symp. Quant. Biol.* **36**, 359 (1971). Copyright 1971 Cold Spring Harbor Laboratory.

[b] 6 reflections with $\sin \theta/\lambda$ < 0.03 weighted 0.

[c] 28 reflections with $\sin \theta/\lambda$ < 0.05 weighted 0.

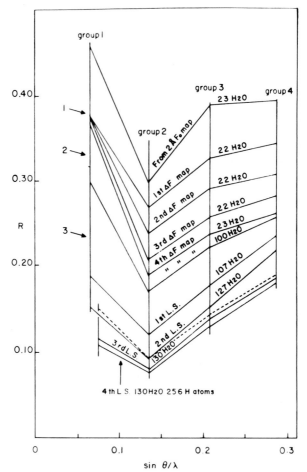

FIG. 14. R factor as a function of $\sin \theta/\lambda$ ($=\frac{1}{2}s$) for different stages of refinement. On the individual curves L.S. represents least-squares refinement and ΔF denotes difference Fourier refinement. The number of water molecules included is also shown. (Reprinted from Watenpaugh *et al.*[54] with permission.)

distributed solvent molecules that are usually ignored. This omission introduces errors in the reflections with low s (low-frequency components, Chapter 6.2). Other factors affect mainly the high-resolution reflections (i.e., large s). They are related to the approximations made during the least-squares refinement which lead to false minima, and to the use of isotropic temperature factors (Section 6.3.1).

Overall, determination of the rubredoxin structure represents a highly successful attempt to obtain a very accurate model of a protein molecule.

In this case it is probable that the extra effort will pay off, as the knowledge of the precise geometry of the Fe–S cluster and of some of the interacting amino acids could become very important when analyzing the function of the protein.

6.5. Fibers

6.5.1. Definition and Description

Fibers form the other major class of biological specimens studied by XRD.[55] They contain periodicities along one axis, called the fiber or z axis, and are nonperiodic in the other two dimensions. They are bundles of thread-like or rodlike objects. For these reasons the theory of fiber diffraction is most conveniently expressed in cylindrical coordinates, where Bessel functions play a prominent role. The broad definition of fibers given above conceals a number of distinct types, which we now discuss.

Biological fibers can be divided into classes depending on the nature of the units that build up the fiber, and also on the way the units pack together to form the fiber. Perhaps the most familiar unit is a polymer chain, arranged in an extended or helical conformation. Frequently, as in most polynucleotides and fibrous proteins, several chains are wound around one another in the unit. The double helical structure of DNA is perhaps the most familiar example of this. These polymers normally have rather few atoms in their monomer and generally come with one of a variety of side groups attached to each monomer. These groups are bases in the case of nucleic acids, and amino acid side chains in the case of fibrous proteins. The side chains normally occur in a random sequence and do not obey the symmetry of the unit. In any structure determination of this type side chains can only be visualized as average masses.

In the other major class of fibers, whole globular proteins, rather than small monomers, are polymerized to form the unit, rodlike object. Monomer–monomer interactions are usually noncovalent. The self-assembly of these units provides interesting systems for physical-chemical study. Rodlike viruses such as the tobacco mosaic virus (TMV) and the tail of the bacterial virus T4 are important examples of this type of fiber. The fibrous proteins actin and myosin, which form the basis of many contractile systems, also belong to this class.

Fibers are also classified by the ways in which the threads are packed side by side to form the specimen. The packing schemes have great variety

[55] B. K. Vainshtein, "Diffraction of X-Rays by Chain Molecules." North-Holland Publ., Amsterdam, 1966.

and complexity, and we mention only a few prototypical cases. In crystalline fibers unit rods pack side by side to form small, truly crystalline bundles. These bundles are then packed with random azimuths (rotations about z) into the fibrous specimen. Such a specimen yields a diffraction pattern composed of discrete spots confined to layer lines normal to the fiber axis. The same pattern would be given by an individual bundle if it were rotated through 360° about z during photography. Certain preparations of DNA are of this type, as are specimens of some simple polymers, such as polyethylene.

In the oriented gel, on the other hand, the unit rods have random azimuths and diffract independently of one another. The diffraction pattern from an oriented gel is the azimuthal average of the pattern from one rod and displays a continuous distribution of intensity along layer lines. The canonical specimen in this group is TMV, which we discuss in detail in Section 6.5.4. It is remarkable that enough information can be recovered from the rotationally averaged diffraction pattern of TMV to generate an image which reveals most of the atoms in the constituent globular protein monomer.

A variety of other specimen types exist, in which various stages of semicrystallinity are present, or in which disorder such as dislocations or irregular alternation of the sense of rods is present.[56] Frequently, great experimental art is required for preparation of fibers suitable for x-ray diffraction study, and a wide variety of techniques, such as extrusion and quick stretching, have been used. Nevertheless, fiber diffraction photographs are usually characterized by arcing, a smearing of the pattern arising from lack of parallelism among threads. The effects of arcing become more serious as one looks farther from the center of the photograph until, finally, layer lines may merge into one another, setting limits on the resolution of the useful data that can be collected. Attempts to deconvolute the smearing from the diffraction photographs have had very promising initial application.[57]

It should be reemphasized that all fibers give rise to azimuthally averaged diffraction patterns. The information contained in these patterns is much more confused than that in the diffraction by single crystals. The problem is especially severe for oriented gels, in which the averaged pattern of a unit rod is observed.

6.5.2. Fiber Diffraction Theory

Fiber diffraction theory was first presented in seminal papers by Cochran et al.[58] and Klug et al.[59] It does not offer any tool as simple as diffraction

[56] S. Arnott, *Am. Crystallogr. Assoc.* **9**, 31 (1973).

[57] L. Makowski, *J. Appl. Crystallogr.* **11**, 273 (1978).

[58] W. Cochran, F. H. C. Crick, and V. Vand, *Acta Crystallogr.* **5**, 581 (1952).

[59] A. Klug, F. H. C. Crick, H. W. Wyckoff, *Acta Crystallogr.* **11**, 199 (1958).

from a spatial plane wave in the crystallographic case, but useful and conceptually simple special cases do exist. We begin by writing down the diffraction pattern of a general object in cylindrical coordinates. From Eq. (6.2.7)

$$\mathbf{F}(X, Y, Z) = \iiint \rho(x, y, z) \exp[2\pi i(xX + yY + zZ)] \, dV.$$

We define polar coordinates by

$$x = r \cos \phi, \qquad y = r \sin \phi, \qquad z = z;$$
$$X = R \cos \psi, \qquad Y = R \sin \psi, \qquad Z = Z. \tag{6.5.1}$$

Then

$$xX + yY + zZ = rR(\cos \phi \cos \psi + \sin \phi \sin \psi) + zZ$$
$$= rR \cos(\psi - \phi) + zZ.$$

Thus

$$\mathbf{F}(R, \psi, Z) = \iiint \rho(r, \phi, z) \exp[2\pi i r R \cos(\psi - \phi)] \exp(2\pi i z Z) r \, dr \, d\phi \, dz. \tag{6.5.2}$$

Using the identity

$$\exp[2\pi i r R \cos(\psi - \phi)] = \sum_{-\infty}^{\infty} J_n(2\pi r R) i^n \exp[in(\psi - \phi)],$$

where J_n is the Bessel function of order n, we find

$$\mathbf{F}(R, \psi, Z) = \sum_n \exp(in\psi) \iiint i^n \rho(r, \phi, z) J_n(2\pi r R)$$
$$\times \exp(-in\phi) \exp(2\pi i z Z) r \, dr \, d\phi \, dz. \tag{6.5.3}$$

Note that $\mathbf{F}$ has merely been expanded in a Fourier series in the periodic angular variable. The Fourier coefficients are the triple integral which multiplies $\exp(in\psi)$. To emphasize this the triple integral is often denoted $\mathbf{G}_n(R, Z)$, so that

$$\mathbf{F}(R, \psi, Z) = \sum \mathbf{G}_n(R, Z) i^n \exp(in\psi). \tag{6.5.4}$$

The factor i^n remains by convention. The relation

$$\mathbf{G}_n(R, Z) = \iiint \rho(r, \phi, z) J_n(2\pi r R)$$
$$\times \exp(-in\phi) \exp(2\pi i z Z) r \, dr \, d\phi \, dz \tag{6.5.5}$$

is a type of Fourier–Bessel transform.

Thus far $\mathbf{F}$ is the diffraction pattern of a perfectly general object in cylindrical coordinates. The period of the fiber can be readily introduced, however. If ρ has period c along z, then $\mathbf{F}$ will be identically zero except on layer lines of spacing l/c—that is, except when $Z = l/c$—where l is any integer. This is a slight generalization of discrete Fourier coefficients for a one-dimensional, periodic function. Thus

$$\mathbf{F}\left(R, \psi, \frac{l}{c}\right) = \sum_n \exp(in\psi) \iiint i^n \rho(r, \phi, z) J_n(2\pi rR)$$

$$\times \exp\left(\frac{ilz}{c}\right) \exp(-in\phi) r \, dr \, d\phi \, dz, \qquad (6.5.6)$$

or, by extension of the previous notation,

$$\mathbf{F}\left(R, \psi, \frac{l}{c}\right) = \sum \mathbf{G}_{n,l}(R) i^n \exp(in\psi). \qquad (6.5.7)$$

The coefficients $\mathbf{G}_{n,l}$ will play a central role in the description of the rotationally averaged pattern.

It is useful and insightful to make an analogous expansion of the electron density function: ρ is now periodic in ϕ and z, so that it can be expanded in a two-dimensional Fourier series

$$\rho\left(r, \phi, \frac{z}{c}\right) = \frac{1}{c} \sum_n \sum_l g_{n,l}(r) \exp(in\phi) \exp\left(-\frac{2\pi ilz}{c}\right). \qquad (6.5.8)$$

The $g_{n,l}$ are slight generalizations of the usual Fourier coefficients and are given by

$$g_{n,l}(r) = \frac{1}{2\pi} \int_0^{2\pi} \int_0^l \rho\left(r, \phi, \frac{z}{c}\right) \exp(-in\phi) \exp\left(\frac{2\pi ilz}{c}\right) d\phi \, dz. \qquad (6.5.9)$$

It can be shown that

$$\mathbf{G}_{n,l}(R) = 2\pi \int_0^\infty g_{n,l}(r) J_n(2\pi rR) r \, dr \qquad (6.5.10)$$

and that

$$g_{n,l}(r) = \frac{1}{2\pi} \int_0^\infty \mathbf{G}_{n,l}(R) J_n(2\pi rR) R \, dR. \qquad (6.5.11)$$

This shows directly that if one can determine the individual $\mathbf{G}_{n,l}$, which are complex, the Fourier coefficients $g_{n,l}$ of the electron density are directly available through Eq. (6.5.11).

It is almost universal that the rodlike elements in a biological fiber are themselves wound into a helix of one sort or another. Helicity in the fiber introduces certain simplifications into the diffraction pattern which permit one (at least in simple cases) to deduce the geometry of the helix, and make possible the generation of an image from a rotationally averaged diffraction pattern. We approach diffraction by helical objects in stages. First we analyze the diffraction pattern from a wire bent in the shape of a helix of radius r_0 and repeat (pitch) c. Such an object can be formally represented as the product of two δ functions

$$\delta(r - r_0)\delta\left(\frac{\phi}{2\pi} - \frac{z}{c}\right), \tag{6.5.12}$$

where we assume that the helix passes through the point $(r_0, 0, 0)$. Inserting this function into Eq. (6.5.6) for the diffraction pattern of a periodic object we find

$$\mathbf{F}\left(R, \psi, \frac{l}{c}\right) = \sum_n \exp(in\psi) \int_0^c \int_0^{2\pi} \int_0^\infty i^n \delta(r - r_0)\delta\left(\frac{\phi}{2\pi} - \frac{z}{c}\right)$$

$$\times J_n(2\pi r R) \exp(-in\phi) \exp\left(\frac{2\pi i l z}{c}\right) r \, dr \, d\phi \, dz. \tag{6.5.13}$$

The integral over r can be done by inspection, giving

$$\mathbf{F}\left(R, \psi, \frac{l}{c}\right) = \sum_n \exp(in\psi) J_n(2\pi r_0 R) i^n$$

$$\times \int_0^{2\pi} \int_0^c \delta\left(\frac{\phi}{2\pi} - \frac{z}{c}\right) \exp(-in\phi) \exp\left(\frac{2\pi i l z}{c}\right) dz \, d\phi. \tag{6.5.14}$$

To integrate over ϕ and z a change of variable is helpful: define

$$u = \frac{\phi}{2\pi} + \frac{z}{c}, \qquad v = \frac{\phi}{2\pi} - \frac{z}{c}.$$

Then

$$\phi = \pi(u + v), \qquad z = \tfrac{1}{2}c(u - v).$$

Substituting

$$\mathbf{F}\left(R, \psi, \frac{l}{c}\right) = \pi c \sum_n J_n(2\pi r_0 R) \exp(in\psi) i^n$$

$$\times \int_0^2 \int_0^2 \delta(v) \exp[-in\pi(u + v)] \exp[\pi i l(u - v)] \, du \, dv, \tag{6.5.15}$$

simplifying,

$$\mathbf{F}\left(R, \psi, \frac{l}{c}\right) = \pi c \sum_n J_n(2\pi r_0 R) \exp(in\psi) i^n$$

$$\times \int_0^2 \int_0^2 \delta(v) \exp[-i\pi v(l + n)] \exp[i\pi u(l - n)]\, du\, dv,$$

$$(6.5.16)$$

and integrating over v, one arrives at

$$\mathbf{F}\left(R, \psi, \frac{l}{c}\right) = \pi c \sum_n J_n(2\pi r_0 R) \exp(in\psi) i^n \int_0^2 \exp[i\pi u(l - n)]\, du. \quad (6.5.17)$$

This integral vanishes unless $n = l$, when

$$\mathbf{F}(R, \psi, l/c) = (2\pi/c) i^l J_l(2\pi r_0 R) \exp(il\psi). \quad (6.5.18)$$

The factor $2\pi c$ is normally omitted.

This pattern is greatly simplified compared with the pattern from a general fiber. First, only a single Bessel function occurs on any layer line, with the order of the Bessel function being equal to the layer line index l. Because of the form of the Bessel functions J_l this gives the diffraction pattern an $\times$-shaped appearance. As shown in Fig. 15, the first maximum of J_0 is at the origin, and the first maxima of the J_n move farther out from the origin as n increases in magnitude. Thus the main bars of the $\times$ are formed by the first maximum ridge of the J_n shown in this figure. Subsidiary maxima of the J_n produce fringes outside the main $\times$. This is illustrated in Fig. 16. The $\times$-shaped form of the pattern, although not its exact equation, can be deduced by a very simple argument from the general shape of the helix. Figure 16b shows that a helix can be roughly expressed as the sum of a set of parallel line segments forming one face of the helix, and a similar set of line segments

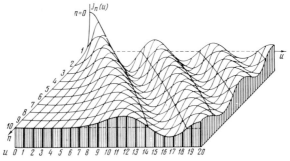

FIG. 15. Bessel functions as a function of order (n) and argument (u). (Reproduced from Vainshtein[55], with permission.)

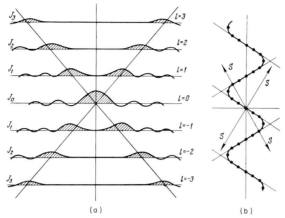

FIG. 16. Diffraction by helices: (a) values of the Bessel function J_l that determine the intensity of the lth layer line in the diffraction pattern of a continuous helix. Note the " $\times$ " drawn through the first maxima of the Js. (b) the $\times$-shaped diffraction pattern is related to structural features in the helix. In projection, the helix is approximated by the sum of two stacks of sloping line segments. Diffraction from these stacks occurs along directions S perpendicular to the stacks, forming the arms of the $\times$. (Reprinted from Vainshtein[55], with permission.)

forming the back. Each set of lines gives rise to a diffraction pattern of a row of points normal to the stack of line segments. The two rows of points comprise the two arms of the $\times$.

A uniform helical wire is not a very useful model for most systems. We will shortly generalize this for a helix made of atoms, by computing the diffraction pattern of a set of beads (atoms) spaced uniformly along a helix. We discuss here a generalization of the helical wire whose diffraction pattern is particularly simple. This parallels the case of crystals, where we showed that by generalizing the usual stack of Bragg planes to a spatial, sinusoidal density fluctuation we could produce an object that could be used to synthesize any specimen, and which also had a very simple diffraction pattern, one Bragg reflection. The existence of a strong reflection therefore implies the existence of a correspondingly strong Fourier component in the object, so that crude ideas about important periodicities can be quickly obtained.

A simple choice for the helical case is suggested by the expansion for $\rho(r, \phi, z/c)$ given in Eq. (6.5.8). We choose

$$\rho(r, \phi, z/c) = J_1(2\pi r R')\exp(i\phi - 2\pi i z/c), \qquad (6.5.19)$$

which represents one term in the expansion, with $\mathbf{g}_{1,1} = J_1$. Substituting for $G_{n,1}$ in Eq. (6.5.10), we find that

$$\mathbf{G}_{1,1}(R) = \int_0^\infty J_1(2\pi r R')J_1(2\pi r R)R \, dR. \qquad (6.5.20)$$

Using the orthogonality conditions for Bessel functions we realize that this integral vanishes unless $R' = R$. Thus

$$\mathbf{G}_{1,1}(R) = \delta(R - R'), \tag{6.5.21}$$

while all other $G_{n,1}$ vanish. This implies that the whole diffraction pattern is

$$\mathbf{F}(R, \psi, 1/c) = \delta(R - R')i^n \exp(i\psi), \tag{6.5.22}$$

which is a ring with periodically varying phase on the first layer. Using this result, very strong reflections on particular layer lines can be related to the existence (in the object) of helical waves or structures of the form chosen in this example.

Note also that the intensity of the diffraction pattern, $\mathbf{FF}^*$, is independent of azimuth in these two cases. Only the phase factor $\exp(i\psi)$ varies. Therefore in this special case the rotationally averaged diffraction pattern from a specimen is identical to the unaveraged one. Since the phase factor is known *a priori* if a helix is assumed, the diffraction pattern from any view of the object is sufficient to reconstruct an image. This is consistent with the idea that, because of helical symmetry, one view of a helix can be used to generate all others.

We now proceed to the example of point masses strung on a massless helical wire.[2] This model (the discrete helix) is of great general utility, since any helical specimen can be represented as a sum of such helices. In DNA, for example, all the phosphates on one strand would form one such helix, all the 3' oxygens another, and so on. These helices will all have the same pitch, but their radii and azimuths will vary. This model can be readily generated from the continuous wire model by multiplying it by a function that has value one on a stack of equally spaced planes normal to the helix axis, and value zero elsewhere. The spacing p between planes becomes the z distance between atoms, and is often termed the rise per residue.

The diffraction pattern of this structure can be generated by use of the convolution theorem, by convolving the Fourier transform of the stack of planes with the Fourier transform (diffraction pattern) of the continuous helix. It is readily shown that the transform of the stack of planes is a set of points spaced $1/p$ apart along the meridian. This convolution amounts to reproducing the $\times$ diagram at each meridional point. This is illustrated in Figs. 17 and 18. For other examples see Harburn *et al.*[60]

To express this result formally some relation between p and P must be assumed. For the case of a helical wire the pitch P was also the true repeat.

[60] G. Harburn, C. A. Taylor, and T. R. Welberry, "An Atlas of Optical Transforms." Cornell Univ. Press, Ithaca, New York, 1975.

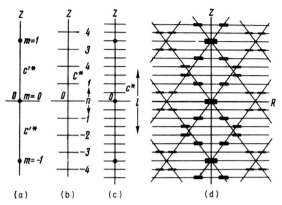

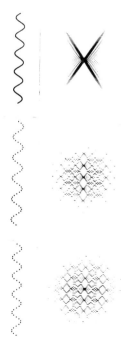

FIG. 17. Formulation of the diffraction pattern of a discontinuous helix: (a) system of points with a repeat distance C'^*, being the transform of a system of planes in real space; (b) system of planes of repeat distance C^*; (c) convolution of the previous two systems, which is a system of planes of repeat distance C^* defining the position of the layer lines; (d) distribution of the first peaks in J_n over the layer planes. The example drawn is for seven residues in two turns. (Reproduced from Vainshtein[55], with permission.)

FIG. 18. Diffraction patterns from a continuous helix and from helices having closely or broadly spaced atoms. This illustrates the convolution of the × pattern with rows of points in reciprocal space.

In a discrete helix, however, there need not be a whole number of atoms per turn. We therefore write the true repeat c

$$c = tP = up, \tag{6.5.23}$$

where t and u are integers. This expression serves to set the relation between the interleaved $\times$s in the diffraction pattern. On layer line l—that is, at $Z = l/c$—only certain J_n can appear, those which fall on this layer line through the convolution process. The allowed values of n are then given by

$$l = tn + um, \tag{6.5.24}$$

where m is in principle any integer; in practice only small values of m generate an n for which J_n is nonzero within the resolution of the photograph.

Given this analysis, we can directly write the diffraction pattern as

$$\mathbf{F}\left(R, \psi, \frac{l}{c}\right) = \sum_n i^n J_n(2\pi r_0 R) \exp(in\psi), \tag{6.5.25}$$

where the sum is over those values of n consistent with the selection rule (6.5.24). The helix in this example was initialized by assuming one atom at the point $r = r_0$, $\phi = 0$, $z = 0$. The full formula for a family of helices with atoms at (r_j, ϕ_j, z_j) is given below, and amounts to the structure factor equation for a helical object instead of a crystal. We can then write

$$\mathbf{F}\left(R, \psi, \frac{l}{c}\right) = \sum_n \sum_j f_j i^n J_n(2\pi r_j R) \exp in\psi\left(- in\phi_j + \frac{2\pi i l z_j}{c}\right), \tag{6.5.26}$$

where the atomic scattering factor f_j has been introduced to give shape to our points, and the phase shifts involving ϕ_j and z_j reflect the positions of the atoms along their individual helices.

It is often convenient to separate the angular and spatial variables in this equation. If we write by analogy with (6.5.5)

$$\mathbf{G}_{n,l}(R) = \sum_j f_j J_n(2\pi r_j R) \exp\left(- in\phi_j + \frac{2\pi i l z_j}{c}\right) \tag{6.5.27}$$

then

$$\mathbf{F}\left(R, \psi, \frac{l}{c}\right) = \sum_n i^n \mathbf{G}_{n,l}(R) \exp(in\psi), \tag{6.5.28}$$

where again the sum is taken only over those values of n that satisfy the selection rule. In general, only one or a few terms contribute significantly to $\mathbf{F}$, because many of the J_n are forbidden by the selection rule. Thus helical structures in general have diffraction patterns much simplified compared with a general fiber, for which all J_n exist on all layer lines.

As mentioned before, the observed diffraction pattern is rotationally averaged. The observed intensity I is

$$I = \langle \mathbf{FF^*} \rangle_\psi = \frac{1}{2\pi} \int_0^{2\pi} (\sum G_{n,l} \exp(in\psi)i^n)(\sum G^*_{n,l} \exp(-in\psi)(-i)^n) \, d\psi.$$

(6.5.29)

The integral vanishes unless $m = n$. Therefore

$$I\left(R, \frac{l}{c}\right) = \sum \mathbf{G}_{n,l}(R)\mathbf{G}^*_{n,l}(R).$$

(6.5.30)

If one wishes to reconstruct an object directly from the diffraction pattern, rather than to guess a model, one must know the phase and amplitude of

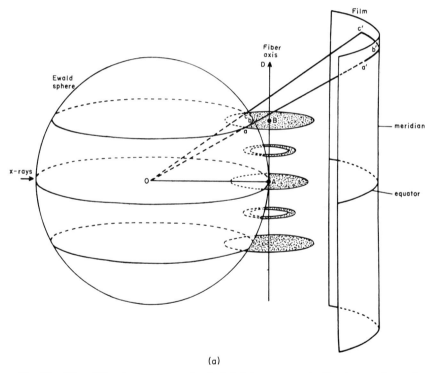

(a)

FIG. 19a. Fiber diffraction geometry: the shaded disks represent the Fourier transform of a fiber. Lines drawn through the intersection of the disks with the Ewald sphere from the center O of the sphere represent the scattered ray directions. The points of intersection of the disk C with the sphere (a, b, and c) give a′, b′, and c′, respectively, on the film. The points a′, b′, and c′ form a layer line. Note that point B does not give rise to any point on the film. (From Holmes and Blow.[2] Copyright © 1966. John Wiley & Sons, Inc. Reprinted by permission of John Wiley & Sons, Inc.)

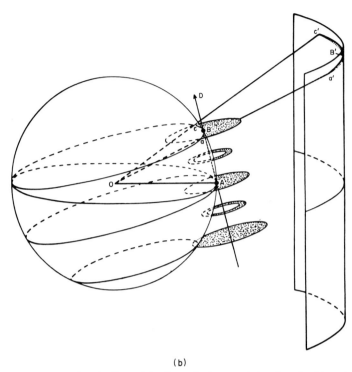

(b)

FIG. 19b. Same as (a) but the fiber axis has been tilted so as to bring the point B into the reflecting position. Note that point B projects onto the meridian at B′ and that the layer line is no longer straight. (From Holmes and Blow.[2] Copyright © 1966. John Wiley & Sons, Inc. Reprinted by permission of John Wiley & Sons, Inc.)

each $\mathbf{G}_{n,l}$. From these one can calculate the full, complex $\mathbf{F}(R, \psi, l/c)$ via Eq. (6.5.28) and then synthesize the object by Fourier transformation of $\mathbf{F}$. Alternatively, one can use Eq. (6.5.11) to generate the $\mathbf{g}_{n,l}$, and use these to generate ρ from Eq. (6.5.8). Since each $\mathbf{G}_{n,l}$ is generated from only one J_n, the problem of recovering the $\mathbf{G}$s from the Is is sometimes referred to as separating the Bessel functions. Remarkably enough, this process has actually been realized in the case of TMV (Section 6.5.4).

A word about how the intensity data are measured is appropriate here. As we have seen, diffraction from fibers is confined to a set of planes in reciprocal space. Since the data are rotationally averaged, there is no variation in intensity as one moves azimuthally in any plane. Thus all the available data are contained in radial scans running from the meridian to the limits of resolution in each of these planes. Figure 19 shows the Ewald sphere construction for a fiber. Note that the sphere intersects each reciprocal space plane along an arc which provides almost all of the desired radial scan. Only a region near the meridian is missing—the blind region. The second panel shows how tilting the fiber can help fill in the missing data.

The required photographs are stills—neither the fiber nor the film moves. If data from the blind region can be ignored, a single photograph provides a complete data set.

6.5.3. Solution of Simple Fiber Structures

We defer discussion of TMV to Section 6.5.4 and examine here the simpler problems that are more characteristic of the field. Problems can be simple because the fiber structure itself is simple (e.g., when the specimen is a polymer having only a small number of atoms in the monomeric unit). They can also be simple because the diffraction patterns are poor and contain so little information that only a crude structure can be determined.

In the case of polymers whose elements are fairly simple monomer units, atomic models are possible. Polypeptides and polynucleotides may fall into this category. The geometry and chemistry of the monomer often provide severe constraints on the model, and, in a sense, only a very small number of the first models may be possible. Watson's description in "The Double Helix"[61] of the thought processes leading to the model of DNA is a readable example of this method. Early on, such models were adjusted by trial and error to give improved agreement with the observed data. Later, Arnott and and others[62] made use of the structure factor formula as an observational equation for least-squares refinement, similar to the process described earlier for crystals. To deal with the relatively small number of data, it was necessary to reduce the number of parameters in the model as much as possible. This was accomplished by using rotations about single bonds as the independent variables, while constraining the lengths of bonds and the angles between them to canonical values. This, the "linked-atom least squares" (LALS) method, has been successful in providing accurate atomic models for a variety of polynucleotides and polysaccharides. Techniques for modeling various types of disorder have also been incorporated into the analysis.

On a much larger scale, modeling techniques have also provided images of specimens in which the repeating unit is a whole protein. In these cases no internal detail is seen in the subunit, which in any case, is often modeled as an ellipsoid or a cylinder. Joint use of electron microscopy is very common for these specimens, so that low-resolution images upon which a model can be based are frequently available.

[61] J. D. Watson, "The Double Helix." Atheneum, New York, 1968.
[62] P. J. Campbell-Smith and S. Arnott, *Acta Crystallogr.*, *Sect. A* **A34,** 3 (1978).

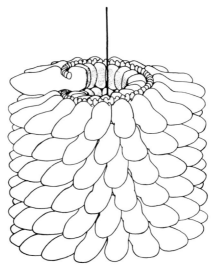

FIG. 20. Drawing of tobacco mosaic virus. Note the Dutch-shoe-shaped protein subunits on the outside, and the RNA wound in the lumen of the central, cylindrical cavity. (From Caspar.[65])

6.5.4. Tobacco Mosaic Virus Structure

As mentioned earlier, an atomic model of TMV has been obtained from fiber diffraction studies. It represents by far the most complex fiber analyzed in this way, and is also the first system in which native protein–ribonucleic acid interactions have been seen in detail.[63] For this reason we present a short review of this project.

TMV is a rodlike virus, some 3000 Å long and 90 Å in radius, with a central hole of radius 20 Å. Identical protein subunits build a helix with 49 subunits in 3 turns. The RNA follows the same helical path and is positioned at the edge of the central hole. Three nucleotides are bound per protein subunit. The pitch of the helix is 23 Å, so that the rise per subunit is only $23 \text{ Å} \times \frac{3}{49} = 1.4 \text{ Å}$. A drawing of the virus, as its structure was understood in the early 1960s,[64] is shown in Fig. 20.[65] Because the helix is so shallow, the opening angle of the × in the diffraction pattern is very small, and it was experimentally difficult to decide whether a reflection was actually meridional, or whether it only appeared to be so because of arcing. Thus the value of 49 residues in 3 turns was in dispute for some time, and was only made certain

[63] G. Stubbs, S. Warren, and K. Holmes, *Nature* (*London*) **267**, 216 (1977).

[64] A. Klug and D. L. D. Caspar, *Adv. Virus Res.* **7**, 225 (1960).

[65] D. L. D. Caspar, *Adv. Protein Chem.* **18**, 37 (1963).

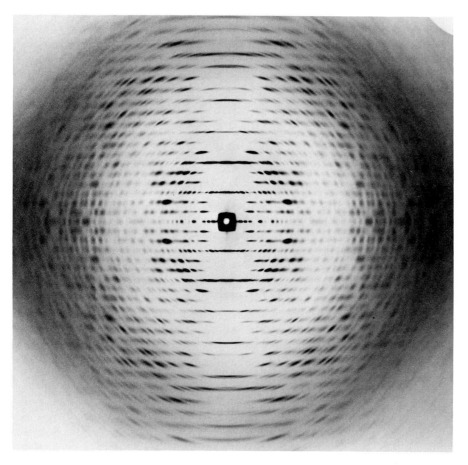

FIG. 21. Photograph of the diffraction pattern of a fiber of tobacco mosaic virus containing data to 3-Å resolution. All the native data needed to reconstruct the fiber are contained in this photograph. (Courtesy of Gerald Stubbs.)

by Franklin and Homes[66] with the use of a heavy-atom derivative. The rods form an oriented gel which gives diffraction patterns of very high quality. The effects of interparticle interference are very small, so that the pattern is essentially the rotational average of the diffraction intensity from a single rod. An example of such a pattern is shown in Fig. 21. It is immediately clear that the repeat is 3 turns, since every third layer line has intensity near the meridian, consisting of the low-order Bessel functions from the central $\times$ in the diffraction pattern. The selection rule is $l = 3n + 49m$. When $l = 0$, on the

[66] R. E. Franklin and K. C. Holmes, *Acta Crystallogr.* **11**, 213 (1958).

equator, the zeroth-order Bessel function appears, and the next is J_{49}, which becomes nonzero at about 10-Å resolution, and mixes hardly at all with J_0, which is only a small fraction of its peak value at this point.

The great wealth of detail in this photograph is readily apparent, and generated the original hope that the atomic structure of TMV might be revealed by it. The final stages of this project were carried out in the laboratory of K. C. Holmes in Heidelberg,[67] where a generalization of the isomorphous replacement method was devised to deal with this problem. We recall that from Eqs. (6.5.28) and (6.5.27) that

$$F\left(R, \psi, \frac{l}{c}\right) = \sum i^n G_{n,l}(R) \exp(in\psi),$$

$$G_{n,l}(R) = \sum_j f_j J_n(2\pi r_j) \exp(-in\phi_j) \exp\left(\frac{2\pi i l z_j}{c}\right),$$

while the observed intensity is

$$I = \langle FF^* \rangle_\psi = \sum_n G_{n,l} G_{n,l}^*.$$

It is sufficient to find the values of the individual $G_{n,l}$, since these lead directly to either F [Eq. (6.5.28)] or ρ [Eq. (6.5.8)]. However, only the sum of G^2 is observed. Imagine now that a heavy atom is bound, as in a protein crystal, to a specific site in TMV, which we can determine. We pass over for the moment how this might be accomplished. This heavy-atom derivative will give rise to a diffraction pattern built from the $G'_{n,l}$, which are simply related to the native $G_{n,l}$. Let

$$G'_{n,l} = G_{n,l} + \gamma_{n,l},$$

$$\gamma_{n,l} = \sum_m f_m J_n(2\pi r_m R) \exp(-in\phi_m) \exp\left(\frac{2\pi i l z_m}{c}\right), \qquad (6.5.31)$$

where m runs only over all heavy atoms, and write as well

$$G = A + iB, \qquad (6.5.32)$$

$$\gamma = a + ib. \qquad (6.5.33)$$

Then

$$I = A_1^2 + B_1^2 + A_2^2 + B_2^2 + \cdots, \qquad (6.5.34)$$

$$I' = (A_1 + a_1)^2 + (B_1 + b_1)^2 + (A_2 + a_2)^2 + (B_2 + b_2)^2 + \cdots. \qquad (6.5.35)$$

[67] A. N. Barrett, J. Barrington Leigh, K. C. Holmes, R. Leberman, E. Mandelkow, and P. von Sengbush, *Cold Spring Harbor Symp. Quant. Biol.* **36**, 433 (1971).

An additional equation like that for I' arises for each heavy-atom derivative. All the as and bs are assumed to be known, since they can be calculated from the known heavy-atom positions. What remains is to calculate the As and Bs; two convenient circumstances make this possible. First, it is characteristic of Bessel functions (Fig. 15) that they remain zero until their argument is about equal to their order. Thus, within a given radius of reciprocal space, say $1/(4\ \text{Å})$, only a limited number of Bessel functions can contribute to the diffraction pattern; the others will all be very small. Second, because of the selection rule for the TMV helix, Bessel functions occurring on the same layer line differ in order by integer multiples of 49, so that very few low-order Js can appear. Given the radius of the virus rod, one can readily show that out to 4-Å resolution at most three Bessel functions contribute at any point along a layer line. This means that Eq. (6.5.34) has only six terms in it—three As and three Bs. A minimum of six heavy-atom derivatives, one for each term, is required to solve these equations.

Solution of these equations is not trivial, and required great insight and care.[68] First, it must be emphasized that Eqs. (6.5.34) and (6.5.35) are for the Is and Gs at one point in the diffraction pattern. They must be solved at a large set of finely spaced sample points to provide the desired information. Further, the difficulties in assigning accurate parameters to the heavy atoms are even greater in fiber diffraction than in protein crystallography, so that the as and bs are subject to considerable error. Equation (6.5.34) is therefore the most accurate of the set. In the technique actually used the heavy-atom equations are solved using (6.5.35) as a constraint. Further, the additional constraint inherent in the smooth variation of the individual Gs was taken into account in a global way, and proved very powerful in eliminating false solutions at certain points.

In the actual TMV case seven heavy-atom derivatives having five independent sites were used. The determination of heavy-atom positions was fraught with difficulty.[67] Consider, for example, that the azimuthal angle ϕ for a single heavy atom cannot be determined from a rotationally averaged diffraction pattern. Only when cross derivatives involving two different heavy atoms are examined can a $\Delta\phi$ between the heavy atoms be calculated.

The proof of all these elaborate methods lies in their ability to produce an interpretable and plausible electron density map. The 4-Å resolution structure of TMV published in 1977[69] makes clear the fundamental soundness of the method. The polynucleotide chain with its electron-dense phosphorus atoms is the most prominent feature in the maps. The coat protein subunits

[68] G. J. Stubbs and R. Diamond, *Acta Crystallogr., Sect. A* **A31**, 709 (1975).

[69] K. C. Holmes, G. J. Stubbs, E. Mandelkow, and U. Gallwitz, *Nature (London)* **254**, 192 (1975).

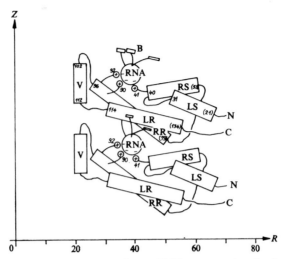

FIG. 22. Schematic drawing of two adjacent TMV coat protein subunits. The cylinders represent helical segments. The interaction of the RNA with the protein is also shown. R and Z represent the radial and fiber axis directions in the structure. (Reproduced from Stubbs *et al.*[63] with permission.)

contain extensive stretches of α-helix, and these made possible the tracing of the protein chain through most of the subunit. For technical reasons the effective resolution of the map drops to about 5.5 Å near the outer 80-Å radius of the virus, and in this area chain tracing was not possible. Fortunately, this portion of the molecule can be visualized in a closely related structure. An intermediate in the assembly of TMV, termed the TMV disk, has been

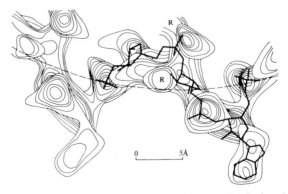

FIG. 23. Perspective drawing of sections of contoured electron density function of TMV. The region shown contains RNA, and the RNA model derived from this density is superimposed. (Reproduced from Stubbs *et al.*[63] with permission.)

crystallized and analyzed in the laboratory of Aaron Klug.[70] The outer portion of the subunits are clearly seen in this structure.

In the virus the RNA chain follows a complex path ranging from 35 to 50 Å from the virus axis. For the first time considerable detail of protein–nucleic acid interaction can be made out. There are three bases associated with each protein subunit, and these all wrap around one segment of helix (termed the left radial) like a roll enclosing a hot dog. A section of the RNA binding site of TMV and a schematic drawing of the structure are shown in Figs. 22 and 23. Additional refinement is now being carried out at 3-Å resolution (G. J. Stubbs, personal communication), and an improved version of the TMV model is expected shortly.

6.5.5. Structure of Microtubules

Microtubules are cellular organelles involved with cell division, transport of materials from one part of the cell to another, and other motile properties of the cell.[71] Like TMV they are hollow rods. The diffraction patterns they produce are also the rotational average of the one-rod pattern, but are much less rich than the pattern from TMV. An x-ray diffraction study of brain microtubules by Mandelkow et al.[72] provides an interesting example of the amount of information that can be gained from a low-resolution study. It also illustrates very useful interplay between electron microscopy and x-ray diffraction. Microtubules are built from two closely related monomers, α- and β-tubulin, which readily form an α–β heterodimer. The dimers readily polymerize into microtubules with increasing temperature. Microtubules have been studied by electron microscopy in the laboratories of Klug[73] and Erickson.[74] These groups showed that the exterior of the tubules was characterized by an array of 13 so-called protofilaments running parallel to the microtubule axis, each staggered axially with respect to its neighbor by about 9 Å. A schematic of this arrangement is shown in Fig. 24. The spacing between tubulin molecules within a protofilament is about 40 Å.

We now consider in a crude and simple way how a microtubule diffracts x rays. Diffraction takes place where there are large fluctuations in electron density. For microtubules this occurs at the inner and outer surface of the hollow rod. Looking at Fig. 24 one can see that there are many different helices which might be drawn to connect the tubulin monomers. In many

[70] J. N. Champness, A. C. Bloomer, G. Bricogne, P. J. G. Butler, and A. Klug, *Nature* (*London*) **259**, 20 (1976).

[71] R. Goldman, T. Pollard, and J. Rosenbaum, eds., "Cell Motility, Book C: Microtubules and Related Proteins." Cold Spring Harbor Lab., Cold Spring Harbor, New York, 1976.

[72] E. Mandelkow, J. Thomas, and C. Cohen, *Proc. Natl. Acad. Sci. U.S.A.* **74**, 3370 (1977).

[73] L. A. Amos and A. L. Klug, *J. Cell Sci.* **14**, 523 (1974).

[74] H. P. Erickson, *J. Cell Biol.* **60**, 153 (1974).

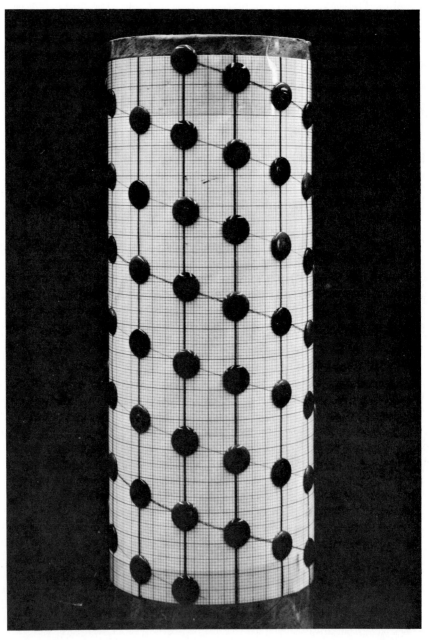

FIG. 24. Surface lattice of the microbule. Each lattice point is represented by a black disk. The vertical protofilaments (13-start helices) and the 3-start helices are drawn in. (Reproduced from Mandelkow *et al.*[72] with permission.)

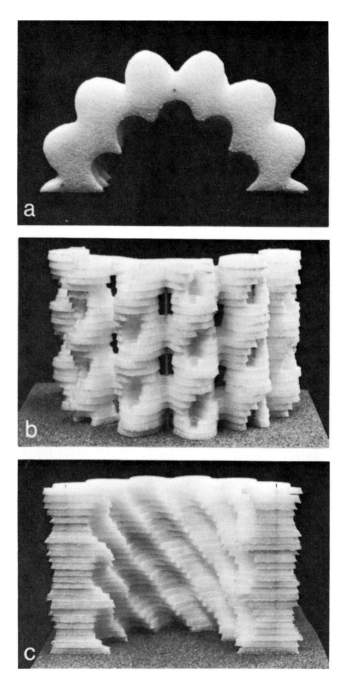

FIG. 25. Styrofoam model of reconstruction of microtubule. (a) View along tubule axis; (b) view from outside of tubule; (c) view from inside of tubule. Note that the high-relief helices are different on the outer and inner surfaces. (Reproduced from Mandelkow *et al.*[72] with permission.)

cases several helical lines have to be drawn if all the subunits are to be included. Each line is said to be a "start," so that the slanted family of lines in Fig. 24 is a three-start helix. In practice some of the helical lines connecting subunits will be in higher relief than others—or the spaces between them will be more deeply etched. It is these particular helices that will contribute most to the diffraction pattern. It is worth emphasizing that this analysis is using as a model the familiar helical wire—only here the wire is a ridge on the surface of the rod.

The inside and the outside of the microtubule, of course, have the same arrangement of subunits, so that the same families of helices can be drawn. However, it is not necessary that the same families be prominent both on the inner and outer surfaces. Because there is very little density fluctuation within the rod wall, Eq. (6.5.27) for the $G_{n,l}$ reduces approximately to two terms: a J_n for the inner tubule radius and a J_n (with the same order) for the outer tubule radius. Such bimodal $G_{n,l}$ are indeed observed.

The analysis began by examining the equator, which contains information about the projection of the tubule down its axis. The investigators concluded that the inner and outer radii of the tubule were 70 and 150 Å, respectively. A reflection corresponding to the spacing between the 13 protofilaments was also observed. Next, strong reflections on the first layer line were examined, using the ideas that went into the derivation of Eq. (6.5.22). Reflections arising from the inner and outer walls of the microtubule were identified by their distance from the meridian. It was found that on the outside (Fig. 24) the predominant feature was the separation of protofilaments, which are themselves clearly divided into subunits with a strong modulation from a ten-start helix (Fig. 25) running at about 45° to the tubule axis. Inside, the major feature is the ten-start helix: the protofilament separation and subdivision are very weak. Further, the ten-start helices on the inside and outside are not in register, at least according to this preliminary model. Rigorous demonstration of this would require a knowledge of the phase of the relevant reflections. Clearly, additional progress on this structure can only come with the use of heavy-atom derivatives.

6.5.6. Conclusion

With these two examples we have tried to illustrate some of the scope and limitations of the fiber diffraction method. In some sense this is a hopeless task, since each problem in fiber diffraction is unique. For lack of a direct way to phase the reflections, most problems are solved by resorting to special-case reasoning, electron microscopy, or other arcane methods. Also, each new problem illuminates and extends some part of helical diffraction theory. Extremely clever deductions from the diffraction patterns are a common feature of the literature. A few generalizations do hold, however. Except for quite small structures, atomic information is difficult to obtain. In addition, lack of good specimens is a limiting factor in many cases. Furthermore, it is actually possible to get the wrong answer in some cases, when very indirect reasoning is used. We find this area of XRD continually entertaining and intriguing.

Acknowledgments

This work was supported by N.I.H. research career development award AI-00271 and N.I.H. research grant AI-14820 (to Eaton E. Lattman), and N.I.H. grant GM-25432 (to L. Mario Amzel).

7. LASER LIGHT SCATTERING

By Ralph Nossal

7.1. Introduction

Light is scattered when it passes through an optically heterogeneous material. Spatial variations in refractive index result in a diffraction pattern which fluctuates with changes in the structure of the scattering medium. The intensity of the scattered light falling upon a small area varies accordingly, at a rate which depends upon the kinematics of the refractive index changes occurring within the sample (see Fig. 1).

The availability of lasers has facilitated measurement and analysis of such fluctuating refractive index distributions. Not only do lasers provide intense, essentially monochromatic sources, but the coherence properties of their emissions allow one to relate the diffraction pattern detected at one time to that detected at a different time. The changes which are measured then can be correlated with molecular reorganizations of the sample.

In Chapter 7.2 we outline the basic theory by which laser inelastic light scattering phenomena are understood. Emphasis primarily is given to fluctuation spectroscopy, but we also discuss some kindred aspects of classical light scattering techniques pertaining to the intensity of the scattered signal. Chapter 7.3 contains a discussion of typical instrumentation and certain common experimental procedures. The remaining chapters are concerned with specific topics, such as measurement of the diffusion coefficients of macromolecules, determination of the electrophoretic mobilities of biological cells, assaying for the swimming speeds of motile microorganisms, detection of changes in blood flow, and measurement of the elastic moduli of soft biological materials.

Our purpose is not to provide an exhaustive review of biomedical uses of laser light scattering but, rather, to familiarize the reader with various *classes* of applications. In so doing we hope to communicate some of the spirit of the physical modeling required to interpret measurements. Recently, some excellent review papers have appeared in which particular biological

METHODS OF EXPERIMENTAL PHYSICS, VOL. 20

applications are described.[1-6] In addition, several books[7,8] and conference proceedings[9-9b] contain extensive discussions of some of the topics that follow. A good bibliography is given in Schurr.[5]

7.2. Theory

When a substance is illuminated by an electromagnetic field, the electrons and nuclei within the material move oppositely to each other and produce an oscillating dipole field. This field is characterized by the macroscopic polarizability $P(r, t)$, which, for weak illumination, is proportional to the incident field $E_0(r, t)$. The proportionality in general is described by a tensor polarizability $\alpha(r, t)$; however, if the scattering medium is optically isotropic the off-diagonal elements of α are zero valued and the diagonal terms are equal.

An oscillating dipole is itself a source of electromagnetic radiation, with a field strength which is proportional to d^2P/dt^2. Unless otherwise stated, we assume the scattering medium to be optically isotropic. In this case the plane of polarization of the scattered light is parallel to that of the incident field, and we say that the scattering is "polarized." "Depolarized" components of scattered light—i.e., those whose polarization is perpendicular to the incident polarization—arise when the off-diagonal elements of the polarizability tensor are different from zero. Depolarized scattering is generally very much lower in intensity than polarized scattering, and thus has not been of much concern in biological applications of quasielastic light scattering.

The intensity of scattered light is a strong function of the scattering angle θ. The latter is defined to lie between the axis of the laser beam and a line drawn

[1] H. Z. Cummins and H. L. Swinney, *Prog. Opt.* **8**, 133 (1970).

[2] N. C. Ford, Jr., *Chem. Scr.* **2**, 193 (1972).

[3] F. D. Carlson, *Annu. Rev. Biophys. Bioeng.* **4**, 243 (1975); V. Bloomfield, *ibid.* **10**, 421 (1981).

[4] B. R. Ware, *in* "Chemical and Biological Applications of Lasers" (C. B. Moore, ed.), Vol. 2, Chapter 5. Academic Press, New York, 1977.

[5] J. M. Schurr, *CRC Crit. Rev. Biochem.* **4**, 371 (1977).

[6] V. A. Bloomfield and T. K. Lim, *Methods Enzymol.* **48F**, 415 (1978).

[7] B. Chu, "Laser Light Scattering." Academic Press, New York, 1974.

[8] B. J. Berne and R. Pecora, "Dynamic Light Scattering with Application to Chemistry, Biology, and Physics." Wiley (Interscience), New York, 1976.

[9] H. Z. Cummins and E. R. Pike, eds., "Photon Correlation and Light Beating Spectroscopy." Plenum, New York, 1974.

[9a] H. Z. Cummins and E. R. Pike, eds., "Photon Correlation Spectroscopy and Velocimetry." Plenum, New York, 1977.

[9b] S. H. Chen, B. Chu, and R. Nossal, eds., "Scattering Techniques Applied to Supramolecular and Nonequilibrium Systems." Plenum, New York, 1981.

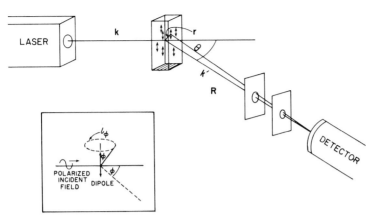

FIG. 1. Schematic of a typical light scattering experiment. Dipoles, which are induced by the incident field, reradiate electromagnetic waves which interfere with emissions from other scattering centers. The phototube views a small part of the resulting diffraction pattern, the intensity of which fluctuates when the scattering centers move. The inset shows the scattering from an individual dipole: l_φ is the locus of directions of equal scattering intensity, which is proportional to sin φ. Usually, the laser is operated in the TEM_{00} mode, and the output is vertically polarized. Dipoles are induced in the vertical plane, and if the detector is in the horizontal plane, the factor sin φ is 1 for any scattering angle θ. If an unpolarized source is used, there is an additional field component which induces dipoles oriented in the horizontal plane. In this case the angles φ and θ are related as $\varphi = \frac{1}{2}\pi - \theta$, for which sin $\varphi = \cos \theta$.

from the phototube surface to the sample through a set of collimating pinholes in the collection optics (Fig. 1). In idealizations of the scattering process the incident electromagnetic wave usually is taken to be a plane wave $E_0(\mathbf{r}, t) \sim E_0 \exp[-i(\omega_0 t - \mathbf{k} \cdot \mathbf{r})]$, and the scattered wave is taken to be spherical. (Each dipole in the sample reradiates energy as a spherical wave, but it is assumed that the detector is positioned sufficiently far from the sample that the resultant wave itself is spherical.)

Within these limitations we now derive an expression to relate the time autocorrelation of scattered light to the space–time correlation of density fluctuations occurring within a sample. The derivation which follows is similar to several which have appeared in the literature,[10–12] various essential features of which first were put forth by Lord Rayleigh (see, e.g., Tanford[13]). Referring to Fig. 1, we see that the detected electromagnetic

[10] R. Pecora, *J. Chem. Phys.* **40**, 1604 (1964).

[11] G. B. Benedek and T. Greytak, *Proc. IEEE* **53**, 1623 (1966).

[12] S. B. Dubin, J. H. Lunacek, and G. B. Benedek, *Proc. Natl. Acad. Sci. U.S.A.* **57**, 1164 (1967).

[13] C. Tanford, "Physical Chemistry of Macromolecules," Chapter 5. Wiley, New York, 1961.

field is the superposition of the fields radiated by each of the oscillating dipoles that have been excited within a sample by the incident electromagnetic field. According to classical electromagnetic theory,[14] this quantity is given as

$$E_s(\mathbf{R}, t) = \int_V \frac{\sin \varphi(\mathbf{r})}{c^2 |\mathbf{R} - \mathbf{r}|} \left(\frac{\partial^2 \mathbf{P}(\mathbf{r}, t')}{\partial t^2} \right)_{t'} d^3 r, \qquad (7.2.1)$$

where $\varphi(\mathbf{r})$ is the angle between the axis of the induced dipole moment at position $\mathbf{r}$ and a line joining that point to the point of observation, c is the speed of light, V is the volume of the effective illuminated region of the sample, and $|\mathbf{R}|$ is the distance from the detector to the origin of the sample. $\mathbf{P}(\mathbf{r}, t)$ is the polarization at point $\mathbf{r}$, and $t' \equiv t - n|\mathbf{R} - \mathbf{r}|/c$ is the "retarded time," where n is the index of refraction of the scattering medium (c/n is the propagation speed of the emitted wave). Because we here restrict ourselves to optically isotropic media, $\mathbf{P}(\mathbf{r}, t)$ can be written as

$$\mathbf{P}(\mathbf{r}, t) = \alpha(\mathbf{r}, t)\mathbf{E}(\mathbf{r}, t), \qquad (7.2.2)$$

where the scalar $\alpha(\mathbf{r}, t)$ is the polarizability and $\mathbf{E}(\mathbf{r}, t)$ is the local electric field within the sample.

If the sample is optically dense, the determination of $\mathbf{E}(\mathbf{r}, t)$ may become quite complicated because of multiple scattering or absorption. Here we assume the relatively simple case that the sample is composed of a dilute collection of nonabsorbing particles which are small enough that an electromagnetic wave traveling through the medium does so without significant reduction in intensity or retardation of phase. In other words, we restrict ourselves here to the Rayleigh–Gans–Debye theory,[13,15,16] which holds for scatterers which are small compared with the wavelength of light or whose refractive index is close to that of the surrounding solvent. (In Chapters 7.5 and 7.7 we discuss scattering from microorganisms and other large particles, and indicate modifications which are necessary in such cases.)

A simple expression relating the scattered field to the structure of the scattering medium can be obtained when the distance R from the sample to the detector is much greater than any of the dimensions of the scattering volume (the "far-field" approximation). For a plane-wave incident field

[14] L. D. Landau and E. M. Lifshitz, "Electrodynamics of Continuous Media," p. 377. Addison-Wesley, Reading, Massachusetts, 1960.

[15] H. C. Van de Hulst, "Light Scattering by Small Particles." Wiley, New York, 1957.

[16] M. Kerker, "The Scattering of Light and Other Electromagnetic Radiation." Academic Press, New York, 1969.

$\mathbf{E} = \mathbf{E}_0 \exp[i(\mathbf{k} \cdot \mathbf{r} - \omega_0 t)]$ we find, from Eqs. (7.2.1) and (7.2.2), that

$$\mathbf{E}_s(R, t) \approx \left(\frac{\omega_0}{c}\right)^2 \frac{\mathbf{E}_0 \sin \varphi \exp(-i\omega_0 t)}{R}$$

$$\times \int_V \alpha(\mathbf{r}, t) \exp\left\{i\left[\mathbf{k} \cdot \mathbf{r} - \omega_0\left(\frac{t - n|\mathbf{R} - \mathbf{r}|}{c}\right)\right]\right\} d^3r$$

$$\approx \left(\frac{\omega_0}{c}\right)^2 \frac{\mathbf{E}_0 \sin \varphi}{R} \exp[i(\mathbf{k}' \cdot \mathbf{R} - \omega_0 t)]$$

$$\times \int_V \alpha(\mathbf{r}, t) \exp[i(\mathbf{k} - \mathbf{k}') \cdot \mathbf{r}] \, d^3r \qquad (7.2.3)$$

To derive the latter we substituted $(\omega_0 n/c)|\mathbf{R} - \mathbf{r}| \approx \mathbf{k}' \cdot (\mathbf{R} - \mathbf{r})$, where the wave vector of the scattered beam $\mathbf{k}'$ is directed along the line of observation but has the same magnitude as that of the incident beam, $|\mathbf{k}'| \approx |\mathbf{k}| = \omega_0 n/c = 2\pi n/\lambda$, where λ is the wavelength of the incident radiation. We also noted that the temporal change of the polarizability is slow in comparison with the oscillation of the electromagnetic field, so terms of $O(\partial^2\alpha/\partial t^2)$ could be neglected.

The quantity $\mathbf{Q} \equiv \mathbf{k} - \mathbf{k}'$ is known as the Bragg scattering vector. From geometric arguments one can show that the magnitude of the Bragg scattering vector is given as

$$|\mathbf{Q}| = |\mathbf{k} - \mathbf{k}'| = 2|\mathbf{k}| \sin \tfrac{1}{2}\theta = (4\pi n/\lambda) \sin \tfrac{1}{2}\theta, \qquad (7.2.4)$$

where θ is the scattering angle. (Note that θ lies in the plane defined by the incident and scattered beams—see Fig. 1.) If the sample is optically homogeneous so that $\alpha(\mathbf{r}, t) = \alpha(t)$, then the integral in Eq. (7.2.3) becomes

$$\int_{V \gg \lambda^3} \alpha(\mathbf{r}, t) e^{i\mathbf{Q} \cdot \mathbf{r}} \, d^3r \simeq \alpha(t)\delta(\mathbf{Q}),$$

where $\delta(\mathbf{Q})$ is the Dirac δ function. In this particular case scattering occurs only in the forward direction, i.e., ordinary transmission.

In general, of course, the polarizability can be expressed as

$$\alpha(\mathbf{r}, t) = \alpha_0(t) + \bar{\alpha}(\mathbf{r}, t), \qquad (7.2.5)$$

and Eq. (7.2.3) thus indicates that the scattered electromagnetic field can be used as a probe of the Fourier components of the spatial variations of the "excess polarizability" $\bar{\alpha}(\mathbf{r}, t)$. For a suspension of particles, $\alpha_0(t)$ might be taken as the polarizability of the solvent and $\bar{\alpha}(\mathbf{r}, t)$ as the difference between the solvent polarizability and that of the particles. In this case we find from Eqs. (7.2.3) and (7.2.5) that the scattered field is given as

$$\mathbf{E}_s(\mathbf{R}, t) = \left(\frac{\omega_0}{c}\right)^2 \frac{\mathbf{E}_0 \sin \varphi}{R} e^{i(\mathbf{k}' \cdot \mathbf{R} - \omega_0 t)} \sum_{\text{scatterers}} \int_{\Omega_i(t)} \bar{\alpha}(\mathbf{r}) e^{i\mathbf{Q} \cdot \mathbf{r}} d^3 r, \quad (7.2.6)$$

where Ω_i is the spatial extent (volume) of the ith scatterer. Therefore the scattering process is sensitive to the optical shape of the scatterers if particle dimensions are comparable to the wavelength of light; however, if the particles are much smaller than λ, one has

$$\bar{\alpha}(\mathbf{r}) \approx \sum_i \bar{\alpha}_i \delta(\mathbf{r} - \mathbf{r}_i),$$

where $\mathbf{r}_i$ and $\bar{\alpha}_i$ are, respectively, the position and excess polarizability of the ith scatterer. We then find

$$\mathbf{E}_s(\mathbf{R}, t) = \left(\frac{\omega_0}{c}\right)^2 \frac{\mathbf{E}_0 \sin \varphi}{R} e^{i(\mathbf{k}' \cdot \mathbf{R} - \omega_0 t)} \sum_i \bar{\alpha}_i e^{i\mathbf{Q} \cdot \mathbf{r}_i}. \quad (7.2.7)$$

It is apparent from Eq. (7.2.7) that the particles will be invisible if their index of refraction and that of the solvent are identical.

To derive Eqs. (7.2.6) and (7.2.7) we considered that scattering caused by local variations in solvent density is insignificant and can be neglected. Also, we implicitly assumed that the incident field is not appreciably disturbed as it passes through the sample and that multiple scattering can be ignored.

7.2.1. Classical Light Scattering

The intensity $\mathscr{I}$ of an electromagnetic field is given as the square of the field amplitude $\mathscr{I} = E^* E$. Early light scattering techniques involved measurement of the time average of the intensity $I = |E_s^* E_s|$, which, for sufficiently long measurement times, can be related to the ensemble average of the spatial distribution of scatterers according to Eq. (7.2.7). We find

$$\frac{I}{I_0} = \frac{\langle E_s^* E_s \rangle}{\langle E_0^* E_0 \rangle} = \frac{16\pi^4 \sin^2 \varphi}{\lambda^4 R^2} \left\langle \sum_i \sum_j \bar{\alpha}_i \bar{\alpha}_j e^{i\mathbf{Q} \cdot (\mathbf{r}_i - \mathbf{r}_j)} \right\rangle \quad (7.2.8)$$

since $\lambda = v/c = \omega_0/2\pi c$. If the particles are uniformly distributed throughout the scattering volume and, also, if they do not interact with one another, the ensemble average $\langle \sum_i \sum_j \bar{\alpha}_i \bar{\alpha}_j e^{i\mathbf{Q} \cdot (\mathbf{r}_i - \mathbf{r}_j)} \rangle$ is simply the sum $\sum_i \bar{\alpha}_i^2$. For identical particles we thus have

$$I/I_0 = (16\pi^4 \sin^2 \varphi/\lambda^4 R^2)(\bar{\alpha})^2 N, \quad (7.2.9)$$

where N is the number of particles in the scattering volume.

Equation (7.2.9) can be put in a somewhat more familiar form by the following manipulations.[13] First, we note that the index of refraction of a medium n can be related to the volume polarizability α according to $n^2 - 1 = 4\pi\alpha$. Thus, if n_0 is the index of refraction of the solvent, we have

$$n^2 - n_0^2 \approx 2n_0(\partial n/\partial\rho)\rho = 4\pi N\bar{\alpha},$$

where ρ is the mass concentration of particles in grams per cubic centimeter. From Eq. (7.2.9) we find

$$\frac{I}{I_0} = \frac{4\pi^2\sin^2\varphi n_0^2(\partial n/\partial\rho)_{n_0}^2\rho^2}{\lambda^4 R^2 N}.$$

Because $\rho = NM/\mathscr{A}$, where M is the molecular weight of a scatterer and $\mathscr{A}$ is Avogadro's number, we have

$$\frac{I}{I_0} = \frac{4\pi^2\sin^2\varphi n_0^2(\partial n/\partial\rho)_{n_0}^2\rho M}{\lambda^4 R^2\mathscr{A}}. \tag{7.2.10}$$

Since all quantities other than M that appear in Eq. (7.2.10) in principle can be independently determined, the molecular weight of the scatterers can be ascertained by measuring the total intensity of scattered light. (If several sizes of particles are present in the sample, the "z-averaged" molecular weight[13] is found.)

Two important assumptions have been invoked which, if relaxed, imply modifications of Eq. (7.2.10). We recall, first, that the particles were assumed to be noninteracting. If the density of scatterers is sufficiently high, interactions can be probed since the expectation $\langle\sum_i\sum_j e^{i\mathbf{Q}\cdot(\mathbf{r}_i-\mathbf{r}_j)}\rangle$ can be expanded in terms of virial coefficients.[17]

Second, when particles are large, we should start with Eq. (7.2.6) rather than Eq. (7.2.7). Consequently, for identical noninteracting randomly oriented particles, the analog of Eq. (7.2.9) is

$$\frac{I}{I_0} = \frac{16\pi^4\bar{\alpha}^2\sin^2\varphi}{\lambda^4 R^2} N\mathscr{S}(Q), \tag{7.2.11}$$

where the structure factor $\mathscr{S}(Q)$, defined as

$$\mathscr{S}(Q) = \left|\int dr\, p(r)e^{i\mathbf{Q}\cdot\mathbf{r}}\right|^2 \tag{7.2.12}$$

is sensitive to the shape of the scatterers. Here, $p(r)$ is the distribution of inducible dipoles within a particle, which generally is taken to be proportional to the mass distribution. For small particles, the only angle dependence

[17] B. H. Zimm, J. Chem. Phys. **16**, 1093 (1948).

of the intensity is through the term $\sin^2 \varphi$, which, when unpolarized light is used as the incident source, is given as [13,17] $1 + \cos^2 \theta$. However, for large particles additional angle dependence affects the intensity through the Q dependence of the structure factor. Consequently, by measuring the angle dependence of the scattered light, information on the size and structure of a particle can be determined. Classical light scattering from bacteria and other biological cells is discussed in papers by Wyatt[18] and Cross and Latimer.[19]

7.2.2. Intensity Fluctuation Spectroscopy

We have just seen that classical light scattering techniques relate the *intensity* of scattered light to certain molecular properties. In contrast, quasielastic light scattering is concerned with extracting information from the *temporal* behavior of the diffraction pattern. By limiting the aperture of the collection optics, fluctuations in the intensity of the light that falls upon the detector can be discerned. These time variations are related to local changes in refractive index of a sample, and, consequently, a measure of the kinetic behavior of the scattering medium can be obtained. The recent availability of amplitude-stabilized lasers—sources of intense, mono-chromatic, and, most importantly, *temporally* and spatially coherent radiation—makes such studies practicable. The coherence time of ordinary light sources is too short to permit resolution of interference fluctuations arising from changes in refractive index of typical biological samples.

For didactic purposes we now consider how the motion of small particles (e.g., protein molecules of molecular weight 80,000 daltons) can be studied with this technique. Other applications are discussed below, where those modifications of the basic theory which are peculiar to each case are indicated.

Two essentially equivalent schemes can be used to extract kinetic information from a time-varying signal: analysis of frequency components or formation of time-delayed "autocorrelations" of that signal. Suppose we have an instrument that can measure the joint expectation of the intensity at one instant of time and of the intensity at a later instant delayed by an increment τ, so that we form the quantity $\langle \mathscr{I}(t)\mathscr{I}(t + \tau) \rangle$. Assuming that the average properties of the sample remain constant, i.e., that the system is in a stationary state, we can determine the autocorrelation function $\langle \mathscr{I}(t)\mathscr{I}(t + \tau) \rangle$ in principle as follows: Multiply the intensity at time t by that at time $t + \tau$, repeat the measurement at some other times t' and $t' + \tau$, and add the product thus obtained to the first; repeat the measurements at yet later times and add those results to the former; finally, divide by the

[18] P. J. Wyatt, *Methods Microbiol.* **8**, 183 (1973).
[19] D. A. Cross and P. Latimer, *Appl. Opt.* **11**, 1225 (1972).

number of additions to the sum. The more often one repeats the measurement, the closer will this experimentally determined quantity be to a theoretical function whose properties derive from the stochastic characteristics of a representative sample. Considerable study has been given to the manner in which finite sampling effects the reliability of deduced kinetic parameters.[20–22]

Because the instantaneous intensity is the square of the field amplitude, we can write for the autocorrelation function of the detected intensity

$$\mathscr{C}(\tau) = \langle \mathscr{I}(t)\mathscr{I}(t + \tau)\rangle = \langle E_{sc}^*(t)E_{sc}(t)E_{sc}^*(t + \tau)E_{sc}(t + \tau)\rangle, \quad (7.2.13)$$

and from Eqs. (7.2.3) and (7.2.4) we find that $\mathscr{C}(t)$ is proportional to the joint expectation of four quantities which contain information about the time dependence of the local polarizability of the sample, namely,

$$\mathscr{C}(\tau) \sim \left\langle \left(\int_V \alpha(\mathbf{r}, t)e^{i\mathbf{Q}\cdot\mathbf{r}}\, d^3r \right)^* \int_V \alpha(\mathbf{r}, t)e^{i\mathbf{Q}\cdot\mathbf{r}}\, d^3r \right.$$
$$\times \left. \left(\int_V \alpha(\mathbf{r}, t + \tau)e^{i\mathbf{Q}\cdot\mathbf{r}}\, d^3r \right)^* \int_V \alpha(\mathbf{r}, t + \tau)e^{i\mathbf{Q}\cdot\mathbf{r}}\, d^3r \right\rangle. \quad (7.2.14)$$

The complicated expression given in Eq. (7.2.14) must next be related to some interesting material parameter via a physical model for the kinematics of the sample. Analysis of the joint expectation of four parameters generally is a difficult matter. Fortunately, if the statistical behavior of the scattering assembly can be characterized as a *Gaussian* random process, an important simplification occurs: If A_1, A_2, A_3, A_4 are Gaussian random variables, it then follows that $\langle A_1 A_2 A_3 A_4 \rangle = \langle A_1 A_2 \rangle \langle A_3 A_4 \rangle + \langle A_1 A_3 \rangle \langle A_2 A_4 \rangle + \langle A_1 A_4 \rangle \langle A_2 A_3 \rangle$. Thus, for a scattering process which obeys Gaussian statistics, Eq. (7.2.14) becomes

$$\mathscr{C}(\tau) \sim i_0^2(\mathbf{Q})[1 + |I_1(\mathbf{Q}, \tau)|^2], \quad (7.2.15)$$

where

$$i_0(\mathbf{Q}) = \left| \int_V \alpha(\mathbf{r}, t)e^{i\mathbf{Q}\cdot\mathbf{r}}\, d^3r \right|^2 \quad (7.2.16)$$

[20] E. Jakeman, E. R. Pike, and S. Swain, *J. Phys. A: Gen. Phys.* **4**, 517 (1971).

[21] E. Jakeman, *in* "Photon Correlation and Light Beating Spectroscopy" (H. Z. Cummins and E. R. Pike, eds.), p. 75. Plenum, New York, 1974.

[22] B. E. A. Saleh and M. F. Cardoso, *J. Phys. A: Math. Nucl. Gen.* **6**, 1897 (1973).

and $I_1(\mathbf{Q}, \tau)$ (the "intermediate scattering function") is defined as

$$I_1(\mathbf{Q}, \tau) = [i_0(\mathbf{Q})]^{-1}\left\langle\left(\int_V \alpha(\mathbf{r}, t)e^{i\mathbf{Q}\cdot\mathbf{r}}\,d^3r\right)^* \int_V \alpha(\mathbf{r}, t+\tau)\exp^{(i\mathbf{Q}\cdot\mathbf{r})}\,d^3r\right\rangle$$

$$(7.2.17)$$

The physical properties of a particular sample are related to acquired data through the quantity $I_1(\mathbf{Q}, \tau)$.

The assumption of Gaussian statistics is oftentimes difficult to substantiate. However, one situation for which it most likely holds is the case when refractive index fluctuations derive from movements of Brownian particles, such as macromolecules in solution. Of course, the concentration of scatterers and the size of the scattering volume must be sufficiently large that fluctuations in the total number of particles in the scattering region are insignificant; if these conditions are not satisfied, additional features will be present in the scattered light spectrum.[23,24] Generally, however, if scattering arises from a large number of statistically independent random events, then by the law of large numbers the sum will be Gaussian distributed and the factorization indicated by Eq. (7.2.15) will hold.[21]

Suppose, now, that the scatterers are indeed undergoing Brownian motion. As before [cf. Eq. (7.2.7)], when the scatterers are identical particles much smaller than the wavelength of the incident radiation, the integral $\int \alpha(\mathbf{r}, t)e^{i\mathbf{Q}\cdot\mathbf{r}}\,d^3r$ can be expressed as $\bar{\alpha}\sum_i e^{i\mathbf{Q}\cdot\mathbf{r}_i(t)}$. We note that the quantity $\sum_i e^{i\mathbf{Q}\cdot\mathbf{r}_i(t)}$ is the Fourier transform of the density operator; thus, from Eq. (7.2.17) one has

$$I_1(\mathbf{Q}, t) = [i_0(\mathbf{Q})]^{-1}(\bar{\alpha})^2\left\langle\sum_i e^{-i\mathbf{Q}\cdot\mathbf{r}_i(t)}\sum_j e^{i\mathbf{Q}\cdot\mathbf{r}_j(t+\tau)}\right\rangle \qquad (7.2.18)$$

$$= [i_0(\mathbf{Q})]^{-1}(\bar{\alpha})^2\langle\delta C^*(\mathbf{Q}, t)\delta C(\mathbf{Q}, t+\tau)\rangle, \qquad (7.2.18')$$

where $\delta C(\mathbf{Q}, t)$ is the "density fluctuation" and the expression within the angular brackets is the "density–density correlation function." Equation (7.2.18') also could have been derived from Eq. (7.2.17) by assuming that the local polarizability $\alpha(\mathbf{r}, t)$ is given as a time-invariant average quantity plus a term whose value depends upon the local deviation of particle density from the statistical average $\alpha(\mathbf{r}, t) = \alpha_0 + (\partial\alpha/\partial C)\partial C(\mathbf{r}, t) + \cdots$; we shall see, however, that for large scatterers, for which intraparticle scattering must be considered, a generalization of the expression given by Eq. (7.2.18) is required (see Chapter 7.5).

[23] D. W. Schaefer and B. J. Berne, *Phys. Rev. Lett.* **28**, 475 (1972).
[24] R. F. Voss and J. Clarke, *J. Phys. A: Math. Gen.* **9**, 561 (1976).

The motions of noninteracting Brownian particles can be described by Fick's diffusion equation. As shown in Chapter 7.4., Eq. (7.2.18′) thus becomes

$$I_1(Q, \tau) = e^{-DQ^2\tau}, \tag{7.2.19}$$

where D is the translational diffusion coefficient. The most common application of laser inelastic light scattering heretofore has been the measurement of the diffusion coefficients of macromolecules and small particles according to Eq. (7.2.19). The photon autocorrelation function derived from Eqs. (7.2.15) and (7.2.19) is shown in Fig. 2a. It has the form of a constant background term plus a simple exponential whose decay is given as $\tau_e = (2DQ^2)^{-1}$. Thus, by measuring the autocorrelation at several scattering angles, the diffusion coefficient would be determined from a representation of data such as that shown in Fig. 2b.

The experimental situation depicted in Eqs. (7.2.13)–(7.2.15) is often called *homodyne* scattering. In this case components of the scattered light field mix with other components of the field at the photodetector surface to provide the fluctuating signal subsequently processed. However, certain kinds of particle motion cannot be detected by this technique. Consider, for example, movements of particles with constant velocity **V**. In such cases the

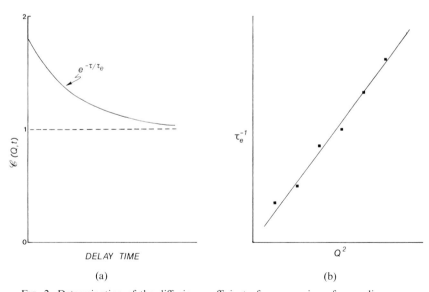

(a)   (b)

FIG. 2. Determination of the diffusion coefficient of a suspension of monodisperse noninteracting Brownian particles: (a) Exponential decay of the time-varying portion of the photon autocorrection function; (b) inverse relaxation time τ_e^{-1}, plotted versus the square of the amplitude of the Bragg scattering vector $Q = 4\pi n\lambda^{-1} \sin \frac{1}{2}\theta$.

position of the ith particle at time $t + \tau$ is related to that at time t as $\mathbf{r}_i(t + \tau) = \mathbf{r}_i(t) + \mathbf{V}\tau$, so that Eq. (7.2.18) becomes

$$I_1(\mathbf{Q}, \tau) = [i_0(\mathbf{Q})]^{-1}(\bar{\alpha})^2 N \, e^{i(\mathbf{Q}\cdot\mathbf{V})\tau}. \tag{7.2.20}$$

Consequently, $|I_1(\mathbf{Q}, t)|^2 = I_1^* I_1 = (\bar{\alpha})^4 N^2/[i_0(\mathbf{Q})]^2$; as indicated by Eq. (7.2.15), the autocorrelation function would not contain any information about the drift velocity of the scatterers.

However, information about $\mathbf{V}$ could be obtained by performing a so-called *heterodyne* experiment in which a portion of the source light does not pass through the sample but is diverted to the detector. Suppose we designate this unscattered field as $E_{LO}(t) = E_{LO}e^{i(\omega_0 t + \varphi)}$, so that the total field at the phototube surface is $E(t) = E_{LO}(t) + E_{sc}(t)$. (The added field is analogous to the local reference oscillator in heterodyne radio detection and therefore sometimes is referred to as the "local oscillator field.") If we assume that an experiment is designed so the amplitude of E_{LO} is much greater than that of the scattered field E_{sc}, then the analog of Eq. (7.2.13) is

$$\mathscr{C}(\tau) \approx \langle (E_{LO}^* E_{LO})^2 \rangle + 2\langle E_{LO}^* E_{LO} \rangle \, \mathrm{Re}\langle E_{sc}^*(t)E_{sc}(t + \tau) \rangle + \cdots \tag{7.2.21}$$

where the right-hand side of the equation also contains a negligible "homodyne" component.[1] From Eq. (7.2.20) we see that $\mathrm{Re}\langle E_{sc}^*(t)E_{sc}(t + \tau) \rangle \sim \cos(\mathbf{Q} \cdot \mathbf{V}\tau)$, so $\mathscr{C}(\tau)$ in this case does indeed provide information on the drift of the scatterers. Finally, we note that heterodyne scattering can always be used to test whether the "Gaussian approximation" given by Eq. (7.2.15) may be invoked to simplify data analysis.

7.3. Instrumentation and General Techniques

7.3.1. Relationship between Photon Autocorrelation and Spectral Analysis

The previous section emphasized autocorrelation techniques for analyzing the fluctuating intensity of scattered light. However, we could as well have considered spectral analysis of that signal, since the information contained in the frequency distribution of a time-varying signal is equivalent to that which can be extracted from autocorrelation. Indeed, by the Wiener–Khintchine theorem, the power spectrum of the scattered light $S(\mathbf{Q}, \omega)$ is related to the autocorrelation function $\mathscr{C}(\mathbf{Q}, t)$ as

$$S(\mathbf{Q}, \omega) = \frac{1}{2\pi} \int_{-\infty}^{\infty} e^{i\omega\tau} \mathscr{C}(\mathbf{Q}, \tau) \, d\tau. \tag{7.3.1}$$

The equivalent nature of these functions is illustrated in Fig. 3.

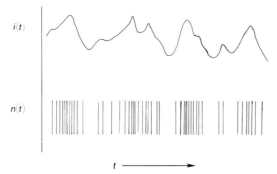

FIG. 3. The average number of detected photons is proportional to the intensity of light falling upon the photodetector. Thus fluctuations in the number of photon detections per unit time [$n(t)$] contain the same information as does spectral analysis of the photocurrent amplitude [$i(t)$].

If the photon autocorrelation function is exponentially decaying, e.g., $I_1^2(\mathbf{Q}, t) \sim e^{-2DQ^2t}$ [see Eqs. (7.2.19) and (7.2.15)], the related spectral power density is Lorentzian,

$$S(\mathbf{Q}, \omega) \sim \frac{2DQ^2/\pi}{\omega^2 + (2DQ^2)^2}. \tag{7.3.2}$$

The frequency width of this function contains the same information as does the decay of the autocorrelation function. Examples of the transformation effected by Eq. (7.3.1) are shown in Fig. 4. Processes whose autocorrelations

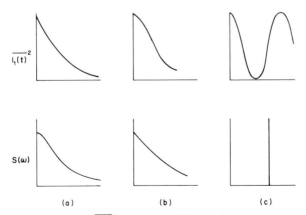

FIG. 4. Correlation functions [$\overline{I_1(t)}^2$] and spectra [$S(\omega)$] are related according to Eq. (7.3.1). Typical results for (a) diffusing macromolecules; (b) subcutaneous blood flow or randomly swimming microorganisms; and (c) particles drifting with uniform velocity.

consist predominantly of a small set of periodic functions are usually best analyzed in terms of spectral components [examples are the motion of charged particles in an electric field (Chapter 7.6) and standing-wave oscillations of soft gels (Chapter 7.10)]. Systems whose scattered fields contain a broad continuum of frequencies are oftentimes most easily related to theoretical models via time–autocorrelation functions [examples are the spatially isotropic swimming of microorganisms (Chapter 7.7) and sub-cutaneous blood flow (Chapter 7.9)]; also, if the scattered signal is very weak or the time scale of the fluctuations very fast, efficient collection and process-ing of primary data presently require photon autocorrelation techniques.

7.3.2. Characteristics of Measured Autocorrelation Functions

Although theoretical analyses based on Eq. (7.2.13) pertain to continuous signals, an instrumental autocorrelator necessarily samples the scattered photon field over discrete intervals of time. Therefore, the correlated com-ponents really are integrated functions, i.e.,

$$\mathscr{C}_{exp}(\tau) = \frac{1}{(\Delta\tau)^2}\left\langle \int_{-\Delta\tau/2}^{\Delta\tau/2} n(t+s)\,ds \int_{-\Delta\tau/2}^{\Delta\tau/2} n(t+\tau+s')\,ds' \right\rangle$$

$$\sim \frac{1}{(\Delta\tau)^2}\left\langle \int_{-\Delta\tau/2}^{\Delta\tau/2} I(t+s)\,ds \int_{-\Delta\tau/2}^{\Delta\tau/2} I(t+\tau+s')\,ds' \right\rangle, \quad (7.3.3)$$

but minimal distortion will occur if the clock interval $\Delta\tau$ is short compared with the characteristic decay time of the true autocorrelation function.

Also, the expression given in Eq. (7.2.13) is only an approximation of the measured autocorrelation function for yet another reason. One really does not directly analyze the field intensity when an autocorrelation analysis is performed on the scattered light. Rather, manipulations are made on the *photocurrent* emitted by the detector (Fig. 3), and theoretical considerations of photoelectron emissions from a photocathode[1] indicate that correlation of electron pulses with themselves gives rise to a shot-noise contribution to the photocurrent correlation function. Thus the photocurrent autocor-relation function $C_i(\tau)$ is given as[1] $C_i(\tau) = e\langle i\rangle\delta(\tau) + \langle i\rangle^2 g_2(\tau)$, where $\langle i\rangle^2 g_2(\tau)$ is proportional to the quantity given by Eq. (7.2.14). [$g_2(\tau)$ is related to the autocorrelation of photopulses $n(\tau)$ as

$$g_2(\tau) = \langle n(t)n(t+\tau)\rangle/\langle n(t)\rangle^2.]$$

In practice however, the signal in the first (zeroth) channel of an autocor-relator is blocked or neglected and the shot-noise term $e\langle i\rangle\delta(\tau)$ ignored. Similarly, when spectral analysis is performed the shot-noise term provides a constant background which is subtracted from the remaining signal.

The factorization $g_2 = 1 + I_1^2$ as shown by Eq. (7.2.15) also is but an idealization. The correlated signal actually appears as

$$\mathscr{C}(\mathbf{Q}, \tau) \sim \{1 + \beta[I_1(\mathbf{Q}, \tau)]^2\}, \tag{7.3.4}$$

where $\beta < 1$. The proportionality term β accounts for statistical errors introduced by clipping (see Section 7.3.3 below) and coherence "losses" at the photodetector surface. The latter arise because the scattering volume "seen" by the photodetector is of finite size so that, even though the diffraction pattern of the scattered light fluctuates, some of the information contained in that pattern is lost. Indeed, if the pinholes through which the sample is viewed by the phototube are very large, the total intensity of light falling upon the photocathode will be almost constant and the ratio of correlated signal to background will be very small, i.e., $\beta \approx 0$. An exhaustive analysis of the ways in which β depends on various instrumental parameters has been performed by Jakeman and associates,[20,21] but such considerations are beyond the scope of this introductory article.

The longer one accumulates a signal, the greater will be the accuracy of the parameters that can be extracted from the data. Of course, the fidelity of $\mathscr{C}(\mathbf{Q}, \tau)$ also depends on the count rate (i.e., the total number of scattering events). Theoretical analyses of the statistical variability of experimental results are quite difficult and have been performed in detail for only a few cases[20-22]; however, one's intuition of this matter can usually be formed from repeated measurements on prototype samples in which instrumental settings and experimental variables such as count rate, channel width, scattering angle, and sample concentration are changed. In the case of diffusion coefficient measurements of monodisperse samples, theoretical and experimental analyses[25] indicate that an accuracy of approximately 1% can be achieved for quite reasonable experimental conditions (e.g., measurement times less than 2 min for a sample composed of 5 mg/ml haemocyanin, using a 50-mW He–He laser).

7.3.3. Equipment

Data can be processed either with a spectrum analyzer or with a pulse autocorrelator. The choice of instrumentation largely depends on the uses anticipated for the spectrometer. Velocimetry experiments, where scatterers move in preferred directions, seem to be most successful when spectrum analyzers are employed (see Chapter 7.6). Applications involving low count rates and short correlation times, such as measurements of the diffusion coefficients of small proteins, are best done in a photon counting mode utilizing autocorrelation analysis.

[25] E. Jakeman, C. J. Oliver, and E. R. Pike, *J. Phys. A: Gen. Phys.* **3**, L45 (1970).

Two kinds of correlators have been used for photon counting applications. One, a "full" autocorrelator, essentially performs the operation indicated by Eq. (7.3.3), i.e., it repeatedly performs multiplications of the actual number of counts appearing in the various time intervals. In the other case, that of a "single-clipped" autocorrelator, the actual number of delayed counts $n(t - j \Delta\tau)$ is replaced by either 0 or 1, depending on whether a preset level is exceeded (e.g., a clipping level of 2 would produce a "1" if the number of counts in the interval is 3 or greater). Clearly, if the average count rate is low (e.g., less than 0.5 pulses per interval $\Delta\tau$), the autocorrelations formed by a clipped autocorrelator operating at zero clipping level will be virtually identical with those formed by a full autocorrelator. Clipped autocorrelators were first used because they afforded finer time resolution and faster operating speeds than other available analyzers and because they were relatively inexpensive to construct. The construction, operation, and theory of such devices has been described by Pike and his associates[26] and Chen et al.[27] Later, an autocorrelator was designed by Asch and Ford[28] which utilizes a four-bit shift register to accomplish an almost exact computation of the full ideal autocorrelation function; it has many of the attributes, such as short sampling times and moderate cost, of clipped autocorrelators. In some laboratory installations the correlator output is fed directly to an electronic calculator for immediate manipulation and storage.[29]

Although an investigator previously had to build his own correlator, several high-quality commercial instruments now are available. Langley–Ford Instruments (Amherst, Massachusetts) markets a full four-bit real-time digital correlator having the capability of choosing the sampling time (clock width) to be as short as 100 nsec; it has many desirable features, including a built-in display, a multiplicity of interfacing options, the ability to simultaneously accumulate correlation functions at different sample times, and an option for up to 1096 channels for resolution of the correlation function. Malvern Scientific Corporation (Ronkonkoma, New York) markets similar signal correlators, as well as a complete laser photon counting spectrometer. Nicomp Instruments (Santa Barbara, California) is producing a spectrometer incorporating a clipped autocorrelator which contains a microcomputer dedicated to diffusion coefficient determinations.

Several firms manufacture excellent real-time spectrum analyzers. Early

[26] R. Foord, E. Jakeman, C. J. Oliver, E. R. Pike, R. J. Blagrove, E. Wood, and A. R. Peacocke, *Nature (London)* **227**, 242 (1970).

[27] S. H. Chen, W. B. Veldkamp, and C. C. Lai, *Rev. Sci. Instrum.* **46**, 1356 (1975); J. P. Yangos and S. H. Chen, *Rev. Sci. Instrum.* **51**, 344 (1980).

[28] R. Asch and N. C. Ford, Jr., *Rev. Sci. Instrum.* **44**, 506 (1973).

[29] M. Holz and S. H. Chen, *Appl. Opt.* **17**, 3197 (1978).

efforts to use laser light beating spectroscopy relied upon swept-frequency analyzers which were slow and cumbersome. The situation is now greatly improved, and for many applications spectrum analyzers are at least as convenient to use as are correlators (see Chapter 7.7). However, correlators are preferable for studying rapidly fluctuating phenomena because the short delay times $\Delta\tau$ available with those instruments correspond to high-frequency ranges which are not accessible when using commercial spectrum analyzers. Nicolet Scientific Corporation (Northvale, New Jersey) and Princeton Applied Research (Princeton, New Jersey) currently offer suitable instruments.

Low-light-level photon counting applications require a phototube which has fast response time, emits narrow pulses, and has a low probability of producing afterpulses. Several tubes are available which have such character-istics, the performance of which has been reviewed in a paper by Gethner and Flynn.[30] Inexpensive photodiodes can be used instead of phototubes for certain applications involving high signal levels and slow fluctuations (see Chapter 7.9). Depending on the particular instrumental configuration, various pulse shapers and/or amplifiers are used to interface the photo-detector and analyzer (see, e.g., Ware[4]).

The detector optics must be designed so that adjustments can be made to delineate the region of the sample which is being observed. Pinholes must be used so that the portion of the detected diffraction pattern will be small enough that changes arising from material fluctuations within the sample will significantly change the amplitude of the signal (i.e., the detected signal must arise from only a few "coherence areas"[1]).

The design of the sample holder or cuvette generally depends on the anticipated application. Electrophoresis experiments (see Chapter 7.6) require particularly sophisticated engineering skills.[4] Good temperature control for the sample is usually advisable, as is vibration isolation.

The laser must have good amplitude and beam direction stability, and should have a polarized output. A variety of He–Ne and argon ion gas lasers have been used. Argon ion sources, because of their generally higher power and shorter emission wavelengths [scattering intensity varies as λ^{-4}; see Eq. (7.2.10)] seem to be favored for research applications where low con-centrations of small macromolecules are being probed; however, in many instances high power can be a nuisance, causing heating artifacts or providing signals which overload the detectors and analyzers unless attenuated with filters. Low-power He–Ne lasers are quite adequate for scattering from structures such as biological cells and tissue (Chapters 7.6–7.10).

[30] J. S. Gethner and G. W. Flynn, *Rev. Sci. Instrum.* **46**, 586 (1975).

7.4. Diffusion Coefficients

The most common biologically related use of laser inelastic light scattering is the determination of the translational diffusion coefficient of a collection of macromolecules or supramolecular assemblies. The results of such a measurement can then be used to obtain other information, such as indications of changes in molecular conformation or of modified interactions between various constituent subunits. Examples which involve measuring particle interactions are studies of the assembly and disassembly of ribosomes[31] and aggregation of antibody-coated microspheres by antigen.[32]

If a sample contains identical (monodisperse) Brownian particles, the intermediate scattering function $I_1(Q, t)$ is given by the expression (7.2.19). The derivation of that expression follows directly from Eq. (7.2.18') and the diffusion equation for the time dependence of the concentration of particles $C(\mathbf{r}, t)$, namely, $\partial C(\mathbf{r}, t)/\partial t = D\nabla^2 C(\mathbf{r}, t)$. Thus, if we subtract a constant term C_0 and define $\delta C \equiv C - C_0$, we find, after taking spatial Fourier transforms, $\partial \delta C(\mathbf{Q}, t)/\partial t = -DQ^2 \delta C(\mathbf{Q}, t)$. The solution is $\delta C(\mathbf{Q}, t) = \delta C(\mathbf{Q}, 0)e^{-DQ^2 t}$, from which an expression for the density–density correlation function then follows as

$$\langle \delta C^*(Q, 0)\delta C(Q, t)\rangle = \langle(\delta C)^2\rangle e^{-DQ^2 t}. \tag{7.4.1}$$

The coefficient D which appears in the exponent of Eq. (7.4.1) is the translational diffusion coefficient. Once D has been determined, information about the size and shape of a particle can be deduced. For example, if the scatterers are spherically symmetric the particle radius can be determined from Stokes's law, i.e.,

$$D = kT/6\pi\eta a, \tag{7.4.2}$$

where k is Boltzmann's constant, T is the temperature, η the viscosity, and a the effective particle radius.

The effective particle size may be somewhat larger than that of a "bare" molecule if solvent is dragged along with the moving scatterer. Another effect of solvent association might be to smooth out the shape of a nonspherical particle, so that Eq. (7.4.2) would be a proper expression for analyzing an experiment even if the bare particle were *not* spherically symmetric. However, if the scatterers are grossly asymmetric a different model is necessary; in particular, the diffusion coefficient for ellipsoidal particles is given by the Perrin equations (see e.g., Tanford,[13] p. 327).

[31] R. Gabler, E. W. Westhead, and N. C. Ford, *Biophys. J.* **14**, 528 (1974).
[32] R. B. Cohen and G. B. Benedek, *Immunochemistry* **12**, 349 (1975).

Particles of many differing types can be studied. Because the intensity varies as the square of the mass, a practical lower limit on the size of the scatterers is presently 2000–3000 daltons, i.e., the size of small peptides. The upper limit on particle size in part is set by one's ability to interpret the effects of intraparticle scattering which occurs when the scatterers are large compared with the wavelength of light (particularly if the sample is not homogeneous in size). Viruses,[33] vesicles of sarcoplasmic reticulum,[34] and small bacteria have been studied. Measurements are frequently performed on particles such as proteins, ribosomes, and phospholipid vesicles, the radii of which fall within the range $15 \text{ Å} \lesssim a \lesssim 10^3 \text{ Å}$. Considerable attention has recently been given to studying large spatially anisotropic molecular assemblies, such as microtubules[35] and actin filaments.[35a]

Complexities occur if the scatterers are not all of the same size. To compute the intermediate scattering function $I_1(\mathbf{Q}, t)$ for a mixed collection of particles, let us again refer to Eq. (7.2.17). The situation for large scatterers is complicated by the fact that intraparticle interference requires integration over the contours of the particles. However, if the particles are small and, also, if it can be assumed that the density of scatterers is so low that interactions between particles are rare, the analog of Eq. (7.2.19) is

$$I_1(Q, \tau) = [i_0(Q)]^{-1} \int \mathscr{P}(a)[\bar{\alpha}(a)]^2 \, e^{-D(a)Q^2\tau} \, da, \qquad (7.4.3)$$

where $i_0(Q) \equiv \int \mathscr{P}(a)[\bar{\alpha}(a)]^2 \, da$ and $\mathscr{P}(a)$ is the relative density of scatterers having radius a. [A generalization of Eq. (7.4.3) is derived in the Chapter 7.5.] If, in Eq. (7.4.3), $\bar{\alpha}$ can be taken to vary directly with the mass of a particle (frequently a good approximation), then a scatterer contributes to the scattering intensity in proportion to the square of its molecular weight.

One procedure which can be used to analyze data from polydisperse samples is the "method of cumulants," first described by Koppel.[36] If $I_1(Q, \tau)$ is represented as $I_1(\tau) = \int_0^\infty G(\Gamma)e^{-\Gamma\tau} \, d\Gamma$ with $\int_0^\infty G(\Gamma) \, d\Gamma = 1$, then the moments of $G(\Gamma)$ are related to the time derivatives of $I_1(\tau)$ as

$$\bar{\Gamma} = M_1 \equiv \int \Gamma G(\Gamma) \, d\Gamma = -\frac{d}{d\tau} [\ln I_1(\tau)]_{\tau=0},$$

$$M_2 = \int (\Gamma - \bar{\Gamma})^2 G(\Gamma) \, d\Gamma = \frac{d^2}{d\tau^2} [\ln I_1(\tau)]_{\tau=0}, \quad \text{etc.}$$

$$(7.4.4)$$

[33] R. D. Camerini-Otero, P. N. Pusey, D. E. Koppel, D. W. Schaefer, and R. M. Franklin, *Biochemistry* **13**, 960 (1974).

[34] J. C. Selser, Y. Yeh, and R. J. Baskin, *Biophys. J.* **16**, 337 (1976).

[35] J. S. Gethner and F. Gaskin, *Biophys. J.* **24**, 505 (1978).

[35a] J. Newman and F. D. Carlson, *Biophys. J.* **29**, 37 (1980).

[36] D. E. Koppel, *J. Chem. Phys.* **57**, 4814 (1972).

Thus, since $\Gamma \equiv DQ^2$, the first derivative of $I_1(Q, \tau)$ in the limit $\tau \to 0$ is proportional to an average diffusion coefficient $\bar{D}$, which, if the particles scatter light in proportion to their molecular weight, is a z-averaged quantity; that is, $\bar{D}$ could be represented as $\bar{D} = \sum_i \rho_i M_i D_i / \sum_i \rho_i M_i$, where ρ_i is the weight concentration of scatterers. The ratio M_2/M_1^2 is a measure of the relative width of the distribution. An analysis of errors expected in values of $\bar{D}$ is given in Koppel's paper.[36]

If only a few particle species are present in the sample, it is often more effective to fit the data to a weighted sum of exponentials. In this case Eq. (7.4.3) becomes

$$I_1(Q, \tau) = \left(\sum_i N_i M_i^2 e^{-D_i Q^2 \tau} \right) \Big/ \sum_i N_i M_i^2. \tag{7.4.5}$$

If the molecular weights of the species are known, the relative concentration of particles can be determined and one can deduce the equilibrium constant for simple reactions, such as protein dimerizations or polypeptide helix–coil transitions. Equation (7.4.5) is also quite useful for analyzing data from samples which contain a small amount of residual dust or other large-sized contaminant, particularly if the time scales for the exponential decays pertaining to sample and dust are widely separated. Generally no more than a few (two or three) exponentials can be used to fit data if parameters are to be unambiguously determined; however, if intensity fluctuation measurements are combined with conventional light scattering measurements, parameter determinations may be facilitated. (Similar considerations arise when frequency analysis is used, in which case data have to be fit to a weighted set of Lorentzians.)

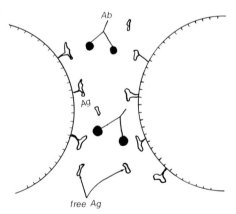

FIG. 5. Immunoassay in the agglutination–inhibition mode. Specific antibody is used to "crosslink" antigen-coated (Ag) spheres. The titre containing the unknown amount of antigen inhibits the agglutination reaction by competing for antibody (Ab) (see von Schulthess et al.[38]).

An interesting and potentially important application of this technique is a laser light scattering immunoassay for detecting antigen–antibody interactions. Changes in $\bar{D}$ are used as an indication of the agglutination of antibody-coated latex microspheres by polyvalent antigen.[32,37] The assay can detect very small amounts of antigen, and in certain instances can approach the sensitivity of radioimmune assays. A modification[38] (the "inhibition" mode) utilizes antigen-coated spheres and probes the degree to which an unknown antigen sample inhibits the agglutination of the spheres by a specific antibody (see Fig. 5).

7.5. Large Particles

Analysis of light scattering experiments generally is more complicated if a sample contains particles whose dimensions are the same order of magnitude as the wavelength of light. Intraparticle interference, by which we mean correlated scattering from several points within a particle, can have two effects. First, the structure factor $\mathscr{S}(\mathbf{Q})$ may have a strong $\mathbf{Q}$ dependence so that in a heterogeneously sized sample the relative contributions to the light scattering spectrum from the various constituents change with the scattering angle. Second, if the particles are not spherical the $\mathbf{Q}$ dependence of the relaxation times of the autocorrelation function may differ from that of spherical particles, even when all scatterers are of identical size and shape.

The effects of size have been of concern ever since light scattering was proposed as a spectroscopic technique almost 100 years ago. Excellent discussions of classical light scattering from large particles may be found in the well-known books written by Kerker[16] and Van de Hulst.[15] In recent years attention has increasingly been given to similar inelastic light scattering situations, but only a few special cases have thus far been analyzed in detail.

We recall that the results derived in the preceding section were obtained after assuming that the scatterers are small compared with the wavelength of light. A generalization to the case of particles whose sizes are, e.g., of the order of a few microns or less (10^4 Å) is easily effected if the so-called Rayleigh–Gans–Debye approximation[15,16] is permissible. The essential assumption in this case is that an incident electric field is not significantly distorted before it interacts with a scattering center within a particle, i.e., that the presence of other scattering centers can be neglected in considering scattering from any particular center. This condition is satisfied if the phase shift experienced by a plane wave transversing the scatterer is small, i.e., if[15] $2(n - n_0)\lambda^{-1} L \ll 1$,

[37] G. K. von Schulthess, R. J. Cohen, N. Sakato, and G. B. Benedek, *Immunochemistry* **13**, 955 (1976).

[38] G. K. von Schulthess, R. J. Cohen, and G. B. Benedek, *Immunochemistry* **13**, 963 (1976).

where $n - n_0$ is the difference between the refractive index of the particle and that of the solvent, L is the maximum dimension of the particle, and λ is the wavelength of light. In general, however, scattering from a large complex particle can be evaluated only by resorting to the solution of Maxwell's equations.[15]

If the Rayleigh–Gans–Debye approximation holds, we see from Eq. (7.2.6) that the scattered field may be written as

$$E_s(\mathbf{R}, t) \sim \sum_{\text{particles}} \int_{\boldsymbol{\Omega}_i(t)} \bar{\alpha}(\mathbf{r}) e^{i\mathbf{Q}\cdot\mathbf{r}}\, d^3r, \tag{7.5.1}$$

where $\boldsymbol{\Omega}_i(t)$ indicates that the integration is to be performed over the volume and orientation of the particles. A useful transformation to a body-fixed coordinate system can be accomplished by writing $\mathbf{r} = \mathbf{R}(t) + \boldsymbol{\rho}$, where $\mathbf{R}(t)$ is the position of the center of mass and $\boldsymbol{\rho}$ is a vector from $\mathbf{R}$ to a point within the particle (see Fig. 6). Thus the expression for $E_s(\mathbf{R}, t)$ can be re-written as

$$E_s(\mathbf{R}, t) \sim e^{i\mathbf{Q}\cdot\mathbf{R}(t)} \int d^3\rho\, P(\boldsymbol{\rho}, t) e^{i\mathbf{Q}\cdot\boldsymbol{\rho}}, \tag{7.5.2}$$

where $P(\boldsymbol{\rho}, t)$ is a weighted distribution of scattering centers, which for optically homogeneous scatterers is simply proportional to the mass distribution within a particle.

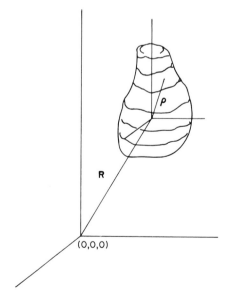

FIG. 6. Transformation to a body-fixed coordinate system.

Let us here assume that the scatterers are identical and noninteracting. Then the intermediate scattering function $I_1(Q, t)$ [cf. Eqs. (7.2.15)–(7.2.18)] can be written as

$$I_1(Q, t) \sim \left\langle e^{i\mathbf{Q} \cdot \Delta \mathbf{R}(t)} \int d^3 \rho' \int d^3 \rho \, P(\boldsymbol{\rho}', 0) P(\boldsymbol{\rho}, t) e^{-i\mathbf{Q} \cdot \boldsymbol{\rho}'} e^{i\mathbf{Q} \cdot \boldsymbol{\rho}} \right\rangle, \quad (7.5.3)$$

where $\Delta \mathbf{R}(t)$ is the displacement of the center of mass at time t. Note that $\mathbf{Q}$ generally is fixed by the geometrical relationship between the source and the detector, so $\int d^3 \rho \, P(\boldsymbol{\rho}, t) e^{i\mathbf{Q} \cdot \boldsymbol{\rho}}$ implicitly depends on the spatial orientation of a particle because of the term $e^{i\mathbf{Q} \cdot \boldsymbol{\rho}}$ in the integrand. Thus, if a particle is nonspherical and, further, if its translational motion is correlated with its spatial orientation, the time dependence of axial reorientation will affect the photon autocorrelation function. This perhaps is clearer if we write Eq. (7.5.3) as

$$I(\mathbf{Q}, t) \sim \langle e^{i\mathbf{Q} \cdot \Delta \mathbf{R}(t)} \mathscr{S}(\mathbf{Q}; \mu_0, \mu_t) \rangle, \quad (7.5.4)$$

where μ_t is the cosine of the angle between the scattering vector $\mathbf{Q}$ and the internal reference axis of a particle; $\mathscr{S}(\mathbf{Q}; \mu_0, \mu_t)$ is the "dynamical structure factor," given by the integrals within expression (7.5.3).

Orientation and translation are strongly correlated when flagellated bacteria such as *E. coli* locomote. These organisms are cigar shaped and have internal structure, so the computation of the expression given in Eq. (7.5.4) is difficult. Various model calculations in this case indicate that the $\mathbf{Q}$ dependence of the *relaxation time* of the autocorrelation function is affected by particle orientation (see Chapters 7.7). If, however, the particles have spherical symmetry, the dynamical structure factor depends only on the magnitude of $\mathbf{Q}$ and not on particle orientation, i.e., $\mathscr{S}(\mathbf{Q}; \mu_0, \mu_t) \rightarrow \mathscr{S}(Q)$, and $I_1(\mathbf{Q}, t) \sim \mathscr{S}(Q) \langle e^{i\mathbf{Q} \cdot \Delta \mathbf{R}(t)} \rangle$. Hence the particle structure in this instance would affect the intensity of the light scattered at a particular angle, but not the $\mathbf{Q}$ dependence of the decay. A similar conclusion holds whatever scheme is used to compute the electromagnetic fields within the scatterer, even if a particle were to be so large that Mie theory (solution of Maxwell's equations for the field within a scatterer[15]) is necessary.

These last comments hold only if all particles are the same size. In a heterogeneous assembly the Q dependence of the structure factor can affect the form of $I_1(Q, t)$. For example, we now consider the diffusion of an assembly of large solid spheres. It is easy to show[15] that in the Rayleigh–Debye–Gans limit, $\mathscr{S}(Q)$ is given as

$$\mathscr{S}(Q) = \{[3/(Qa)^3](\sin Qa - Qa \cos Qa)\}^2 \quad \text{(spheres)}, \quad (7.5.5)$$

where a is the radius. Thus the analog of Eq. (7.4.3) is

$$I_1(Q, \tau) = [i_0(Q)]^{-1} \int \mathcal{P}(a)\mathcal{S}(Qa)[\tilde{\alpha}(a)]^2 e^{-D(a)Q^2\tau}\, da. \qquad (7.5.6)$$

In the limit that a is small, $\mathcal{S}(Qa) \to 1$ and the expression given in Eq. (7.4.3) is recovered; however, $\mathcal{S}(Qa)$ decreases markedly for large values of Qa. Thus, for a heterogeneously sized collection of scatterers, the signal from larger particles will be relatively less significant as the scattering angle is increased; consequently, since large particles usually move more slowly than small particles, contributions to the autocorrelation function which decay slowly will be deemphasized and the apparent decay rate will increase at large scattering angles In this way, artifactual scattering arising from dust contaminants in a sample can in part be eliminated. For example, when a sample is illuminated with light from a He–Ne laser (red), the relative intensity of the scattering from a 0.5-μm particle decreases by a factor of almost 300 as the scattering angle is increased from 20° to 90°; in contrast, for a 0.05-μm particle, the intensity varies by only 15% over that range.

The structure factor for spherical shells has been derived by Tinker[39] and by Pecora and Aragon[40]:

$$\mathcal{S}(Q) = \{[3/Q^3(a^3 - a_i^3)](\sin Qa - \sin Qa_i - Qa\cos Qa + Qa_i\cos Qa_i)\}^2$$
$$\text{(shells)}, \qquad (7.5.7)$$

where a is the outside radius of the shell and a_i is its inside radius. This expression is important in analysis of the scattering from phospholipid vesicles[41,42] and membraneous structures such as particles formed from sarcoplasmic reticulum.[34] Chen et al.[43] have derived a similar expression for coated ellipsoids and have investigated relationships for dynamic light scattering which arise when the integrations indicated by Eq. (7.5.1) are carried out.

The situation for rodlike scatterers is more complicated. Pecora[44] was the first to recognize that, under certain circumstances, reorientations of long, thin molecules would contribute features to the scattered light spectrum from which information about the rotational diffusion coefficients of the scatterers could be extracted. Pecora's expression, which was derived for rigid, vanishingly thin rods of length L, is

$$I_1(Q, t) = B_0(QL)e^{-Q^2Dt} + B_1(QL)e^{-(Q^2D + 6\Theta)t} + \cdots, \qquad (7.5.8)$$

[39] D. O. Tinker, *Chem. Phys. Lipids* **8**, 230 (1972).
[40] R. Pecora and S. R. Aragon, *Chem. Phys. Lipids* **13**, 1 (1974).
[41] F. C. Chen, A. Chrzeszczyk, and B. Chu, *J. Chem. Phys.* **64**, 3403 (1976).
[42] E. P. Day, J. T. Ho, R. K. Kunze, Jr., and S. T. Sun, *Biochim. Biophys. Acta* **470**, 503 (1977).
[43] S. H. Chen, M. Holz, and P. Tartaglia, *Appl. Opt.* **16**, 187 (1977).
[44] R. Pecora, *J. Chem. Phys.* **48**, 4126 (1968).

where $B_0(x)$ and $B_1(x)$ are defined as

$$B_0(x) = \left(\frac{2}{x} \int_0^{x/2} \frac{\sin y}{y}\, dy\right)^2 \tag{7.5.9}$$

and

$$B_1(x) = 5\left(\frac{1}{x} \int_0^{x/2} \frac{\sin z}{z}\, dz - 3j_1\left(\frac{x}{2}\right)\right)^2, \tag{7.5.10}$$

where j_1 is the spherical Bessel function (of the second kind) of order 1. Almost all of the scattered intensity is represented by the B_0 term if QL is less than 3, in which case a single exponential characterizes the autocorrelation function. However, for large scattering angles and particle lengths (e.g., such that $QL > 5$), the intensities associated with higher-order terms in the expansion are quite significant. If the correlation times for translation and rotation are comparable, information about Θ can be obtained. Cummins et al.[45] used light scattering techniques to measure the rotational diffusion coefficient of tobacco mosaic virus, the length of which is approximately 3000 Å, and for which diffusion coefficients are found to be[45] $D \simeq 2.8 \times 10^{-8}$ cm^2 sec^{-1} and $\Theta \simeq 320$ sec^{-1}.

In addition to the requirement that the scatterers be rigid, several other assumptions were made to obtain expression (7.5.8). First, it was assumed that particle orientation and direction of translational diffusion are uncorrelated. Calculations performed[45] for particles so large that their rotational correlation times greatly exceed those of translational diffusion (so that the rotational terms essentially contribute only a flat background term to the correlation function) indicated that translational diffusion, if correlated with particle orientation, leads to deviations from the simple exponential decay given in the first term of Eq. (7.5.8). Second, Rayleigh–Gans theory was used to compute the form factor and results were found to be different[46] if particles are so large that Mie theory[15] must be used (e.g., sizes of the order of several microns).

Additional features in the autocorrelation function may arise from molecular configurational fluctuations if the scatterers are not rigid. Internal modes of motion such a stretching and bending may thus be discernible when long, flexible molecules are examined.[47,48] The internal dynamics of DNA have been studied by quasielastic light scattering,[49] but molecular entanglements and hydrodynamic interactions which occur in congested solutions of such molecules are a complication.[50]

[45] H. Z. Cummins, F. D. Carlson, T. J. Herbert, and G. Woods, *Biophys. J.* **9**, 518 (1969).

[46] M. Kotlarchyk, S. H. Chen, and S. Asano, *Appl. Opt.* **18**, 2470 (1979).

[47] S. Fujime, M. Murayama, and S. Asakura, *J. Mol. Biol.* **68**, 347 (1972).

[48] S. Fujime and M. Murayama, *Macromolecules* **6**, 237 (1973).

[49] M. Caloin, B. Wilhelm, and M. Daune, *Biopolymers* **16**, 2091 (1977).

[50] W. I. Lee, K. S. Schmitz, S. C. Lin, and J. M. Schurr, *Biopolymers* **16**, 583 (1976).

7.6. Determination of Electrophoretic Mobilities

In the preceding chapters we have assumed that the translational movements of the scatterers were spatially isotropic. If, however, the scatterers move preferentially in certain directions, the scattering function $I_1(\mathbf{Q}, t)$ will have special characteristics. This feature forms the basis of a laser inelastic light scattering assay for electrophoretic response in which particles move along an electric field vector and thus impart a corresponding Doppler shift to the spectrum.

A kinetic equation to describe movement of particles in an electric field can be obtained from the continuity equation $\partial C/\partial t = -\mathbf{V} \cdot \mathbf{J}$, with the current $\mathbf{J}$ now taken as the sum of a "diffusion current" $\mathbf{J}_D = -D\mathbf{V}C$ and a drift term $\mathbf{J}_E = u\mathbf{E}C$, where $\mathbf{E}$ is the electric field acting upon the particles and u is the electrophoretic mobility. We thus have

$$\partial C/\partial t = D\mathbf{V}^2 C - u\mathbf{E} \cdot \mathbf{V}C. \qquad (7.6.1)$$

When the effects of boundaries are neglected, the solution of the above equation is

$$\delta C(\mathbf{Q}, t) = \delta C(\mathbf{Q}, 0)e^{-(DQ^2 + iu\mathbf{E}\cdot\mathbf{Q})t}, \qquad (7.6.2)$$

where $\delta C \equiv C - C_0$ is the fluctuation of scatterers about an equilibrium value C_0 which satisfies $\partial C_0/\partial t = 0$.

In a *homodyne* experiment, in which only components of scattered light are mixed on the surface of the photodetector (see Chapter 7.2), the measured photon autocorrelation function $\mathscr{C}(\tau)$ is related to the square of the density–density correlation function $|\delta C^*(Q, 0)\delta C(Q, t)|^2$ [see Eqs. (7.2.15) and (7.2.18′)]. Thus, were an electrophoresis experiment to be performed in this configuration, we would obtain the result normally seen for diffusing particles, namely, $\mathscr{C}(\tau) \sim 1 + \beta e^{-2DQ^2\tau}$. In order to extract information related to electrophoretic drift a fraction of the light emitted by the laser source must be diverted to the photodetector before it passes through the sample. In this situation the spectrometer is operated in a *heterodyne* mode and, as shown in Eq. (7.2.21), the useful signal is then given as

$$I_1(\mathbf{Q}, t) \sim \mathrm{Re}\,\langle \delta C^*(\mathbf{Q}, 0)\delta C(\mathbf{Q}, t)\rangle \sim e^{-DQ^2 t}\cos \mathbf{v}_E \cdot \mathbf{Q}t, \qquad (7.6.3)$$

where $\mathbf{v}_E$ is defined as

$$\mathbf{v}_E = u\mathbf{E}. \qquad (7.6.4)$$

Analogous expressions are realized for other processes in which drift is superimposed upon spatially isotropic motion (e.g., the chemotactic response of bacteria[51]—see Chapter 7.7).

If a collection of diffusing particles all were to have the same drift velocity, the autocorrelation function would be given by Eq. (7.6.3). The spectrum of scattered light would be given as a Lorentzian shifted from the origin by an amount $\omega_s = \mathbf{Q} \cdot \mathbf{v}_E$, i.e.,

$$S(\omega) \sim 1/[(\omega - \mathbf{Q} \cdot \mathbf{v}_E)^2 + (DQ^2)^2]. \tag{7.6.5}$$

Note that as the diffusion coefficient decreases so does the spectral width. Thus, for large particles such as biological cells, were the scatterers to have identical surface charge the spectrum would appear as a sharp spike (cf. Fig. 4c).

In fact, though, the particles in a sample generally do not all have identical charge on their surfaces, nor are they all of the same size. Extraction of electrophoretic parameters from the data can be somewhat complicated if several Lorentzians overlap. However, if diffusion can be ignored the entire distribution of electrophoretic mobilities of a sample can in principle be determined quite easily. In this case the relative height of the spectrum at a given frequency ω_s is directly proportional to the intensity of light scattered by particles whose electrophoretic mobilities are consistent with $\omega_s = \mathbf{Q} \cdot \mathbf{v}_E$. One practical application of this technique has been an investigation of the electrophoretic mobilities of lymphocytes and monocytes obtained from the blood of leukemic patients[52] (see Fig. 7). Several other studies of cell surface charge are discussed in Smith.[53]

Recent applications of electrophoretic light scattering include an assay for antigen–antibody reactions,[54] and a study of the fusion of secretory vesicles with membrane receptors.[55] Ware has recently reviewed work in this area emanating from his own and other investigators' laboratories.[4] Certain difficulties not present in other applications must be resolved when engineering an electrophoretic light scattering experiment. For example, although we implicitly assumed that the electric field is time independent, a constant applied electric field can cause a sample to overheat, resulting in convection. It also might cause electrode polarizations. Consequently, a specially designed pulse generator must be used to cycle the electric field in coordination with data accumulation by the spectrum analyzer. Furthermore, the scattering cell and electrodes must be carefully designed in order to obtain a determinable electric field in the scattering region, and the sample holder

[51] R. Nossal and S. H. Chen, *Opt. Commun.* **5**, 117 (1972).
[52] B. A. Smith, B. R. Ware, and R. S. Weiner, *Proc. Natl. Acad. Sci. U.S.A.* **73**, 2388 (1976).
[53] B. A. Smith, Ph.D. Thesis, Harvard University, Cambridge, Massachusetts (1977).
[54] E. E. Uzgiris, *J. Immunol. Methods* **10**, 85 (1976).
[55] D. P. Siegel and B. R. Ware, *Biophys. J.* **25**, 304a (1979).

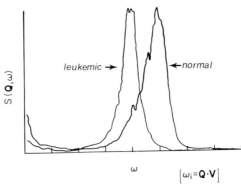

FIG. 7. Electrophoretic light scattering spectrum of mononuclear cells obtained from the blood of leukemic patients. (Adapted from Smith.[53])

should be cooled to effect adequate temperature control. Certain applications require high field strengths, in which case one runs the risk that particles will be leached off the surfaces of the electrodes and contribute an artificial signal. Finally, wave front matching to accomplish adequate signal heterodyning must be done carefully, and, if measurements are made on protein solutions, very small ($\lesssim 5°$) scattering angles must be employed in order to reduce spectral broadening relative to the frequency shifts arising from electrophoretic drift. These considerations are discussed at length in Smith's thesis.[53]

7.7. Motility Measurements

When a suspension of motile microorganisms is illuminated by laser light, the diffraction pattern that results fluctuates at a rate dependent upon the swimming speeds of the scatterers. In many cases the tracks of motile microorganisms can be represented as essentially straight lines with intermittent stops, at which times changes in speed or direction take place. These straight-line trajectories occur over distances generally much greater than a length which is the inverse of the Bragg scattering vector Q, and the organisms' velocities seem to be essentially constant over similarly long distances (see Fig. 8). In this instance a relatively simple model can be used to compute $I_1(\mathbf{Q}, t)$, the implication of which is that the root-mean-square (rms) speed or, in some circumstances, the entire distribution of swimming speeds of a sample can be deduced.

We first assume that the scatterers are homogeneous in size, do not interact, and are *spherically symmetric insofar as their light scattering properties*

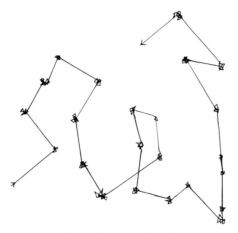

FIG. 8. Approximate pattern of the trajectories of motile microorganisms. Certain strains of flagellated bacteria stop and "twiddle" between successive straight-line portions of their tracks (see Berg and Brown[56]).

are concerned. For the model illustrated in Fig. 8 it follows from Eq. (7.2.18) that $I_1(Q, t)$ is given as

$$I_1(Q, \tau) \sim \left\langle \sum_{\text{scatterers}} e^{-i\mathbf{Q}\cdot\mathbf{r}(t)} e^{i\mathbf{Q}\cdot\mathbf{r}(t+\tau)} \right\rangle \qquad (7.7.1)$$

$$\sim \alpha\langle e^{i\mathbf{Q}\cdot\mathbf{V}\tau} \rangle + (1 - \alpha)e^{-DQ^2\tau}, \qquad (7.7.2)$$

where α is the fraction of the population of microorganisms locomoting at the instant that the sample is probed; $1 - \alpha$ therefore represents the resting fraction. The first term in expression Eq. (7.7.2) results from an assumption that during locomotion the trajectories are given simply as $\mathbf{r}(t + \tau) = \mathbf{r}(t) + \mathbf{V}\tau$, where $\mathbf{V}$ is constant. The second term follows if we assume that movement occurs only by diffusion during the resting phase. For motile bacteria, movements during the rest period are in fact generated by the microorganism itself. These "twiddling" movements[56] result from uncoordinated oscillations of flagellar bundles and do not necessarily lead to a correlation function of the form given in the second term of Eq. (7.7.2); however, an analysis of the twiddle movements predicts functions which are similar and in good agreement with data.[57]

[56] H. C. Berg and D. A. Brown, *Nature (London)* **239**, 500 (1972).
[57] M. Holz and S. H. Chen, *Biophys. J.* **23**, 15 (1978).

If no external forces act upon the microorganisms, so that the distribution of velocities is spatially isotropic, the term $\langle e^{i\mathbf{Q}\cdot\mathbf{V}t} \rangle$ in Eq. (7.7.2) takes the following form:

$$\langle e^{i\mathbf{Q}\cdot\mathbf{V}t} \rangle = \int_0^\infty \frac{\sin QVt}{QVt} W(V)\, dV, \qquad (7.7.3)$$

where $W(V)$ is the distribution of swimming *speeds* of the scatterers. [The distribution $W(V)$ is related to the *velocity* distribution of the microorganisms $P(\mathbf{V})$ by integrations of the latter over direction,[58] e.g.,

$$W(V) = 4\pi V^2 \int_0^{\pi/2} \int_{-\pi/2}^{\pi/2} P(\mathbf{V}) \sin \varphi\, d\varphi\, d\theta. \Bigg]$$

The most striking property of Eq. (7.7.3) is the Q dependence of the predicted autocorrelation function: if data taken at different scattering angles are plotted as a function $X = Qt$, the autocorrelation functions associated with locomotion $\langle e^{i\mathbf{Q}\cdot\mathbf{V}t} \rangle$ will superimpose. In contrast, the signal arising from diffusing particles scales as $Y = Q^2 t$. Consequently, a mixed correlation function such as that given in Eq. (7.7.1) can be analyzed by scaling the experimental autocorrelation functions according to $X = Qt$ and subtracting data taken at one scattering angle from those taken at a different angle.[58]

If Eq. (7.7.3) is written in terms of the scaled variable $X = Qt$ as

$$I^l(X) = \int_0^\infty \frac{\sin XV}{XV} W(V)\, dV, \qquad (7.7.3')$$

we see that the swimming speed distribution can in principle be obtained by Fourier transformation as

$$W(V) = \frac{2V}{\pi} \int_0^\infty \sin XV [X I^l(X)]\, dX. \qquad (7.7.4)$$

But, unless data for $I^l(X)$ can be fitted by an analytic function, the procedure indicated by Eq. (7.7.4) may be impractical because of uncertainties related to truncating the transformation. Consequently, other procedures such as spline fitting[59] or assumption of a specific analytic form for the swimming speed distribution[57,60,61] have been employed to deduce features of $W(V)$. A useful parameter of motion can sometimes be obtained by examining the

[58] R. Nossal and S. H. Chen, *J. Phys. (Orsay, Fr.)* **33-Cl**, 171 (1972).
[59] G. Stock, *Biophys. J.* **22**, 79 (1978).
[60] B. J. Berne and R. Nossal, *Biophys. J.* **14**, 865 (1974).
[61] F. R. Hallett, T. Craig and J. Marsh, *Biophys. J.* **21**, 203 (1978).

initial decay of the autocorrelation function,[62] which from Eq. (7.7.3) is seen to be related to the rms swimming speed as

$$\langle V^2 \rangle \equiv \int_0^\infty V^2 W(V)\, dV \approx 6X^{-2}[1 - I(X)]|_{X \to 0}. \qquad (7.7.5)$$

Many assumptions have been employed in order to derive the relationships given in Eqs. (7.7.1)–(7.7.5). Perhaps the most severe is that of spherical symmetry. Generally, swimming microorganisms have the shapes of cylinders or oblate spheroids, and they are so large that structural factors must be considered (cf. Chapter 7.5). Various model calculations have been performed in which such spatial asymmetries are taken into account. If it is assumed that the direction of locomotion of a microorganism is strongly correlated with its orientation (e.g., that motion occurs parallel to the major axis of the scatterer), then significant deviations from simple Qt scaling as indicated by Eq. (7.7.3) are predicted.[43,47,59,60] For example, when a continuous ellip-soidal mass distribution is assumed for the scatterer, the half-decay point of the autocorrelation function is found to deviate from that predicted for motile spheroids by an ever larger amount as the scattering angle is increased. The discrepancy between the autocorrelation functions for spherical and ellipsoidal particles increases with the size of the scatterers or degree of asymmetry. For solid particles having the size and shape of *E. coli* bacteria, differences by a factor of two or greater might be observed,[60] whereas a model of a prolate ellipsoid coated with a thin membrane leads to periodic deviations from Qt scaling.[43] Recent experimental data on *Chlamydomonas*, though, exhibit scaling of the half-decay point over a wide range of scattering angles.[62a]

Because the dimensions of microorganisms are large compared with the wavelength of light, movements other than simple translation should also contribute to the light scattering spectrum. Theoretical analyses[63,64] indicate that wobbling motions and rotation should increase the rate of decay of $I_1(Q, t)$. However, these perturbations are unimportant at very small scattering angles, where only translational motions are measurable.[59] Thus, by careful experimental design it should be possible to distinguish and quantitate certain complex movements of the scatterers. Unfortunately, samples of microorganism such as flagellated bacteria are amenable to al-most unlimited possibilities of data fitting, as they generally contain translat-ing but perhaps wobbling scatterers, some rotating or twiddling individuals, a

[62] R. Nossal and S. H. Chen, *Nature (London) New Biol.* **244**, 253 (1973).

[62a] T. J. Racey, R. Hallett, and B. Nickel, *Biophys. J.* **35**, 557 (1981).

[63] D. W. Schaefer, G. Banks, and S. S. Alpert, *Nature (London)* **248**, 164 (1974).

[64] J. P. Boon, R. Nossal, and S. H. Chen, *Biophys. J.* **14**, 847 (1974).

fraction of resting or diffusing particles, and members which collide with each other. Thus, using laser light scattering as an assay for *relative* changes in motion may be more appropriate than employing it to determine absolute values of parameters associated with detailed movements. On the other hand, self-consistent computer-assisted data fitting gives reasonably good agreement with certain simple theoretical models.[57,59]

One potentially significant biomedical use of laser light scattering is in studies of spermatozoan motility. Attempts to correlate laser light scattering data with visual observations of sperm viability suggest that light scattering can form the basis of a rapid and objective assay.[61] The motion of sperm cells through extracts of cervical mucus has also been studied.[65,66] By orienting the sperm in narrow capillaries,[65] the entire swimming speed distribution could be determined from the Doppler-shifted components of the spectrum (see Chapter 7.6, particularly Fig. 7). However, several points must still be resolved concerning the interpretation of light scattering data from cells as large as spermatozoa, among which is the importance of head rotations as compared with translational motion.[66a]

7.8. Applications in Cell Biology

Laser light scattering will probably receive increased use in studies of the dynamic behavior of biological cells and tissues. Unfortunately, though, the more ambitious the applications, generally the more complex and uncertain will be the physical models needed to explain experimental observations.

A particularly suitable application is the study of protoplasmic streaming in large algae[67,68] and giant slime molds.[69,70] Protoplasm contains inclusions ranging in size from that of mitochondria down to that of submicroscopic particles. However, using techniques similar to those developed for measuring the speeds of charged particles responding to an electric field (Chapter 7.6), one can determine the velocity distributions of cytoplasmic constituents quickly and objectively.

[65] M. DuBois, P. Jouannet, P. Bergé, and G. David, *Nature (London)* **252**, 711 (1975).

[66] W. I. Lee, P. Verdugo, R. J. Blandau, and P. Gaddum-Rose, *Gynecol. Invest.* **8**, 254 (1977).

[66a] T. Craig, F. R. Hallett, and B. Nickel, *Biophys. J.* **28**, 457 (1979).

[67] R. V. Mustacich and B. R. Ware, *Biophys. J.* **16**, 373 (1976); **17**, 229 (1977).

[68] K. H. Langley, R. W. Piddington, D. Ross, and D. B. Sattelle, *Biochim. Biophys. Acta* **444**, 893 (1976).

[69] R. V. Mustacich and B. R. Ware, *Rev. Sci. Instrum.* **47**, 108 (1976).

[70] S. A. Newton, N. C. Ford, K. H. Langley, and D. B. Sattelle, *Biochim. Biophys. Acta* **496**, 212 (1977).

Another interesting use of laser light scattering is as a probe of ciliary activity on the surfaces of cells and epithelia. Lee and Verdugo[71,72] have used intensity correlation spectroscopy to measure the periodic frequency of ciliary beat in cells of the rabbit oviduct[71] and have shown that their results agree with data obtained by phase microscopy and high-speed cinematography. However, cinematographic methods are expensive and cumbersome, and generally cannot be used to study tissues *in situ*. Light scattering instruments based on fiber optics techniques can potentially be used for clinical investigation of ciliary activity in the respiratory tract.[72] A theoretical model of ciliary beat phenomena used to explain light scattering data[72] treats the ciliated epithelium as an oscillating rough surface, the movements of which are considered to be the sum of a periodic ciliary oscillation and a modulation arising from the coordinated motions of clusters of cilia ("metachronal coordination"). Even for this simplified model, the theoretical analysis is rather complicated and several approximations have to be made. Nevertheless, the theory gives a good representation of the prominent features of the measured autocorrelation function.

A third instance where laser light scattering has been adopted for investigations of phenomena of cellular physiology is in studies of resting and contracting muscle.[73,74] This undertaking is especially challenging due to the complex structure of muscle tissue.[74a]

7.9. Blood Flow

Instruments based on laser light scattering can be designed to monitor local blood flow in a noninvasive manner. Potential applications include detecting the onset of shock in patients confined in a critical-care facility, determining blood flow in tissues chosen for biopsy, investigating the effectiveness of flow in skin transplants, and assessing the extent of peripheral vascular disease. Compact instruments currently are being developed[75,76] for specialized clinical applications which utilize flexible fiber optic probes

[71] W. I. Lee and P. Verdugo, *Biophys. J.* **16**, 1115 (1976).

[72] W. I. Lee and P. Verdugo, *Ann. Biomed. Eng.* **5**, 248 (1977).

[73] F. D. Carlson, R. Bonner, and A. Fraser, *Cold Spring Harbor Symp. Quant. Biol.* **37**, 389 (1972); R. F. Bonner and F. D. Carlson, *J. Gen. Physiol.* **65**, 633 (1975).

[74] R. C. Haskell and F. D. Carlson, *Biophys. J.* **33**, 27 (1981).

[74a] S. Fujime and S. Yoshino, *Biophys. Chem.* **8**, 305 (1978).

[75] R. F. Bonner, P. Bowen, R. L. Bowman, and R. Nossal, *in* "Proceedings of the Electro-Optics/Laser '78 Conference," p. 539. Industrial and Scientific Conference Management, Inc., Chicago, Illinois, 1978.

[76] R. W. Wunderlich, R. C. Folger, B. R. Ware, and D. B. Giddon, *Rev. Sci. Instrum.* **51**, 1258 (1980).

to collect light scattered from tissues such as skin or gums. Photodiodes are used as detectors, and specially designed low-cost electronic analyzers are used to extract flow parameters from the data.

Suppose that laser light is shined on a subject's extremity (such as a finger) and detected as backscattered radiation. The motion of red cells in the microvasculature causes Doppler broadening of the incident light.[77,78] This process is difficult to model analytically because of possible multiple scattering from the erythrocytes, uncertain knowledge of the spatial pathways for flow through the vascular bed, and scattering from surrounding tissue. Nonetheless, various studies[75,78–79] have indicated that changes in the width of the Doppler-shifted spectrum can be correlated well with changes in flow, even though the precise value of the proportionality coefficient is hard to determine theoretically.

One model which shows how the spectral bandwidth scales with flow rate is as follows.[75,78a] Assume that, before interacting with an erythrocyte, photons make many collisions with the surrounding tissue and are randomly diffused. Consequently, even though light is seemingly incident upon the tissue from a specific direction, at the time of the first interaction of a photon with an erythrocyte illumination from all directions is equally probable. Assume further that photons, once scattered by a red cell, do not interact with other moving targets (i.e., other blood cells) before emerging from the tissue. We need not consider the details of photon trajectories subsequent to interaction with the erythrocytes because, although several collisions with static tissue elements might occur, only the *net* path between a red cell and the fixed detector determines the phase shift of the scattered light.

For simplicity, we assume that the scatterers are spherically symmetric, having an effective radius a. We also presume that the cells move in random directions in the capillary bed, with velocities which are constant over distances large compared with the wavelength of light. Thus, by Eqs. (7.5.4) and (7.7.3), the intermediate scattering function can be written as

$$I(t) \sim \int_0^\pi \mathcal{S}(Q(\theta)a) \left\langle \frac{\sin[Q(\theta)Vt]}{Q(\theta)Vt} \right\rangle \sin\theta \, d\theta, \qquad (7.9.1)$$

where the θ integration accounts for all possible scattering angles, assuming a fixed detector and 4π incident illumination ($\sin\theta$ is a geometric factor). The brackets indicate averaging over speeds, and $\mathcal{S}(Qa)$ is the structure factor which, in the Rayleigh–Gans limit, is given above in Eq. (7.5.5).

[77] M. D. Stern, *Nature (London)* **254**, 56 (1974).

[78] M. D. Stern, D. L. Lappe, P. D. Bowen, J. E. Chimosky, G. A. Holloway, Jr., H. R. Keiser, and R. L. Bowman, *J. Physiol. (London)* **232**, H441 (1977).

[78a] R. Bonner and R. Nossal, *Appl. Opt.* **20**, 2097 (1981).

[79] M. D. Stern, P. D. Bowen, R. Parma, R. W. Osgood, R. L. Bowman, and J. Stein, *Am. J. Physiol.* **231**, F80 (1979).

A good analytical approximation to Eq. (7.5.5) which facilitates the integration in Eq. (7.9.1) is[78a]

$$\mathscr{S}(x) = e^{-2\alpha x^2}, \tag{7.9.2}$$

where $\alpha = 0.1$. If the speed distribution is Gaussian, it then follows[60] that the expression in the angular brackets is

$$\langle \sin[QVt]/[QVt] \rangle = e^{-Q^2\langle V^2 \rangle t^2/6}, \tag{7.9.3}$$

where $\langle V^2 \rangle^{1/2}$ is the root-mean-square value of the velocity.

Using in Eq. (7.9.1) the expressions given in Eqs. (7.9.2) and (7.9.3), we can express the result of the integration in terms of reduced variables of length and time, which are, respectively, $L = 4\pi n \lambda^{-1} a$ and $T = \langle V^2 \rangle^{1/2} t / \sqrt{6} a$. When $2\alpha L^2$ is large (as for red blood cells) the resulting expression for $I(t)$ effectively becomes a function only of T,

$$I(t) = 2\alpha/(2\alpha + T^2). \tag{7.9.4}$$

The half-decay point of $I(t)$ is proportional to $\langle V^2 \rangle^{-1/2}$ through the relationship $t_{1/2} = (\sqrt{6} a/\langle V^2 \rangle^{1/2})T_{1/2}$. For example, if we set $\alpha = 0.1$ and numerically evaluate Eq. (7.9.4), we find the half-decay of the autocorrelation function to be related to the rms flow speed as $\langle V^2 \rangle^{1/2} = 0.71 a t_{1/2}^{-1}$. The equivalent relationship to the half-decay of the power spectrum is $\langle V^2 \rangle^{1/2} = 4.1 a v_{1/2}$. A more elaborate derivation also accounts for multiple Doppler scattering from moving blood cells[78a] and provides information on the manner in which spectral shifts depend on blood volume in addition to red cell velocity.

Tanaka et al.[80] have shown that inelastic light scattering can be used to study blood flow noninvasively in human retinal vessels. An instrument based on their prototype studies now is being used for clinical diagnosis.[81] Also, blood flow in arteries of experimental animals has been measured with a fiber-optic catheter,[82] although the necessity for surgical intervention limits the usefulness of this particular procedure.

7.10. Gels and Solutions of Fibrillous Proteins

Various biological materials, either *in situ* or as reconstituted models, have the mechanical characteristics of amorphous solids or highly viscous fluids. Examples are the lens and *vitreous humor* of the eye, extracts of cell

[80] T. Tanaka, I. Ben-Sira, and C. Riva, *Science* (*Washington, D.C.*) **186**, 830 (1974).

[81] G. T. Feke and C. E. Riva, *J. Opt. Soc. Amer.* **68**, 526 (1978).

[82] T. Tanaka and G. Benedek, *Appl. Opt.* **14**, 189 (1975).

cytoplasm, gels formed of fibrin and other blood constituents, cervical mucus, and lubricants of joints such as sinovial fluid. Although these substances typically contain many constituents, in essence they can be characterized as aqueous solvents containing a relatively high concentration of long polymeric molecules. If the interaction between the polymer chains is sufficiently strong, the material is resistant to deformation and has gel-like characteristics. In other instances the material can flow if it is subjected to steady stress, and it exhibits viscous behavior.

Laser light scattering can be used as a probe of the viscoelastic properties of such materials. In certain instances the noninvasive quality of laser light scattering is essential, particularly if a sample is easily perturbed by the stresses imposed by conventional rheometers. In general, the less rigid the material, the more easily it is disturbed if classical rheological techniques are used for measurement. (For example, long molecules tend to align in the stress field between the plates of a viscometer.) Another advantage of certain light scattering techniques is that they can be used to study local properties of a sample, whereas conventional rheological techniques probe bulk moduli.

In Chapter 7.4 we discussed procedures for probing the motions of dilute molecules in solution. Clearly, however, physical models invoked for diffusing macromolecules are unsuitable for analysis of viscoelastic materials whose properties derive mainly from frequent strong interactions between polymer chains. It is preferable to consider a gel or dense polymer solution as a continuum, in which case light scattering fluctuations are linked to local displacements of material within the sample.

We here assume that our sample is optically homogeneous on a length scale comparable to the wavelength of light. It then follows that the diffraction pattern varies because of local changes in density. In a continuum the deviation of the density ρ from local equilibrium is given by the continuity equation

$$\delta\rho(\mathbf{r}, t) = -\rho(\mathbf{r}, t)\mathbf{V} \cdot \mathbf{U}(\mathbf{r}, t), \qquad (7.10.1)$$

where $\mathbf{U}(\mathbf{r}, t)$ is the displacement vector of a differential volume element whose equilibrium position is at $\mathbf{r}$. Upon taking spatial Fourier transforms we find

$$\delta\rho(\mathbf{Q}, t) \simeq -i\rho_0 \mathbf{Q} \cdot \mathbf{U}(\mathbf{Q}, t), \qquad (7.10.2)$$

where ρ_0 is the equilibrium density. Thus, by Eq. (7.2.18') we see that the autocorrelation of photons would be sensitive to autocorrelations of various components of the Fourier transform of the displacement vector.

To compute $\mathbf{U}(\mathbf{r}, t)$ one first needs to assume a physical model. Tanaka et al.[83] considered the question of light scattering from a gel and took the

[83] T. Tanaka, L. O. Hocker, and G. B. Benedek, *J. Chem. Phys.* **59**, 5151 (1973).

polymer lattice to move as an elastic continuum subject to frictional interactions with surrounding fluid. Thus they considered movements of a gel lattice *against* the fluid. Their assumed kinetic equation has the form[83]

$$\rho \partial^2 \mathbf{U}/\partial t^2 = \mu \nabla^2 \mathbf{U} + (K + \tfrac{1}{3}\mu)\nabla(\nabla \cdot \mathbf{U}) - f \partial \mathbf{U}/\partial t, \quad (7.10.3)$$

where K and μ are taken to be the compressibility and shear moduli of the fiber network alone and f is a friction coefficient which is proportional to the viscosity of the gel liquid. Solution of this equation, assuming infinite spatial boundaries, yields waves which propagate or are damped depending on the values of K, μ, ρ, and f. Nonpropagating modes are predicted for soft gels, yielding[83] the following expression for the autocorrelation of the component of the displacement which is parallel to $\mathbf{Q}$:

$$\langle U_z(Q, 0)U_z(Q, t)\rangle = \langle \overline{U_z(Q)}^2\rangle \exp\{-[(K + \tfrac{4}{3}\mu)/f]Q^2 t\}. \quad (7.10.4)$$

Thus the photon autocorrelation function would decay with a time constant whose Q dependence is the same as for Brownian motion [cf. Eq. (7.4.1)], but with a proportionality coefficient given as $(K + \tfrac{4}{3}\mu)/f$. The basic Q^2 dependence of the decay time has been verified in several laboratories,[84,85] although deviation from a single exponential decay also has been reported.[86]

In these studies the driving forces for the displacements are thermal excitations, and the term $\langle U_z(Q)^2\rangle$ is evaluated by thermodynamic arguments.[83] The field autocorrelation function for polarized scattering then derived from Eq. (7.10.4) is given as[83]

$$\langle E^*(Q, 0)E(Q, t)\rangle = \frac{I_0}{c}\left(\frac{\omega_0}{c}\right)^4 \frac{\sin^2 \varphi}{\pi R^2} n^2 \left(\frac{\partial n}{\partial \rho}\right)_T^2$$

$$\times \rho^2 \frac{LkT}{(K + \tfrac{4}{3}\mu)Q^2} \exp\left(-\frac{K + \tfrac{4}{3}\mu}{f}Q^2 t\right),$$

$$(7.10.5)$$

where k is Boltzmann's constant, T the temperature, and L the illuminated length of the scattering volume. The other terms in the expression have the same definitions as in Eqs. (7.2.3) and (7.2.10). Note that, although the parameter which can be obtained from the decay of the autocorrelation is the ratio of the elastic moduli to the friction coefficient, only the moduli appear in the term which multiplies the exponential. Thus, by measuring the intensity of scattered light and knowing instrumental and material

[84] J. P. Munch, S. Candau, J. Herz, and G. Hild, *J. Phys. (Orsay, Fr.)* **38**, 971 (1977).
[85] E. Geissler and A. M. Hecht, *J. Phys. (Orsay, Fr.)* **39**, 955 (1978).
[86] K. L. Wun and F. D. Carlson, *Macromolecules* **8**, 190 (1975).

parameters, one can obtain values of $K + \frac{4}{3}\mu$ and f separately. This procedure has been used by Tanaka et al.[87] to study phase transitions in polyacrylamide gels and, in a somewhat analogous fashion, to examine the formation of cataracts in excised bovine lens.[88] A similar technique has been used by Geissler and Hecht[85] to study the concentration dependence of the compound modulus $K + \frac{4}{3}\mu$ of various polyacrylamide samples.

Recently a light scattering scheme has been developed which yields values of the bulk shear modulus G. This modulus pertains to the solvent and lattice acting together, but because of the low shear resistance of the solvent, it closely approximates μ for the lattice alone. The technique involves driving a sample with a weak external mechanical field and relating observed resonances in the light scattering spectrum to the standing-wave frequencies of gel displacements.[89,90]

Attempts have been made to use scattering schemes based on thermal excitation to study actin gels.[91] Significant deviations from an exponentially decaying form have been observed for the autocorrelation function, but heterogeneities in the sample have made it difficult to obtain consistent results. Similarly, discrepancies between experimental results and the expression given by Eq. (7.10.4) have been observed in studies of the properties of cervical mucus,[66] and it may be that a modification of the basic kinetic equation (7.10.1) should be used to describe such fluid like, though highly viscous, materials.

[87] T. Tanaka, S. Ishiwata, and C. Ishimoto, *Phys. Rev. Lett.* **38**, 771 (1977).

[88] T. Tanaka, C. Ishimoto, and L. T. Chylack, Jr., *Science (Washington, D.C.)* **197**, 1010 (1977).

[89] R. A. Gelman and R. Nossal, *Macromolecules* **12**, 311 (1979).

[90] R. Nossal, *J. Appl. Phys.* **50**, 3105 (1979).

[91] F. D. Carlson and A. B. Fraser, *J. Mol. Biol.* **89**, 273 (1974); **95**, 139 (1975).

8. SMALL-ANGLE SCATTERING TECHNIQUES FOR THE STUDY OF BIOLOGICAL MACROMOLECULES AND MACROMOLECULAR AGGREGATES

By Peter B. Moore

8.1. Introduction

8.1.1. The Nature of the Technique

When short-wavelength radiation (λ about 1 Å) strikes amorphous materials, diffuse scattering results. Observed under proper conditions, the elastic component of this diffuse scatter can yield useful information about the structure of materials at the molecular level. Measurements of this kind have contributed significantly to our understanding of the structures of gases, liquids, and amorphous solids. The variant of this experiment of most interest to biologists, the study of the diffuse scatter given by solutions of biological macromolecules, is the topic of this article.

In biological experiments the scattering studied is the incremental contribution to the total scatter of a solution due to macromolecular solutes. Analysis of this contribution can lead to accurate estimates of molecular weight and volume, molecular size and shape, and knowledge of some aspects of internal structure. Since macromolecules are large compared to the wavelength of the radiation generally used in such experiments (10–100 versus 1 Å), most of the macromolecular scatter is confined to regions close to the direction of the incident radiation. The technique, therefore, is called "small-angle" or "low-angle" scattering, or sometimes simply "solution scattering."

Small-angle scattering commends itself to the biophysicist for three reasons. First, it is easy to prepare samples. All that is needed is a monodisperse solution of the molecule or macromolecular aggregate in question. Crystallization, for example, is not a prerequisite. In addition, sample sizes are small; a few tenths of a milliliter of a 1 or 2 % solution can often produce interesting data. Second, the information obtained relates to the molecule as it exists free in solution. Thus the potential for artifacts is minimal and the

METHODS OF EXPERIMENTAL PHYSICS, VOL. 20

data correspondingly likely to be relevant physiologically. Third, data collection is fast; a few days often suffice.

8.1.2. Historical Comments

Guinier's paper[1] on the measurement of the radius of gyration of particles in suspension from their diffuse scatter is a convenient landmark from which to date the birth of the small-angle field. At the time of introduction, small-angle techniques were virtually unique in offering a direct means for estimating the size and shape of biological macromolecules. Although the method has always attracted adherents, it has never become widely practiced despite its virtues. There are undoubtedly many reasons for this, but historical factors have certainly played a role. By the mid 1950s, when the scientific world was completing its recovery from World War II, two new techniques which could provide similar information had come of age, electron microscopy and x-ray crystallography. The interpretation of the molecular images generated by an electron microscope can be difficult, but the pictures it produces are visual and direct. However misleading it may be, the appeal of the micrograph is often overwhelming when compared with a model inferred from scattering data, by what is to many a nonintuitive process. The attraction of crystallographic analyses of macromolecular structures lies in their richness of detail and accuracy. The effort required is very large, but the results cannot be equaled by any other technique.

8.1.3. The Sources of Current Interest

Crystallography and microscopy, the competition, are widely used and become more effective every year. Yet, paradoxically, the small-angle field is as vigorous now as it has ever been. Four important innovations have contributed to its current strength.

First, a new kind of radiation has become available for small-angle work, thermal neutrons. The small-angle field grew up using x radiation as its standard tool. While this radiation is effective, neutron radiation has many advantages (see below), and its use has significantly broadened the range of problems approachable by solution methods. The first neutron facilities suitable for biological work came on line in the mid to late 1960s, notably at the Julich, Harwell, and Brookhaven laboratories. The completion of the splendid installation at Institut Laue–Langevin in France in the early 1970s gave tremendous impetus to the field.

Second, a new kind of radiation detector has been introduced into the field, the position-sensitive detector. Detectors can be constructed to count

---

[1] A. Guinier, *Ann. Phys.* (*Paris*) [5] **12**, 161 (1939).

either x rays or neutrons and to determine where in the sensitive volume counting events take place. Both one- and two-dimensional detectors are feasible. They can be built to count all the scattering angles of interest in a small-angle experiment simultaneously. These devices replace the traditional single-detector, step-scan mode of data collection and increase the speed of data acquisition roughly 100-fold.

Third, the influence of the computer has been felt in the small-angle field as in so many others. There are many steps in the analysis of small-angle data which require numerical processing beyond the capabilities of hand computation. The widespread availability of computers has encouraged the development of new and powerful algorithms for processing small-angle data, and their use is now routine. As a result the sophistication of data analysis in the small-angle field has improved and along with it the reliability of results obtained. Computers have also materially simplified the task of data collection.

Fourth, in the past few years, a new kind of ultrahigh-flux x-ray source has become available; its potential is still under investigation. Synchrotron radiation sources can provide x rays at all wavelengths useful in scattering experiments and at fluxes orders of magnitude higher than obtainable with conventional laboratory equipment. Among other distinctive features, a synchrotron radiation source is a pulsed radiation source rather than a continuous one, as is the case with ordinary x-ray generators. There is a definite possibility that the unique properties of synchrotron radiation sources may engender some new applications of small-angle techniques, including time-resolved experiments.

The net result of these developments is that a variety of experiments which could hardly be considered using the equipment of 15 or 20 years ago are now practical. In addition, the terms of trade between experimenter's time expended and data acquired have improved 100–1000-fold. There is hardly any physical technique that would not enjoy strong interest in re-sponse to stimuli such as these.

8.1.4. Purpose

The objective of this article is to provide the reader with an introduction to the macromolecular small-angle scattering field. This article will describe the kinds of information that can be obtained about a biological macro-molecule or macromolecular aggregate by small-angle scattering using both x rays and neutrons. The experimental problems faced will be discussed in general terms, as will the approaches generally taken in data analysis. The applications of these methods will be illustrated from the recent literature, and sufficient references provided so that an interested reader can complete his study of the field.

8.1.5. Earlier Reviews

The small-angle field has been reviewed regularly over the years and a number of useful works are available. Much of the theory of the method was developed in the field's early years with the result that the older reviews retain their utility to a degree which is unusual in the biochemical and biophysical literature. For example, the book by Guinier and Fournet[2] remains invaluable despite the fact that it has reached its 25th anniversary. Many other reviews can be cited which deal with the x-ray small-angle literature.[3-7] In addition, the proceedings of a number of symposia on small-angle topics have been published over the years (e.g., Brumberger[8]); two symposium proceedings have appeared in the Journal of Applied Crystallography (1974, 1978). The proceedings of a recent symposium dealing with all aspects of the application of neutron scattering to biology are available.[9] Jacrot[10] has reviewed the neutron small-angle field.

Background information on x-ray techniques may be obtained from James[11] and the International Tables for X-Ray Crystallography.[12] Bacon[13] has provided an excellent work for those wishing an introduction to neutron diffraction. A rigorous theoretical development of the theory of thermal neutron scattering can be found in Marshall and Lovesey.[14]

[2] A. Guinier and G. Fournet, "Small-Angle Scattering of X-Rays." Wiley, New York, 1955.

[3] W. W. Beeman, P. Kaesberg, J. W. Anderegg, and M. B. Webb, *Handb. Phys.* **32**, 321 (1957).

[4] O. Kratky and I. Pilz, *Q. Rev. Biophys.* **5**, 481 (1972).

[5] H. Pessen, T. F. Kumosinski, and S. N. Timasheff, *Methods Enzymol.* **27**, 151 (1973).

[6] I. Pilz, *in* "Physical Principles and Techniques of Protein Chemistry" (S. J. Leach, ed.), Part C, pp. 141–243. Academic Press, New York, 1973.

[7] O. Kratky and I. Pilz, *Q. Rev. Biophys.* **11**, 39 (1978).

[8] H. Brumberger (ed.), "Small-Angle X-Ray Scattering" (Proceedings of Conference on Small-Angle X-Ray Scattering, June 1965, Syracuse, New York). Gordon & Breach, New York, 1967.

[9] B. P. Schoenborn, (ed.), *Brookhaven Symp. Biol.* **27** (1976).

[10] B. Jacrot, *Rep. Prog. Phys.* **39**, 911 (1976).

[11] R. W. James, "The Optical Principles of the Diffraction of X-rays." Cornell Univ. Press, Ithaca, New York, 1962.

[12] K. Lonsdale, ed., "International Tables for X-Ray Crystallography." Kynoch Press, Birmingham, England, 1968.

[13] G. E. Bacon, "Neutron Diffraction," 3rd ed. Oxford Univ. Press (Clarendon), London and New York, 1975.

[14] W. Marshall and S. W. Lovesey, "Theory of Thermal Neutron Scattering." Oxford Univ. Press, London and New York, 1971.

8.2. The Experimental Problem

8.2.1. The Scattering Signal

If the scatter given by a single isolated molecule could be observed over a period of time as it diffused rotationally, a time-averaged scatter $I(R)$ would be measured which is directly related to its structure[15]:

$$I(R) = \sum_i \sum_j f_i f_j \frac{\sin 2\pi R r_{ij}}{2\pi R r_{ij}}. \tag{8.2.1}$$

Here $R = (2 \sin \theta)/\lambda$, where 2θ is the angle between the incident and the scattered radiation and λ the wavelength of the radiation (Fig. 1); r_{ij} is the distance between the ith and jth atoms in the molecule and both sums are taken over all atoms; f_i and f_j are the scattering lengths of the two atoms appropriate for the type of radiation used.

For x rays, $f_i = (e^2/mc^2) f_i(R)$, where $f_i(R)$ is the atomic structure factor for atom i which can be thought of as the Fourier transform of the electron density distribution of that atom.[11] [The polarization correction will be very small for low scattering angles. It can be ignored in many cases or included in $f_i(R)$ as a small correction factor.] In the neutron literature, where there is no physical equivalent of electron density, atomic scattering properties are tabulated directly as scattering lengths, b_i being the scattering length of the ith atom; hence $f_i = b_i$.[13] $I(R)$ is the number of photons (neutrons) scattered by a single molecule per steradian of solid angle per unit time, at a given R, if the molecule is experiencing an incident radiation flux of one photon (neutron) per unit time. Flux is measured in photons (neutrons) per unit area per unit time, where the area is measured in the same units as the scattering lengths.

Equation (8.2.1) describes the scatter given by gases at low pressure, where molecules are well separated and behaving in an uncorrelated manner. It is also the correct basis for understanding macromolecular contributions to the scattering of a dilute solution. The scattering of the macromolecules in solution, however, will not be identical to that which they would give *in vacuo*. Equation (8.2.1) uses f_is, which are measured relative to a vacuum which has an average f per unit volume equal to zero. Solvent viewed at low resolution (i.e., small scattering angles) can be regarded as a continuum having an average f per unit volume, or scattering-length density, which is not zero. The effect of being immersed in this continuum can be taken into account to a good first approximation by replacing each f_i in Eq. (8.2.1) by

[15] P. Debye, *Ann. Phys. (Leipzig)* [4] **46**, 809 (1915).

$f_i - v_i\rho$, where v_i is the volume made inaccessible to solvent by the ith atom and ρ the solvent's average scattering-length density.

The experimental objective of virtually all macromolecular small-angle studies is to measure $I(R)$, the macromolecular contribution, under conditions in which Eq. (8.2.1) holds, with appropriately evaluated f_is. The reader should recognize that, although knowledge of a molecule's structure at atomic resolution enables one to compute $I(R)$, the reverse operation is not possible: a knowledge of $I(R)$ will not give back the structure. The rotational averaging involved in the production of the signal, reflected in the cylindrical symmetry of $I(R)$ about the direction of the direct beam, leads to the loss of all information about the relative orientations of the interatomic vectors whose lengths r_{ij} appear in Eq. (8.2.1). However, as is made clear below, a number of lower-resolution aspects of the molecule's size and structure can be elucidated once $I(R)$ is known (see Chapter 8.3).

8.2.2. Isolation of the Macromolecular Signal

All components in a solution contribute to its scattering. The separation of macromolecular contributions from those of the solvent is usually done in a simple-minded manner. A sample of pure solvent and a sample of the solution are measured in the same apparatus under the same conditions. The two measured profiles, $I_{sol}(R)$ and $I_{solv}(R)$, which are simply the counts registered by the detector at each scattering angle, are scaled to normalize for any differences in the total amount of radiation used in the collection of the two profiles and $I(R)$ found as follows:

$$I(R) = I_{sol}(R) - (v/v_t)I_{solv}(R), \qquad (8.2.2)$$

where v is the volume of solvent in an aliquot of volume v_t of the solution. Note that knowledge of $\bar{v}$, the partial specific volume of the solute, is required, as is knowledge of concentration of the solute, in order to carry out the subtraction indicated.

This subtraction is justified provided there are no correlations between the positions of the molecules of the solvent and those of the macromolecular solute, averaged over the time of observation.[16] If these positions do not correlate, the scattered intensities due to solute and solvent will simply add to give the total measured signal, and Eq. (8.2.2) will hold. If solvent molecules bind to the macromolecules or organized regions of solvent exist around them, problems can arise.[16-19] Equation (8.2.2), therefore, is to be

[16] J. Goodisman and H. Brumberger, *Monatsh. Chem.* **104**, 598 (1973).
[17] H. Eisenberg and H. Cohen, *J. Mol. Biol.* **37**, 355 (1968).
[18] S. N. Timasheff, *Adv. Chem. Ser.* **125**, 327 (1973).
[19] A. S. Hyman, *Macromolecules* **8**, 849 (1975).

regarded as a first-order approximation for estimating $I(R)$ from experimental data.

The macromolecules themselves may aggregate or otherwise correlate in solution. Distortions in $I(R)$ due to these effects can be dealt with by making measurements at several concentrations and extrapolating to zero concentration. In the dilute limit, the macromolecular contribution will be proportional to $I(R)$, the profile for a single molecule.

The only condition besides diluteness which must be met by a sample in order that the scattering described by Eq. (8.2.1) be measured is that the sample be thin. If the sample is thick, a significant fraction of the radiation scattered by the average macromolecule will be scattered again by others before emerging from the sample. Equation (8.2.1) does not hold for multiple scattering (see Beeman et al.[3]).

8.2.3. Small-Angle Apparatus: General Features

All small-angle instruments consist of the same set of elements: a radiation source, a monochromator, a collimator, a sample holder, and a detector (Figs. 1 and 2). The purpose of these elements is to provide a monochromatic, appropriately collimated beam of radiation directed at a specimen held in a fixed position. The radiation scattered by the specimen is measured by the detector at some distance from it. The design of these elements is dictated by three factors: (i) the type of radiation being used, (ii) the kind of radiation source available, and (iii) the type of measurement to be made. While it is not appropriate to discuss instrument design in great detail, some general comments are in order.

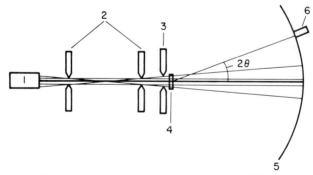

FIG. 1. Schematic diagram of a low-angle scattering apparatus. Critical components and positions in the instrument are numbered: (1) radiation source, (2) collimation slits, (3) guard slits, (4) specimen, (5) plane of registration, and (6) detector. The plane of registration is more properly a surface of registration and is the region of space explored by the detector. Its low-angle region will be well approximated as a plane; hence its name. The scattering angle 2θ is the angle between the optic axis of the instrument, defined by the beam center, and a line drawn between the center of the beam's intersection with the sample and the detector axis.

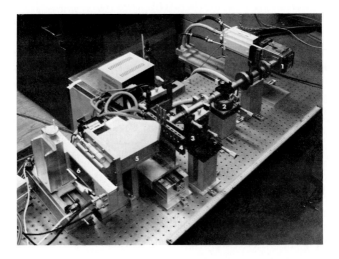

FIG. 2. X-ray small-angle instrument equipped with a linear position-sensitive detector. The instrument depicted here is in operation at Yale University. It was designed by D. M. Engelman and constructed with the assistance of J. Johnson. The numbered elements are (1) x-ray tube, (2) focusing-mirror monochromator, (3) quard slit assembly, (4) sample holder, (5) flight path chamber, and (6) position-sensitive detector. The tubes connecting the source to the mirror and the mirror to the guard slits are all evacuated as is the flight path chamber. This instrument has a monitor counter (the cylindrical object on top of the flight path chamber at its detector end) which measures backscatter from the beam stop.

8.2.4. The Small-Angle Compromise

The principal problem in measuring diffuse scatter stems from the fact that the signal being measured is continuous. In order to measure such a profile, the ideal small-angle instrument should have a source–monochromator–collimator combination which delivers a perfectly monochromatic, infinitely fine pencil of radiation which is directed through an infinitesimally thin specimen. The scattered radiation should then be detected using a device with an infinitesimally small acceptance aperture. The reason this kind of geometry is desirable is that if the detector aperture is finite, it will admit radiation scattered over a range of angles and at any given time will not measure the signal scattered at a single angle only. If the beam is a finite size, even an ideal detector will be seeing scatter from particles whose angular relationships to the detector vary. Failure to achieve perfect monochromatization of the beam has a similar effect. In all cases the signal measured at some nominal scattering angle 2θ will not be $I(2\theta)$ but some kind of weighted average of $I(2\theta)$ over a range of angles $2\theta \pm \Delta$. The two functions $I(2\theta)$, given by Eq. (8.2.1), and the weighted average of $I(2\theta)$, measured by instruments which deviate from the ideal, are different; the profile measured is a distorted version of the profile desired. This instrumental distortion effect is called "slit smearing."

Given the finite brightness of real radiation sources, the collimated beam must have finite size, the detector used must have a finite aperture, and the monochromatization must deviate from perfection in order to ensure a practical rate of data acquisition. Thus some degree of deviation between the ideal measured profile and that actually observed has to be accepted in order to make any measurement whatever. The practical problem is compounded by the fact that the signal of interest is often quite weak.

The size and shape of the beam pose a particularly difficult problem since it is usually highly desirable to measure I at angles very close to zero. The larger the physical size of the beam, the larger are the angular regions around zero where no measurements can be made. This situation is not entirely without hope, however. As a general rule, the smaller the molecule, the larger the smallest angle at which measurement must be made is. Since small molecules scatter more weakly than large molecules on a weight for weight basis, beam size can be adjusted upwards as the molecular weight of the material being studied falls so as to keep the data acquisition rate and the degree of slit smearing approximately constant.

The deviation of the measured from the ideal signal depends on the properties of the ideal signal, which is to say, on the structure of the molecule being studied, as well as on the properties of the instrument making the measurement. While computational correction of measured signals for instrument smearing effects *a posteriori* is possible (see Section 8.2.19), it is obviously desirable to minimize smearing as far as possible by careful instrument design; the smaller the magnitude of the correction which needs to be made, the more accurate the final result is. The interactions between the characteristics of the instrument being used and the structure of the material being studied, as well as the additional constraints imposed by the purposes to which the data will be put, ensure that there is no single "best" instrument for small-angle purposes.

8.2.5. X Rays

For the foreseeable future, most x-ray small-angle experiments will be done using conventional laboratory x-ray generators, as has always been the case. (A full discussion of synchrotron sources and their possible application to small-angle work may be found in a recent review.[20]) These sources work by exciting the characteristic x-ray emission lines of the metallic elements. An electron beam having an energy of a few tens of kilovolts is directed against a metal target. The beam knocks electrons out of the low-lying orbitals of

[20] H. B. Stuhrmann, *Q. Rev. Biophys.* **11**, 71 (1978).

the atoms. The refilling of these holes with electrons from higher-lying orbitals gives rise to sharp emission lines. In a typical generator roughly half the emitted x-ray energy will be contained in these lines. The other half shows up as a white background continuum which has a maximum cutoff energy equal to the electron beam energy. The favorite target material for biological studies is copper, and its K_α emission line ($\lambda = 1.54$ Å) is the one usually used.

The area of the target illuminated by the electron beam is the x-ray source. There are a variety of generator designs on the market differing in size and shape of the illuminated area. In general the source areas are 0.1–1 mm^2 in area and rectangular in shape. Where generators differ most significantly is in their brilliance, the number of photons they emit per area of source. The more brilliant the source, the faster the rate of data acquisition and/or the smaller the beam one can use is. High brilliance is achieved by raising the amperage of the electron beam falling on the source. High intensity requires elaborate expedients to prevent the destruction of the target due to melting, which leads in turn to high initial cost and maintenance problems. The result is that someone considering the acquisition of an x-ray small-angle apparatus should bear in mind the possibility that the savings in cost and ease of operation obtained by the purchase of a simple, low-intensity generator may more than balance the gains to be made in data collection rate by the purchase of a high-intensity source.

8.2.6. Monochromatization of X Rays

Conventional x-ray sources have a partially monochromatic character due to the way they generate radiation. A variety of steps can be taken to "clean up" their output to the extent required for small-angle work.

The simplest technique for monochromatization takes advantage of the fact that the linear absorption coefficient of metals for x rays is strongly wavelength dependent.[21] Of particular importance is the fact that the linear absorption coefficient for each metal shows a sharp discontinuity (its absorption edge) at a wavelength determined by its atomic number. Thus near the absorption edge, metal foils will act as bandpass filters, passing radiation of wavelengths longer than the critical value and strongly absorbing radiation of shorter wavelengths. For Cu K_α radiation, nickel has an appropriately positioned absorption edge. It has a high transmittance for Cu K_α, but strongly absorbs Cu K_β radiation, which has a somewhat shorter wavelength and which is the second most important emission line from a copper target. The radiation emerging from a nickel filter of appropriate thickness will be

[21] B. W. Roberts and W. Parrish, in "International Tables for X-Ray Crystallography" (K. Lonsdale, ed.), Vol. III, p. 73. Kynoch Press, Birmingham, England, 1968.

effectively free of Cu K_β contributions and also be significantly reduced in shorter-wavelength white radiation.

The Cu K_α contributions to a measured scattering profile can be more rigorously identified using the two-filter technique of Ross.[22] This technique employs two metals, one whose edge is at shorter wavelength than Cu K_α and the other whose edge is at a wavelength longer than Cu K_α. One measures the scattering profile first with one of the filters in place and again with the other filter in place and then takes the difference of the two profiles.[23]

A useful degree of monochromatization can also be imposed by the detector. Both proportional detectors and scintillation detectors have energy-discriminating properties which allow them to "tune out" events due to radiation of energies different from that of Cu K_α, for example. The discriminating power of these detectors is not very high, but sufficient to contribute to making a scattering experiment effectively monochromatic. The combination of nickel filters and detector energy discrimination often suffices.

For applications where better monochromatization is required, two further techniques can be employed. X-ray beams can be totally reflected when they strike smooth surfaces ("mirrors") at glancing angles of incidence. The critical angle for total reflection is a function of wavelength. A surface placed in a beam at the critical angle for Cu K_α will not reflect Cu K_β or any of the other components in the beam of a shorter wavelength. The reflected rays, which include Cu K_α and all longer-wavelength radiation from the source, can then be used for scattering purposes. The ultimate expedient involves monochromatization by directing the beam onto a crystal. The radiation scattered in a strong Bragg reflection, taken at the correct angle, can provide an excellent source of virtually pure Cu K_α radiation. Both mirrors and crystals can be bent to make the total reflection or Bragg reflection process focus the beam. A thorough discussion of the properties of monochromators of both kinds, both focusing and otherwise, can be found elsewhere.[24]

8.2.7. Collimation

X-ray small-angle instruments stand or fall on the performance of their collimating systems. The purpose of the collimating system is to define the size and shape of the beam incident on the specimen. A collimator often consists of a sequence of slits which shape the beam and control its angular divergence. The size, shape, and divergence of the beam must be such that at

[22] P. A. Ross, *Phys. Rev.* **28**, 425 (1926).

[23] O. Kratky, *Naturwissenschaften* **31**, 325 (1943).

[24] J. Witz, *Acta Crystallogr., Sect. A* **A25**, 30 (1969).

the plane of registration the detector can measure scatter at the smallest angles required without interference from the components of the beam that have been transmitted through the specimen without scattering. (Note that the transmitted beam is generally a large fraction of the total radiation incident in the specimen.)

Conventional generators are not bright enough to permit one to collimate the beam to a round or square profile and achieve a geometry which permits measurement at the necessary low angles in an instrument of convenient size. Therefore most small-angle instruments use a beam with a narrow rectangular profile. This profile is convenient since most x-ray generators are set up to illuminate a long rectangular source which, viewed from a low angle, appears as a line. An appropriately oriented set of collimator slits will deliver a slit-shaped beam at the specimen which receives contributions from the whole illuminated area of the source and is therefore as bright as possible. In addition, if a slit-shaped beam is used, the detector can be brought very close to the beam center in its narrow direction; measurement can thus begin at very small angles. Slit-shaped beams derived from conventional sources have sufficient numbers of photons in them to give good counting rates with ordinary specimens. The penalty paid for the count rate is serious slit smearing in the lowest-angle regions of the measured scattering profile.

Collimator design would appear to be a simple problem in geometric optics; indeed, this is the way it is usually approached. However, x-ray collimators never perform up to expectation. The problem is a phenomenon called parasitic scatter. The slit edges that are put in the x-ray beam to define its size and shape scatter x rays powerfully, regardless of their composition. The result is that instead of creating a beam with sharp, well-defined edges, a simple collimation system will produce a beam whose edges trail out over regions much larger than expected from geometric optics. The "sloppy" edges of the beam due to parasitic scatter have intensities which are large compared to the signal seen in solution scattering experiments; they cannot be ignored. There are basically two strategies for controlling parasitic scatter. The first and simplest is to mount the detector on a moveable arm whose axis is coincident with the position of the beam as it passes through the specimen. Between the specimen and the detector a second collimator is placed coaxial with the arm and restricting the detector's field of view to the point where it only "sees" x rays scattered from the illuminated region of the specimen and not those scattered from slit edges.[25] The second approach involves the use of auxiliary slits (or equivalent elements) called "guard slits." These slits are placed between the sample and the collimator and are

[25] W. W. Beeman, *in* "Small-Angle X-Ray Scattering" (H. Brumberger, ed.), p. 197. Gordon & Breach, New York, 1967.

positioned to intercept the parasitic scatter due to preceding elements in the collimator without themselves being illuminated by any part of the main beam. The first strategy is appropriate for devices that collect data using a single detector in the step-scan mode. The second can be used for single-detector instruments; it is also appropriate for devices that make use of position-sensitive detectors. It should be emphasized that control of parasitic scatter is probably the single most important aspect of the design and construction of a small-angle instrument.

8.2.8. Flight Path

Samples are usually placed immediately following the guard slit, or the equivalent if guard slits per se are not used. The thin-walled glass capillaries used for conventional x-ray crystallography are adequate sample containers, although more elaborate sample holders can be used if necessary.

Between the sample and the detector there will be a flight path. There are two problems which must be dealt with in this part of the instrument. The first is air scatter. Air scatters x rays like everything else. This scattering has two undesirable aspects. On the one hand it attenuates the signal seen at the detector, and on the other, air scatter of the portion of the beam which passes through the specimen can contribute substantially to background. The simplest solution to both problems is to evacuate the space between the sample and the detector. If for some reason the instrument cannot accommodate a vacuum, flushing of these spaces with helium will reduce the problem. The second issue to be considered is the undeflected beam. Both as a protection to the detector and for the safety of the people working in the area of the instrument it is essential to place a stop in the beam path between the sample and the detector. A small piece of lead placed in the appropriate position is adequate for this purpose. Obviously the beam stop should be as small as possible and positioned to shade the minimum angular region possible in the plane of registration. Placement of the stop inside the vacuum chamber prevents the air scatter that would occur if the beam were stopped outside.

8.2.9. Detection

There are many ways to measure scattered x-ray intensities.[26] Over the years the small-angle scattering fraternity has relied primarily on digital electronic counting devices of one kind or another, mainly because they give

[26] K. Lonsdale, D. H. MacGillavry, H. J. Milledge, P. M. de Wolff, and W. Parrish, *in* "International Tables for Crystallography" (K. Lonsdale, ed.), Vol. III, p. 133. Kynoch Press, Birmingham, England, 1968.

direct numerical output. The traditional small-angle instrument is equipped with a single electronic counter which is supplied with some kind of aperture limiting the range of angles over which it can accept radiation. These apertures are often narrow slits of dimensions comparable to that of the beam at the plane of registration. The long dimension of the detector aperture is set parallel to the long axis of the beam profile and the detector scans the scattered radiation along an arc whose center is located at the specimen and whose periphery is tangent to the line perpendicular to the beam's long axis, passing through the beam center. In modern instruments the scanning operation is computer controlled. It has usually been the practice to use either a scintillation or a proportional counter in order to take advantage of their energy discrimination properties.

Recently linear position-sensitive x-ray detectors have become available.[27,28] Linear x-ray detectors typically have sensitive areas which are roughly 5 mm high and 5–10 cm long. Spatial resolution in these detectors is of the order of 0.1 mm. Devices of this kind can be placed with their long axis along the line one would scan with a single detector. Their dimensions and resolutions are such that they will cover the entire angular range of interest with adequate resolution if they are placed at about the same distance from the sample as the single detector device they ordinarily replace. Two-dimensional x-ray detectors can also be constructed.[29] The position-sensitive detectors in common use are proportional counters which retain the energy discrimination properties characteristic of all devices of that kind.

The gain in data acquisition rate obtained by the use of these detectors has already been mentioned. They have an additional virtue: fluctuations in beam intensity during the measurement of a single scattering profile will not influence the profile shape obtained. Fluctuations obviously do affect the profile obtained with a single detector.

8.2.10. Beam Monitors

In some experiments it is useful to be able to normalize data sets for differences in the total flux passing through the specimen during the time of data collection, caused by instabilities in the source, for example. Accurate comparison of data sets, such as might be required when studying conformational changes in macromolecules, virtually requires such normalization.

The way normalization is usually handled is to equip the instrument with an auxiliary counter which counts a small fraction of the incident or transmitted beam and accumulates a running total which is proportional to the

[27] C. J. Borkowski and M. K. Kopp, *Rev. Sci. Instrum.* **39**, 1515 (1968).

[28] W. R. Kuhlmann, K. H. Lauterjung, B. Schimmer, and K. Sistemich, *Nucl. Instrum. Methods* **40**, 118 (1966).

[29] R. W. Hendrichs, *J. Appl. Crystallogr.* **11**, 15 (1978).

total number of photons that have been involved in a given experiment. This can be done by mounting the monitor counter so that it measures back-scatter off the beam stop. A more elegant variation of this approach is to place a small crystal on the back of the beam stop and monitor one of its Bragg reflections.[30]

In instruments using position-sensitive detectors a transmitting beam can be used. The stop is constructed to transmit a small fraction of the direct beam which leads to the presence of an attenuated image of the beam in the data accumulated by the detector. Its integrated area provides the required measure of relative intensity from experiment to experiment. This approach has the virtue that it compensates both for changes in source strength and for changes in detector sensitivity.

Monitors have not been widely used in x-ray small-angle work in the past. Conventional laboratory x-ray sources, especially low-intensity x-ray sources, are so stable that the flux averaged over periods of minutes to hours will be quite reproducible from one experiment to the next, provided the equipment is maintained in a properly air-conditioned environment.

8.2.11. Apparatus Designs

Scores of different x-ray small-angle instruments have been built and used over the years. For the purposes of discussion one might categorize these instruments ("cameras") into several different classes according to the most striking aspects of their design. The following classes are described in the literature: (i) focusing crystal cameras,[31-33] (ii) total-reflection cameras,[34,35] (iii) the Kratky camera and its variants,[36-39] (iv) collimated-detector cameras,[25,40,41] (v) position-sensitive detector cameras,[29,42,43] and (vi)

[30] O. Kratky, in "Small-Angle X-Ray Scattering" (H. Brumberger, ed.), p. 63. Gordon & Breach, New York, 1967.

[31] V. Luzzati, Acta Crystallogr. **13**, 939 (1960).

[32] V. Luzzati, J. Witz, and R. Baro, J. Phys. (Paris), Suppl. **24**, 141A (1963).

[33] H. Pessen, T. F. Kumosinski, S. N. Timasheff, R. R. Calhoun, Jr., and J. A. Connelly, Adv. X-Ray Anal. **13**, 618 (1970).

[34] W. Ehrenberg and A. Franks, Nature (London) **170**, 1076 (1952).

[35] G. Damaschun, Exp. Tech. Phys. **13**, 224 (1965).

[36] O. Kratky, Z. Elektrochem. **58**, 49 (1954).

[37] O. Kratky and Z. Skala, Z. Elektrochem. **62**, 73 (1958).

[38] G. Damaschun, G. Kley, and J. J. Mueller, Acta Phys. Austriaca **28**, 233 (1968).

[39] R. W. Hendrichs, J. Appl. Crystallogr. **3**, 348 (1970).

[40] H. P. Thomas, Z. Phys. **208**, 338 (1967).

[41] H. Wawra, Z. Naturforsch., C.: Biosci. **31C**, 635 (1976).

[42] J. Schelten and R. W. Hendrichs, J. Appl. Crystallogr. **8**, 421 (1975).

[43] A. Tardieu, L. Mateu, C. Sardet, B. Weiss, V. Luzzati, L. Aggerbeck, and A. M. Scanu, J. Mol. Biol. **101**, 129 (1976).

special-purpose cameras.[44,44a] The designs cited are chosen to represent what can be done; this list of references is by no means all inclusive. References to earlier designs may be found in Guinier and Fournet.[2]

A few of the designs are available from commercial manufacturers as off-the-shelf items. Satisfactory instruments can also be assembled on an optical bench using commercially available elements (e.g., detectors or monochromators) and some ingenuity (e.g., the instrument illustrated in Fig. 2). For someone wishing to set up an apparatus for measurements of this kind with minimum trouble, the purchase of a Kratky camera is probably advisable. The Kratky design has collimation properties that would be very hard for a first-time designer and builder to match. It can be readily adapted for use with monochromators (see above) and position-sensitive detectors,[45] and the analysis of the data it puts out is well understood.

8.2.12. Neutrons

A neutron in motion has wave properties and will act, under conditions where its wave properties are being expressed, as if it had a wavelength which can be calculated from its mass and velocity from the de Broglie relationship. It is easy to calculate the kinetic energy of a neutron whose de Broglie wavelength is in the neighborhood of 1 Å. When this is done it is discovered that neutrons whose wavelengths are in that range have kinetic energies roughly equal to $\frac{3}{2}kT$, where T is approximately room temperature.[13] Slow, or thermal, neutrons will be diffracted by materials in much the same way as x rays of the same wavelength.

The only devices available today which can generate thermal neutrons at the flux levels required for biological-structure work are nuclear fission reactors. From the standpoint of a biophysicist the move from x-ray experiments to neutron experiments constitutes a move to a new world. An x-ray apparatus can be operated within the context of an ordinary laboratory as "private" equipment. Fission reactors on the other hand are facilities which because of their size and cost must be used on a shared basis by large groups of investigators. The scattering equipment which is attached to such devices to permit their use for structural studies is similarly large and costly. There are only a handful of research facilities of this kind in the whole world.

A fission reactor consists of an assembly of uranium elements (the core) surrounded by a jacket full of a moderator substance, usually a fluid; this region is in turn surrounded by a heavy layer of radiation shielding for safety

---

[44] V. Bonse and M. Hart, in "Small Angle X-Ray Scattering" (H. Brumberger, ed.), p. 121. Gordon & Breach, New York, 1967.

[44a] O. Kratky and H. Leopold, *Makromol. Chem.* **133**, 181 (1970).

[45] D. Atkinson, personal communication.

reasons. Fission events in the core produce energetic neutrons in quantities beyond those needed to sustain the chain reaction. These excess neutrons escape into the moderator substance, where they lose kinetic energy by collision with moderator molecules. The neutrons equilibrated with the moderator will have the Maxwell–Boltzmann distributions of energies appropriate for the temperature at which the moderator is maintained. In research reactors it is customary to keep the moderator at around room temperature. This yields a neutron energy distribution with a peak wavelength in the neighborhood of 1 Å. Neutrons from the neutron "gas" which exist in the moderator region can be tapped for experimental purposes outside the reactor by insertion of a suitable pipe through the shielding.

The thermal spectrum of the neutrons derived from the reactor is manipulable. In the past decade it has become apparent that for biological applications and some applications in chemistry and physics it would be desirable to have a distribution of neutron energies which peaks at longer (i.e., lower-energy) wavelengths. This spectral shift is brought about by inserting a device into the moderator to produce a strongly refrigerated region within it. Such devices, known as "cold sources," commonly operate near absolute zero, and neutrons taken from these refrigerated regions have an energy distribution which peaks in the neighborhood of 6–10 Å.

The design of neutron instruments, as compared to corresponding x-ray instruments, is dictated by the relatively low fluxes available from neutron sources. When the spectra given by the most powerful available neutron sources are measured, it is found that the neutron flux available per unit wavelength at any wavelength in the spectrum is orders of magnitude less than that available as Cu K_α emission from a conventional laboratory x-ray source.[13] The probability of a scattering event occurring when a neutron or photon passes through a sample is given by the scattering cross sections of the material contained in the sample. (The total cross section for a macromolecule is $\sim 4\pi \sum f_i^2$.) It turns out that the scattering cross sections for biological macromolecules are about the same for x rays and neutrons. It is clear then that one cannot simply transfer an x-ray camera to an appropriate reactor and hope to collect any data at all; the beam its collimating system would produce would be much too weak to produce data at a useful rate.

8.2.13. Monochromatization of Neutrons

The neutron beam that comes out of a reactor beam core contains neutrons of all energies. In order to perform diffraction experiments it must first be monochromatized. Bragg reflection of the beam from a crystal is just as suitable a means of monochromatization for neutrons as it is for x rays; the same principles apply. A number of crystalline materials have been used for this purpose in the past, but in recent years pyrolytic graphite has come to be the favored material.[46,47] Pyrolytic graphite surfaces have very high reflectivity; there is little beam intensity lost as a result of Bragg reflection. The surfaces used are artificially made and can be formed on curved surfaces, opening the possibility of focusing optics. In addition the surfaces have a mosaic spread which is also under control during manufacture. In general it is advisable to use pyrolytic graphite surfaces with as large a mosaic spread as possible to *increase* the bandwidth of the wavelength region reflected by the crystal into the collimating system. Reasonable small-angle measurements can often be made with beams whose wavelength distributions are quite broad ($\Delta\lambda/\lambda \sim 0.1$). The broader the band used, the greater the number of neutrons available for experimentation. For these devices the theories developed for the corresponding x-ray apparatus are fully transferrable.[24] Because the spectrum of the beam incident on such monochromators is white, it is necessary to place filters in the reflected beam to reduce contamination with $\frac{1}{2}\lambda$ radiation.

A second method of monochromatization in common use is velocity selection.[48] Neutrons of wavelength in the angstrom range have velocities of the order of 1000 m/sec. This speed is low enough to permit neutrons to be sorted on the basis of velocity by mechanical means. A cylindrical rotor is placed with its axis parallel to the beam and set so that the beam is intercepted by its edge. Helical channels are provided along the cylinder's outside wall and pitched so that when the cylinder rotates in an appropriate direction at an appropriate speed neutrons having the desired velocity will be able to follow the helical grooves without colliding with the walls. The width of the grooves controls the bandwidth of the velocity distribution passed by the selector, and its rate of rotation controls the average velocity and hence wavelength.

Crystal monochromators are cheap, simple, and require no maintenance. They have three disadvantages, however. First, alterations of the wavelength at which a given piece of apparatus is operated require repositioning and realigning all the apparatus downstream from the monochromator. Given the size and weight of neutron instruments, adjustments of this kind can be a major undertaking. Second, the spacings of the crystals available sets significant limits on the maximum wavelength that can be obtained. (The plane-to-plane spacing used in pyrolytic graphite is about 6 Å). Third, the monochromatization achieved using crystals is often "too good" for small-angle applications because of a smaller than optimal mosaic spread, which

[46] R. W. Gould, S. R. Bates, and C. J. Sparks, *Appl. Spectrosc.* **22**, 549 (1968).
[47] T. Riste and K. Otnes, *Nucl. Instrum. Methods* **75**, 197 (1969).
[48] B. Buras, K. Mikke, B. Lebech, and J. Leciejewicz, *Phys. Status Solidi* **11**, 567 (1965).

costs the experimenter usable flux. Some relief from both the spacing and bandwidth problems appears likely in the near future because of the development of multilayer monochromators.[49,50] A multilayer monochromator is effectively a one-dimensional synthetic crystal produced by depositing thin films of metals of alternating scattering lengths on an appropriate substrate. These Bragg-reflect incident neutrons just like a normal crystal. They have very large plane-to-plane spacings and their reflectivity can be controlled to permit them to reflect a very broad wavelength band, if required.

Velocity selectors are costly to build. However, they can be adjusted over a wide range of wavelengths simply by changing the rotor speed. Such adjustments require no alterations in the optical track of the rest of the instrument. Their efficiency increases with the width of the wavelength band they pass.

8.2.14. Pulsed Reactors

The neutron sources discussed so far produce a steady output of thermal neutrons. Most of the experimental work done to date has made use of "dc" sources of this kind. However, some fission reactors and virtually all neutron sources which depend on particle accelerators ("spallation sources") produce pulses of neutrons rather than a continuous stream. It seems likely that many of the high-flux neutron sources built in the future will be this kind. Pulsed sources are best used with no monochromator. As a pulse of thermal neutrons travels through space, it spreads out in the direction of flight owing to differences in the velocity of the neutrons it contains. Its leading parts will consist of short-wavelength, high-velocity neutrons and its trailing portions will contain nothing but long-wavelength, low-velocity neutrons. The motion of the pulse monochromates it. There is a whole class of instruments called "time-of-flight" instruments which exploit this property of neutron pulses. In this case one might place a detector to view the scatter given by the sample at a fixed angle and monitor the output of this detector as a function of time. A scan of the detector's output in time corresponds to a scan in R as each pulse goes by since the neutrons producing the events seen at each time are constantly changing in λ and hence R, since $R = (2 \sin \theta)/\lambda$. It is possible to construct time-of-flight instruments suitable for biological small-angle work. A description of the design and properties of such devices can be found elsewhere.[51,52]

---

[49] B. P. Schoenborn, D. L. D. Caspar, and O. F. Kammerer, *J. Appl. Crystallogr.* **7**, 508 (1974).

[50] A. M. Saxena and B. P. Schoenborn, *Acta Crystallogr., Sect. A* **A33**, 805 (1977).

[51] L. Cser, *Brookhaven Symp. Biol.* **27**, VII-3 (1976).

[52] D. F. R. Mildner, *Nucl. Instrum. Methods* **151**, 29 (1978).

8.2.15. General Design: Collimation

The primary fact about neutron sources is their low flux. If the flux an instrument can deliver is low, the only way a significant number of scattering events can be observed per unit time is to increase the amount of material placed in the beam.

Some relief from the problem of low flux can be obtained by taking a broad cut from the thermal spectrum, as already mentioned. Relaxation of the degree of monochromatization of the beam will increase the flux entering the experiment. The penalty paid is an increase in smearing; the amount of broadening that can be tolerated depends on the type of data being collected.

A second factor which assists the neutron experimenter is the low absorption coefficient of most biological materials for neutrons. Most of the apparent attenuation of a neutron beam as it passes through a sample results from high-angle scattering, especially incoherent scattering if the sample is rich in hydrogen. This is in contrast with the x-ray situation, where much of the attenuation is due to the absorption of photons by the material in the sample. The way the cross sections for these kinds of events work out, samples of about 2 mm thickness can be tolerated for water solutions and thicknesses of 0.5–1 cm can be accommodated for materials suspended in D_2O. The maximum tolerable specimen thickness for an x-ray experiment is about 1 mm. Thus, in a neutron experiment, one can place two to five times as much material in the beam as in an x-ray experiment, simply because of differences in sample thickness.

The third and most important way the amount of radiated material is increased in neutron experiments is by setting up the collimating system to deliver a beam with a very large (by x-ray standards) cross-sectional area. The size of the neutron source is set fundamentally by the diameter of the beam pipe through which the particles leave the reactor. Typical pipes deliver beams which are 2×2 cm. In order to take full advantage of that area, or some large fraction of it, the collimator must be very long. Large beams can be tolerated only if their angular divergence is small, and the only way angular divergence can be controlled if the aperture areas are big is by placing successive slits at very large distances from each other. If angular divergence is not controlled, the direct beam at the plane of registration will have an angular size which forbids measurements at the low angles required for small-angle work. The second aspect of the apparatus which follows from the large beam is the requirement for a long flight path, to ensure that the angular area corresponding to the direct beam due to its cross-sectional area at the final aperture is small compared to the contribution due to its angular divergence.

The result of the operation of these factors is that neutron small-angle instruments are typically many meters from monochromator to detector, whereas most x-ray instruments are of the order of half a meter. The weight and cost of the equipment goes up with size.

Once the need for a physically large instrument is accepted, square or round beam profiles become practical. The use of such beam profiles means that the smearing effects due to beam size are roughly equal in all directions in the plane of registration. This has two practical consequences. First, it will be useful to collect data over the entire plane of registration rather than along one line, as is the case with slit-beam x-ray devices where the degree of slit smearing varies radically with direction in the plane of registration. Second, if the angular size of the beam is small, slit correction may often be close to negligible. A point which should be made about neutron collimators is that parasitic scatter tends to be much less of a problem in neutron optics than in x-ray optics. Slits made with cadmium absorb neutrons so efficiently that parasitic scatter is negligible and thus collimation will be about as predicted from geometric optics.

The net effect of the differences in design between neutron and x-ray equipment can be appreciated by considering the volume of solution illuminated by the beam. For x rays this volume is generally of the order of half a microliter; for neutrons it is commonly 100 microliters or greater. It is this difference in illuminated sample volume that makes neutron small-angle experiments feasible despite the low flux levels available.

8.2.16. Flight Path and Detection

The flight path in a neutron instrument is effectively a scaled-up version of its x-ray counterpart. The flight path should be evacuated if possible, or failing that, flushed with dry helium in order to reduce air scatter. The path must also contain a beam stop composed of a strong neutron absorber, e.g., boron carbide.

Detectors for neutrons consist of a chamber full of a gas including a strong neutron-absorbing species, such as ^{10}B as BF_3, or ^3He. A nuclear reaction takes place after the neutrons are absorbed by these atoms and the products of those reactions are ionizing.[13] These ionizing events are detected just as they would be for x rays. When the field started out, single-detector, step-scan instruments were commonly used. These have been rendered obsolete by newer two-dimensional position-sensitive detectors.[53-56] For instruments having round or square beams the scattered

[53] J. L. Alberi, J. Fisher, V. Radeka, L. Rogers, and B. P. Schoenborn, *IEEE Trans. Nucl. Sci.* **NS-22**, 255 (1975).

[54] R. Allemand, J. Bourdel, E. Roudant, P. Convert, K. Ibel, J.-P. Cotton, and B. Farnoux, *Nucl. Instrum. Methods* **126**, 29 (1975).

[55] J. Schelten, P. Schlect, W. Schmatz, and A. Mayer, *J. Biol. Chem.* **247**, 5436 (1972).

[56] R. Klesse and G. Kostorz, *Crystallogr. Comput. Tech., Proc. Int. Summer Sch., 1975* p. 383 (1976).

profile will be cylindrically symmetric about the beam axis or nearly so. In this case an area detector acts like a piece of digital film. The signal ascribed to each scattering angle will be the average at locations on the detector face within an annular ring of appropriate radius and width. This mode of data collection insures that *all* neutrons scattered over the entire angular region of interest contribute to the measured data. The gain in data acquisition rate is roughly 1000-fold compared with single-detector step-scan strategy. The importance of these detector developments in increasing the rate at which work can be done cannot be overstated.

Monitors are routine in neutron instruments. Data collection often takes days; changes in reactor temperature, state of the fuel cycle, demand due to other users, etc., can all affect the flux available at a given wavelength. Such alterations can only be allowed for by use of a monitor.

8.2.17. Available Instruments

The instrumentation of a neutron beam port for scattering experiments is a major undertaking both because of the physical size and cost of the components required and because of the elaborate electronics often associated with such devices. Most of those doing biological experiments will find themselves traveling to a central facility to make use of an already existing, general purpose instrument on a shared basis. The properties of the instrument will usually have to be accepted as given. A number of instruments have been constructed for dc sources. Two closely related designs for simple, Söller slit cameras have been used.[57,58] The possibility of constructing focusing cameras based on curved multilayers has been explored,[49,59] and it seems likely that instruments of this kind will soon be available. Two large instruments based on velocity-selector monochromatization and two-dimensional position-sensitive detectors have been built.[55,60] An instrument based on crystal monochromatization and a two-dimensional position detector of high spatial resolution is available at Brookhaven.[61] Time-of-flight designs can be found in the references given in Section 8.2.14.

8.2.18. Absolute Intensity Measurements

For determination of molecular weights and volumes by both x rays and neutrons, it is useful to be able to place measured intensities on an *absolute*

[57] A. C. Nunes, *Nucl. Instrum. Methods* **108**, 189 (1973).

[58] B. C. G. Haywood and D. L. Worcester, *J. Phys. E* **6**, 568 (1973).

[59] A. C. Nunes and G. Zaccai, *Brookhaven Symp. Biol.* **27**, VII-49 (1976).

[60] W. Schmatz, T. Springer, J. Schelten, and K. Ibel, *J. Appl. Crystallogr.* **7**, 96 (1974).

[61] B. P. Schoenborn, J. Alberi, A. M. Saxena, and J. Fisher, *J. Appl. Crystallogr.* **11**, 455 (1978).

scale. What is meant by absolute scale in the small-angle field is the determination of what fraction of the incident beam is represented by a given measured intensity. What makes absolute-scale measurements difficult in practice is that the intensity scattered in any direction by a typical specimen is orders of magnitude less than the intensity of the incident beam. Indeed, most detectors used for measuring scattered intensities would be damaged by exposure to the direct beam.

For neutrons, absolute scale is reasonably easy to achieve. A set of attenuators of known thickness can be prepared and beam intensity measured as a function of total attenuator thickness, with the same detector used for measuring scattering profiles. A semilog plot of the data permits extrapolation to the unattenuated intensity at zero thickness. Lucite is a satisfactory attenuator. Thicknesses of several centimeters are required so that the preparation of a set of blocks of well-characterized thickness is easy. Foils of Cd can also be used. The thickness of the sample, the total integrated beam intensity, and the solid angle accepted by the detector suffice to set measurements on an absolute scale. A second method in wide use is to employ a standard scatterer and calibrate in terms of the scatter it gives. H_2O is excellent for these purposes. Almost all the scattering from water is incoherent scatter owing to its ^1H content.[13] Thus it gives a strong scattering profile which is virtually constant with angle. For scatterers for which $I(R)$ is constant there is no slit smearing effect; hence the observed $I(R)$ is the true $I(R)$. The cross section of water is well known.[10] Provided the volume of the sample in the beam is also known, the relationship between total sample cross section and measured intensity follows, and once again the absolute scale is set.

Absolute scale is harder to establish for x rays. For an x-ray apparatus in which the beam is fully monochromatized (by Bragg reflection), attenuation with filters is useful.[31,32,62-64] For beams which are not rigorously monochromatized filters cannot be used because the linear absorption coefficient of matter for x rays is strongly wavelength dependent with short wavelength being more penetrating. (Such strong wavelength dependence is not seen with neutron attenuators.) In such circumstances attenuation will not be logarithmic and accurate extrapolation to zero attenuation impossible. Two other approaches have been used, the simplest being a resort to standards whose scattering can be calculated *a priori*, just as is done with H_2O in the neutron field.[25,63] Examples of substances useful in this regard include sulfur hexafluoride and octafluorocyclobutane, which are gases of known structure[25] whose scatter can be calculated using Eq. (8.2.1). The second

[62] V. Luzzati, J. Witz, and A. Nicolaieff, *J. Mol. Biol.* **3**, 367 (1961).

[63] D. L. Weinberg, *Rev. Sci. Instrum.* **34**, 691 (1963).

[64] G. Damaschun and J. Mueller, *Z. Naturforsch., A* **20A**, 1274 (1965).

alternative is to prepare precalibrated secondary standards. Kratky and his collaborators have used mechanical devices to attenuate the beam in order to estimate its intensity[65-67] in conjunction with the calibration of standards. Further discussions of these problems may be found in Batterman et al.[68] A useful theoretical critique of the various methods of determining absolute intensity has been provided by Hendrichs,[69] and a report has recently appeared on the accuracy with which profiles can be measured.[70]

For either x-ray or neutron experiments, the option is always open of determining molecular weights by comparison of unknowns with molecules of the same chemical type of known molecular weight, under equivalent experimental conditions. The knowns in this case are used as secondary standards.

It should be recognized that the beam intensity of interest in any scattering experiment is the intensity *after* passage through the specimen. Passage through the specimen attenuates both x rays and neutron beams because of absorption and high-angle scatter. It is easy to show that for low scattering angles, the beam and the scattered radiation of interest are equally affected by attenuation. Hence one always wishes to compare measured scatter with the attenuated-beam intensity.

8.2.19. Data Correction

Before data sets are analyzed consideration must be given to the impact of slit smearing (see Section 8.2.4). In general it is advisable to attempt to correct for these effects computationally. Slit smearing alters *both* the shape of a profile, on which estimates of molecular size and shape depend, *and* its absolute magnitude, from which estimates of molecular weight and volume come. What the analytical correction techniques attempt to do is to estimate what the data would have been if a detector of the same solid angle of acceptance as that actually used had measured the scatter given by the sample irradiated with a beam of the same integrated intensity as that used, under conditions of ideal geometry and monochromatization.

The effect of smearing on the data can be thought of as a convolution. $O(R)$, the curve actually measured, is given by the following expression:

$$O(R) = \int_{-\infty}^{+\infty} I(R - R')W(R, R')\, dR', \qquad (8.2.3)$$

[65] O. Kratky and H. Wawra, *Monatsh. Chem.* **94**, 981 (1963).

[66] O. Kratky, I. Pilz, and P. J. Schmidt, *J. Colloid Interface Sci.* **21**, 24 (1966).

[67] H. Stabinger and O. Kratky, *Makromol. Chem.* **179**, 1655 (1978).

[68] B. W. Batterman, D. R. Chipman, and J. J. De Marco, *Phys. Rev.* **122**, 68 (1961).

[69] R. W. Hendrichs, *J. Appl. Crystallogr.* **5**, 315 (1972).

[70] Commission on Crystallographic Apparatus, *J. Appl. Crystallogr.* **11**, 196 (1978).

where $I(R)$ is the ideal profile and $W(R, R')$ the weighting function representing the averaging effects of the instrument.[71,72] W is generally a function of R; the amount of averaging varies with angle. $W(R, R')$ has two components: (i) a component due to the size and shape of the beam and the detector aperture, and (ii) a component due to the spectrum of the radiation used in the experiment. The first of these can either be calculated from the geometry of the beam collimator and the detector's properties, or measured directly. What is required is the size and shape of the beam at the plane of registration, as seen by the detector. For x-ray experiments wavelength effects are usually small. In the case of neutron experiments, the beam's spectrum will usually be known and will be provided to the experimenter by those running the facility.

Equation (8.2.3) is an integral equation in which both $W(R, R')$ and $O(R)$ are known. There are a number of techniques for solving such equations to obtain $I(R)$. What keeps the problem from being straightforward is the fact that $O(R)$ is usually measured only at a finite number of points over a limited range of R. In addition $O(R)$, as measured, includes some error, from counting statistics if nothing else. The problem is to find an estimate of $I(R)$ in the face of these limitations in the information given.

In the past ten years a number of algorithms have been proposed which accept a reasonably wide range of scattering profiles and scattering geometries and find solutions which take the noise in the data into account. One might single out for mention the iterative algorithms,[73,74] the spline algorithm of Schelten and Hossfeld,[75] and the indirect transformation method of Glatter.[76] Deutsch and Luban[77] have recently proposed a new method for dealing with slit height correction. An extensive bibliography and discussion of techniques has been prepared by Schmidt,[78] and some practical comparisons of algorithm performance have been performed.[79]

Someone starting out in the field would be well advised to write the authors of the algorithms he finds appealing to obtain listings of the computer programs that have been written to implement them. The art of data correction is in a fairly satisfactory state at present. However, further progress may be expected in the future; the ideal algorithm is yet to be discovered.

[71] F. W. Jones, *Proc. R. Soc. London, Ser. A* **166**, 16 (1938).

[72] R. Spencer, *Phys. Rev.* **38**, 618 (1931).

[73] J. A. Lake, *Acta Crystallogr.* **23**, 191 (1967).

[74] O. Glatter, *J. Appl. Crystallogr.* **7**, 147 (1974).

[75] J. Schelten and F. Hossfeld, *J. Appl. Crystallogr.* **4**, 210 (1971).

[76] O. Glatter, *Acta Phys. Austriaca* **47**, 83 (1977).

[77] M. Deutsch and M. Luban, *J. Appl. Crystallogr.* **11**, 87, 98 (1978).

[78] P. W. Schmidt, *Crystallogr. Comput. Tech. Proc. Int. Summer Sch., 1975* p. 362 (1976).

[79] G. Walter, R. Kranold, J. J. Mueller, and G. Damaschun, *Crystallogr. Comput. Tech., Proc. Int. Summer Sch., 1975* p. 383 (1976).

8.3. Data Analysis

Suppose that full data have been collected on the scattering of a given macromolecule in solution. What information can be extracted from those data? In practice "full data" means that the incremental contribution of the macromolecule to the scattering of a solution has been measured at several solute concentrations and the data placed on an absolute scale. The data are then used to construct a scattering profile corrected for concentration effects *and slit smearing*. For such a corrected scattering profile $I_{exp}(R)$, it will be true that

$$I_{exp}(R) = NI(R), \tag{8.3.1}$$

where N is the number of solute molecules contributing to the measured profile and $I(R)$ the scatter of a single molecule as given by Eq. (8.2.1).

8.3.1. Molecular Weight and Partial Specific Volume

The first aspect of the corrected profile that can be examined with profit is its value at $2\theta = 0$, $I(0)$. Because of the direct beam, $I(0)$ cannot be measured directly. However, provided data have been collected at R sufficiently close to zero, extrapolation to $R = 0$ is straightforward. (The range of R over which data must be collected in order to permit extrapolation depends on the size of the molecule. Criteria are discussed in Beeman et al.[3]) Guinier[1] demonstrated that the lowest-angle portions of the scattering curve of any molecule are Gaussian in shape. Thus the lowest-angle data plotted as $\ln[I_{exp}(R)]$ versus R^2 will be linear, and extrapolation to $R = 0$ follows.

The interest in $I(0)$ lies in its relationship to molecular weight and molecular volume.[31,62,80] Equation (8.2.1) at $R = 0$ becomes

$$I(0) = \left(\sum_i f_i - \rho_0 v \right)^2,$$

where $v = \sum_i v_i$, is the total volume of the molecule inaccessible to solvent and ρ_0 the scattering-length per unit volume, or scattering length density of the solvent. Defining $\bar{\rho} = (\sum_i f_i)/v$, then taking the incident flux and experimental geometry into account, one obtains (Luzzati[31]; see Kratky and Pilz[4])

$$I_{exp}(0) = (A_d/d^2)I_i AtC(MW/N_0)(\bar{\rho} - \rho_0)^2\bar{v}^2. \tag{8.3.2}$$

Here I_{exp} and I_i are measured in photons or neutrons per second, I_i is the flux in the beam measured *after* its passage through the sample; A_d is the area of the aperture of the detector and d the distance from the sample to

---

[80] O. Kratky, *Prog. Biophys. Biophys. Chem.* **13**, 107 (1963).

the detector; A is the area of the sample illuminated by the beam and t its thickness; MW is the solute's molecular weight, N_0 Avogadro's number, $\bar{v}$ the partial specific volume of the solute, and C its concentration (mass/volume). Then

$$MW = I_{exp}(0)d^2N_0/A_d ItC[\bar{\rho} - \rho_0]^2\bar{v}^2, \qquad (8.3.3)$$

where $I = I_i A$, the total number of neutrons or photons in the beam. The partial specific volume of the solute must be measured. The scattering-length density of the solvent can be obtained from its chemical composition and density. The scattering-length density of the solute is generally obtained from knowledge of its atomic composition and $\bar{v}$. If MW is known, $\bar{v}$ follows; if $\bar{v}$ is known, MW follows.

Equation (8.3.2) suggests a method for obtaining $\bar{v}$ from scattering data. ρ_0 can be varied (see Section 8.3.8). In a neutron experiment ρ_0 can be altered over a wide range by mixing D_2O into the solvent (see Section 8.3.12). In an x-ray experiment low-molecular-weight solutes of high electron density, such as salts, glycerol, or sugars, can be added to the solution. $[I_{exp}(0)/C]^{1/2}$ is proportional to $\rho - \rho_0$. Thus data collected on the same macromolecule at different ρ_0s will permit determination of the ρ_0 for which $\rho = \rho_0$. This procedure can be carried out on data that are not on an absolute scale but simply normalized for differences in transmitted beam. Given the atomic composition of the molecule as obtained, for example, from its amino acid or base composition, $\bar{v}$ follows from ρ. These values can be used to calculate MW [Eq. (8.3.3)], leading to a determination dependent only on scattering and atomic composition data. Note that the molecular volume emerging from a neutron study using H_2O–D_2O mixtures will be the anhydrous molecular volume, with the proviso about hydrogen exchange explained below (see Section 8.3.11). The x-ray volume will be that inaccessible to the solute used to manipulate ρ_0 and will be a *hydrated* volume.

While Eq. (8.3.3) is often convenient, it is necessary to be aware that its theoretical justification requires something more rigorous than Eq. (8.3.1). The scattering of solutions near $R = 0$ must be considered in thermodynamic terms. When $R \sim 0$, volumes much larger than that of a single molecule are being examined; it is the fluctuations in solute concentration within these volumes that give rise to the scattering seen.[2] Thus the molecular weight value obtained from a low-angle experiment is a thermodynamic quantity. The Einstein–Debye theory developed first for light scattering[81] is correct in this case as well; its extension to x-ray or neutron scattering may be found in Eisenberg and Cohen[17] or Timasheff.[18,82]

[81] C. Tanford, "Physical Chemistry of Macromolecules." Wiley, New York, 1961.

[82] S. N. Timasheff, *in* "Electromagnetic Scattering" (M. Kerber, ed.), p. 337. Pergamon, Oxford, 1963.

In practice published values for $\bar{v}$ for macromolecules (within a given chemical type variation is small) are often used in estimating $(\rho - \rho_0)\bar{v}$ for use in Eq. (8.3.3). This usage assumes that the scattering-length density of the solution is the sum of the scattering-length densities of both solute and solvent. Additivity will not be observed if solute and solvent interact, as they do with polyelectrolytes.[83] In principle one should measure $\delta\rho/\delta C$, the change in scattering-length density with change in solute concentration at constant temperature with the chemical potentials of all species except that of the macromolecule held constant, and use it in place of $(\rho - \rho_0)\bar{v}$. In practice $\delta d/\delta C$, where d is the weight density of the solution, can be measured under the same conditions for the solute–solvent combination of interest and $\delta\rho/\delta C$ estimated from it (for details, see Eisenberg and Cohen[17]). The buffer sample whose scattering profile is used for background subtraction [Eq. (8.2.2)] should be equilibrated by dialysis with the solute solution of interest. Under these conditions, a correct molecular weight will be found. Similar precautions are routine in the determination of molecular weight by other thermodynamic methods; they are equally necessary here.

In practice the molecular weights obtained from x-ray or neutron forward scattering have been found to be of excellent accuracy, comparable to sedimentation equilibrium except for a few situations involving highly charged polyelectrolytes where the precautions mentioned above were not observed. The advantage of small-angle molecular weight determination over its light scattering analog is that sample preparation is much easier. Dust, which is the bane of light scattering studies, scatters almost entirely in the forward direction when observed with radiation of small λ; it does not interfere with this technique.

Recognizing the thermodynamic nature of $I(0)$, it can be understood that a study of $I(0)$ as a function of solute concentration will give information about the dependence of the chemical potential of the solute on concentration. Thus the virial coefficients for the molecule can be obtained from such an analysis just as they can be derived from light scattering or osmotic pressure data.

8.3.2. Molecular Size and Shape: Radius of Gyration

As already pointed out, the scattering curve of a macromolecule at low R is Gaussian in shape; different molecules differ only in the breadth of this Gaussian. It can be shown by expansion of Eq. (8.2.1) as a power series in $2\pi Rr$ and considering the first two terms not only that the curve is Gaussian,

[83] A. Hyman and P. A. Vaughan, *in* "Small-Angle X-Ray Scattering" (H. Brumberger, ed.), p. 477. Gordon & Breach, New York, 1967.

but that the breadth of the Gaussian reports the molecule's radius of gyration.[1] For small R,

$$I(R) = I(0) \exp(-\tfrac{4}{3}\pi^2 R_g^2 R^2).$$ (8.3.4)

Hence the slope of the ln $I(R)$ versus R^2 plot used for estimating $I(0)$ will be $-\tfrac{4}{3}\pi^2 R_g^2$, where R_g is the radius of gyration of the molecule;

$$R_g^2 = \sum (f_i - v_i\rho_0)r_i^2 \Big/ \sum (f_i - v_i\rho_0),$$

where r_i is the distance from the ith atom to the center of mass of the scattering length of the particle, i.e., to the location $\bar{r}_0$,

$$\bar{r}_0 = \sum (f_i - v_i\rho_0)\bar{r}_i \Big/ \sum (f_i - v_i\rho_0),$$

where $\bar{r}_i$ is the vector from an arbitrary coordinate origin to the ith atom. The sums in all cases run over all atoms.

The scattering radius of gyration gets its name from its analogy with the mechanical radius of gyration. The value found by x-ray methods for a protein in water solution or determined by neutron scattering for a protein in D_2O solution will be close to the mechanical value for the molecule in question and provides a useful indication of the molecule's size. (The relationship between molecular structure and radius of gyration is considered at greater length in Section 8.3.9.)

Radius of gyration is the sole fact about a macromolecule which can be deduced from its solution scattering profile without resort to ancillary data. Furthermore, radii can be obtained from data which are not on an absolute scale. Because this parameter is extractable from low-angle data in such a direct fashion, virtually every small-angle paper ever published includes radius of gyration estimates.

8.3.3. Quantities Related to the Radius of Gyration

Long, thin particles or thin plates will often have radii of gyration which are very large indeed. When small-angle data are collected on such objects, it will often prove hard or impossible to measure those radii because of the narrowness of the central Gaussian. Useful information can be obtained, however, on the cross-sectional distribution of scattering length in the case of long, thin objects, or on the transverse distribution in the case of flat plates.[84,85]

[84] G. Porod, *Kolloid-Z.* **124**, 83 (1951).
[85] G. Porod, *Kolloid-Z.* **125**, 51 (1952).

For long, thin objects it will be observed that plots of $\ln[RI(R)]$ versus R^2 are linear over an appreciable range of R. The slope of this linear region will be $-2\pi^2 R_x^2$ where R_x is the radius of gyration of the cross section of the object. If the density of the object is projected onto a plane normal to its long axis and the radius of gyration of the projected density calculated, R_x is the result. For platelike objects, the analogous plot is $\ln[R^2 I(R)]$ versus R^2. Once again a linear region will be discerned whose slope is $-4\pi^2 R_t^2$, where R_t is now the radius of gyration of the density of the plate projected onto a line normal to its surface. For a uniform plate, $R_t\sqrt{12}$ is the thickness of the plate.

Note that the existence of a linear region in a $\ln[RI(R)]$ versus R^2 plot or in a $\ln[R^2 I(R)]$ versus R^2 plot is diagnostic for a long, thin shape or a thin plate shape as the case may be.

The value of RI or $R^2 I$ found when the linear regions of one of these plots is extrapolated to $R = 0$ is useful. We shall call this value "$I(0)$" for this purpose. If measurements have been made on an absolute scale, for the long rod $2I(0)$ can be substituted into Eq. (8.3.3) in place of $I_{exp}(0)$ and an estimate of the molecular weight per unit length of the object will result. For the thin plate, similar use of $2\pi I(0)$ will give an estimate of the molecular weight per unit area.

The possibility of measurement of membrane thicknesses and mass per unit area by the method just suggested has seldom been taken advantage of. Mass per unit length (M/L) and radius of gyration of the cross section, however, have often been measured for elongated structures. The classic material for studies of this kind is DNA and its various complexes with histones and other protein (i.e., chromatin). The early work on pure DNA by Luzzati and collaborators[86] and by Bram[87] is evaluated in the paper by Eisenberg and Cohen[17] with reference to the polyelectrolyte problem alluded to above. Evaluation of M/L and the radius of gyration of the cross section for chromatin-related structures has been reported by several groups.[88–90] Zipper and Bunemann[91] have recently described the alterations in DNA structure due to drug binding in a paper which includes a careful explanation of the data analysis used in M/L and R_x investigations. For indefinitely long structures like these, M/L and diameter information are valuable parameters since they can constrain model-building studies, and they are hard to obtain by other means. These same techniques are useful for the analysis of far less elongated molecules; they materially assisted

[86] V. Luzzati, F. Masson, A. Mathis, and P. Saludgian, *Biopolymers* **5**, 491 (1967).

[87] S. Bram, Ph.D. Thesis, University of Wisconsin, Madison (1968).

[88] S. Bram, *Biochimie* **57**, 1301 (1975).

[89] V. I. Aleksanyan, V. I. Vorobev, and B. A. Federov, *Biopolymers* **14**, 1133 (1975).

[90] L. Sperling and A. Tardieu, *FEBS Lett.* **64**, 89 (1976).

[91] P. Zipper and H. Bunemann, *Eur. J. Biochem.* **51**, 3 (1975).

Pilz *et al.*[92] in their recent analysis of the tRNA structure from small-angle data.

8.3.4. Length Distributions

Equation (8.2.1), the expression for $I(R)$, is written as a double sum over all atoms of terms of the form $(\sin x)/x$. In computing $I(R)$, one could equally well make a list of all atomic distances and collect these distances into groups of identical value. The contribution of the group to $I(R)$ is simply

$$\sum_{\substack{\text{all pairs} \\ r_{ij}=r}} f_i f_j \frac{\sin 2\pi rR}{2\pi rR}$$

and

$$I(R) = \sum_{\text{all } r} \left(\sum_{r_{ij}=r} f_i f_j \right) \frac{\sin 2\pi rR}{2\pi rR}.$$

Let us define a new function $p(r)$:

$$p(r) = \sum\sum_{r_{ij}=r} \frac{f_i f_j}{f_{av}^2},$$

where $f_{av} = (\sum f_i)/N$ and N is the number of atoms in the molecule; then

$$I(R) = f_{av}^2 \sum p(r) \frac{\sin 2\pi rR}{2\pi rR},$$

where the sum is over all distances r.

Often it is more convenient to consider $p(r)$ a continuous function; then

$$I(R) = \bar{\rho}^2 \int_0^\infty p(r) \frac{\sin 2\pi rR}{2\pi rR} \, dr, \tag{8.3.5}$$

where $\bar{\rho}$ is the average scattering-length density of the object; $p(r) \, dr$ is the (weighted) number of vectors in the structure having lengths between r and $r + dr$. It follows from Eq. (8.3.5) that $p(r)$ can be obtained from $I(R)$ by spherical Patterson inversion[2,93,94]:

$$p(r) = \frac{8\pi r}{\bar{\rho}^2} \left(\int_0^\infty R I(R) \sin(2\pi Rr) \, dR \right). \tag{8.3.6}$$

Note that knowledge of $p(r)$ is equivalent to knowledge of $I(R)$; the two are Fourier mates. *Thus the total information contained in a solution scattering*

[92] I. Pilz, O. Kratky, F. Cramer, F. von der Haar, and E. Schlimme, *Eur. J. Biochem.* **15**, 401 (1970).

[93] P. Debye and M. H. Pirenne, *Ann. Phys. (Leipzig)* [5] **33**, 617 (1938).

[94] P. Debye and A. M. Bueche, *J. Appl. Phys.* **20**, 518 (1949).

profile is the distribution of interatomic vector lengths in the structure under study.

The radius of gyration of the object can be obtained from the length distribution[2]:

$$R_g^2 = \frac{1}{2} \int_0^\infty r^2 p(r)\, dr \left/ \int_0^\infty p(r)\, dr \right. ;$$

r_{max}, the largest vector in the structure or its longest chord or maximum linear dimension, can also be estimated.

A function closely related to $p(r)$ is Porod's[84,85] "characteristic function" $\gamma_0(r)$, the probability that a point a distance r from a given point in the structure will also be within the particle, averaged over all points. For structures of uniform internal scattering-length density,

$$p(r) = 4\pi r^2 v \gamma_0(r), \tag{8.3.7}$$

where v is the volume of the structure.

8.3.5. Characteristic Function Parameters

From the definition of $\gamma_0(r)$ it is clear that it refers to a structure of the same shape and volume as the molecule being studied, but one in which all volume elements have the same scattering-length density, which is, of course, not true of a real molecule. However, when the molecule is observed at low resolution such that $R^{-1} \gg a$, where a is the average distance between nearest-neighbor atoms, $I(R)_{exp}$ will approximate the profile predicted by the characteristic function abstraction of the true molecule.

Several interesting aspects of $\gamma_0(r)$ can be readily demonstrated. First, $d\gamma_0(r)/dr = -S/4v$ at $r = 0$, where S is the surface area of the molecular volume.[84,85] From this relationship it can be shown that the trend of $I(R)$ at large values of R must be given by:

$$I(R) \propto \bar{\rho}^2 S/8\pi^3 R^4. \tag{8.3.8}$$

For volumes of uniform density, this $1/R^4$ dependence of $I(R)$ holds rigorously. For real molecules, when R gets large $I(R)$ begins to reflect the inhomogeneities of the structure with respect to scattering-length density and may deviate from the $1/R^4$ trend. For structures with gross fluctuations in scattering-length density (e.g., nucleoproteins), the trend may be significantly obscured.

In practice this relationship is used by plotting the data as $R^4 I(R)$ versus R^4. The large-R data should show a linear trend. In fact it should be flat, provided background subtraction has been done perfectly and internal fluctuation contributions are not too severe. The intercept of the trend line at $R = 0$ yields an estimate of S. Once S and, correspondingly, the trend of the high-angle data are established, Porod's invariant Q can be evaluated:

$$Q = 4\pi \int_0^\infty R^2 I(R) \, dR = (\Delta\rho)^2 v, \tag{8.3.9}$$

where v is the molecular volume and $\Delta\rho$ the difference in scattering-length density between solvent and solute, on average. Without knowlege of the trend of the data this integration would be impossible since data are never taken at large R. Note that $I(0) = (\Delta\rho)^2 v^2$. Thus[84,85,95]:

$$I(0)/Q = v. \tag{8.3.10}$$

The beauty of this relationship is that it permits v to be estimated from data which are *not* on an absolute scale since the scaling factors for $I(0)$ and Q are identical.

8.3.6. Molecular Shape

In addition to permitting one to evaluate the parameters just described, the analysis of scattering data can lead to an estimate of the shape of the molecule which gives rise to it. These shape estimates have traditionally been arrived at by a model-building process whose goal is to discover the shape of uniform scattering-length density whose (calculated) scatter most closely matches the data. This shape is then taken as a description of the molecule. Model building is, of course, guided by the numerical information whose derivation is described above as well as by whatever information may be available from other sources.

Model building is not straightforward. Real molecules do not have uniform internal scattering-length density so that even a correct model will predict a scattering profile which deviates from the data, especially when R is large. This being the case, no hard rules can be given for how well a model must match the data to be useful. Unhappily, models differing significantly in morphology can give rather similar scattering curves. One is then placed in the position of having to decide whether a deviation between measured and model data is significant or whether it simply reflects contributions due to internal variations within the true structure. Finally, where choices between alternatives are made on the basis of other data, one is at the mercy of their reliability.

The simplest kind of model building is equivalent ellipsoid determination. Procedures exist for comparing measured data to the profiles of ellipsoids of the same radius of gyration which differ in axial ratio.[3] The "best ellipsoid"

[95] V. Luzzati, A. Tardieu, L. Mateu, and H. B. Stuhrmann, *J. Mol. Biol.* **101**, 115 (1976).

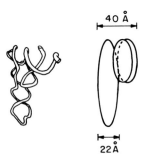

FIG. 3. Comparison of the small-angle model of yeast tRNA[Phe] and its crystallographic structure. On the right is the ellipsoid model of Pilz *et al.*[92] arrived at from small-angle analysis, with its principal dimensions indicated. On the left, drawn to the same scale, is a view of the same molecule as deduced from x-ray crystallographic data.[96a] The trace of the polynucleotide backbone is shown. [Reproduced with permission from Kratky and Pilz, "A comparison of x-ray small-angle scattering results to crystal structure analysis and other physical techniques in the field of biological macromolecules," *Q. Rev. Biophys.* **11**, 39 (1978). Published by Cambridge Univ. Press, London and New York.]

so found constitutes an unsophisticated model for a macromolecule whose main virtue is that its limitations are well understood by physical chemists. Such models can be useful, however. The equivalent ellipsoid models for *E. coli* ribosomes arrived at by Hill *et al.*[95a] from small-angle data have guided work in that area for many years. Recent studies of animal ribosomes should be equally profitable.[96]

More elaborate efforts involve modeling in terms of assemblies of spheres and/or ellipsoids and explorations of more complex shapes. A number of studies of this kind have been reported in the past few years. Perhaps the most remarkable recent success was the determination of a shape for tRNA from solution scattering data by Pilz *et al.*[92] There is an extremely good correspondence between their model and the structure of tRNA determined crystallographically several years afterwards (see Kratky and Pilz,[7] and Figure 3).[7,92,96a] The Pilz model was derived entirely from solution scattering data.

Two other recent model-building studies by the same group were guided electron microscopic results. Studies on G-type immunoglobulins in solution demonstrated a T shape for the molecule in solution, rather than a Y shape as indicated by electron microscopy, a finding recent crystallographic

[95a] W. E. Hill, J. D. Thompson, and J. W. Anderegg, *J. Mol. Biol.* **44**, 89 (1969).

[96] G. Damaschun, J. J. Mueller, H. Bielka, and M. Bottger, *Acta Biol. Med. Ger.* **33**, 817 (1974).

[96a] S. H. Kim, G. J. Quigley, F. L. Suddath, A. McPherson, D. Sneden, J. J. Kim, J. Weinzierl, and A. Rich, *Science (Washington, D.C.)* **179**, 285 (1973).

results have tended to confirm.[7,97,97a] A less useful effort was that made on the DNA-dependent RNA polymerase from *E. coli*.[98] In this case, the model building may have been biased by spurious electron microscopic results. Some of the more complex model-building studies in macromolecular aggregates are discussed below (Section 8.3.16).

8.3.7. Spherical Objects

If an object is spherical or nearly so, this fact will be readily apparent from its solution scattering profile. Not only will the ellipsoid of axial ratio unity fit its low-angle data best, but also there will be secondary maxima at higher angles at predictable positions. The classic biological spherical scatters are the small isometric viruses (e.g., Fishbach *et al.*[99]; for review, see Kratky and Pilz[4]). These viruses are in fact icosahedral in symmetry, but at low resolution they are effectively spherical.[100]

Because of their quasispherical symmetry, the low-resolution data from viruses can be analyzed in a more sophisticated manner than is normal in the low-angle field. For objects of this kind the rotationally averaged intensity profile observed by solution scattering is the same as the intensity profile not averaged by rotation. Thus $I(R)^{1/2} = |F(R)|$, where $|F(R)|$ is the modulus of the Fourier transform of the object's scattering-length density distribution. In addition, it will be true that the maxima in the profile have real phases and alternate in sign. In most cases phase assignments can be made. Thus the intensities measured in solution can be inverted to give the scattering-length density of the object as a function of distance from the object's center, i.e., its radial density distribution. The reader should appreciate that direct statements about the distribution of material in an object result from small-angle data in no other case. The problems of extending such an analysis to resolutions higher than those where strict spherical symmetry hold have been discussed by Jack and Harrison.[101]

Viruses are complexes of protein and nucleic acid which sometimes include lipid. Contrast variation methods, discussed in the next sections, can be used to decide where the different chemical components of the virus reside. Both x-ray[100] and neutron experiments[102,103] of this kind have been carried out

[97] I. Pilz, O. Kratky, A. Licht, and M. Sela, *Biochemistry* **12**, 4998 (1973).

[97a] L. Cser, I. A. Gladkikh, Z. A. Kozlov, R. S. Nezlin, M. M. Ogievetskaya, and Yu. M. Ostanevich, *FEBS Lett.* **68**, 283 (1976).

[98] I. Pilz, O. Kratky, and D. Rubassay, *Eur. J. Biochem.* **28**, 205 (1972).

[99] F. A. Fishbach, P. M. Harrison, and J. W. Anderegg, *J. Mol. Biol.* **13**, 638 (1965).

[100] S. C. Harrison, *J. Mol. Biol.* **42**, 457 (1969).

[101] A. Jack and S. C. Harrison, *J. Mol. Biol.* **99**, 15 (1975).

[102] B. Jacrot, C. Chauvin, and J. Witz, *Nature (London)* **266**, 417 (1977).

[103] C. Chauvin, J. Witz, and B. Jacrot, *J. Mol. Biol.* **124**, 641 (1978).

and have provided useful information on the arrangement of the different chemical species in small viruses.

8.3.8. Contrast

During the early years small-angle scattering studies mainly concerned themselves with determination of molecular parameters of the kinds discussed above with the object of discovering the shape and dimensions of the object of uniform scattering density which best approximates the molecule of interest. In the past 15 years there has been a significant shift of emphasis in both theory and practice towards consideration of the influence of density fluctuations within a molecule on its scattering behavior, and of the effect of variation in solvent scattering-length density on scattering profiles. These developments have led in two directions: (i) toward better solutions of the classical problems of molecular size and shape, and (ii) toward a characterization of the internal structure of macromolecules and macromolecular aggregates.

Contrast is a concept used above without calling it by name. The contrast $\Delta\rho$ between a molecule and its environment is simply the difference between its average scattering-length density $\bar{\rho}$ and that of the environment (e.g., solvent), ρ_0; i.e., $\Delta\rho = \bar{\rho} - \rho_0$. Contrast may also exist between different regions within the same molecule. The significance of contrast lies in the fact that the visibility of chemically identifiable regions within a scattering specimen, i.e., the intensity of the scattering signal they produce, is proportional to $\Delta\rho^2$.

In the classic analysis, a molecule is regarded as a shape with a constant $\bar{\rho}$, the argument being that for small enough R, internal fluctuations will average out. In fact this approximation is a good deal *less* useful than often thought. The prediction it makes is that scattering profiles collected on the same molecule in solvents of different ρ_0 will differ *only* in amplitude, not in shape. In fact, however, this prediction does not hold. Even the initial slope of a Guinier plot of $I(R)$, from which R_g is determined, varies significantly with ρ_0 in molecules whose internal density fluctuations are as modest as those of a globular protein.[104] It follows that a correct estimate of a molecule's shape from its solution scattering profile may not result unless these factors are taken into account.

8.3.9. The Dependence of Radius of Gyration on Contrast

In real molecules significant fluctuations in ρ exist which can affect the radius of gyration manifested by the molecule under any set of experimental

[104] H. B. Stuhrmann, *J. Appl. Crystallogr.* 7, 173 (1974).

conditions (see, e.g., Mateu et al.[105]). A satisfactory theory for these effects has been developed by considering macromolecules as having an average scattering-length density $\bar{\rho}$ on top of which is superimposed a fluctuation in scattering-length density $\rho^1(r)$. The sum of the two functions $\rho(\bar{r}) = \bar{\rho} + \rho^1(\bar{r})$ then describes the scattering-length density of the whole molecule. Substituting this expression into the definition of radius of gyration, one can show that

$$R^2 = R_c^2 + \alpha/\Delta\rho - \beta/\Delta\rho^2, \tag{8.3.11}$$

where R is the radius of gyration observed when the contrast is $\Delta\rho$, and R_c the radius of gyration of the object at infinite contrast. R_c is also the radius of gyration of the uniform solid the size and shape of the whole molecule.

$$\alpha = \frac{1}{v}\int_0^\infty \rho^1(\bar{r})(r)^2 \, dr, \qquad \beta = \frac{1}{v^2}\left(\int_0^\infty \rho^1(\bar{r})\bar{r} \, dr\right)^2, \tag{8.3.12}$$

where α will be recognized as the second moment of the fluctuations in density relative to the center of gravity of the uniform shape and β the square of the first moment of these fluctuations[95,106–110]; v is the volume of the molecular shape.

For an object composed of two regions of uniform but different scattering-length density ρ_1 and ρ_2, the following expressions hold:

$$\alpha = (v_1/v)[(R_1^2 - R_2^2) + (\bar{r}_1^2 - \bar{r}_2^2)](\rho_1 - \bar{\rho}),$$
$$\beta = ([v_1/v][(\bar{r}_1 - \bar{r}_2)(\rho_1 - \bar{\rho})])^2, \tag{8.3.13}$$

where R_1 and R_2 are the radii of gyration of the two regions considered separately and $\bar{r}_1$ and $\bar{r}_2$ vectors from the center of the uniform solid for the whole object to the centers of the two regions whose volumes are v_1 and v_2[111,112]:

$$\bar{\rho} = (v_1\rho_1 + v_2\rho_2)/(v_1 + v_2).$$

Thus for such an object β gives information about the separation of the centers of mass of the two regions and α reports primarily their differences in radius of gyration.

[105] L. Mateu, A. Tardieu, V. Luzzati, L. Aggerbeck, and A. M. Scanu, J. Mol. Biol. 70, 105 (1972).

[106] H. Benoit and C. Wippler, J. Chim. Phys. 57, 524 (1960).

[107] H. B. Stuhrmann and R. G. Kirste, Z. Phys. Chem. 46, 247 (1965).

[108] H. B. Stuhrmann and R. G. Kirste, Z. Phys. Chem. 56, 335 (1967).

[109] J. P. Cotton and H. Benoit, J. Phys. (Orsay, Fr.) 36, 905 (1975).

[110] D. K. Carpenter and W. L. Mattice, Biopolymers 16, 67 (1977).

[111] C. Sardet, A. Tardieu, and V. Luzzati, J. Mol. Biol. 105, 383 (1976).

[112] R. R. Crichton, D. M. Engelman, J. Haas, M. H. J. Koch, P. B. Moore, R. Parfait, and H. B. Stuhrmann, Proc. Natl. Acad. Sci. U.S.A. 74, 5547 (1977).

8.3.10. Dependence of the Extended Scattering Profile on Contrast

In studying the full scattering profile of a molecule, the same decomposition of the molecule's scattering-length density distribution into constant and fluctuation parts which leads to Eq. (8.3.11) retains its utility. When $I(R)$ is developed from Eq. (8.2.1) using this premise, it is found that:

$$I(R) = \Delta\rho^2 I_c(R) + \Delta\rho I_{cs}(R) + I_s(R), \qquad (8.3.14)$$

where $I(R)$ is the observed data. $I_c(R)$ is the profile given by a uniform volume of the size and shape of the molecule, $I_s(R)$ is the scatter due to the internal fluctuations within the molecule, and $I_{cs}(R)$ the component due to interference between shape and fluctuation scatter.[83,95,107,113] The three components of $I(R)$ are its "characteristic functions." If data are collected at several $\Delta\rho$, one can fit the values observed for each R with a quadratic expression in $\Delta\rho$ with second-order coefficient I_c, first-order coefficient I_{cs}, and the constant term I_s.

One obvious point in favor of such a decomposition is that $I_c(R)$, like R_c, relates to the shape of uniform density most closely related to the molecule's structure. *All* the classical methods for the analysis of scattering profiles are rigorously valid for data of this kind. It is clearly the profile to use as a basis for shape determination.

8.3.11. Absolute Scale and Contrast Variation

Luzzati *et al.*[95] have recently pointed out that Porod's invariant Q, like virtually everything else connected with small-angle profiles, depends on contrast and is affected by scattering-length density fluctuations within the molecule under investigation. Plots of $Q/I(0)$ versus $1/\Delta\rho^2$ should be linear, and the value at $1/\Delta\rho^2 = 0$ should be $1/v$, where v is the volume of the molecular shape. $\Delta\rho$ is a calculable quantity once the contrast match point is known, using the chemical composition and density of the solvent. The contrast match point can be established from data which are *not* on an absolute scale (see Section 8.3.1). As already observed (Section 8.3.5), v determination likewise does not require scaled data. However, once v and $\Delta\rho$ are known, absolute scale is set since these two parameters allow one to predict the absolute scatter expected from a given specimen in a specified incident flux. Thus contrast variation combined with a consideration of Porod's invariant leads both to a setting of absolute scale and a value for particle molecular weight independent of measurements of beam intensities, solid angles, etc., all of which are sources of uncertainty in ordinary

[113] H. B. Stuhrmann, *Z. Phys. Chem.* (*Wiesbaden*) **72**, 177 (1970).

absolute-scale determinations. A practical application of these ideas to the analysis of the structure of lipoproteins from small-angle data has been given by Tardieu et al.[43]

8.3.12. Neutrons versus X Rays

As has been emphasized repeatedly in the past,[9] the primary reason for studying biological structures using neutron radiation is the sensitivity of neutrons to ^1H and the large difference in b between ^1H and ^2H.[13] The scattering length of ^1H is -3.74 F (1 F $= 10^{-15}$ m), whereas that of ^2H is $+6.67$ F. Since biological macromolecules, not to mention water, are H-rich substances, large alterations in $\bar{\rho}$ can be brought about by substitution of ^2H and ^1H with minimal effects on the chemical and physical properties of substances in question. Table I gives some representative values for $\bar{\rho}$ for normal ^1H materials and their ^2H counterparts.

TABLE I. Neutron and X-Ray Scattering Length Densities for Biological Materials[a]

	ρ (cm$^{-2} \times 10^{10}$)	
Substance	Neutron[b]	X ray
Fully H substituted		
H_2O	-0.55	$+9.3$
protein	$+3.11$	$+12.4$
nucleic acid	$+4.44$	$+16.0$
fatty acid	-0.01	$+8.2$
carbohydrate	$+4.27$	$+14.1$
Fully D substituted		
D_2O	$+6.36$	$+9.3$
protein	$+8.54$	$+12.4$
nucleic acid	$+7.44$	$+16.0$
fatty acid	$+6.89$	$+8.2$
carbohydrate	$+8.07$	$+14.1$

[a] Reproduced, with permission, from D. M. Engelman and P. B. Moore, *Annual Review of Biophysics and Bioengineering*, Volume 4. © 1975 by Annual Reviews, Inc.

[b] Neutron scattering length densities are calculated assuming that all exchangeable hydrogen positions are fully deuterated.

Inspection of Table I reveals that ρ_{H_2O} is less than and ρ_{D_2O} greater than the scattering length of any ^1H biological material. Thus a wide range of $\Delta\rho$s for protonated macromolecules can be explored, including the condition of contrast match, $\Delta\rho = 0$, by use of solvents which are H_2O/D_2O mixtures. Also, $\bar\rho$ of a macromolecule can be altered over a wide range in a nontraumatic manner by isotopic labeling, opening the possibility of experimental manipulation of the distribution of densities *within* a macromolecule. Manipulations of this kind lead to isomorphous replacement experiments for the study of internal structure.

Alteration of $\bar\rho$ in an x-ray experiment corresponds to alteration of electron density. Unfortunately, the chemical and physical properties of substances inevitably alter with changes in electron density. Moreover, the range of contrasts available is limited. Lipids cannot be contrast matched in an aqueous environment since $\rho_l < \rho_{H_2O}$. Carbohydrate materials (e.g., glycerol, glucose) can be used to raise solvent densities into the region of ρ_{prot}, but can never produce solutions such that $\rho_0 = \rho_{RNA}$. A wider range can be explored with salts, but their presence in high molarity is often incompatible with the requirement that a monodisperse solution be examined. It is not surprising, therefore, that contrast variation has not been popular in the past in x-ray solution work. The advent of neutrons for study of biological structure has stimulated interest in contrast techniques enormously.

The volume explored by addition of small-molecule solutes is the hydrated volume. (An interesting illustration of the difference between the anhydrous and hydrated volumes of a macromolecule was recently given by Stuhrmann and his collaborators,[114] who collected contrast variation data on the same object using both D_2O and deuterated glucose as the contrast modifying agents.)

8.3.13. The Hydrogen Exchange Problem

In most contrast variation experiments it is the scattering-length density of the solvent, ρ_0, which is manipulated. An "ideal" experiment is one in which alterations in ρ_0 do not effect ρ, the scattering-length density of the macromolecule, and do not perturb its structure.[95] In this connection it should be noted that the D_2O/H_2O method for ρ_0 adjustment is not ideal. The problem is that all biological macromolecules include significant numbers of hydrogens which exchange at appreciable rates with water protons. When D_2O is included in the solvent, these positions deuterate, raising ρ. The

[114] H. B. Stuhrmann, M. H. J. Koch, R. Parfait, J. Haas, K. Ibel, and R. R. Crichton, *J. Mol. Biol.* **119**, 203 (1978).

exchange process has a big enough effect on ρ that it can be followed quite accurately by measuring $I(0)$ as a function of time following the mixing of a protein in water, for example, with an aliquot of D_2O.[104]

As a general rule, in small organic molecules hydrogens bonded to C are nonexchangeable and those bonded to O, N, S, etc., do exchange. In the few situations in biological small molecules related to macromolecules where proximity effects labilize C-bonded H (e.g., the C_8 proton in guanine) the rates of exchange are negligible at neutral pH and room temperature. In a biological macromolecule, therefore, knowledge of its monomer composition leads not only to an accurate overall atomic composition, but also to an accurate estimate of the total number of hydrogens that can exchange. Not all chemically labile hydrogens in a macromolecule will exchange, however, owing to steric constraints imposed by molecular folding. In most cases, the great majority of the labile hydrogens do ($\sim 75\%$ or more).[115,116] In order to make full use of neutron data, it is important to know what fraction of the potentially exchangeable hydrogens do in fact exchange. There are a number of ways to obtain this information. First, tritium exchange methods can be used.[116] Second, $\bar{v}$, determined by solution physical-chemical methods, can be combined with chemical composition data to predict $\bar{\rho}$, assuming no exchange and that value compared to ρ_0 at contrast match. The difference should be due to exchange, which can then be evaluated. The justification for this approach is that in many cases $\bar{v}$ determined from neutron contrast matching experiments for molecules where the exchange level was known in advance has been identical to $\bar{v}$ determined by classical methods, within experimental error.[117,118] Third, the molecule in question can be deuterated. If a macromolecule can be obtained both in deuterated and protonated form, then comparison of the density match points for the two forms gives both $\bar{v}$ and the fraction of the protons undergoing exchange.[118]

The formalism for the perturbations exchange in neutron contrast variation experiments has been worked out to some extent.[10,112,117] Firm estimates for molecular weight and volume can be obtained provided the extent of exchange is properly taken into account. The radius of gyration found at infinite contrast, R_c, [Eq. (8.3.11)] will include contributions due to the distribution of exchangeable protons. If they are distributed uniformly through the molecule's volume, no perturbation will be seen. If the distribution is nonuniform, the perturbation can be significant. α will be altered only if the center of gravity of the exchange distribution differs from that of the

[115] A. Hvidt and S. O. Nielsen, *Adv. Protein Chem.* **21**, 288 (1966).

[116] S. W. Englander, N. W. Downes, and H. Teitelbaum, *Annu. Rev. Biochem.* **41**, 903 (1972).

[117] K. Ibel and H. B. Stuhrmann, *J. Mol. Biol.* **93**. 255 (1975).

[118] P. B. Moore, D. M. Engelman, and B. P. Schoenborn, *J. Mol. Biol.* **91**, 101 (1975).

whole structure, and β will be unaltered.[117] I_c, in the ideal case the characteristic function for the uniform volume, will now correspond to a volume of nonuniform ρ whose nonuniformities relate to the distribution of exchangeable protons.

It should be emphasized that in practice exchange effects are not overwhelming. They are not large enough to upset the qualitative validity of the results of neutron contrast variation experiments. They should be regarded more as a source of frustration in that were they not present, H_2O/D_2O contrast variation would be the perfect answer for many small-angle problems, rather than merely a very good one.

8.3.14. Contrast Variation in Practice: Radius of Gyration

The simplest contrast variation experiment is the measurement of radius of gyration as a function of contrast. This is a particularly useful experiment when applied to structures which are aggregates of two kinds of material differing significantly in ρ. For such an object a contrast variation experiment should yield the radius of gyration of the whole structure, the radii of gyration of the regions occupied by the two species, and an estimate of the separation of the centers of mass of the two regions.

Three experimental approaches to this kind of information have been used in the past: (i) solvent density manipulation, for which many examples may be cited (for review, see Jacrot[10]; e.g., Tardieu et al.,[43]); (ii) solute density manipulation[112,118]; and (iii) use of several types of radiation to measure the same object.[119,120] The operation of the first two kinds of experiment is fairly obvious from what has already been said. The third type of contrast variation experiment deserves further comment. It exploits the fact that the physics of the scattering of x rays, neutrons, and light differ significantly. The scattering length densities of different chemical species vary according to the type of radiation with which they are studied, and the changes are different for each species. Thus the internal contrast between chemical species within a particle will change with radiation type as will the contrast of the whole particle with its surrounding solvent. Radii of gyration observed will differ with radiation and give information about the issues in question.

The now classic neutron solvent variation method involves measurement of the radius of gyration of the molecule of interest in solvents of different D_2O/H_2O composition. $\Delta\rho$ for each data set is evaluated from the observed dependence of $[I(0)/C]^{1/2}$ on solvent composition. A straight-line relationship will be observed for homogeneous samples.[104] From the solvent composition at which $\Delta\rho = 0$, $\bar{\rho}$ is estimated in absolute terms with due allowance for exchange. $\Delta\rho$ can then be calculated for all other solvent conditions from

[119] I. N. Serdyuk and B. A. Federov, *J. Polym. Sci.* **11**, 645 (1973).
[120] I. N. Serdyuk, *Brookhaven Symp. Biol.* **27**, IV49 (1976).

knowledge of solvent composition and density, i.e., $\Delta\rho = \bar{\rho} - \rho_0$. There are a number of ways of plotting the data to evaluate their trends. Perhaps the most convenient is to plot R^2 versus $1/\Delta\rho$ (see Stuhrmann[104]). The intercept at $1/\Delta\rho = 0$ is R_c^2. The slope of the plot $1/\Delta\rho = 0$ gives α and its curvature β. If α is positive, high ρ regions in the particle must lie towards its exterior; if α is negative, the opposite is true. β curvature will always be concave down. A nonzero β implies a lack of coincidence in the positions of the centers of mass of the high and low density regions within the particle.

In general, in "two phase" aggregates, prior knowledge of chemical compositions and partial volumes is precise enough so that the chemical identity of the high and low density regions is certain. In these situations the qualitative statements that can be made about their relative distributions in the aggregate from an analysis of the type described are unambiguous. R_1 and R_2 are obtained from the plot by picking off the values for R at the contrasts where phase 2 and phase 1, respectively, are contrast matched. Thus the quantitative estimation of R_1, R_2, and $|\bar{r}_1 - \bar{r}_2|$ [see Eq. (8.3.13)] from α and β will only be as good as the available information about the volumes occupied by the phases *in situ*. The quality of these estimates can be greatly improved by studying the structure in a variety of deuterium-labeled forms. Such an investigation can lead to accurate estimation of the volume fraction occupied by each phase (i.e., $\bar{v}$ *in situ*) as well as values for R_1, R_2, and $|\bar{r}_1 - \bar{r}_2|$.[112]

Neutron contrast variation studies have been done on a variety of two-phase structures including ferritin,[121] nucleosomes and other chromatin related materials,[122–127] ribosomes,[112,128,129] rhodopsin in detergent micelles,[130,131] and fibrinogen.[132]

[121] H. B. Stuhrmann, J. Haas, K. Ibel, M. H. Koch, and R. R. Crichton, *J. Mol. Biol.* **100**, 399 (1976).

[122] J. P. Baldwin, P. G. Boseley, and E. M. Bradbury, *Nature (London)* **253**, 245 (1975).

[123] S. Bram, G. Butler-Browne, P. Baudy, and K. Ibel, *Proc. Natl. Acad. Sci. U.S.A.* **72**, 1043 (1975).

[124] J. F. Pardon, D. L. Worcester, J. C. Wooley, K. Tatchell, K. E. Van Holde, and B. M. Richards, *Nucleic Acids Res.* **2**, 2163 (1975).

[125] P. Baudy, S. Bram, D. Vastel, J. Le Pault, and A. Kitzes, *Biochem. Biophys. Res. Commun.* **72**, 176 (1976).

[126] B. P. Hjelm, G. G. Kneale, P. Suau, J. P. Baldwin, E. M. Bradbury, and K. Ibel, *Cell* **10**, 139 (1977).

[127] S. Bram, P. Baudy, J. Le Pault, and P. Hermann, *Nucleic Acids Res.* **4**, 2275 (1978).

[128] H. B. Stuhrmann, J. Haas, K. Ibel, B. De Wolf, M. H. J. Koch, R. Parfait, and R. R. Crichton, *Proc. Natl. Acad. Sci. U.S.A.* **73**, 2379 (1976).

[129] P. Beaudry, H. O. Petersen, M. Grunberg-Manago, and B. Jacrot, *Biochem. Biophys. Res. Commun.* **72**, 391 (1976).

[130] M. Yeager, *Brookhaven Symp. Biol.* **27**, 1113 (1976).

[131] H. B. Osborne, C. Sardet, M. Michel-Villaz, and M. Chabre, *J. Mol. Biol.* **123**, 177 (1978).

[132] G. Marguerie and H. B. Stuhrmann, *J. Mol. Biol.* **102**, 143 (1976).

The nucleosome studies alluded to are notable in providing direct physical evidence for the now widely accepted view that the basic subunit of chromatin structure is DNA wound around the outside of a histone core particle. The ribosome studies have produced the only data available on the relative distributions of bulk protein and RNA in these structures (see also Ser-dyuk[120]). The 30 S subunit structure has been recognized as a reasonably homogeneous mixture of RNA and protein, while the 50 S subunit shows considerable segregation of the two species, with protein tending to form the exterior surface of the particle.

Recent extension of such studies to 70 S ribosomal couples in which one subunit is deuterium labeled has led to determination of the intersubunit distance and evidence that couple formation is accompanied by little conformational change on the part of the individual subunits.[133] Giege et al.[134] have done an interesting study on the formation of complexes between tRNA and tRNA ligase by contrast-matching the protein. Under these conditions it was possible to show that at low ligase concentrations a 2 tRNA: 1 ligase complex forms.

With the increase in interest in contrast variation methods, some extensive x-ray studies have also been undertaken on two-phase systems. A number of lipoproteins have been investigated[43,105,135,136] (for a review see Laggner and Mueller[137]), and recently the rhodopsin–detergent system has been studied by these same means.[111] The paper of Tardieu et al.[43] and its companion theory paper[95] contain a particularly detailed exposition of the analysis of two phase systems.

8.3.15. Contrast Variation in Practice: Extended Scattering

The analysis of contrast variation data sets into characteristic functions is still a relatively new method. As already emphasized, $I_c(R)$, the shape scattering profile or at least an approximation to it, is the best possible input into a shape analysis. Shapes can be derived from $I_c(R)$ by model building just as described in Section 8.3.6. Recently, however, a second, more deductive approach has been suggested by Stuhrmann[137a] which requires $I_c(R)$ as its input. The concept behind the approach is to regard the density distribution

[133] M. H. J. Koch, R. Parfait, J. Haas, R. R. Crichton, and H. B. Stuhrmann, *Biophys. Struct. Mech.* **4**, 251 (1978).

[134] R. Giege, B. Jacrot, D. Moras, J.-C. Thierry, and G. Zaccai, *Nucleic Acids Res.* **4**, 2421 (1977).

[135] K. Mueller, P. Laggner, O. Kratky, G. Kostner, A. Holasek, and O. Glatter, *FEBS Lett.* **40**, 213 (1974).

[136] K. Mueller, P. Laggner, O. Glatter, and G. Kostner, *Eur. J. Biochem.* **82**, 73 (1978).

[137] P. Laggner and K. W. Mueller, *Q. Rev. Biophys.* **11**, 371 (1978).

[137a] H. B. Stuhrmann *Brookhaven Symp. Biol.* **27**, IV-3 (1976).

of a molecule as the sum of spherical-harmonic density waves.[100,101,137b] A unique decomposition into spherical harmonics exists for any object, and its solution scattering profile directly reflects the magnitudes of the coefficients of this expansion. If the object in question has uniform internal scattering-length density, as is true for the $I_c(R)$ object, considerable restrictions are placed on the values these coefficients may have.[137a] Taking these into account one can deduce the coefficients from the data within certain limitations and can then calculate models. Only a few structures have been dealt with in this way. One of the first was lysozyme,[138] for which a shape quite close to that of the crystal structure was obtained. Fibrinogen has also been analyzed,[132] as have the subunits of the E. coli ribosome. The 50 S structure[139] obtained resembles the shape the object appears to have in the electron microscope.

The results of this kind of analysis are intriguing, but its limitations are still unclear. There are two problems with the analyses done so far. First, the impact of error in the input data on the final result has not been worked out. Second, most of the data used have been neutron data, which are affected by exchange. For an object undergoing exchange, the $I_c(R)$ obtained corresponds to an object of nonuniform density, contrary to the assumption of uniform density which guides the choice of coefficients for harmonics in model building. For a discussion of the impact of these problems see Stuhrmann et al.[139]

8.3.16. Macromolecular Aggregates: Synthesis Methods

In the past decade an increasing amount of low-angle work has been done on macromolecular aggregates. In this class of structure one would include multisubunit enzymes, multienzyme complexes, viruses, ribosomes, nucleosomes and other chromatin-related structures, and lipoproteins. At some level of description many of these can be regarded as two-phase problems and approached by contrast variation as already indicated. Investigations of the arrangement of the two phases can hinge either on naturally existing differences in scattering-length density or on artificially created distinctions such as can be produced by deuteration.[112,118] For structures with more than two subunits, these methods are not readily applicable and other means must be used.

Two methods have evolved for studying multisubunit structures which might be described as (i) "synthesis" and (ii) "distance finding." Synthesis

[137b] H. B. Stuhrmann, Acta Crystallogr., Sect. A 26, 297 (1970).

[138] H. B. Stuhrmann and H. Fuess, Acta Crystallogr., Sect. A A32, 67 (1976).

[139] H. B. Stuhrmann, M. H. J. Koch, R. Parfait, J. Haas, K. Ibel, and R. R. Crichton, Proc. Natl. Acad. Sci. U.S.A. 74, 2316 (1977).

is the only approach possible for aggregates containing many copies of only one or two subunits, whereas distance finding comes into its own for structures containing single copies of several different kinds of subunits. These two techniques are thus suitable for the analysis of different kinds of structures. The objective of these analyses, in either case, is to determine the disposition of the subunits within the aggregate.

A simple example of synthesis would be the work of Damaschun and his colleagues on the trypsin–trypsin inhibitor complex.[140,141] R_gs were measured for the two molecules separately and for their complex, and their distance apart on the complex inferred using the parallel-axis theorem:

$$R^2_{comp} = f_1 R^2_{trp} + f_2 R^2_{inh} + f_1 f_2 \Delta^2, \tag{8.3.15}$$

where f_1 is the fraction of the total scattering length of complex relative to solvent contributed by trypsin, f_2 the corresponding quantity for the inhibitor, and Δ the distance between their centers of mass. The parallel-axis theorem is the proper formula for combining radii of gyration, and for two subunit systems gives results equivalent to those obtainable by contrast variation [Eq. (8.3.12)]. One could imagine a third protein in a complex with the other two, measure its R_g separately and the R_g of the ternary complex, and so work up to a description of the whole complex in terms of subunit radii of gyration and locations based on relative distances.

One can also study the shapes of the subunits and those of the subcomplexes they may form, as well as that of the whole aggregate, using extended scattering curves. Model building is then done in which the objective is to match the whole structure's scattering profile by assembling aggregates of its substructures. It is clear that an important premise of such an analysis is that the shapes of subunits and subassemblies do not change during degradation and/or reassembly of the complete structure. The analysis performed on haemocyanin, which is an aggregate of 200 subunits, by Pilz and colleagues is a fine example of synthesis.[142] The work of Stöckel et al.[143] on the hemoglobin of Tubifex can also be cited.

It should be noted that deuteration of subunits followed by appropriate contrast matching offers the possibility of testing the assumption that subunit radii of gyration are unaltered by complex formation, but even this approach is of limited applicability.[144]

[140] G. Damaschun, P. Fichtner, H. V. Purschel, and J. G. Reich, Acta Biol. Med. Ger. 21, 309 (1968).

[141] G. Damaschun and H. V. Purschel, Acta Biol. Med. Ger. 21, 865 (1968).

[142] J. Berger, I. Pilz, R. Witters, and R. Lontie, Eur. J. Biochem. 80, 79 (1977).

[143] P. Stöckel, A. Mayer, and R. Keller, Eur. J. Biochem. 37, 193 (1973).

[144] P. B. Moore, J. A. Langer, and D. M. Engelman, J. Appl. Crystallogr. 11, 479 (1978).

8.3.17. Macromolecular Aggregates: Distance Finding

Over the years the possibility of measuring distances between points within a large structure by x-ray solution scattering has been repeatedly pointed out theoretically and demonstrated in model systems.[145-148] The points in question are labeled with heavy-metal atoms by chemical means and solution scattering profiles obtained for the doubly labeled species $I_{12}(R)$, both of the possible singly labeled species $I_1(R)$ and $I_2(R)$, and for the unlabeled structure $I(R)$. Let

$$A = \sum_i \sum_j f_i f_j \frac{\sin 2\pi R r_{ij}}{2\pi R r_{ij}},$$

$$B = 2f_1 \sum_j f_j \frac{\sin 2\pi R r_{1j}}{2\pi R r_{1j}},$$

$$C = 2f_2 \sum_j f_j \frac{\sin 2\pi R r_{2j}}{2\pi R r_{2j}},$$

$$D = f_1^2,$$

$$E = f_2^2,$$

$$F = 2f_1 f_2 \frac{\sin 2\pi R r_{12}}{2\pi R r_{12}},$$

where f_1 and f_2 are the scattering lengths of the heavy atoms and the sums run over the unlabeled atoms. Then

$$I_{12}(R) = A + B + C + D + E + F,$$
$$I_1(R) = A + B + D,$$
$$I_2(R) = A + C + E,$$
$$I(R) = A.$$

Thus if profiles are normalized to equal numbers of molecules and total flux,

$$I_x(R) = [I_{12}(R) + I(R)] - [I_1(R) + I_2(R)] = F.$$

F will be recognized as the interference contribution in the scatter of the doubly labeled structure due to the presence of the two heavy atoms. F is a damped sinusoidal ripple whose nodes occur at regular intervals $R = n/2r_{12}$ in R. Thus measurement of $I_x(R)$ leads to determination of r_{12}, the distance between the two labeled points.

[145] O. Kratky and W. Worthmann, *Monatsh. Chem.* **76**, 263 (1947).
[146] B. K. Vainshtein, N. I. Sosfenov, and L. A. Feigin, *Dokl. Akad. Nauk SSSR* **190**, 574 (1970).
[147] W. Hoppe, *Isr. J. Chem.* **10**, 321 (1972).
[148] W. Hoppe, *J. Mol. Biol.* **78**, 581 (1973).

Hoppe[147,148] pointed out a unique feature of this type of measurement, namely that it can be carried out with specimens at very high concentration. If the doubly labeled and unlabeled samples are mixed in equivalent amounts, the profile of the mixed samples obtained at high concentration, and the same done with the two singly labeled samples at equally high concentration, the subtraction will still yield $I_x(R)$. The intermolecular vectors that make normal solution scattering measurements useless at high sample concentrations will be the same in both samples and their contributions will cancel when the difference is taken.

Simultaneously with these developments it was observed that this labeling strategy was likely to work using deuterium as a label for whole subunits within a macromolecular aggregate and, of course, neutron radiation.[149] The signal-to-noise ratio for such an experiment is 10 to 100 times better than in its x-ray analogue. The possibility was envisioned of mapping the positions of subunits within aggregates by triangulation based on distances. So far, the practical applications of the distance method to macromolecules have all involved the mapping of subunit locations by the neutron version of the experiment rather than its x-ray counterpart.

The biochemical requirements for such measurements are stringent. The structure must come from an organism which will grow in the presence of high concentrations of D_2O; only biosynthesis leads to levels of D incorporation into macromolecules sufficiently high to make labeling useful. So far this requirement has limited work to structures from bacterial sources. In addition, the structure must be reconstitutable, i.e., it must be possible to disassemble it into its subunits and cause it to reassemble, in order that the required placement of deuterated subunits be achieved, another difficult step. Nevertheless, two systems of considerable biological importance have been worked on, the ribosome and the DNA-dependent RNA polymerase. At the time of this writing, over 30 distances have been measured within the 30 S ribosomal subunit of E. coli,[150,150a] leading to the determination of the placement of 9 of its 21 protein subunits; the DNA-dependent RNA polymerase, which is a four-subunit structure, has been fully analyzed.[151] The results are generally in good agreement with other data available on these structures.

Some comments about data interpretation are in order for these experiments. When whole subunits are labeled rather than simple points, $I_x(R)$

[149] D. M. Engelman and P. B. Moore, Proc. Natl. Acad. Sci. U.S.A. 72, 3888 (1972).

[150] J. A. Langer, D. M. Engelman, and P. B. Moore, J. Mol. Biol. 119, 463 (1978).

[150a] D. E. Schindler, D. M. Engelman, and P. B. Moore, J. Mol. Biol. 134, 595 (1979).

[151] P. Stöckel, R. May, I. Strell, Z. Cjeka, W. Hoppe, H. Heumann, W. Zillig, H. L. Crespi, J. J. Katz, and K. Ibel, J. Appl. Crystallogr. 12, 176 (1979).

becomes (for neutrons)

$$I_x(R) = 2(b_D - b_H)^2 \sum_1 \sum_2 \frac{\sin 2\pi R r_{ij}}{2\pi R r_{ij}},$$

where b_D and b_H are the scattering lengths of ^2H and ^1H, respectively, and the sums run over all deuteriums in subunit 1 and all deuteriums in subunit 2. This function looks superficially like a damped sinusoidal ripple, just as in the two-atom case, but is not. The periodicity of its ripples is irregular, being influenced by subunit shapes and relative orientations. Two methods for obtaining distances commend themselves. First, $I_x(R)$ can be inverted to obtain its corresponding $p(r)$. $p(r)$ in this case will be the length distribution of all vectors joining d in the two subunits, and d_{12}, the distance between subunit centers of mass, will be [152]:

$$d_{12} \sim \bar{r} = \int_0^\infty r p(r) \, dr.$$

Alternatively, one can make use of the fact that

$$\bar{r}^2 = \int_0^\infty r^2 p(r) \, dr = d_{12}^2 + R_1^2 + R_2^2,$$

where R_1 and R_2 are the radii of gyration of the two subunits as they are configured *in situ*.[144,151] For large structures like the ribosome this relationship can be used as the basis for solving sets of distance data both for subunit positions and radii of gyration.[152a] For less complex structures the possibility exists of deuterating single subunits, contrast-matching out the protonated parts and measuring the necessary R_is directly. The possibility of analyzing such data for subunit shape and relative orientation has been pointed out, but no general procedure has yet been discovered.[152]

8.3.18. Solution Scattering Studies on Molecules of Known Structure

So far we have discussed small-angle methods primarily from the viewpoint of an investigator seeking initial information about the shape and properties of a molecule of unknown structure. Although this is the most common

[152] P. B. Moore, J. A. Langer, B. P. Schoenborn, and D. M. Engelman, *J. Mol. Biol.* **112**, 199 (1977).

[152a] P. B. Moore and E. Weinsiein, *J. Appl. Crystallogr.* **12**, 321 (1979).

application of these techniques, they have also been used from time to time to study macromolecules whose structures are known at high resolution. The purposes of such studies have been several: (i) as benchmark investigations to establish the validity of small-angle analysis, (ii) to study the correspondence between the crystallographic structure of a molecule and its configuration in solution, and (iii) to investigate conformational changes connected with ligand binding or alterations in environmental conditions.

The favorite objects for benchmark studies have been hemoglobin and myoglobin, with lysozyme coming in a distant third. A critical step in such investigations is the computation of an expected scattering curve from the crystallographic data. This is not a trivial computation for a number of reasons: (i) Eq. (8.2.1) is cumbersome to apply for molecules in which the number of atoms ≥ 1000, as is the case with proteins; (ii) surface polar residues are often poorly located in a crystallographic structure because of intrinsic disorder; and (iii) there is the solvent problem. The last is particularly difficult because the structure of the solvent at the surface of the molecule will affect its apparent size and the shape of the boundary between the molecule and the solvent. A number of investigators have suggested approaches for making these calculations.[153–156] Given the problems to be overcome in such a calculation, it is perhaps not surprising that calculated radii of gyration commonly differ from measured values by 0.5–1.0 Å in 25 Å. These problems lead to an interaction between goals (i) and (ii) of the preceding paragraph.

In general the results obtained have led to the conclusions that (i) the crystallographic structures of proteins are very close to those found for the same proteins in solution, and (ii) x-ray and neutron results on the same molecule are consistent. Myoglobin has been investigated by Beeman[25] and Stuhrmann.[117] In the latter investigation the myoglobin characteristic functions were determined and the intriguing finding made that although the fit of calculated to observed scatter is good at low and intermediate angles, the high-resolution solution scattering curve shows much *less* structure than anticipated from the crystal structures. This suggests that myoglobin in solution may be a more flexible structure than it appears to be in the crystalline state. The lysozyme work of Stuhrmann and Fuess[138] has been described

[153] H. C. Watson, *in* "Small-Angle X-Ray Scattering" (H. Brumberger, ed.), p. 267. Gordon & Breach, New York, 1967.

[154] J. Soler, *Crystallogr. Comput. Tech. Proc. Int. Summer Sch. 1975*, 376 (1976).

[155] Yu. A. Robbin, R. L. Kayushina, L. A. Feigin, and B. M. Shchedrin, *Kristallografiya* 18, 701 (1973).

[156] W. L. Mattice and D. K. Carpenter, *Biopolymers* 16, 81 (1977).

above. Krigbaum and Kugler[157] have also examined lysozyme by x-ray methods as well as chymotrypsin and chymotrypsinogen.

There have been a whole series of solution scattering experiments done on hemoglobin, most notably by the Julich group. Both x-ray and neutron data have been collected[55,158-162] (see also Damaschun et al.[163]). The neutron data are significant because they are among the first ever obtained on a protein. In addition to verifying the possibility of collecting neutron data and showing the close correspondence between the solution and crystal structures, they also demonstrated a substantial alteration in radius of gyration of hemoglobin upon oxygenation and deoxygenation. The change observed is once again close to that anticipated from crystal data.

The idea of using small-angle methods to monitor conformation changes is likely to be a significant application of the method in the long term. In addition to hemoglobin, other studies of the same kind have been done on lipoproteins[164] tRNA,[165,166] and antibodies,[97] to name a few other systems. Durchschlag et al.[167] have demonstrated a small volume change in yeast glyceraldehyde-3-phosphate dehydrogenase as a function of saturation with NAD. Small changes in structure have also been observed in malate dehydrogenase upon substrate binding.[168] Efforts have been made to develop an ultrasensitive difference method for detecting small changes,[169] and a useful critique of the information one can derive from studies of changes has been prepared.[170] The point is that for molecules of known structure,

[157] W. R. Krigbaum and F. R. Kugler, *Biochemistry* **9**, 1216 (1970).

[158] V. H. Conrad, A. Mayer, S. Schwaiger, and R. Schneider, *Hoppe–Seyler's Z. Physiol. Chem.* **350**, 845 (1969).

[159] V. H. Conrad, A. Mayer, H. P. Thomas, and H. Vogel, *J. Mol. Biol.* **41**, 225 (1969).

[160] J. Schelten, A. Mayer, W. Schmatz, and F. Hossfeld, *Hoppe–Seyler's Z. Physiol. Chem.* **350**, 851 (1969).

[161] R. Schneider, A. Mayer, W. Schmatz, B. Kaiser, and R. Scherm, *J. Mol. Biol.* **41**, 231 (1969).

[162] R. Schneider, A. Mayer, H. Eicher, P. Stöckel, W. Schmatz, and J. Schelten, *Hoppe–Seyler's Z. Physiol. Chem.* **351**, 1499 (1970).

[163] G. Damaschun, H. Damaschun, J. J. Mueller, K. Ruchpaul, and M. Zinke, *Stud. Biophys.* **47**, 27 (1974).

[164] L. Mateu, T. Kirchauser, and G. Camejo, *Biochemistry* **17**, 1436 (1978).

[165] J. Ninio, V. Luzzati, and Y. Moshe, *J. Mol. Biol.* **71**, 217 (1972).

[166] I. Pilz, R. Malnig, O. Kratky, and F. von der Haar, *Eur. J. Biochem.* **75**, 35 (1977).

[167] H. Durchschlag, G. Puchwein, O. Kratky, I. Schuster, and K. Kirschner, *Eur. J. Biochem.* **19**, 9 (1971).

[168] P. Zipper and H. Durchschlag, *Eur. J. Biochem.* **87**, 85 (1978).

[169] I. Simon, *Eur. J. Biochem.* **30**, 184 (1972).

[170] H.-V. Purschel, *Stud. Biophys.* **69**, 119 (1978).

minor alterations in small-angle scatter which in the absence of other information might appear totally mysterious are often interpretable. The formation of complexes between macromolecules can also be usefully followed by small-angle methods.[171-173]

It will undoubtedly become common to test hypotheses about the conformational properties of macromolecules by small-angle scattering. One of the failings of spectroscopic methods for studying these kinds of changes has always been the difficulty of relating alterations in spectral properties to alterations in structure. Small-angle data, of course, report structure directly and offer unique opportunities for studying the subtleties of macromolecular action in solution.

8.4. Concluding Remarks

8.4.1. Summary

Table II briefly summarizes the most important kinds of information one can extract from small-angle data on macromolecules or macromolecular aggregates. In the case of uncharacterized macromolecules it is clear that small-angle techniques can lead to a rather complete characterization in terms of size and overall shape. In addition to the parameters listed in Table II there are a number of more esoteric quantities which can be measured[2] but are not described above for reasons of space. In the case of aggregates it is sometimes possible to go beyond this "whole molecule" description to specify how its subunits are arranged.

The resolution of the description produced by a small-angle analysis corresponds roughly to the best electron microscopy can do with favorable macromolecular specimens. The strength of the small-angle approach lies in the accuracy of the size and dimensional information it produces. In electron microscopy there are always problems of visibility of molecular edges, distortions during sample preparation, etc., which limit its accuracy in this regard. On the other hand, if a molecule has a highly unusual shape, its shape is more likely to be identified correctly from electron microscopic data than from small-angle results, although the small-angle data must necessarily be consistent with the correct shape, whatever it is.

Finally, it should be remembered that one of small angle's primary virtues is that it characterizes a molecule's structure as it exists free in solution. This

[171] R. Osterberg, *J. Mol. Biol.* **99**, 394 (1975).

[172] W. Folkhard, I. Pilz, O. Kratky, R. Garrett, and G. Stoeffler, *Eur. J. Biochem.* **59**, 63 (1975).

[173] R. Osterberg, B. Sjöberg, L. Rymo, and V. Lagerkvist, *J. Mol. Biol.* **77**, 153 (1973).

TABLE II. Principal Molecular Parameters Deducible from
Small-Angle Data

A. *All macromolecules*

    1. molecular weight
    2. molecular volume
    3. radius of gyration
    4. distribution of interatomic distances
    5. surface area
    6. molecular shape, as equivalent ellipsoid or better
    7. low-resolution picture of the internal distribution of
       high and low scattering-length density regions

B. *Macromolecular aggregates only*

    1. position of the centers of mass of subunits
    2. subunit radii of gyration
    3. subunit shapes and orientations

fact makes small-angle techniques useful as a way of testing models and assessing conformational changes.

8.4.2. Critique

Low-angle scattering, as already emphasized, is a low-resolution technique. A typical macromolecule contains 10^3–10^5 atoms and requires $3n - 6$ coordinates and n scattering lengths for its full description. The rotational averaging inescapably connected with solution experiments results in the loss of much of the coordinate information otherwise available in the single-molecule diffraction pattern, and the effective lumping of vectors into classes according to length does the same for the scattering-length information. A well-collected solution data set might permit one to specify a length distribution as 15 or 20 independent numbers. A full-contrast series might increase the number of specifiable parameters by three. Internal labeling methods, which involve large increases in experimental effort, may raise that number severalfold more. In any case nothing approaching a full description of a macromolecule is ever possible.

The numerical data given by small-angle experiments constitute solid information. Models deduced for molecular shapes, especially the more complex ones, must be approached with caution, however. In a crystallographic study the covalent structure of the molecule under investigation eventually emerges. Its correspondence with the molecule's chemistry verifies the correctness of the analysis. No comparable internal control is available for the models deduced from small-angle data, a failing this technique shares with other low-resolution methods. There is an additional shortcoming common to all low-resolution techniques: most biochemical problems ultimately require structural information at atomic resolution.

The question must then be asked: under what circumstances can the limited information available be made to yield a biochemically significant result? Four kinds of profitable situations can be recognized. First, a small-angle opportunity exists at the very outset of an investigation of a new biological macromolecule. Small-angle methods can rapidly yield a molecular weight and good estimates of size and shape, traditionally valuable information. Second, there are a number of large biological structures which are simply not suitable for crystallography at this time, for reasons of size or lack of crystals, or both. For some of these it is also true that low-resolution information is relevant to the kinds of questions biochemists are asking about them, and in such situations the means and the ends are well balanced. The ribosome is a prime example of such an object. Third, small-angle methods have no rival as a means for testing models for structures. Many macromolecules have very eccentric shapes when viewed, for example, in the electron microscope. The validity of these shapes as a description of the molecules in solution can be examined by small-angle means and by virtually no other. Similarly, small-angle methods are a sensitive means for examining the relationship between solution and crystal structures for biological macromolecules. The fourth opportunity is the study of conformational change and binding interactions of well-understood structures. It is in this last area that one anticipates the most growth in the years to come. It is conceivable that synchrotron radiation sources and high-speed detectors may prove of great value for investigations of this kind.

Acknowledgments

I wish to acknowledge my indebtedness to those who, both consciously and unconsciously, have contributed to this work by furthering my education in small-angle scattering: D. M. Engelman, B. P. Schoenborn, and V. Luzzati. This work was supported by a grant from the National Science Foundation (PCM78-10361).

9. ELECTRON MICROSCOPY

By Robert M. Glaeser

9.1. Electron Microscopy as a Tool for Structure Determination

The electron microscope has by now become such a routine tool for the cell biologist that one might suppose that electron microscopy had long ago passed out of the realm of experimental physics. Such is by no means the case; the development of new experimental methods of electron microscopy that are suitable for molecular structure determination and the development of new physical methods that are suitable for application to special problems in cell biology still represent lively fields of research. In the field of electron microscopy there still remain many challenges for the experimental physicist, and particularly for the biophysicist.

The purpose of this article is first to describe some of the current developments in methods of electron microscopy, placing much of the emphasis on the new methods that have been developed for molecular structure determination. As will be seen, these experimental methods are in many cases still at an incomplete stage of development. Therefore, a second objective of this paper will be to indicate areas where there is still a need for continued development of experimental methods of a physical nature in the biophysical applications of electron microscopy.

The electron microscope is an optical instrument[1,2] which permits (but does not guarantee) the direct visual observation of structure at very high resolution. High-resolution images can only be obtained, however, provided that the specimen materials are prepared in a suitable form. Among the requirements that must be met are that the specimens should be very thin, that they can be placed into a vacuum of 10^{-6} Torr or better, and, finally, that the specimen structure should not change during the period of electron irradiation while the image is being recorded.

[1] C. E. Hall, "Introduction to Electron Microscopy," 2nd ed. McGraw-Hill, New York, 1966.

[2] R. D. Heidenreich, "Fundamentals of Transmission Electron Microscopy." Wiley (Interscience), New York, 1964.

METHODS OF EXPERIMENTAL PHYSICS, VOL. 20

Electron microscopes have traditionally been designed to be operated at an energy of 100 keV. At this energy the electron has a relativistic wavelength of $\lambda = 0.037$ Å. This very short wavelength is two orders of magnitude smaller than the actual resolution limit of the electron microscope. It is the spherical and chromatic aberration of the objective lens and not the electron wavelength which limit the resolution, because the electron optical lens aberrations cannot be corrected in the same way as is possible in light optical systems. Electron microscopes are also commercially available which operate at an energy of 1 MeV ($\lambda = 0.0087$ Å), and a variety of research instruments going to energies as high as 3 MeV have been in operation for several years. Such high-voltage electron microscopes were initially designed to make it possible to work with thicker specimens than can be used at 100 keV. It has only recently been the case that specially designed high-voltage electron microscopes have begun to surpass 100-keV instruments in terms of improved instrumental resolving power.

The instrumental resolving power is best specified in terms of the resolution at which the contrast transfer function of the instrument has its first zero, under conditions of optimal defocus. For 100-keV instruments this resolution limit is approximately 3.5 Å. In some microscopes, the contrast transfer function continues with appreciable magnitude out beyond the point of the first zero, under proper conditions of instrumental operation. Improved "resolution" can also be obtained with tilted (nonaxial) illumination and in the dark-field image mode. However, in all cases the images so obtained are frequently subject to ambiguous or even misguided interpretation, and it is prudent to assume that the limit of interpretable contrast is given by the expression $\frac{2}{3}(C_s\lambda^3)^{1/4}$, where C_s is the coefficient of spherical aberration and λ is the electron wavelength. At 100 keV and with a realistic value $C_s = 1.0$ mm, this resolution limit works out to be about 3.2 Å.

One limitation on the specimen thickness is given by the depth of field of the microscope. The depth of field is, of course, not just a single value, but depends upon the resolution at which the image is being examined. If the objective aperture is optimized for the resolution that is of interest, then the depth of field is given by the expression $\Delta Z = d^2/2\lambda$, where d is the resolution. For a resolution of 3.5 Å and an electron energy of 100 keV, the depth of field works out to be approximately 165 Å. This sort of specimen thickness is not a limiting factor for biological applications if the image resolution is not as good as 3.5 Å, but it is important to have this factor in mind when going for very high resolution.

In practice the resolution experimentally obtained with biological specimens is limited in most cases to approximately 25–30 Å, or worse. This limitation in the practical results is due to numerous different factors which may be of varying relative importance according to the specific problem under investigation. First, there may be the problem of poor specimen preservation. This limitation is not directly due to any instrumental factors, but rather is indirectly a result of the dual requirements for having thin specimens and for placing these specimens in the vacuum of the microscope. Protein denaturation and other types of destruction of the specimen material can be expected to occur whenever the preparation methods require chemical fixation and dehydration, not to mention possible embedding and sectioning. A second factor which can limit the resolution attainable with biological materials is the serious problem of radiation damage. This damage is an inescapable consequence of the inelastic scattering events that are produced by a fraction of the electrons that pass through the specimen. When there are a large number of inelastic scattered electrons, the energy spread produced in the transmitted electron beam may become large enough that chromatic aberration can become a significant limiting factor in determining the resolution. The effect is particularly troublesome in the case of thick specimens, where the spread in energy loss values can be very great and where multiple scattering events involving one or more elastic processes, together with one or more inelastic processes, cause the inelastic electrons to be scattered out to quite wide angles.

With sufficiently thin specimens and careful control of the radiation exposure, a resolution of approximately 10–15 Å is possible on certain stained and unstained biological specimens. Even higher resolution can be achieved on unstained, hydrated specimens, provided that the electron exposure is kept extremely low and provided that extensive spatial averaging of the image is used in order to compensate for the shot-noise fluctuations produced in the image at low electron exposures. There is no reason, in principle, why the spatial averaging technique cannot be used to obtain experimental results down to the instrumental limit of resolution. However, numerous technical difficulties still stand in the way of this achievement. Overcoming these technical difficulties represents an area of active research development at the present time.

The inelastically scattered electrons in themselves carry a great deal of useful information about the object structure. This information is essentially of a spectroscopic nature; that is to say, the examination of the electron energy loss spectrum at each point in the specimen allows one to carry out spectroscopic observations with extremely small volumes of specimen material and to record differences in the spectrum from point to point. The imaging of inelastically scattered electrons therefore provides structural information that is not obtainable from the images produced by elastically scattered electrons. In discussing the information available in the images produced with inelastically scattered electrons, one must remember that the spatial resolution is once again limited by radiation damage. However, the resolution limitations that are caused by radiation damage have not yet been so carefully analyzed in the case of images produced with inelastically scattered electrons as is the case for elastically scattered electron images.

The preponderent number of inelastic scattering events occur at energy losses of 3–50 eV. The scattering events are "delocalized" in the sense that excitation and ionization of a molecule can occur quite easily due to the passing of an incident electron at some distance away from the "target" molecule. The amount of delocalization of such excitations is in the range of 10–20 Å. As a result, images produced with these inelastically scattered electrons cannot, in principle, provide the localization of the "chromophore" to a resolution better than 10–20 Å.[3] Higher-resolution images can in principle be obtained with electrons that have suffered much larger energy losses, for example, energy losses that would be produced in the excitation of inner-shell electrons. But while such high-loss events are quite well localized, the cross section for such events is smaller by about a factor of a thousand than is the case for the lower-energy-loss events. The reduced frequency of such high-loss events greatly reduces the signal-to-noise ratio in the measurements, which may in turn impose a severe limitation on the attainment of high resolution in the image.

In terms of instrumental approaches, it can be said that inelastic electron images are best obtained with a scanning transmission electron microscope equipped with a suitable energy loss spectrometer. If a resolution of only 20–30 Å is desired, the instrumental features needed for operation in the scanning transmission mode with energy filtering can be provided as an attachment to a conventional transmission electron microscope. For a resolution in the 3–5 Å range it is still necessary to use a dedicated scanning transmission electron microscope (STEM) which has been built exclusively for operation in the STEM mode. It is true that an energy filter lens can, in principle, be developed for the normal imaging mode in the transmission electron microscope. However, a great deal of physics and electron optics must still be worked out in order that "filter lenses" can be regarded as practical devices in conventional microscopes. In fact, there also remains considerable need and opportunity for improvement in the design of spectrometers for scanning transmission electron microscopes.

Other areas of major instrumental improvement also deserve mention. The continued development and refinement of high-voltage electron microscopes offers a promising direction for the achievement of higher resolving power. At the same time, high-voltage electrons have a distinct advantage in overcoming the multiple scattering processes that can seriously confound image interpretation. An alternative direction to take for improving instru-

[3] M. Isaacson, J. P. Langmore, and H. Rose, *Optik* **41**, 92 (1974).

mental resolution is the development of devices for correcting lens aberrations. Although the theoretical possibilities have been quite clearly defined for some time, their experimental realization remains still as a great challenge. Finally, it should be mentioned that the resolution limitations of present-day electron microscopes could be quite directly overcome by the use of image processing techniques. In particular, the resolution limitation normally associated with the spherical aberration of the objective lens is, in fact, only a systematic error, one which can be easily corrected for provided that the image formation occurs under conditions of very high coherence. The experimental task confronted in this case is therefore to provide an electron source of high brightness and unusually small energy spread for use in the conventional transmission electron microscope.

9.2. Image Formation in the Electron Microscope

In discussing image formation in the electron microscope, and in the subsequent discussion of contrast transfer function theory and methods of three-dimensional reconstruction, we shall restrict our attention to a description of images that are formed by the elastically scattered electrons. The use of images formed by inelastically scattered electrons will be dealt with again in Section 9.7.3. The theoretical justification for leaving out the inelastically scattered electrons in the description of "normal" image formation in electron microscopy has not been well developed in the literature. That the question is complex and cannot be easily treated in general terms is in itself not a sufficient excuse to avoid a discussion of image formation of inelastically scattered electrons. Fortunately, one can now say, on the basis of considerable experimental experience, that the theory of image formation by elastically scattered electrons provides an excellent representation of the actual image intensities that one normally records in transmission electron microscopy. On this basis, therefore, we shall proceed to "forget about" the inelastically scattered electrons.

9.2.1. A Simplified Picture: The Mass Thickness Approximation

The earliest ideas of image contrast in electron microscope studies with biological materials were based upon the idea that a fraction of electrons are always scattered at angles sufficiently wide that they are stopped by the objective aperture. In this view of the origin of image contrast, the effects that can be seen in the image intensity are the same as if the electrons had actually been absorbed in the specimen. Thus, where the specimen is thicker or has greater density, there is more scattering and therefore more "absorption," and the image is correspondingly darker.

This whole idea of "pseudoabsorption" or scattering contrast was translated into terms of practical use to biologists by Zeitler and Bahr,[4] who derived the result that, under suitable conditions, the image intensity would vary exponentially with the local "absorption coefficient." The electron image intensity is thus expected to obey a law that is identical to Beer's law in spectroscopy or the exponential attenuation law in the absorption of x-ray radiation. According to the approximations of Zeitler and Bahr, the image intensity at a point $\mathbf{x}$ in the image is given by

$$I(\mathbf{x}) = I_0 \exp[-\mu(\mathbf{x})],$$

where $\mathbf{x}$ is a vector in the plane of the image. In this expression the effective absorption coefficient $\mu(\mathbf{x})$ is proportional to the projection of the mass density in the direction of propagation of the electron beam, that is,

$$\mu(\mathbf{x}) = k \int \rho(\mathbf{x}, z) \, dz \equiv k\rho'(\mathbf{x}),$$

where $\rho(\mathbf{x}, z)$ is the mass density at each point in the object and z is the direction of propagation of the electron beam. The parameter k is an instrument-dependent constant which depends upon the objective aperture semiangle and the incident electron energy, but which does not depend upon the properties of the specimen itself. Clearly, if the specimen is sufficiently thin and the mass density is correspondingly small, the exponential function can be expanded to give

$$I(\mathbf{x}) \cong I_0[1 - k\rho'(\mathbf{x})].$$

In this limiting case, the image intensity can be seen to be a linear function of the projection of the structure; more precisely, a linear function of the projected mass density of the structure.

The conditions of validity of the mass-density approximation described above have been carefully and accurately described by Zeitler and Bahr in terms of the specimen thickness for which the approximations apply in materials of different atomic number. In addition, it should be stressed that the derivation assumes that the object structure is amorphous, in the sense that the angular dependence of the scattering intensity should be the incoherent sum of the individual atomic scattering intensities. Furthermore, the derivation implicitly assumes that the image is taken in the "in-focus condition," such that phase contrast image effects are negligible. Other possible sources of intensity variation in the transmitted electrons, such as

---

[4] E. Zeitler and G. F. Bahr, *Exp. Cell Res.* **12**, 44 (1957).

inelastic scattering processes, are also neglected. In practice, the assumption of incoherent scattering and lack of phase contrast effects means that the "mass-density" approximation is limited to rather low-resolution detail and "close-to-focus" imaging conditions. Nevertheless, this picture of image contrast is useful for the average cell biologist, even under conditions when its derivation is clearly no longer valid. However, the mass-density approximation is not at all suitable for modern applications of the electron microscope in the area of molecular structure analysis, particularly at very high resolution.

9.2.2. The "Weak Phase Object" Model

A more rigorous treatment of image formation in the electron microscope must proceed from a physical optics point of view, making use of Abbé's "diffraction theory" of image formation, which tends now to be more fashionably referred to as "Fourier optics" (see, for example, Goodman[5] or Glaeser[6]). The wave optical approach to image formation automatically takes into consideration the effects of lens aberrations, image defocus, and the "absorption" of electrons by the objective aperture. This treatment, therefore, automatically includes the "mass thickness effect" described above, as well as all other factors that may affect image contrast by the elastically scattered electrons. However, it is still necessary for us to neglect the effects of inelastically scattered electrons.

As a first step in describing the images formed by elastically scattered electrons we shall make the assumption that the electron wavelength is vanishingly small and that the intensity of the scattered portion of the wave is very much smaller than the intensity of the unscattered wave. Under these conditions, the transmitted wave Ψ_t can be written as the product of the incident wave Ψ_0 and a transmission function which is linear in the scattering potential (see, for example, Jap and Glaeser[7]):

$$\Psi_t(\mathbf{x}) = \Psi_0(\mathbf{x})[1 - (i/\hbar c\beta)eV'(\mathbf{x})].$$

In the equation above, $\hbar$ is Planck's constant divided by 2π, c is the velocity of light, β is the relativistic electron velocity divided by c, and e is the electron charge. $V'(\mathbf{x})$ is the projection of the Coulomb potential within the object in the direction of propagation of the incident beam, that is,

$$V'(\mathbf{x}) = \int_{-\infty}^{\infty} V(x, y, z)\, dz.$$

[5] J. W. Goodman, "Introduction to Fourier Optics." McGraw-Hill, New York, 1968.
[6] R. M. Glaeser, *J. Microsc. (Oxford)* **17**, 77 (1979).
[7] B. K. Jap and R. M. Glaeser, *Acta Crystallogr., Sect. A* **A34**, 94 (1978).

The approximations described in the paragraph above are commonly known in the field of high-resolution electron microscopy as the "weak phase object" approximation.

If we further assume that the incident wave is a plane wave, that is, if we make the approximation of perfect coherence, then the application of Abbé's diffraction theory of image formation[5] leads to the simple result that the image wave Ψ_i is given by

$$\Psi_i(\mathbf{x}) = [1 - (i/\hbar c\beta)eV'(\mathbf{x})] * h(\mathbf{x}),$$

where the $*$ symbol indicates the convolution product (integral) and the function $h(\mathbf{x})$ is the inverse Fourier transform of the function describing the effects of wave aberrations and the objective aperture. More explicitly,

$$h(\mathbf{x}) = \mathscr{F}^{-1}\{\exp[-i\gamma(\mathbf{s})]A(\mathbf{s})\},$$

where $\mathbf{s}$ is a vector in Fourier space and is called a "spatial frequency," $\gamma(\mathbf{s})$ incorporates the effects of lens aberrations and defocus, the function $A(\mathbf{s})$ is unity in the open part of the aperture and zero in the opaque part of the aperture, and $\mathscr{F}^{-1}$ indicates the inverse Fourier transform. A more detailed and explicit discussion of the very useful weak phase object case is given in Chapter 9.3 below. Before proceeding to that discussion, however, we pause to introduce some of the complications that can invalidate the approximations which we have so far employed.

9.2.3. Complications

A number of complications can arise that make the weak phase object approximation an invalid description of image formation. These problems include the following:

1. Situations where the scattering is too strong for the weak phase object approximation to apply. These strong scattering situations are effectively the same as those which would lead us to declare the failure of the first Born approximation.

2. Situations in which the specimen thickness is so great that we can no longer ignore Fresnel diffraction effects, which can occur even as the wave propagates through the specimen. Such a situation can, on the one hand, be regarded as the failure of the "zero-wavelength" approximation. At the same time it is also useful to recognize that the problem in this case is essentially a problem of inadequate depth of field in the image forming processes.

3. Situations when the effects of inelastic scattering cannot be ignored. At present there are two effects that are of concern to us. The first is the fact that the electrons that are inelastically scattered may just as well have been totally absorbed, so far as a description of the elastically scattered, coherent

wave is concerned. At the same time the inelastically scattered electrons are not truly absorbed in the specimen, but pass through it, and many of them continue down to the image plane, where they form their own separate image intensity.

As a first step, we can consider what happens if we keep the zero-wavelength approximation but allow strong interaction and failure of the first Born approximation to occur. Assuming once again that the incident electrons are described as a plane wave, we find that the transmitted wave retains a simple but no longer linear relationship to the projected Coulomb potential;

$$\Psi_t(\mathbf{x}) = \exp[(-i/\hbar c \beta)eV'(\mathbf{x})].$$

This nonlinear description of the transmittance function of the object is known in the electron microscope literature as the "strong phase object" approximation. A rigorous derivation of its validity, using the summation of the infinite Born series under the stationary phase approximation, was first given by Schiff,[8] albeit in quite a different context. It is worth pointing out that in the exponential approximation the specimen is still a perfect phase object, in the sense that the transmitted wave has constant intensity and there are no amplitude (i.e., intensity) variations in the wave transmitted through the object. The relationship between the strong phase object approximation and the weak phase object approximation discussed above is obvious upon expansion of the exponential to first order in the projected Coulomb potential.

Relaxation of the zero-wavelength approximation, which means essentially the admission of wave propagation and Fresnel diffraction within the specimen, leads to rather considerable complications in the description of image formation. The simplest case to discuss is one where the weak phase object approximation is valid for a thin slice of material, at each level of the specimen. One can then picture the electrons as penetrating without interaction to a certain depth in the specimen, where a small fraction are scattered. The scattered and unscattered electrons then proceed to be transmitted through the remainder of the specimen with no further interaction. The final transmitted wave can no longer be written as the product of the incident wave and a transmittance function of the object, but instead the transmitted wave must be written as the sum of waves transmitted through different layers of the object; each such transmitted wave contributes to image formation with its own value of "image defocus." This simple model is described here not with any recommendation of its accuracy or validity, but only to give a first idea of the type of complication that is involved under conditions where the zero-wavelength approximation is no longer valid. It should

[8] L. I. Schiff, *Phys. Rev.* **103**, 443 (1956).

perhaps be said that the difficulty in this case has to do not so much with our being able to calculate what the transmitted wave would be, and therefore what the image intensity would be, for an object of known structure; this is an easy task for modern digital computers. Rather, the difficulty is that if we are given only the final image intensity, it is quite impossible to work backward to find out what the object structure is. Of course, a trial and error approach can be taken, where different three-dimensional models of the object structure can be used to calculate the expected image intensity. However, this trial and error approach can become so formidably long and expensive as to be impossible to use as a practical method.

A fairly rigorous description of the transmitted wave requires not only that one take into consideration the Fresnel propagation (i.e., defocus) of the individual transmitted waves, but also requires that allowance be given for a scattered wave to subsequently interact with and be further scattered by the specimen as it continues to propagate to the lower surface of the specimen. Such a description of the transmitted wave was first given by Cowley and Moodie,[9] and this description is commonly referred to as the multislice dynamical theory of electron transmission through moderately thin specimens. In the Cowley–Moodie multislice formulation, propagation through individual, very thin slices is approximated by the phase object approximation. The wave that is transmitted through one slice is then imagined to propagate by Fresnel diffraction to the next slice, where it serves as the next "incident wave." This idea is expressed mathematically by

$$\Psi_n(\mathbf{x}) = \left[\Psi_{n-1}(\mathbf{x}) \exp\!\left(\frac{-i}{\hbar c \beta} \int_{Z_{n-1}}^{Z_n} eV(\mathbf{x}, z)\, dz \right) \right] * \frac{i}{\lambda\, \Delta Z_n} \exp\!\left(\frac{-i\pi x^2}{\lambda\, \Delta Z_n}\right),$$

where $\Psi_n(\mathbf{x})$ is the wave function entering the nth splice and

$$\Delta Z_n = Z_n - Z_{n-1}.$$

Exactly the same expression for the transmitted wave function can also be derived on the basis of the Feynman path integral formulation of quantum mechanics. The path integral formulation has the advantage of showing clearly what portions of the electron–specimen interaction are taken into consideration and what portions are neglected.[7] The Cowley–Moodie multislice formulation is sufficiently rigorous for any practical application, provided that the slice is thin enough. But it has the great disadvantage that it is no longer possible to describe the transmitted wave as a product of the incident wave and a simple function representing the transmission function of the object.

[9] J. M. Cowley and A. F. Moodie, *Acta Crystallogr.* **10**, 609 (1957).

Fresnel propagation and spreading of the electron wave front, within the finite thickness of the specimen, introduces a further important effect in our description of the transmitted wave which does not appear in the zero-wavelength, "pure phase object" approximation. This effect arises from the fact that the intensity of the transmitted wave is no longer constant in space when Fresnel diffraction is included, contrary to the results in the simple phase object approximation. Since there can be amplitude variations in the transmitted wave, these can be faithfully imaged under ideal, "in-focus" imaging conditions. It is therefore important to keep in mind that amplitude variations in the elastically transmitted wave can arise from Fresnel propagation effects as well as from inelastic scattering effects.

9.3. Contrast Transfer Function Theory

In this chapter we treat the process of image formation in the electron microscope according to the Fourier-optical theory of "linear systems." That is, we show that the image intensity is related to the object structure, under appropriate conditions, by a set of linear equations. The performance of a linear system is most elegantly described in terms of the "transfer function" of the system. The power of the transfer function lies in the fact that it tells us how a sinusoidal variation in the image intensity, for a given periodicity or "spatial frequency," is related to a sinusoidal variation, of the same periodicity, in the transmittance function of the object.

This "Fourier optics" treatment of contrast in the electron microscope is essential to an understanding of the focus-dependent contrast effects that are seen in high-resolution applications. We also discuss at some length the restrictions on the imaging conditions under which linear transfer theory may be applied with validity, and at the end of the chapter describe some of the complications that may be encountered in practice when the strict conditions of valid use are not observed.

9.3.1. Origin of Phase Contrast in Electron Microscopy

Phase contrast effects in electron microscopy are generated as a result of the phase distortion function $\exp[i\gamma(\mathbf{s})]$, where $\mathbf{s}$ is a general spatial frequency. The phase distortion $\gamma(\mathbf{s})$ is due in the first instance to any sort of wave aberration associated with the lens, of which the spherical aberration effect is the primary example. For nearly all situations of practical interest in electron microscopy, the wave aberration can be described by the function

$$\gamma(\mathbf{s}) = 2\pi(\tfrac{1}{4}C_s\lambda^3s^4 - \tfrac{1}{2}\Delta Z\lambda s^2)\},$$

where **s** is the spatial frequency vector and s its magnitude, C_s is the coefficient of spherical aberration, ΔZ is the instrumental defocus, and λ is the electron wavelength.

The wave function in the image plane is given by

$$\Psi_i(\mathbf{x}) = [1 - (i/\hbar c\beta)eV'(\mathbf{x})] * \mathscr{F}^{-1}\{\exp[-i\gamma(\mathbf{s})]A(\mathbf{s})\},$$

if we assume that the specimen is a weak phase object.

The image intensity is obtained simply by squaring the image wave function. We thus have

$$I(\mathbf{x}) \cong 1 - [(2/\hbar c\beta)eV'(\mathbf{x})] * \mathscr{F}^{-1}\{\sin[\gamma(\mathbf{s})]A(\mathbf{s})\}.$$

In the spirit of the weak phase object approximation, we have dropped from further consideration the quadratic image terms. This is consistent with the fact that quadratic terms in the object potential were already neglected when approximating the object's transmittance function by the linear expansion of

$$\exp[(-i/\hbar c\beta)eV'(\mathbf{x})]$$

It is important to realize that certain approximations and special conditions have been embedded in the procedure used to obtain the expression for the image wave function above. First, we have assumed that the illumination is axial, that is, that the wave propagation vector is parallel to the optical axis. We have further assumed that the objective aperture is axially symmetric. Finally, we have assumed that we can ignore the curvature of the Ewald sphere and that the Fourier transform of the transmitted wave satisfies Friedel symmetry.† These last two assumptions are already guaranteed by the approximations that we have had to make in describing the transmitted wave in terms of the projection of the structure.

Our next task is to consider the image intensity as a linear superposition of many different sinusoidal functions. This, of course, involves nothing more than looking at the Fourier-transform representation of the image intensity. Taking the Fourier transform of the right-hand side of the equation for the image intensity, we obtain

$$\hat{I}(\mathbf{s}) = \delta(\mathbf{s}) - (2e/\hbar c\beta)\hat{V}'(\mathbf{s}) \sin \gamma(\mathbf{s})A(\mathbf{s}).$$

In this equation, $\hat{I}(\mathbf{s})$ is the Fourier transform of the image intensity, $\hat{V}'(\mathbf{s})$ is the Fourier transform of the projected scattering potential, and we have invoked the convolution theorem of Fourier transforms to obtain the

---

† The Ewald sphere is the locus of all vectors **s** such that $2\pi\mathbf{s} = \mathbf{k} - \mathbf{k}_0$, where **k** is the scattered wave vector. Friedel symmetry applies when the scattered wave at $-\mathbf{s}$ is the complex conjugate of the scattered wave at $+\mathbf{s}$.

result that the convolution product in real space is converted to a direct multiplication in Fourier space. From this result we can see that there is a simple, one-to-one relationship between any one Fourier component of the image intensity and a Fourier component, of the same period, of the projected potential. In the image intensity we find that the amplitude of each sinusoidal (Fourier) component of the potential is modulated by an amount $2 \sin[\gamma(\mathbf{s})]$, and that, furthermore, the amplitude undergoes a change in sign whenever the function $\sin[\gamma(\mathbf{s})]$ goes through zero. We thus find that the various sinusoidal components of the object structure are each transferred with an individual modulation of contrast, and depending upon the defocus value, some sinusoidal components may even be transferred with reverse contrast. For these reasons, the function $\sin[\gamma(\mathbf{s})]$ is commonly referred to as the "phase contrast" transfer function in electron microscopy.

9.3.2. The Envelope Function: Partial Coherence

The result given in the previous section for the weak phase object is valid only under conditions of perfectly coherent illumination. In the real case, several factors can reduce both the degree of temporal coherence and the degree of spatial coherence. Temporal coherence is determined primarily by the monochromaticity of the illumination. Thus a finite energy spread in the illuminating electrons must lead to a corresponding reduction in the degree of coherence. The energy spread can arise from the physics of the emission process at the filament (source), from instabilities in the accelerating voltage, and from electron–electron interactions at points where a beam crossover may be formed. It is also appropriate to include, in the factors leading to degradation of temporal coherence, the effect of focus instability in the objective lens, which is associated with ripple in the objective-lens current. Although this does not directly involve the incident electrons, the effect that objective-lens current instability has on the image is indistinguishable from the effect of a corresponding energy spread in the incident beam (given a perfectly stable objective-lens current). The spatial coherence is determined primarily by the angular convergence (or divergence) of the illumination. The finite size of the primary electron source is obviously a factor in determining the divergence, but also of importance is the condenser-lens aperture size and the conditions of focus of the condenser lens that are used.

The effect that partial coherence will have on the transfer function can be derived from the perfectly coherent case by considering the linear superposition of image intensities, each calculated for the appropriate wavelength and propagation vector, the superposition being weighted according to the distribution of energies and angles of incidence.

In the case in which the object's transmittance function is approximated by the weak phase object approximation, it has been shown that the effect of partial coherence can be represented as a multiplicative envelop function according to

$$\hat{I}(\mathbf{s}) = \delta(\mathbf{s}) - (2e/\hbar c\beta)\hat{V}'(\mathbf{s}) \sin[\gamma(\mathbf{s})]A(\mathbf{s})E(\mathbf{s}). \qquad (9.3.1)$$

The envelope function $E(\mathbf{s})$ takes on values between 0 and 1. The exact functional form of the envelope function depends upon a detailed description of the factors influencing the partial temporal and spatial coherence, and depends also upon the defocus value.[10,11] Under normal circumstances the envelope function of a modern, high-resolution electron microscope has values close to unity out to a resolution of 4–5 Å. At very large defocus values, the envelope function may also affect lower-resolution details.

9.3.3. Image Restoration

The term "image restoration is used here to refer to the process by which the projection of the object structure, $V'(\mathbf{x})$, is retrieved from the observed image intensity $I(\mathbf{x})$. Stated more specifically, image restoration is the process by which one corrects for the systematic errors that exist in the image due to the fact that the function $\sin[\gamma(\mathbf{s})]$ is not always equal to one and does not always have the correct sign; in addition, one must correct for the fact that the envelope function is not always equal to unity.

Now it is clear that in order to obtain the projected potential $V'(\mathbf{x})$, we need only obtain the Fourier transform $\hat{V}'(\mathbf{s})$. But examining Eq. (9.3.1) above, we can see that a formal solution to the restoration problem for all spatial frequencies except $s = 0$ would be to divide the Fourier transform of the image intensity by the product $\sin[\gamma(\mathbf{s})]E(\mathbf{s})$. That is, the formal solution is given by

$$V'(\mathbf{x}) = \mathscr{F}^{-1}[\hat{V}'(\mathbf{s})],$$

where

$$-(2e/\hbar c\beta)\hat{V}'(\mathbf{s}) = \hat{I}(\mathbf{s})/[\sin \gamma(\mathbf{s})]E(\mathbf{s}). \qquad (9.3.2)$$

Needless to say, this approach cannot be used where $A(\mathbf{s}) = 0$, so we have suppressed the aperture function. In addition, the function $V'(\mathbf{x})$ obtained according to Eq. (9.3.2) excludes any zero-frequency (constant) term, and thus lacks the correct "average value" of $V'(\mathbf{x})$.

In practice this approach to the image restoration problem leads to

---

[10] J. Frank, *Optik* **38**, 519 (1973).
[11] R. H. Wade and J. Frank, *Optik* **49**, 81 (1977).

noise amplification whenever the product $\sin[\gamma(\mathbf{s})]E(\mathbf{s})$ becomes rather close to zero. This condition can arise in three regions of the Fourier spectrum of the object. At low spatial frequencies, $\sin[\gamma(\mathbf{s})]$ can never take on appreciable values except for very large defocus. At intermediate spatial frequencies, $\sin[\gamma(\mathbf{s})]$ can oscillate, particularly for moderate values of defocus. And finally, at very high resolution the envelope function must gradually tend to zero, while, of course, at the same time $\sin[\gamma(\mathbf{s})]$ will be continuing to undergo oscillations.

In the low- and intermediate-resolution regions, we can take advantage of the fact that $|\sin[\gamma(\mathbf{s})]|$ takes on values close to one for different spatial frequencies, depending upon the value of defocus. It follows, therefore, that if $|\sin[\gamma(\mathbf{s})]|$ is close to zero in an image taken with one defocus value, one can always find another defocus value such that $|\sin[\gamma(\mathbf{s})]|$ is close to unity, at the spatial frequency of concern. In this way it is possible to patch together data from many different images, each taken with a different defocus, such that the product $\sin[\gamma(\mathbf{s})]E(\mathbf{s})$ is sufficiently close to unity at all spatial frequencies in one or another micrograph. The main limitation to image restoration will then occur either at low resolution or at very high resolution, in the latter case the problem being that the envelop function has become too small.

It is enlightening to see that the envelope function is the real villain when it comes to considering the image resolution that can be obtained, provided that one is prepared to engage in the image restoration task. It is commonly supposed, and not without justification, that the objective-lens spherical abberation is the main factor limiting image resolution. However, the envelope function is not very sensitive to the value of the spherical aberration coefficient, while at the same time the envelope depends quite strongly upon the chromatic aberration coefficient and the energy spread. The use of transfer function theory therefore teaches us that the real limit to image resolution is associated with the energy spread and the chromatic aberration effect and not really with the spherical aberration effect.

By way of a practical example, it is worthwhile to insert some numbers into the analytic functions[11] representing the envelope function. A typical value for the energy spread of a 100-keV microscope is approximately 1.5 eV, and the coefficient of chromatic aberration normally has a value not smaller than 1.5 mm. The practical limit on the illumination angle for conditions of imaging at high magnification is approximately 10^{-3} rad for short exposures and 10^{-4} rad for long exposures. Using these figures, we find that the envelope function has decreased to a value of ~ 0.25 at a spatial frequency of ~ 0.35 Å^{-1}, that is, at a resolution of ~ 2.8 Å. This resolution limitation is in fairly good agreement with the practical limit for useful image restoration (~ 3.5 Å) that is actually realized in present-day electron microscopes.

9.3.4. Complications

The discussion has dealt thus far with the electron microscope contrast transfer function for the special case of a specimen which satisfies the defining conditions for a weak phase object. However, it is not at all uncommon that one wishes to work with specimens that do not accurately satisfy the conditions of the weak phase object approximation. A slightly more complicated case than the weak phase object is obtained by admitting a weak amplitude variation in the transmitted wave, in addition to a weak phase variation. In this case the object transmittance function becomes

$$T(\mathbf{x}) = 1 - i\eta(\mathbf{x}) - \mu(\mathbf{x}),$$

where $T(\mathbf{x})$ is the transmitted wave function when the incident wave is a plane wave.

To understand the effect that the amplitude variation has in the contrast transfer that is obtained in the image, it is convenient to consider first the case of a pure-amplitude object, but still making the assumption of weak scattering. The same analysis as was used in the previous section leads to the result that the image intensity is given by

$$I(\mathbf{x}) = 1 - 2\mu(\mathbf{x}) * \mathscr{F}^{-1}\{\cos[\gamma(\mathbf{s})]\}.$$

Once again taking the Fourier transform of both sides of the equation, we obtain

$$\hat{I}(\mathbf{s}) = \delta(\mathbf{s}) - 2\mathscr{F}[\mu(\mathbf{x})]\cos[\gamma(\mathbf{s})].$$

Thus for a pure amplitude object the contrast transfer function is found to be $\cos[\gamma(\mathbf{s})]$, in contrast to the weak phase object where the transfer function is $\sin[\gamma(\mathbf{s})]$.

The image intensity for a weak, mixed phase and amplitude object is once again easily derived by the procedures above. The result of such analysis is that the Fourier transform of the resulting image intensity is given by

$$\hat{I}(\mathbf{s}) = \delta(\mathbf{s}) - 2\mathscr{F}[\eta(\mathbf{x})]\sin[\gamma(\mathbf{s})] - 2\mathscr{F}[\mu(\mathbf{x})\cos[\gamma(\mathbf{s})], \qquad (9.3.3)$$

which is the linear combination of the phase contrast and amplitude contrast effects. We note now that an essential property of the concept of a transfer function has been lost in this description of the relationship between the image intensity and the object structure. According to Eq. (9.3.3), it is no longer possible to represent the Fourier transform of the image intensity as the product of the Fourier transform of the object function and a phase-and-amplitude "modulating function." Nevertheless, the result remains exceedingly simple and convenient to use, since the phase contrast term is characterized by one transfer function while the amplitude contrast term is characterized by another transfer function. The mechanism for carrying out image restoration remains as clear and straightforward as it is in the case of the pure phase object. The question of the effects of partial temporal and spatial coherence in the case of the mixed weak phase and weak amplitude object have not yet received attention in the literature. However, it seems quite likely that an envelope function expression similar to that obtained in the case of the weak phase object would continue to be a valid expression for use in the case of the mixed object.

The conditions under which the weak phase and amplitude object might be a valid representation of the object transmittance function have not yet been carefully explored in the literature. There is no good theoretical indication as to whether one might expect to find a wide range of types of specimens for which the weak phase object approximation fails, but for which the weak phase and amplitude object provides an accurate representation. Nevertheless, there is some experimental evidence that the mixed object representation is a practical representation, even though theoretical considerations might not encourage one to accept its validity. A fairly extensive analysis of the defocus-dependent contrast transfer characteristics of uranyl-acetate-stained catalase was carried out by Erickson and Klug.[12] In this work the relative weights of the phase contrast and the amplitude contrast terms were allowed to vary continuously at each spatial frequency. These relative weights were then obtained from a series of micrographs taken at different values of defocus. Thus the relative proportions of phase and amplitude contrast, which could be obtained from just two micrographs, were used to determine whether the contrast transfer could be correctly predicted at other defocus values. For all but a few spatial frequencies, the experimental fit was found to be moderately good.

The analysis of Erickson and Klug might have been an overly simplified approximation of the complete problem, however, in the sense that these authors sought only to determine the magnitude of the Fourier coefficients of the phase object and amplitude object terms in the transmittance function. A third parameter, which perhaps should not be ignored, is the possible difference in the phase origin between the phase-object Fourier coefficients and the amplitude-object Fourier coefficients. To illustrate this point, let us suppose that at a particular spatial frequency the Fourier transform of $\mu(\mathbf{x})$ is 90° out of phase with the Fourier transform of $\eta(\mathbf{x})$. Let us make the further, perhaps pathological, assumption that the two Fourier coefficients have equal magnitude. As can be easily seen from Eq. (9.3.3), the magnitude of the Fourier transform of the image intensity will not vary with changes in defocus. More general cases are easily evaluated by the same sort of analysis.

---

[12] H. P. Erickson and A. Klug, *Philos. Trans. R. Soc. London, Ser. B* **261**, 105 (1971).

The effects of the other complications, which have been described in Section 9.2.3, are not so easily discussed in terms of contrast transfer function theory. One can only say that it is prudent to be aware of the existence of these complications and to consider effects such as Fresnel diffraction within the finite thickness of the specimen and dynamical interaction of the electrons with the specimen as sources of possible discrepancy between the obtained experimental results on the one hand and the mathematical representation of contrast transfer function theory on the other.

One further complication merits brief mention at this point. We have so far neglected the effects of the quadratic terms obtained when squaring the image wave function in order to obtain the image intensity. In those cases where the image contrast is quite large, one cannot possibly hold to the assumption that the magnitude of the scattered wave is negligibly small in comparison to the unscattered wave. In these cases we might truly expect to find nonnegligible contributions from the quadratic image terms. These are actually identical to the dark-field image intensity that would be produced by stopping out the central, unscattered beam in the objective lens's back focal plane. The effect that the quadratic terms have on the Fourier transform of the image intensity is to add an expression of the form

$$\hat{I}_d(\mathbf{s}) = (e/\hbar c\beta)^2 [\hat{V}'(\mathbf{s})e^{-i\gamma(\mathbf{s})}] * [\hat{V}'(\mathbf{s})e^{-i\gamma(\mathbf{s})}], \qquad (9.3.4)$$

where $\hat{I}_d(\mathbf{s})$ is the Fourier transform of $I_d(\mathbf{x})$, the dark-field image intensity, and the asterisk indicates the autocorrelation product (integral). This result follows directly from the application of the convolution theorem of Fourier transforms, which is applicable in the case of perfectly coherent illumination. Inclusion of the quadratic terms necessarily prevents the use of a transfer function in the description of the image intensity.

As may be clearly seen from the autocorrelation (self-convolution) representation in Eq. (9.3.4), the operation of squaring the image wave function generates spurious coefficients for spatial frequencies representing all possible combinations of differences of the true spatial fequencies of the object. The interesting point that is worth emphasizing in this connection is the fact that the "new" function generated by admission of the quadratic terms is identical to the dark-field image intensity that would be obtained by blocking the unscattered beam. In practical electron microscopy, it is not possible to use a central beam stop to produce a dark-field image. Dark-field images are usually produced with either a displaced aperture or tilted illumination. Nevertheless, the coherent dark-field image, however obtained, must necessarily be of the same general form as the quadratic terms we normally wish to eliminate from our description of the contrast transfer theory, under coherent bright-field illumination conditions.

9.4. Three-Dimensional Reconstruction

The expression "three-dimensional reconstruction" is used here to refer to the class of methods in electron microscope (EM) structure analysis by which information obtained from several different EM images is used to construct a three-dimensional model of the sturcture of the object. More precisely, the phrase "three-dimensional reconstruction" refers to the task of obtaining a three-dimensional model of the object from many different two-dimensional projections, each obtained at a different tilt angle of the specimen.

There exist many difficulties in the task of three-dimensional reconstruction, and many different approaches have been taken in an attempt to overcome these difficulties. One can distinguish two general classes of methods for three-dimensional reconstruction. One class of methods attempts, first, to recover the three-dimensional Fourier transform of the object, after which the three-dimensional structure is obtained directly by a three-dimensional inverse Fourier transform. This class of methods has a great kinship to the well-developed methods of macromolecular crystallography. The second class of methods works directly in real space and does not resort to the use of Fourier transforms except for steps in which the Fourier transform might present a particular convenience in the correction of systematic errors. As an example, the Fourier transform would still be used during the steps of two-dimensional image restoration; as another example, the Fourier transform provides the easiest way of obtaining the correct weighting of the different spatial frequency components in the "filtered back projection" method of three-dimensional reconstruction, which is still fundamentally a "real-space" method.

9.4.1. Fourier (Crystallographic) Methods

The Fourier transform methods of three-dimensional reconstruction are in many respects similar to the x-ray crystallographic methods of macro-molecule structure analysis. One important difference, of course, is contained in the fact that the structural phases are obtained directly from images in electron microscopy. There is no need in electron microscope structure analysis to obtain phases by the use of isomorphous heavy-atom derivatives of the native structure, as long as the image resolution extends to sufficiently high spatial frequencies.

That the structural phases can be obtained from the images is a direct result of the fact that, under suitable conditions, the image intensity is a linear function of the projection of the object's structure. The conditions needed to provide this linear relationship are, of course, the conditions of validity for the weak phase object approximation. For a weak phase object we need only to apply the projection theorem of the Fourier transform to obtain the result

$$\mathscr{F}_{2D} I(x, y) \propto \hat{V}(S_x, S_y, 0),$$

where $\hat{V}(S_x, S_y, S_z)$ is the three-dimensional Fourier transform of the object potential. The projection theorem thus states that the Fourier transform of a projection is a central section of the three-dimensional Fourier transform of the object. Since we assume here that the Fourier transform of the image intensity can be calculated numerically, we do not have the usual problem encountered when a diffraction pattern is produced physically by some form of radiation. In the latter case, only the modulus of the Fourier transform can be obtained from experimental measurements, whereas in the case of numerical calculations both the magnitude and the phase can be computed.

The net result is that the Fourier transform of one image gives a two-dimensional slice through the three-dimensional Fourier transform of the whole object. Thus, if we have many different projections of the object, each taken at a different angle, then by computing the Fourier transforms of each of them we can obtain many different two-dimensional, central sections through the three-dimensional Fourier transform. The idea of the Fourier methods of three-dimensional reconstruction is then to obtain a sufficient number of different central sections that they will "fill in" all of the three-dimensional Fourier transform of the object. Once the three-dimensional Fourier transform $\hat{V}(s)$ has been computed, it is a straight-forward matter to obtain the three-dimensional Coulomb potential by inverse Fourier transformation, that is,

$$V(\mathbf{r}) = \mathscr{F}_{3D}^{-1}[\hat{V}(s)].$$

The "crystallographic" Fourier transform method described above is completely general in nature and approach. In addition, it has some special advantages in those cases when the object has some form of known symmetry.

One of the most favorable instances in which symmetry facilitates a three-dimensional reconstruction is provided in the case of cylindrical or helical structures. Such structures are assembled from identical subunits (e.g., proteins), which are arranged in a repetitive pattern around and along a cylindrical (helical) axis. Considering for the moment only one repeating segment of the structure, it is evident that a single projection in a direction perpendicular to the axis will be composed of a superposition of many different, independent projections of the individual subunits. Thus a single projection of a cylindrical or helical structure contains as much information about the basic subunit as could be obtained from a large number of equivalent projections of a single subunit. A cylindrical or helical structure is, in effect, its own "tilting stage" in the electron microscope.

In the case of a helical structure with a large number of subunits per repeat, a single projection is sufficient to give a good three-dimensional reconstruction, as was first demonstrated by DeRosier and Klug[13] in their study of the structure of the T4 bacteriophage tail. Helical structures with only a few subunits per repeat, and all cylindrical (nonhelical) structures require two or more independent projections with a known angle of rotation about the symmetry axis. The method of helical reconstruction has now been applied to a wide variety of biological specimens, including the stacked-disk protein of tobacco mosaic virus,[14] ribosomes,[15] microtubules,[16] myosin-decorated actin filaments,[17] and helical fibers of proteins such as glutamine synthetase[18] and sickle cell hemoglobin.[19]

Other symmetries besides helical symmetry can also be of great help in recovering the three-dimensional Fourier transform from a limited number of "views" or projections. In the case of the icosahedral ("spherical") viruses, a single projection again provides many different views of the identical protein subunits that lie at the surface of the virus.[20,21]

A limited amount of work has also been done on the use of Fourier methods for three-dimensional reconstruction of individual macromolecules, in which case the object lacks any special symmetry relationships. A severe limitation on the resolution that can be obtained in three-dimensional reconstructions of individual objects necessarily accompanies the fact that a large number of successive micrographs must be taken of the same object, one for each time the tilt angle is changed. For this reason it has been necessary to do such work only with stained, dehydrated specimens. As a consequence, the amount of biological information that can be obtained is rather limited, and this approach is not to be recommended.

High-resolution three-dimensional reconstructions can only be achieved at the present time if the specimen can be obtained in the form of rather large, two-dimensional periodic arrays of the fundamental, asymmetric unit.

[13] D. J. DeRosier and A. Klug, *Nature (London)* **217**, 130 (1968).

[14] P. N. T. Unwin and A. Klug, *J. Mol. Biol.* **87**, 641 (1974).

[15] J. A. Lake and H. S. Slayter, *J. Mol. Biol.* **66**, 271 (1972).

[16] H. P. Erickson, *J. Cell Biol.* **60**, 153 (1974).

[17] P. B. Moore, H. E. Huxley, and D. J. DeRosier, *J. Mol. Biol.* **50**, 279 (1970).

[18] T. G. Frey, *in* "Electron Microscopy 1978" (J. M. Sturgess, ed.), Vol. III, p. 107. Microscopical Society of Canada, Toronto, 1978.

[19] G. Dykes, R. H. Crepeau, and S. J. Edelstein, *Nature (London)* **272**, 506 (1978).

[20] R. A. Crowther, L. A. Amos, J. T. Finch, D. J. DeRosier, and A. Klug, *Nature*, **226**, 42 (1970).

[21] R. A. Crowther, *Philos. Trans. R. Soc. London, Ser. B* **261**, 221 (1971).

CYTOPLASMIC SIDE

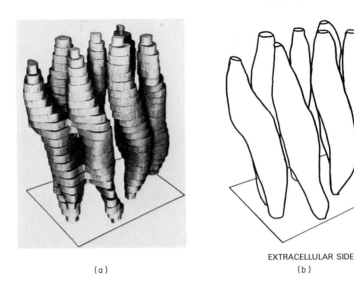

EXTRACELLULAR SIDE

(a)   (b)

FIG. 1. Three-dimensional model of the structure of bacteriorhodopsin at about 7-Å resolution. Seven α-helical rods are shown according to the model published by Henderson and Unwin,[22] while the identification of the cytoplasmic and extracellular surfaces of the protein was carried out independently by Hayward et al.[23] and by Henderson et al.[24] (a) Structural model built from the 3D contour maps. (From Henderson and Unwin.[22]) (b) Line-drawing idealization of the model with the cytoplasmic and extracellular surfaces identified. (Courtesy of Dr. Steven Hayward.)

The requirement for two-dimensional arrays exists strictly to overcome the limitations arising from radiation damage. These limitations are described in greater detail in Chapter 9.5. The requirement for two-dimensional arrays has nothing to do with providing additional crystallographic symmetry, and objects with two-dimensional redundancy must be treated in the same way as general, asymmetric objects as far as the three-dimensional reconstruction problem is concerned.

The present state of the art in the area of three-dimensional reconstruction in high-resolution electron microscopy is represented by the work of Henderson and Unwin[22] on the structure of bacteriorhodopsin. Bacteriorhodopsin is a membrane protein of the halophilic, photosynthetic bacterium *Halobacterium halobium*. This protein acts as a photon-driven proton pump and thereby establishes a pH gradient across the cell membrane when the bacterium is exposed to light. Bacteriorhodopsin is particularly well suited for electron microscope structure analysis since the protein naturally

---

[22] R. Henderson and P. N. T. Unwin, *Nature (London)* **257**, 28 (1975).

occurs in two-dimensional crystalline arrays, which can be biochemically isolated and which are referred to as "purple membrane." These membrane patches occur in pieces as large as one micron in diameter which contain more than 10^4 protein molecules per patch. Unwin and Henderson first demonstrated that two-dimensional projections of the structure could be obtained from electron microscope images out to a resolution of about 7 Å. In subsequent work Henderson and Unwin[22] collected image data over a range of tilt angles of $\pm 60°$ from the normal. The three-dimensional Fourier transform could then be "filled out" from two-dimensional central sections over all of reciprocal space except for a "hollow cone" of semiangle equal to 30° about the direction perpendicular to the membrane plane. Inverse Fourier transform of the sampled data yielded a three-dimensional model, shown in Fig. 1,[22–24] of the structure of the protein with a resolution of approximately 7 Å in the plane of the membrane and approximately 14 Å perpendicular to the plane of the membrane. From these studies it was learned that the bacteriorhodopsin molecule has seven α-helical regions of polypeptide chain that run nearly perpendicular to the plane of the membrane, and which appear to span the full thickness of the membrane.

9.4.2. Direct-Space Methods

The Fourier transform (crystallographic) method is by no means the only one by which three-dimensional reconstruction can be accomplished. There exist quite a number of other approaches which also represent formal solutions to the problem of obtaining a three-dimensional model from many different two-dimensional projections. Whether one real-space method is better than another, and whether real-space methods are better than the Fourier transform method are questions which have sometimes been the subject of very hot debate. The factors by which the various methods might differ from one another include (i) the computer time and costs required for completion of the three-dimensional reconstruction, (ii) the uniqueness of the solution that might be obtained, (iii) the stability of the reconstruction procedure with respect to experimental noise and errors in the measurements and (iv) the ability of the method to achieve the correct solution when one has available only "incomplete data." This last problem will be treated at further length in Section 9.4.3.

One class of direct-space reconstruction methods is built upon the idea of producing a "back projection" from many different two-dimensional projections. The basic idea of the back projection is to project the density of

[23] S. B. Hayward, D. A. Grano, R. M. Glaeser, and K. A. Fisher, *Proc. Natl. Acad. Sci. U.S.A.* **75**, 4320 (1978).

[24] R. Henderson, J. S. Jubb, and S. Whytock, *J. Mol. Biol.* **123**, 259 (1978).

the object backward along a straight line perpendicular to the plane of the projection itself. Thus a single projection is used to generate a three-dimensional density function in real space, parallel sections through which are everywhere identical to the original projection. When the back projections from many different images are superimposed, the true features of the object are reinforced at points of intersection, while unwanted density makes only a general background underlying the actual object. Because this background level is rather high, the use of the simple back projection method is not a very satisfactory approach to the three-dimensional reconstruction problem.

It can be shown in a straightforward way that the naive back projection approach produces a function which is actually the convolution product of the object structure and the inverse Fourier transform of $1/s$. The unwanted convolution is easily removed by computing the Fourier transform of the naive back projection, multiplying this Fourier transform by s, and then carrying out an inverse Fourier transformation. This simple modification of the back projection procedure is referred to as the "filtered back projection." It is to be emphasized that the filtered back projection approach is still philosophically a real-space method, and the use of the Fourier transform to execute the spatial filtering is only incidental to the philosophy of the approach. Complications to the filtered back projection are encountered first in handling the abrupt cutoff at the edge of the "filter function," as must be done in any real, numerical implementation of the approach. Another complication is that the theorem regarding the point-spread function (that is, the inverse Fourier transform of $1/s$), becomes invalid if the projections are not equally spaced, and if they are not very "dense" in reciprocal space.

Another class of real-space three-dimensional reconstruction methods is based on what one might almost call trial and error methods of fitting an object structure to the experimentally obtained projections. These approaches represent essentially the "refinement" of a real-space model, and therefore have a certain approach in common with the reciprocal-space refinement methods that are now becoming commonplace in x-ray crystallography of macromolecules. The basic approach to the three-dimensional reconstruction problem can be described as follows: Pointwise density values of a model structure are permitted to vary so as to achieve a best fit between the calculated projections of the model structure and the actual (experimental) projections that have been obtained from electron micrographs. A common approach to this method is to vary the density parameters of the model in an iterative procedure. One algorithm allows for the simultaneous optimization of the density values to all projections in each iterative step.[25] The optimization of the model to the experimental projections can also be set up as a

---

[25] P. Gilbert, *J. Theor. Biol.* **36**, 105 (1972).

least-squares fitting problem which can then be solved by any of the variety of methods that are possible for the solution of least-squares problems.

The direct-space model refinement procedures may appear, to some extent, to be more attractive than they really have a right to be. The model fitting procedures have no rigorous mathematical foundation by which one might evaluate the limitations of the methods. Their sensitivity to noise and their sensitivity to problems that might arise when the projections are not obtained at equally spaced tilt angles are not easily analyzed except through numerical simulations. The lack of a rigorous analytical foundation might tend to seduce some investigators into believing that the real-space model fitting approaches are less troubled by such difficulties than is the case for the Fourier, crystallographic approach. It is always possible that such a favorable result could actually be the case, but it must be said that there is little mathematical justification to believe that it is true. In fact, numerical simulations by Crowther and Klug[26] have shown that the simultaneous iterative reconstruction procedure does not lead to as accurate a three-dimensional reconstruction from the data as does the Fourier method, when the tilt views are equally spaced in angle. This is particularly true when noise has been added to otherwise perfect data.

A rather special version of the real-space model fitting procedures is the so-called maximum-entropy technique.[27] The rationale in this case is to find the set of "quantified" density values m_i in real space which serve to maximize the sum

$$S = -\sum_i m_i \ln(m_i),$$

while giving projections of the object structure that are consistent, within experimental errors, with the observed data. At this time, no study has yet been published on the suitability of the maximum-entropy method for three-dimensional reconstruction in electron microscopy.

Perhaps the most direct approach to three-dimensional reconstruction in real space would be the formal solution of the simultaneous linear equations that relate the pointwise density values, as unknowns, to the observed projections, that is, to the observed summation of density values in individual columns. A formal solution of the problem is obtained by application of Cramer's rule or by other, more rapid methods of matrix inversion. Matrix inversion solutions are also formally available when the linear equations are overdetermined, giving in this case the least-squares solution. There are two difficulties confronted by the approach of directly solving the linear equations. The first difficulty has to do with the size of the matrices that are involved

[26] R. A. Crowther and A. Klug, *Nature (London)* **251**, 490 (1974).
[27] S. F. Gull and G. J. Daniell, *Nature (London)* **272**, 686 (1978).

for problems of interest in high-resolution structure analysis. It would not be at all uncommon to expect as many as a thousand simultaneous linear equations. The size of the corresponding matrix equation makes the numerical calculations in the formal solution quite formidable. In addition, the matrix inverse can easily be very poorly conditioned. The consequence of poor conditioning is that the presence of the least amount of experimental error makes it quite worthless to compute the matrix inverse, even if one has an unlimited amount of free computer time. We elaborate on this point in more detail in the next section.

9.4.3. The Hollow Cone Problem

The task of three-dimensional reconstruction in electron microscopy faces a very special problem which is due to the fact that projections of the structure can only be obtained over a limited range of tilt angles, for example, $\pm 60°$. The effect of being able to work with only a limited range of views is to produce a three-dimensional reconstruction in which the resolution is not equally good in all directions. The need to find an optimal solution to this problem is very great, since the extension of electron-microscopic structural studies to atomic resolution might not be possible without an adequate solution.

The reason why we refer to this difficulty as the "hollow cone problem" is most clearly seen in the context of the Fourier, crystallographic method of three-dimensional reconstruction. As already described in Section 9.4.1, the Fourier transform of each two-dimensional projection corresponds to a two-dimensional central section in reciprocal space. It therefore follows that a continuous distribution of projections at different angles in real space serves to fill out continuously the three-dimensional Fourier transform of the object over the entire range of angles that can be realized in real space. But by the same token the limited range of tilt angles make a conical section of reciprocal space inaccessible to direct experimental observation, as is illustrated in Fig. 2. The existence of a hollow cone in the sampled data is therefore a general problem in three-dimensional reconstructions which can be avoided only in those happy instances when the object itself has some additional symmetry as, for example, in the case of helical or cylindrical objects.

It is always possible to compute an inverse Fourier transform of just the data accessible to experimental measurement. The object structure retrieved in this way will certainly have some relationship to the real object structure, but at the same time there will necessarily be some differences between the reconstructed object and the original object. The precise nature of the relationship can be seen by application of the convolution theorem. Thus the experimentally obtained data are related to the full three-dimensional

Real Space

Fourier Space

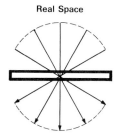
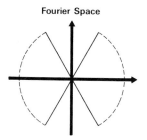

An arbitrarily large number of different
projections within a restricted domain
of tilt angles

Fourier coefficients known to arbitrary
accuracy within a restricted domain
of reciprocal space

FIG. 2. The hollow cone problem is illustrated by the fact that, on the left, a family of projections obtained over a limited angular range (say $\pm 60°$) in real space corresponds, on the right, to a family of "central sections" (planes) over the same angular range in Fourier space. The diagrams should be interpreted as figures of revolution about the vertical axis. The limited range of views leaves a cone about the vertical axis in Fourier space which is inaccessible to experimental measurement.

Fourier transform by a product between the real Fourier transform and a three-dimensional "aperture function." The three-dimensional aperture function $A(\mathbf{s})$ is unity everywhere within the domain of observation and is zero within the region of the hollow cone. Thus we have

$$\hat{V}_{obs}(\mathbf{s}) = \hat{V}(\mathbf{s})A(\mathbf{s}).$$

The inverse Fourier transform of this product is, of course, the convolution product (integral) of the real object structure and a three-dimensional point-spread function. Thus we have

$$\mathscr{F}^{-1}[\hat{V}_{obs}(\mathbf{s})] = \mathscr{F}^{-1}[\hat{V}(\mathbf{s})] * \mathscr{F}^{-1}[A(\mathbf{s})] = V(\mathbf{r}) * a(\mathbf{r}).$$

From this equation it is quite clear that the resolution in the z direction must be substantially degraded by the three-dimensional point-spread function $a(\mathbf{r})$. The primary effect of the hollow cone can therefore be seen, by the use of the convolution theorem, to result in a loss of resolution in the z direction.

The success of interpolation and extrapolation procedures in solving the hollow cone problem is, unfortunately, rather limited. Extrapolation is traditionally known to be a very refractory mathematical problem, and the application to the three-dimensional reconstruction problem is no exception to that generalization. The method most commonly used for interpolation and extrapolation in electron microscopy has involved the representation of the sampled Fourier transform in terms of cylindrical Bessel functions for helical objects, and in terms of spherical Bessel functions for general objects. The limitations of the safe application of Fourier–Bessel reconstruction have been described in some detail.[28,29] This analysis has given the general rule of thumb that the highest isotropic resolution d that can be obtained with m equally spaced projections is given by $d \propto 2D/m$, where D is the "diameter" of the object under investigation. Other "basis function" methods are certainly to be admitted as candidates for interpolation and extrapolation. The question of what basis set would be optimal for the reconstruction problem has been addressed by Zeitler.[30]

The hollow cone problem of three-dimensional reconstruction cannot be really said to have yet been solved as well as it might be, within the context of the Fourier crystallographic method of reconstruction. In addition, a formal evaluation of the effects of the hollow cone problem in the case of three-dimensional reconstruction by direct-space methods has apparently not been dealt with in analytical terms in a way that is equivalent to the state of analysis for the Fourier crystallographic methods. Nor has there been any effort to apply any compensations or corrections, such as interpolation and extrapolation, to the original data before the real-space three-dimensional reconstruction procedures are started. A general assumption seems to run through the literature of real-space three-dimensional reconstructions, that these methods are not sensitive to the same problems felt by the Fourier crystallographic method. Evidence on this point is not yet compelling and is confined mainly to numerical simulations of the effects caused by using data that are obtained with unequal projection angles.

9.5. Radiation Damage

Radiation damage is, without question, the fundamental limitation in the overall effort to obtain high-resolution images of biological materials. This limitation is particularly severe when one wishes to work with hydrated, unstained biological materials. Organic materials in general undergo considerable changes in structure and properties as a result of exposure to very large doses or fluxes of electron irradiation, and biological materials are no exception. The incident electron beam must, after all, be considered to be a flux of ionizing radiation, and there is therefore no way to escape the damage that must occur as a result of prolonged exposure to such radiation. With stained biological materials it is true that a stable specimen can normally be obtained which provides biologically significant information at the level of 25–50 Å resolution. For unstained hydrated specimens the object

[28] R. A. Crowther, D. J. DeRosier, and A. Klug, *Proc. R. Soc. London, Ser. A* **317**, 319 (1970).

[29] A. Klug and R. A. Crowther, *Nature (London)* **238**, 435 (1972).

[30] E. Zeitler, *Optik* **39**, 396 (1974).

is far less stable to electron irradiation, and the limitations on meaningful resolution at very high exposures are far worse than for stained specimens. On the other hand, if great care is taken with the microscopy and if a minimum-exposure technique is used, object detail as small as 20–30 Å can be imaged in noncrystalline objects, in the frozen hydrated state. Even higher resolution can be obtained, and the radiation damage problem can be further circumvented, if periodic specimens are imaged with very low electron exposures and if spatial averaging techniques are then used to superimpose the individual, statistically noise images in the periodic array.

9.5.1. Radiation Physics and Radiation Chemistry

The problem with radiation damage in organic materials arises primarily from the inelastic scattering of the incident electrons. So-called knock-on collisions, which are elastic scattering processes in which an atom is torn out of its molecular context by a head-on collision with the incident electron, have a far smaller probability or cross section and can generally be ignored.[31]

The inelastic scattering processes can be classified as individual molecular excitations, collective molecular excitations, valence electron ionizations, and inner-shell ionizations. The collective excitations can be regarded phenomenologically as having their origin in the imaginary part of the dielectric constant; that is, they are a kind of "dielectric loss" process that is characteristic of the material as a whole and cannot be understood as inelastic scattering processes involving only a single molecule. In the case of metals and other conductors, the theory of such collective losses becomes relatively simple, and the inelastic scattering processes are then referred to as "plasmon" excitations. In nonconductors the "collective" excited state decays rather rapidly into individual, localized molecular excitations. Since the collective losses often involve as much as 30–50 eV, it is not infrequent that the decay of the collective excitation results in multiple ionizations, all within a close distance of one another. The inner-shell electron excitations, which for organic materials means K-shell ionizations, also deserve some special attention. The K-shell ionizations are often ignored in discussions of radiation damage since their cross section is again quite small in comparison to the cross sections for other inelastic scattering events. However, the K-shell excitations can be particularly damaging processes, since for low-atomic-number elements a K-shell ionization is most commonly followed by a second ionization through the Auger process. The molecule therefore winds up with a double vacancy in the valence electron orbitals, and this invariably results in some form of molecular damage, even in the more radiation-resistant aromatic materials.

[31] I. Dietrich, F. Fox, H. G. Heide, E. Knapek, and R. Weyl, *Ultramicroscopy* 3, 185 (1978).

The spectrum of energy losses exhibited by the inelastically scattered, transmitted electrons normally shows three main regions of interest. In the region from zero- to 10-eV energy losses, the inelastic scattering processes are primarily due to valence shell excitation and possibly ionization; the features of the energy loss spectrum in this region can be well correlated with the known electronic absorption spectrum of the molecule. In the region from 10 to 40 or 50 eV, there normally is a broad, featureless absorption curve which shows little difference from one organic material to another. This broad peak represents the superposition of individual molecular ionization processes and collective excitation processes. Approximately 90 % of the inelastic scattering events occur within this general region of energy loss.[32] On the whole one can think of the lower-energy-loss region from ~ 5 to 20 eV as being dominated by individual molecular ionizations, while the higher-energy-loss region is dominated by collective excitations. The third region of interest of the energy loss spectrum is due to the K-shell ionization processes, which occur at energies of 300–400 eV for organic materials. The K-shell edges exhibit an interesting fine structure which is characteristic of the chemical bonding and molecular structure of the parent molecule. In addition, an extended fine structure, which is also a characteristic of high-resolution x-ray absorption spectra, is manifest in the electron energy loss spectrum at energies slightly above the K-shell edge.

It can be taken as a reliable rule that nearly every molecular ionization will result in some form of bond disruption in biological molecules. One reason for this is simply the fact that removal of one of the valence electrons from a molecule results in a purely repulsive potential-energy curve for the vibrational motion of the nuclei. Thus, in the time that it takes for the nuclei to travel 1 or 2 Å, the molecule is bound to separate into two pieces. Since this time is on the order of 10^{-14} sec, there is virtually no hope of the ionized molecule capturing an electron from its environment and reconstituting itself into a stable, electronically excited state. Many highly excited (neutral) molecule electronic states are also characterized by purely repulsive potential-energy curves. As a result, many bond-scission events can occur either as the result of direct excitation or as the result of recombination of an electron with a positive molecular ion. Negative molecular ions, which can be formed by neutral molecule capture of secondary electrons, are frequently unstable to bond scission as well.

For completeness, however, it is worth pointing out that aromatic materials are highly resistant to effects of radiation damage. This resistance can be ascribed to the delocalized nature of the molecular orbitals of the valence electrons in aromatic materials. The delocalization of the valence electrons

[32] M. Isaacson, D. Johnson, and A. V. Crewe, *Radiat. Res.* **55**, 205 (1973).

in aromatic molecules serves to distribute the loss of chemical bonding that accompanies ionization over the entire molecular network. Thus a repulsive intramolecular potential may not occur in aromatic materials as easily as it does in the case of aliphatic materials, where the loss of the valence electron is confined to a single chemical bond.

Also worth mentioning is the phenomenon of resonance energy transfer, which can be seen both in complex biological macromolecules and in simpler chemical systems that are "doped" with small molecules that can serve as an "energy sink." The effect of resonance energy transfer, which is by nature a quantum-mechanical phenomenon, is that energy absorbed by one part of the molecule can be transferred to other parts of the molecule or even to other neighboring molecules. The excited state thus "hops" around from place to place, and in this process the excitation energy may eventually find a somewhat lower energy state and become trapped at that location. Alternatively, the excited state may move about until it lands at a particularly labile or easily dissociated bond, at which time the energy may be dissipated as heat, through bond rupture. Thus two effects of resonance energy transfer are (i) an increased amount of fluorescence from aromatic moities in the specimen, which is due to the fact that aromatic groups tend to be traps or sinks for excited states created elsewhere in the molecule, and (ii) a large cross section for the rupture of specific, labile bonds is seen which is much larger than can be accounted for by random direct hits at those particular sites within the molecule.

The processes of radiation damage are by no means over with once the primary steps of molecular ionization and bond rupture have occurred. These primary steps lead to the formation of a variety of molecular ions and molecular radicals which have enhanced chemical reactivity with other molecules in their environment. These products therefore engage in a host of secondary chemical reactions, which in some cases can go through a considerable sequence of steps involving many neighboring molecules. In most systems there is a large number of different pathways that can be taken by the secondary products, and there is no simple scheme or generalization that can describe all of the effects that can occur.

For our purposes, it is sufficient to take an empirical approach in order to understand the expected consequences of electron-beam-induced radiation damage. This empirical approach is based upon the extensive published studies of radiation chemistry in which the disappearance of parent molecules and the appearance of a variety of product molecules have been measured.

The results of empirical studies in radiation chemistry are traditionally presented in terms of the "G value" for the particular reaction being considered. Thus there are G ($-$parent molecule) values to describe the disappearance of the original molecule and G (product) values that describe the creation of a particular product molecule. These G values represent the number of molecules created or destroyed per 100 eV of energy absorbed from the incident radiation. While it is true that these G values can differ depending upon the quality and type of radiation involved, the majority of the radiation chemical literature has dealt with the use of energetic x or gamma rays, and for these the radiation chemistry is virtually identical to that of high-energy incident electrons.

Thus, to compare the radiation damage effects observed in electron microscopy to direct radiation chemical studies carried out on similar specimens, it is necessary to calculate the energy deposited in a specimen for a given amount of electron exposure in the electron microscope. The statistically expected value of the deposited energy can be calculated from stopping-power theory, which is a macroscopic representation of the inelastic scattering processes. According to stopping-power theory the average amount of energy lost per unit path length by a charged particle traveling in matter is almost inversely proportional to the square of the velocity of the particle.[33] For thin specimens, the incident electron suffers a negligible change in velocity over the whole specimen thickness, and the stopping power can be taken to be constant throughout the thickness of the specimen. The average amount of energy deposited per unit volume is then given directly as the product of stopping power times thickness times electron exposure, the latter being expressed in units of electrons per unit area. Extensive tables of stopping power for electrons of different energy in a variety of different materials are given by Berger and Seltzer.[34] The energy deposited in the specimen is often expressed in units of rads, one rad corresponding to 100 ergs per gram. The rad dose is again easily calculated from the stopping power by first calculating the energy deposited per unit volume as above and then dividing by the mass density of the material.

9.5.2. Empirical Studies of the Radiation Damage Effect under Electron Microscope Conditions

A number of factors concerning the conditions of electron irradiation in electron microscopy are rather different from the conditions that exist in radiation chemical experiments. While none of these factors are such as to make the situation seem tremendously more encouraging in electron microscopy than would be inferred from previous radiation chemical studies, the conditions are still sufficiently special in nature as to make their description worthwhile, and the microscopist should always be conscious of these

[33] F. Rohrlich and B. C. Carlson, *Phys. Rev.* **93**, 38 (1954).

[34] M. J. Berger and S. M. Seltzer, *in* "Studies in Penetration of Charged Particles in Matter," pp. 205–268. Natl. Acad. Sci.–Natl. Res. Council Publ. 1133, Washington, D.C. (1964).

differences. First, the specimens used in electron microscopy are necessarily exceedingly thin. This means that many of the secondary electrons created by primary impact may in fact escape from the specimen. Thus part of the energy which is lost by the incident beam, and which would be calculated to be deposited in the specimen according to the use of stopping-power theory, will in fact escape from the specimen without doing any damage. In addition, the secondary electrons that do escape cannot form negative molecular ions, which would themselves undergo chemical reactions including bond scission. A second difference regarding the electron microscope conditions of specimen irradiation has to do with the fact that the experiments are most commonly carried out with the specimen in a vacuum. For anhydrous materials one would not expect any change in the radiation chemistry of specimens in air or in vacuum. For specimens that are kept in the hydrated state, it is clear that the surrounding vacuum will not have any effect. Thus, in either case, it is unlikely that the vacuum conditions in the electron microscope will have any significant effect on the radiation damage problem. A third factor is the exceedingly high dose rates that can commonly be achieved in electron microscopy. These high dose rates open up the possibilities of introducing dose-rate-dependent effects. Put another way, it becomes possible, under very high-dose-rate conditions, to introduce chemical reactions that obey second- or higher-order kinetics. Finally, and most importantly, the doses received by specimens in the electron microscope are quite commonly many orders of magnitude larger than the total doses that can be realistically achieved by radiation chemists. Thus, for comparison, a dose of 10^6 rad is a rather high dose as far as the radiation chemist is concerned, while a dose of 10^{12} rad is not at all uncommon in electron microscopy at high magnification. It is for all of these reasons, but particularly to explore the effects at exceedingly high doses, that it has been necessary for electron microscopists to carry out empirical studies on the damage that is produced in their specimens under normal electron-microscopic conditions.

One of the most informative methods of observing radiation damage, from a structural point of view, has been to observe the changes that occur in the electron diffraction patterns of crystalline organic materials. One initially sees a general fading of the diffraction spot intensities, and in many materials the higher-resolution Bragg reflections are seen to fade more rapidly than do the low-resolution reflections. Eventually, the crystalline diffraction pattern disappears completely, and all that remains is a continuous, diffuse scattering pattern, which is highly characteristic of amorphous materials. The fading of the diffraction pattern is often characterized in a semiquantitative way in terms of the electron exposure that results in a "complete loss" of the crystalline diffraction pattern. A somewhat more quantitative characterization involves the measurement of the electron exposure at which the diffraction intensity of a given reflection falls to e^{-1} of its initial value. This latter measure, however, has sensible meaning only if the change in intensity is exponential with time, and, unfortunately, this can often prove to not be the case.

It is a fairly accurate rule of thumb to say that the diffraction patterns of crystalline biological materials are normally destroyed by electron exposures in the range from 0.5 to 5.0 electrons/$Å^2$. This range of values reflects not so much the differences that occur from one type of specimen to another as it does the differences in end point that might be chosen to describe "destruction," and particularly the differences in the resolution of the Bragg spacing that is being monitored as the end point for following specimen destruction. Biological specimens that contain aromatic molecules are noticeably more resistant to radiation damage; for example, crystalline adenosine can tolerate an electron exposure approximately ten times greater than can be tolerated by the crystalline amino acid l-valine. Other aromatic organic compounds, such as the aromatic hydrocarbons, can stand even higher doses, and some aromatic dyes can tolerate a dose as high as 1000 electrons/$Å^2$. The most radiation-resistant organic material known in the field of electron microscopy is chlorine-substituted copper phthalocyanine, for which the lower-resolution Bragg reflections will tolerate an exposure of 10,000 electrons/$Å^2$.

Changes in the chemical bonding and structure within molecules can also be accurately detected by observing changes in both the electronic absorption spectrum and the molecular vibrational absorption spectrum. The electronic spectrum is most easily observed by optical spectroscopy in the visible and the ultraviolet, but optical spectroscopy is rather impractical to perform for specimens as thin as those used in electron microscopy. Fortunately, the development of energy loss spectroscopy techniques applied to the inelastically scattered electrons has provided a valuable alternative to optical spectroscopy. Electron energy loss spectroscopy is chiefly useful for molecules that have aromatic ring structures or other, extensively conjugated double-bond systems. This is because the electronic transitions in such molecules involve energy losses of a few electron volts or more, which fall into the range of a high-quality electron energy loss spectrometer. Electronic excitation in aliphatic compounds involves rather higher energies, and the electronic absorption spectrum becomes lost within the continuum of ionization and collective excitations which sets in at about 10 eV.

The study of electronic excitation spectra shows the expected result that the aromatic (chromophore) portion of the molecule is rather more radiation resistant than is the rest of the molecule. As an example, in studies of the amino acid phenylalanine, the electron diffraction pattern was seen to fade at a rate which was much more rapid than the rate at which changes occurred in the absorption spectrum.[35] The total destruction of aromatic biological moieties is normally found to occur at electron doses in the range 10–1000 electrons/$Å^2$.

Changes in the vibrational spectrum cannot be observed in energy loss spectroscopy because the energy resolution required to see such changes greatly exceeds the energy resolution normally available in a spectrometer. Thus changes in the infrared spectrum have previously been studied by irradiating rather large specimens and then examining the "bulk" specimen in a conventional infrared spectrometer.[36] More recently, a technique involving the irradiation of very thin films and subsequent examination in a specimen device that utilizes total internal reflection has permitted the observation of changes in the infrared spectrum of thin-film aliphatic molecules.[37] The results of both types of studies have shown that changes in the infrared spectrum occur at very much the same rate as do changes in the electron diffraction pattern or some of the other changes in the specimen structure that are reflected as a mass loss.

Perhaps one of the most easily observed damaging effects of electron beam irradiation is the phenomenon of mass loss. Once again, from previous radiation chemical research we know that bond scission will result in the formation of molecular fragments of reduced molecular weight. For example, the irradiation of crystalline amino acids results in the formation of ammonia, carbon dioxide, methane, and other low-molecular-weight species. The more volatile of these products will be lost immediately to the vacuum of the microscope, and some of the less volatile products may be readily dispersed by surface diffusion. The net result of mass loss is an immediately detectable "thinning" of the specimen. Specimen thinning has been studied both in terms of the increased transparency of the specimen and the decrease in electron scattering that may be detected in the dark-field image mode. The amount of mass loss produced varies tremendously with the chemical composition of the specimen material. Thus proteins are commonly stated to lose from 20 to 40% of their mass with an electron exposure in the range from 1 to 5 electrons/$Å^2$. Nucleic acids lose considerably less mass. Long-chain paraffins lose only about 5% of their mass, and a rather similar value might be expected for biological lipids. Appreciable beam-induced thinning can easily be seen in collodion (cellulose nitrate) films, from which one might

[35] S. D. Lin, *Radiat. Res.* **59**, 521 (1974).

[36] K. Stenn and G. F. Bahr, *J. Ultrastruct. Res.* **31**, 526 (1970).

[37] W. Baumeister, U. P. Fringeli, M. Hahn, F. Kopp, and J. Seredynski, *Biophys. J.* **16**, 791 (1976).

conclude that carbohydrates in general probably undergo a rather substantial mass loss.

Mass loss has also been measured by autoradiographic techniques. Those experiments, while elegant in concept, have proven to be exceedingly difficult in execution. Their principal virtue is that they permit the measurement of mass loss of specifically labeled chemical groups as opposed to a total, averaged mass loss.[38] The mass loss values recorded from autoradiographic experiments have tended to be rather less than those determined from electron scattering measurements.[39]

It is worth mentioning that the effects of mass loss can be greatly diminished if the specimen is irradiated at low temperature. At helium temperature, the amount of mass lost, even in rather labile systems, is reduced to an amount less than can be measured.[39] The reduction in mass loss by the use of low temperature is no doubt valuable, but one should not be misled by the thought that the absence of mass loss might mean the absence of radiation damage of any kind. Unfortunately, the diffraction patterns of crystalline biological materials are still destroyed by very low electron exposures in specimens that are irradiated at low temperature.[40,41]

The extensive changes that occur in biological structure (i.e., changes in electron diffraction patterns, changes in chemical bonding, and mass loss) all suggest that the original ultrastructure is not likely to be seen in images of biological materials for which the electron exposure received by the specimen greatly exceeds the characteristic dose that leads to complete destruction. Some authors believe, however, that dark-field images of individual molecules which have been taken with a total specimen exposure of more than 100 electrons/$Å^2$ to 1000 electrons/$Å^2$ can still be interpreted in terms of the native molecular structure down to a resolution of 5 Å or so.[42] The suggestion is made that the damaged residue of the original molecule collapses onto the support film in a state which closely resembles the projection of the original structure. The *a priori* likelihood of this occurring is not only remote, but one can also say that in crystalline biological materials such a hopeful outcome does not occur at all. If a facsimile of the original structure did remain in crystalline materials, then some features of the initial electron diffraction pattern would persist indefinitely. This is not seen to occur except in the case of stained biological macromolecules.

[38] R. E. Thach and S. S. Thach, *Biophys. J.* **11**, 204 (1970).

[39] J. Dubochet, *J. Ultrastruct. Res.* **52**, 276 (1975).

[40] R. M. Glaeser, V. E. Cosslett, and U. Valdré, *J. Microsc.* (*Paris*) **12**, 133 (1971).

[41] R. M. Glaeser and L. W. Hobbs, *J. Microsc.* (*Oxford*) **103**, 209 (1975).

[42] F. P. Ottensmeyer, D. P. Bazett-Jones, H. P. Rust, K. Weiss, F. Zemlin, and A. Engel, *Ultramicroscopy* **3**, 191 (1978).

9.5.3. Shot-Noise Limitation for High-Resolution Electron Microscope Work

From the empirical studies of radiation damage described above, it would seem prudent and necessary that high-resolution images should be recorded with a total exposure to the specimen that is significantly less than the exposure that results in complete destruction. It is clear, however, that if the safe dose for microscopy is on the order of magnitude of 1 electron/$\AA^2$, then very large statistical fluctuations in the actual number of electrons that are recorded in each picture element will occur as the size of the resolved element becomes smaller and smaller. It is intuitively reasonable that the true variations in image intensity will ultimately be completely obscured and therefore become invisible because of the large statistical fluctuations in the number of incident electrons.

The problem of visual recognition of simple features at very low statistical definition was studied many years ago by Rose and other authors in connection with the development of television display devices.[43] Rose in particular was responsible for the development of a quantitative psychophysical law which says that the natural contrast of the object must be at least five times greater than the expected fluctuations in intensity that will arise due to shot noise in order for the object to be seen. This law can be expressed mathematically in the form

$$d \geq 5/C\sqrt{N},$$

where d is the object size, C is the contrast, and N is the exposure in electrons/area. A quantitative relationship of this type is useful in that it gives us some guideline or estimate of the smallest object feature that can be visually detected in images of single molecules when the electron exposure is kept to a safe value. For example, when the expected contrast is about 10% and the safe electron exposure is only 1 electron/$\AA^2$, we can immediately see that shot-noise considerations will limit the perceptible resolution to approximately 50 Å. Attempts to overcome the shot-noise limitation by increasing the electron exposure can only lead to disaster, in the sense that one will obtain a statistically well-defined image of nothing but a thoroughly destroyed specimen.

9.5.4. Solution of the Shot-Noise Problem by Image Superposition

The only known possibility for overcoming the shot-noise limitation on image resolution under safe conditions of irradiation is to superimpose a

[43] A. Rose, *Adv. Electron.* **1**, 131 (1948).

large number of images of identical objects. From the Rose equation one might suggest that somehow improving the image contrast would be another method to improve the shot-noise-limited resolution. This idea fails because of the fact that all methods for improving contrast, such as the use of the dark-field image mode, result in a reduced efficiency of utilizing the incident electrons and in a concomitant increase in the shot noise. In a similar vein, efforts to improve the shot-noise-limited resolution by greatly increasing the amount of the electron dose that can be safely tolerated by the specimen are apparently doomed to failure, since the primary steps of radiation damage involve inexorable physical processes over which the electron microscopist seems to have no experimental control.

In the case of crystalline biological specimens, it is fortunate that the process of superimposing many images is not a difficult one at all. The advantage of working with crystalline specimens is, of course, that one knows *a priori* that the motif in one unit cell is repeated exactly in all other unit cells. Furthermore, one has precise *a priori* information about the orientation and translation of the contents of one unit cell with respect to the contents of all other unit cells. In practice, the superposition of the noisy images of the individual unit cells of crystalline specimens is most easily done by first computing the Fourier transform of the image of the periodic object. If an inverse Fourier transform is then computed, using only the values on the reciprocal lattice and setting to zero all values off of the reciprocal lattice, the result can be shown to be formally equivalent to a perfect superposition of all unit cells in the array.

The feasibility of using spatial averaging to overcome the shot-noise limitations on resolution was first demonstrated in a numerical simulation by Glaeser et al.[44] The practical application of this principle to real, biological specimens had to await the development of methods for obtaining images of hydrated biological macromolecules under high-resolution conditions. The first successful application of the principle of spatial superposition was carried out by Unwin and Henderson using glucose-embedded "purple" membranes from the photosynthetic prokaryote *Halobacterium halobium*. In that first study and in the subsequent three-dimensional structure analysis, the image resolution was limited to approximately 7Å.[45] The limiting factor was not the degree of preservation of the hydrated structure, since single patches of glucose-embedded purple membrane will give electron diffraction patterns extending to 2.64 Å.[46] Similarly, the

[44] R. M. Glaeser, I. Kuo, and T. F. Budinger, *Proc.—Annu. Meet., Electron Microsc. Soc. Am.* **29**, 466 (1971).

[45] P. N. T. Unwin and R. Henderson, *J. Mol. Biol.* **94**, 425 (1975).

[46] S. B. Hayward and R. M. Glaeser, *Ultramicroscopy* **5**, 3 (1980).

limiting factor was not the resolving power of the instrument, since a resolution of 3.5 Å is commonly achieved in modern electron microscopes.

The actual limitations have still to be worked out and demonstrated, but it seems likely that these have to do with the need to record images at a rather low magnification (for example, 40,000 magnification) in order to produce sufficient blackening of the photographic emulsion, when the electron exposure at the specimen is kept in the range of approximately 1 electron/Å^2. The use of somewhat more sensitive photographic emulsions and taking advantage of a moderate improvement in the safe electron dose achieved at low temperature may ultimately result in the resolution of the spatial averaging method being extended to a value that is close to the true resolving power of the microscope. As part of that overall process it may be necessary to compute Fourier transforms of rather larger array sizes than those used in the work of Unwin and Henderson[45] and the related study of Hayward et al.[23] One should be advised that systematic correction of the image for the effects of the pincushion and spiral distortions of the projector lens may be a necessary first step before executing the large-array Fourier transform. An increase in the size of the area being averaged will also put increasingly tight restrictions on the degree of "specimen flatness" that must be produced during specimen preparation.

Superimposing images of identical molecules when the molecules are not in a periodic array is certainly a more formidable task than it is for crystals, but it is by no means impossible. We start with the assumption that a large number of molecules are all attached to the support film with a common axis perpendicular to the film, but with random translational displacements and with random azimuthal rotations about that axis. We can then address the question of whether the images of these projected structures can be successfully superimposed if they have been recorded under low-dose, high-shot-noise conditions. The mathematical question was first addressed by Saxon and Frank,[47] who explored the use of cross correlation methods for carrying out the superposition. Saxon and Frank developed an analytical expression similar in nature to the Rose equation which says that the minimum electron exposure required in order to obtain a successful superposition is given by $N \geq 3/C^2 Dd$, where N is the exposure in electrons/area, C is the contrast, D is the size of the entire object, and d is the attainable resolution. The feasibility of the cross correlation and superposition of individual molecules was further demonstrated by numerical simulations using artificially noise-degraded images of single particles located randomly in a field of view. These simulations confirm the analytical result and give strong encouragement that the superposition of individual molecules may in fact be a practical

[47] W. O. Saxon and J. Frank, *Ultramicroscopy* 2, 219 (1977).

method of improving the resolution attainable with individual, free-standing molecules. The practical application of this approach is continuing, and preliminary progress has been reported by Frank and collaborators[48] with images of negatively stained glutamine synthetase.

9.6. Specimen Hydration within the Vacuum of the Instrument

9.6.1. The Need for Maintaining the Hydrated State with Complex Biological Structures

It is well known from x-ray diffraction studies that biological macromolecules must be kept in a wet, hydrated state in order to preserve their native, well-ordered structure. Nearly all complex biological materials that give diffraction patterns in the hydrated state will fail to give diffraction patterns when they are partially or completely dehydrated. Electron diffraction studies with complex biological materials have only served to confirm what was already known from x-ray studies. In fact, electron diffraction studies have further shown that all conventional methods of specimen preparation for electron microscopy destroy, to varying degrees, the well-ordered native structure of biological macromolecules. Included in these methods are the conventional procedures for fixation and embedding, the use of negative stains, and even the freeze-drying technique. Apparently, specimen preparation by critical-point drying has not yet been evaluated by the use of electron diffraction, but there is no reason to believe that critical-point drying will be any better than freeze drying when it comes to preservation of high-resolution features of ordered macromolecular structures. The failure of the freeze-drying technique to preserve high-resolution order is perhaps a bit surprising, since a large number of macromolecules retain their native structure and function when they are rehydrated from the freeze-dried state.

9.6.2. Difficulty of the Problem

It is difficult to maintain a hydrated state for biological specimens when they are observed in the electron microscope because the electron-optical path must be maintained at the relatively high vacuum of 10^{-5} Torr or better. The high vacuum is needed in order to prevent excessive scattering of the electrons by gas molecules along the optical path. For the same reason, it is not practical to enclose a wet specimen between thin windows, because

---

[48] J. Frank, W. Goldfarb, D. Eisenberg, and T. S. Baker, *Ultramicroscopy* **3**, 283 (1978).

even a relatively thin window can produce such excessive electron scattering as to obscure both the electron diffraction pattern and the image of the specimen. Very thin windows, for example those in the range of a few hundred angstroms in thickness, might conceivably be of some use. However, the experience of many investigators, going back to the early years of development of the electron microscope, has consistently been that windows of such thickness are extraordinarily fragile and invariably burst under vacuum. This seems to be particularly the case when the windows are exposed to a rather intense electron beam. Because of these difficulties, the hydration problem for biological specimens went unsolved until the early 1970s, even though the need for its solution was appreciated almost as soon as the electron microscope was first applied to the examination of biological materials.

9.6.3. Technical Solutions to the Hydration Problem

The first practical solution of the hydration problem was developed by Parsons and his collaborators.[49] The approach taken by this group was to insert the specimen between two small apertures and continuously supply water vapor to the region of the specimen at a rate which exactly compensated for the escape of water vapor through the small apertures. This approach avoids completely the problems associated with the use of the thin-window specimen cell. The method has been demonstrated to work sufficiently well to allow one to collect electron diffraction patterns of hydrated protein crystals. The method is not without its difficulties, however, in that the water flow must be adjusted to precisely the correct rate such that the specimen is neither dehydrated because of an insufficient supply of water vapor, nor "flooded" by an excessive amount of water vapor. At the present time, the technique has not yet been demonstrated to be suitable for high-resolution imaging applications.

A second method of preserving hydrated structure has recently been demonstrated in which thin hydrated specimens are rapidly frozen and the frozen hydrated specimen then observed at low temperature in the microscope.[50,51] It was not obvious that long-range order would be preserved during rapid freezing, particularly since attempts to freeze specimens for x-ray diffraction studies apparently have been quite unsuccessful. Nevertheless, Taylor and Glaeser demonstrated that very high-resolution electron diffraction patterns could be obtained from frozen catalase crystals, an

[49] V. R. Matricardi, R. C. Moretz, and D. F. Parsons, *Science (Washington, D.C.)* **177**, 268 (1972).

[50] K. A. Taylor and R. M. Glaeser, *J. Ultrastruct. Res.* **55**, 448 (1976).

[51] H. G. Heide and S. Grund, *J. Ultrastruct. Res.* **48**, 259 (1974).

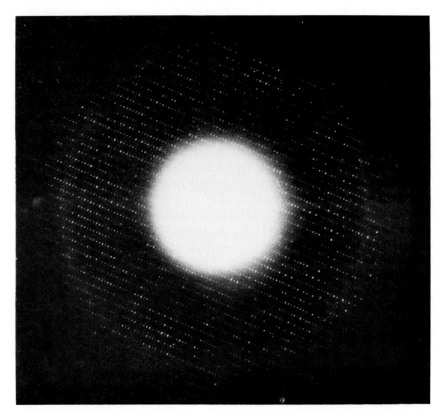

FIG. 3. Electron diffraction pattern of a frozen hydrated crystal of unfixed, unstained beef liver catalase. The unit cell dimensions are approximately 170×68 Å, and spots are clearly visible out to a Bragg spacing of about 3.5 Å on the print. (Courtesy of Dr. Kenneth Taylor.)

example of which is shown in Fig. 3. On the basis of that work, it would seem that the frozen hydrated-specimen technique can provide an approach generally quite useful in solving the hydration problem. This technique is also not without its difficulties, foremost of which are the need to improve the mechanical stability of low-temperature stages for high-resolution imaging studies, and the need to avoid condensation of water vapor on the face of the low-temperature specimen.

More recently, it has been demonstrated by Unwin and Henderson,[45] in their study of purple membrane and of catalase, that the hydrated structure of biological specimens can apparently be preserved if the specimen is air dried from a dilute glucose solution. This technique now has been demonstrated to work exceedingly well for a variety of thin protein crystals. Other hydrophilic molecules besides the simple sugars also seem to work just as well as glucose. The use of iodinated and phosphorylated sugar derivates opens up the possibility for "contrast matching" experiments, both in electron diffraction and in electron imaging. Indeed, glucose already very nearly matches the contrast of protein at low resolution. The great advantage of the glucose embedding technique is that it allows one to examine the specimens on the conventional stage of the microscope; thus there are no additional problems that interfere with high-resolution imaging other than the always-present radiation damage problem.

9.7. Recent Innovations in Experimental Methods

9.7.1. Dark-Field Methods

A number of different experimental procedures can be used to produce a dark-field image in the electron microscope. The common feature of all dark-field images, of course, is the fact that unscattered electrons must in some way be eliminated during image formation. The various dark-field image modes differ from one another according to the portion of the distribution of scattered electrons that is used, the efficiency of utilization of the scattered electrons, and the degree of partial coherence of the image formation process.

The dark-field image mode offers several advantages that are worthwhile to consider in certain cases. The dark-field image is expected to have much higher contrast than the bright-field image for very thin specimens. The increased contrast of the dark-field image mode is achieved because of the removal of the unscattered electrons, which produce a strong "background" intensity in the bright-field case. In the case of crystalline specimens, the elastically scattered electrons are mainly confined to the sharp, Bragg diffraction spots. Dark-field images can be easily formed using a selected number of these diffracted beams. This "preselection" of data in the diffraction pattern results in images that possess highly restricted yet very specific information, and such a restriction in information can often be very useful.[52]

In considering the potential advantages of any dark-field imaging mode, one must also consider certain possible disadvantages. The electron utilization efficiency of all dark-field image modes is inherently low, since the unscattered electrons are necessarily not utilized. In the bright-field image mode, the unscattered electrons serve the function of being a "reference wave" (in the case of thin, weak phase objects), and in this way the unscattered electrons

---

[52] R. M. Glaeser and G. Thomas, *Biophys. J.* **9**, 1073 (1969).

can contribute useful information. Thus, even though the dark-field image has higher contrast, the signal-to-noise ratio of the bright-field image is essentially the same as that of the dark-field image for the same electron exposure. Removal of the unscattered wave also means that the dark-field image intensity must be quadratic in the scattered wave function. This quadratic dependence on the scattered wave function can introduce unwanted "frequency doubling" artifacts in the dark-field image. In addition, the loss of the linear dependence of image intensity upon object structure can make image restoration extremely difficult. While this linear dependence is lost in coherent dark-field images, it can be regained if highly incoherent dark-field image formation conditions can be achieved.

The difficulties of interpretation that can arise in the use of coherent dark-field images are easily appreciated from a simple comparison of the image wave functions and the image intensities for bright- and dark-field images. If we represent the image wave function in the bright-field case as two separate terms, one due to the unscattered electrons and one due to the scattered electrons, then we have

$$\Psi_i(\mathbf{x}) = \Psi_u(\mathbf{x}) + \Psi_s(\mathbf{x}),$$

where Ψ_i, Ψ_u, and Ψ_s designate the image wave, the unscattered wave, and the scattered wave. The dark-field image wave function will be simply

$$\Psi_i(\mathbf{x}) = \Psi_s(\mathbf{x}).$$

The image intensities in the two cases will then be

$$I_b(\mathbf{x}) \cong |\Psi_u(\mathbf{x})|^2 + 2\,\mathrm{Re}[\Psi_u(\mathbf{x})\Psi_s(\mathbf{x})]$$

for weak scattering objects, and

$$I_d(\mathbf{x}) = |\Psi_s(\mathbf{x})|^2$$

for all objects.

From these equations one can easily see that a positive deviation in $\Psi_s(\mathbf{x})$ (i.e., from its average value) cannot be distinguished in the dark-field image from an equivalent negative deviation. A sinusoidal variation in $\Psi_s(\mathbf{x})$ produces both the fundamental and the first harmonic in $I_d(\mathbf{x})$.

A particularly useful insight into the interpretation of dark-field images is obtained by considering the Fourier transform of the image intensity. Thus for the bright-field image

$$\hat{I}_b(\mathbf{s}) \simeq \delta(\mathbf{s}) + 2\hat{\Psi}_s(\mathbf{s}),$$

where $\Psi_u(\mathbf{x})$ has been set to unity for simplicity, while for the dark-field image

$$\hat{I}_d(\mathbf{s}) = \hat{\Psi}_s(\mathbf{s}) * \hat{\Psi}_s(\mathbf{s}).$$

In the equations above, $\hat{\Psi}_s(\mathbf{s})$ represents the Fourier transform of the scattered wave.† As is shown in this last equation, the different spatial frequencies of $\Psi_s(\mathbf{s})$ get "scrambled together" by the convolution operation in the dark-field case.

In the conventional transmission electron microscope (CTEM), dark-field images are obtained by means of a suitable aperture placed in the back focal plane of the objective lens. The ideal aperture for dark-field image formation would consist of a central beam stop, but this geometry cannot be realized in practice since the beam stop must be supported in some way. Several authors have experimented with the use of a single thin wire suspended across the hole of a normal objective aperture. In practice, the wire beam stop is plagued by problems related to the buildup of a contamination layer, which leads in turn to electrostatic charging and its associated image defects. Exceedingly small wire diameters are also required if the beam stop is not to introduce an unwanted reduction in the image resolution.[53]

One of the easiest methods for obtaining dark-field images in the CTEM is to simply displace the normal objective-lens aperture to one side, so that the edge of the aperture intercepts the unscattered beam. While this is the most simple method, it is not suitable for high-resolution work, since the electrons going through the aperture do not utilize the central portion of the objective lens. As a result, the image resolution in the dark-field image is more severely affected by the lens aberrations.

A better method of dark-field image formation consists of tilting the incident illumination by a large enough angle (typically about 10^{-2} rad) such that the unscattered electrons are no longer brought to focus on the optical axis, but instead are intercepted by the edge of the (centered) objective aperture. In this way the scattered electrons that pass through the objective aperture still use the central portion of the objective lens, and full image resolution is preserved. However, it must be cautioned that electrostatic charging at the aperture is always a potential problem and can degrade image resolution well below the theoretically expected valued.

All of the dark-field image methods described above will produce coherent dark-field images, provided that the incident illumination is nearly coherent. Since interference effects can be quite troublesome in dark-field images, it is sometimes worthwhile to strive for incoherent image formation conditions.

---

[53] W. Chiu and R. M. Glaeser, *J. Microsc.* (*Oxford*) **103**, 33 (1975).

---

†For clarity we mention more explicitly that $\Psi_s(\mathbf{x})$ represents the deviation of the transmitted wave from the incident wave, and in that sense represents the "scattered wave" at the specimen plane. $\hat{\Psi}_s(\mathbf{s})$ is the Fourier transform of $\Psi_s(\mathbf{x})$ and thus represents the scattered wave in the "far-field" or Fraunhofer limit.

A good degree of partial incoherence can be achieved by the use of an annular condenser aperture. With such a condenser aperture the illumination can be focused at the specimen, producing a conical shell of illumination. The specimen is then effectively illuminated by electrons that have a constant tilt angle relative to the optical axis and a continuous distribution azimuthally about the axis. A disadvantage of the annular-aperture method is associated with the fact that the condenser aperture can be "ideally matched" to the objective aperture at only one setting of the condenser lens. Since the condenser lens setting is normally varied in order to adjust the illumination intensity at the specimen, the annular-aperture dark-field method imposes an undesirable restriction on the operation of the microscope. An equivalent condition of conical illumination can be produced by setting up the tilted-beam dark-field conditions without an annular aperture, and then electronically precessing the tilt angle about the optical axis.[54] In this way a condition of "dynamical" conical illumination is produced which has much greater flexibility for the operator than does the annular-condenser-aperture method.

The scanning transmission electron microscope (STEM) is particularly well suited as an instrument for producing nearly incoherent dark-field images. Results with the first high-resolution STEM were reported by Crewe and collaborators in 1970.[55] The principle of operation in the STEM is to scan a well-focused beam in a raster pattern over a selected area of the specimen. The interaction of the incident electrons with the specimen is monitored by a suitable detector, the output of which is used to modulate a synchronously scanned display tube. In the simplest dark-field image mode, an annular detector lets the unscattered electrons pass through a central hole while the majority of the scattered electrons are collected on the annulus. If the scattered electron current is summed together without regard to the position on the annular detector where the electrons had fallen, then the resulting image is nearly a completely incoherent dark-field image. Of course, other detector configurations are not precluded, including the use of separate detectors in the bright-field cone. A further practical advantage of the STEM is the fact that an energy spectrometer can be used to analyze the distribution of energy losses in the transmitted electron beam. Many different scattered electron signals can be collected and displayed simultaneously. Certain composite images can be of great value, such as the incoherent dark-field image minus a scale factor times the inelastic scattering image.

The dark-field STEM mode has proven to be far superior to any other approach for producing images of isolated, single, heavy atoms. A representative example of the sort of single-atom images that can now be obtained

[54] W. Krakow, *Ultramicroscopy* **3**, 291 (1978).

[55] A. V. Crewe, J. Wall, and J. Langmore, *Science* (*Washington, D.C.*) **168**, 1338 (1970).

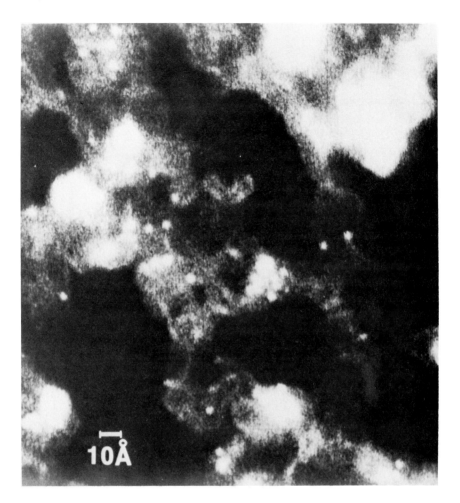

FIG. 4. Individual gold atoms deposited on a very thin carbon film, as they are visualized with the scanning transmission electron microscope in the dark-field image mode. (From Ohtsuki *et al.*[55a])

with the STEM is shown in Fig. 4.[55a] The superior performance of the STEM is due to several factors. First, the STEM dark-field image is nearly incoherent, and as a result the "amorphous" structure of a thin carbon support film generates far less of a disturbance than is the case in the partially coherent dark-field images that are produced in the CTEM. Second, the collection efficiency of the STEM is nearly 80%, assuming a 2.0-Å probe size, while the

[55a] M. Ohtsuki, M. S. Isaacson, and A. V. Crewe, *in* "Scanning Electron Microscopy," Vol. II, p. 375. SEM, Inc. AMF O'Hare, Illinois, 1979.

collection efficiency of a tilted-beam CTEM image would be only about 20%.[56] Finally, the ability to collect the inelastically scattered electrons separately from the elastically scattered electrons, and in general the ability to manipulate the collected signals, can be a considerable advantage of the STEM. The clarity of the images of single heavy atoms that can be produced with the STEM has now been confirmed by other laboratories besides Crewe's.[57,58] While numerous other laboratories have claimed to produce images of single heavy atoms on thin carbon films with the CTEM, the results are not nearly as distinct and unambiguous, in this author's opinion, as those produced with the STEM.[59]

9.7.2. Single-Sideband Images and Holography

We have already mentioned that the strong interaction of electrons with matter can quickly lead to the breakdown of the weak phase object approximation. A major advantage of electron microscopy, namely the preservation of phase information, can then be lost. Because of this fact, the possibility of a holographic recording of the scattered-electron wave function has been investigated for many years. While none of the proposed schemes has yet been demonstrated to be highly successful, the so-called single-sideband image mode is beginning to emerge as a very promising possibility.

The single-sideband image mode is achieved by the use of an asymmetrically shaped objective aperture which lets the unscattered beam and a half-plane of the scattered electrons through, but which blocks the other half-plane of scattered electrons. The diffraction-limited image resolution that is permitted by the half-plane aperture is the same as that of the (normal) double-sideband aperture. However, only half as many scattered beams contribute to the formation of the image in the single-sideband case. The formal possibility exists, therefore, that both the magnitude and the phase of the diffracted wave function, within the open half-plane, can be obtained from the single-sideband image intensity alone. As a parenthetical comment, it should be pointed out that the magnitude and phase of the diffracted wave can be obtained from the (normal) double-sideband image intensity of a weak phase object, since the diffracted beams of a weak phase object obey Friedel symmetry, i.e.,

$$\hat{V}'(-\mathbf{s}) = [\hat{V}'(\mathbf{s})]^*.$$

[56] J. P. Langmore, J. Wall, and M. S. Isaacson, *Optik* **38**, 335 (1973).

[57] M. D. Cole, J. W. Wiggins, and M. Beer, *J. Mol. Biol.* **117**, 387 (1970).

[58] J. S. Wall, J. F. Hainfeld, and J. W. Bittner, *Ultramicroscopy* **3**, 81 (1978).

[59] M. Isaacson, D. Kopf, M. Utlaut, N. W. Parker, and A. V. Crewe, *Proc. Natl. Acad. Sci. U.S.A.* **74**, 1802 (1977).

Under the conditions of Friedel symmetry the use of a double-sideband aperture does not introduce any more parameters or "unknowns" into the image formation equations than are present in the single-sideband case.

The holographic nature of the single-sideband image mode was first noted by Hanssen.[60] Algorithms for the retrieval of both the phase information and the amplitude information have been proposed, notably by Misell and Greenaway.[61] Downing has pointed out that the weak, mixed amplitude and phase object presents a situation of special interest, since the sum of two complementary single-sideband images is equal to the amplitude term, while the difference of the complementary single-sideband images is equal to the phase term. Up until the present time, the electrostatic charging of the edge of the half-plane aperture has proven to be a severe experimental difficulty which limits the quality and the resolution of the single-sideband images that can be obtained.[62]

9.7.3. Image Formation with Inelastically Scattered Electrons

The ability to completely remove the inelastically scattered electrons from the image and the ability to form images using only a selected portion of the inelastically scattered electrons represent two aspects of electron microscopy which have not yet been fully developed and exploited.

The complete removal of the inelastically scattered electrons would seem to be an experimentally useful thing to do, since the quantitative interpretation of image is based upon a theoretical treatment in which the inelastically scattered electrons are neglected. Removal of the inelastically scattered electrons would be expected to improve the image resolution in the case of thicker specimens, since the chromatic aberration effect can be severe for energy losses of 10 eV or more and for scattering angles of 5×10^{-3} rad or more.

The formation of images using only electrons that have experienced a specific energy loss represents a method of considerable potential value. One such example would be to form successive images with electrons whose energy losses corresponded to the characteristic K-shell excitation energies of the various elements in the specimen, thereby giving images of the distribution of the different elements within the specimen. Another possibility is to select electrons that have suffered an energy loss characteristic of the optical excitation of a specific type of molecule[63]; this sort of approach is equivalent

[60] K. J. Hanssen, *Z. Naturforsch. A* **24a**, 1849 (1969).

[61] D. L. Misell and A. H. Greenaway, *J. Phys. D* **7**, 832 (1974).

[62] K. Downing, *Ultramicroscopy* **4**, 13 (1979).

[63] J. Hainfeld and M. Isaacson, *Ultramicroscopy* **3**, 87 (1978).

to the introduction of stains (chromophores) in optical microscopy. In all of these approaches, as in the more conventional EM studies, one must, of course, consider the effects of radiation damage in formulating a realistic experimental design.

A related experimental possibility is to image a small portion of the specimen onto the entrance slit of an energy spectrometer. Analysis of the energy loss spectrum allows one to perform spectroscopic studies on the specimen material at an extremely small size scale. Once again, radiation damage normally limits the smallest size that can be used, rather than other factors such as the electron optics or the signal-to-noise ratio due to the finite beam current.

A variety of energy dispersing systems have been devised for use in the conventional transmission electron microscope. If energy dispersion at only one point of the image is desired, or if it is satisfactory to deflect the image across the entrance aperture of a spectrometer for point-by-point analysis, then a wide variety of spectrometer designs can be used. On the other hand, one often wishes to have filtered (no-loss) electron images, or specific energy loss images, in which all image points pass through the spectrometer simultaneously. Simultaneous filtering is an obvious advantage if one must avoid any unnecessary (or unproductive) irradiation of the specimen. An optical scheme involving a magnetic prism and an electrostatic mirror was devised by Castaing and Henry[64] for simultaneous energy filtering of all image points. A modified version replaces the electrostatic mirror with a 180° magnetic deflection.[65] This latter scheme allows one to use the Castaing image filter for higher-energy electrons. Needless to say, these complex optical elements are difficult to align so as to simultaneously provide good energy resolution and good spatial resolution in the image, and they have so far not been usable for high-resolution electron microscope applications.

The scanning transmission electron microscope (STEM) is much better suited to energy filtering and energy analysis than is the CTEM. Since only one point of the specimen is irradiated at a time in the STEM, there is no need to use a "full-image" spectrometer, as in the CTEM. The problems of spatial resolution and energy resolution are decoupled in the STEM, which represents a major advantage. While the problem of energy analysis is greatly simplified in the STEM, there is still room for improvement in the physical design of spectrometers.[66] An energy resolution of 0.25 eV (or better) at an energy of 100,000 eV is one of the "ideal" design goals. At the same time one would like to use a large entrance aperture to ensure a high collection effici-

[64] R. Castaing and L. Henry, *C. R. Hebd. Seances Acad. Sci.* **255**, 76 (1962).

[65] G. Zanchi, J. P. Perez, and J. Sevely, *Optik* **43**, 495 (1975).

[66] D. E. Johnson, *Ultramicroscopy* **3**, 361 (1978).

ency, and to analyze the energy loss spectrum of electrons that have been scattered at angles as large as 10^{-2} rad.

Crewe and his collaborators have demonstrated a wide range of applications of energy loss spectroscopy in the STEM, and have thereby opened the way to a very promising aspect of electron beam microanalysis. The experiments from Crewe's laboratory have shown that it is possible to obtain high-resolution energy loss spectra from a variety of organic compounds. The spectral features that have been observed span the range from optical transitions in the visible and the ultraviolet all the way out to the excitation and ionization of K-shell electrons. The energy resolution of the K-shell excitations can be sufficient to show a characteristic fine structure at the leading edge of the transition, and the extended fine-structure features at the tailing edge of the transition.[67,68] Inner-shell energy loss spectroscopy has been shown to be sufficiently sensitive to detect the iron within the core of individual ferritin molecules—about 5×10^{-19} g or 5000 atoms. In most applications, however, radiation damage can be expected to limit the smallest acceptable sample size to much larger volumes of material.[68]

9.7.4. High-Voltage Electron Microscopy

Up until this point we have implicitly assumed that the electron microscope operates at an accelerating voltage of 100 kV or less. In that context, the phrase "high-voltage electron microscopy" would refer to any instrument operating above 100 kV. In practice, when the expression "high-voltage electron miscroscopy" is used, one normally thinks in terms of instruments that operate at energies of 500 keV or higher. Most "high-voltage" electron microscopes are designed to operate up to energies of 1 MeV or a bit more. However, one instrument has been built in Toulouse, France, which is routinely operated at an energy of 3 MeV.

The major advantage of existing high-voltage microscopes is their ability to give high-quality images of thick specimens. The improved image quality with thick specimens is due to the reduction in chromatic aberration, which in turn is due to a reduction in inelastic scattering at high energies and a reduction in the average scattering angle as the energy increases. An obvious advantage of using thicker specimens is the ability to view and comprehend the full three-dimensional relationships that may exist between different structural elements. To fully appreciate the three-dimensional structural relationships, it is usually necessary to tilt the specimen and record successive images for subsequent viewing as stereo pairs. The practical limit on specimen

[67] M. Isaacson and D. Johnson, *Ultramicroscopy* 1, 33 (1975).
[68] M. S. Isaacson and M. Utlaut, *Optik* 50, 213 (1978).

thickness tends to be between 0.25 and 1.0 μm, depending upon the complexity of the structure. The thin margins of whole-mounted tissue-culture cells can be viewed without embedding and sectioning, as is seen by the example shown in Fig. 5.[69] For thicker cells and whole tissues, it is still necessary to use conventional embedding and sectioning procedures. However, the use of thick (0.25–0.5 μm) sections can greatly facilitate the task of cutting and collecting serial sections that go through the entire thickness of a cell, or related tissue structure.

Another feature associated with higher electron energies is the fact that the depth of field increases as the wavelength (and the average scattering angle) decreases. By depth of field, we mean the thickness of the specimen—or range of instrumental defocus—over which an object feature of size d remains in sharp focus. The depth of field varies with the wavelength according to the expression

$$\Delta Z = d^2/2\lambda,$$

where d is the resolution and λ is the electron wavelength. Some typical values of the electron wavelength are 0.037 Å (100 keV), 0.0142 Å (500 keV), 0.00872 Å (1.0 MeV), and 0.00637 Å (1.5 MeV). The increased depth of field can be an advantage in high-resolution electron microscopy of thick specimens, and also in high-resolution microscopy of thin specimens viewed at a high tilt angle.

The use of higher accelerating voltages also offers the possibility of improved instrumental resolving power. The point-to-point (Scherzer) limit of resolution is proportional to $(C_s\lambda^3)^{1/4}$, where C_s is the coefficient of spherical aberration and λ is the electron wavelength. A good rule of thumb is that $C_s\lambda$ remains constant as the voltage is changed. Thus the instrumental resolution can be improved in proportion to $\sqrt{\lambda}$ as the voltage is increased. The reduction in chromatic aberration already mentioned is another factor that favors the use of high voltages for high-resolution work. In this context it is convenient to think of the reduced chromatic aberration in terms of an improved temporal coherence, which in turn will result in the envelope function (of the contrast transfer function) extending to higher resolution. In practice this sort of improvement has not yet been fully realized on high-voltage microscopes because of the increased difficulties in obtaining good electrical and mechanical stability with the bigger machines.

Finally, it is worth mentioning that the reduced scattering cross section at high voltages has the effect of making biological specimens fit more accurately the description of being weak phase objects. This is clearly an advantage of a very fundamental nature, since the quantitative interpretation

[69] J. J. Wolosewick and K. R. Porter, *J. Cell Biol.* **82**, 114 (1979).

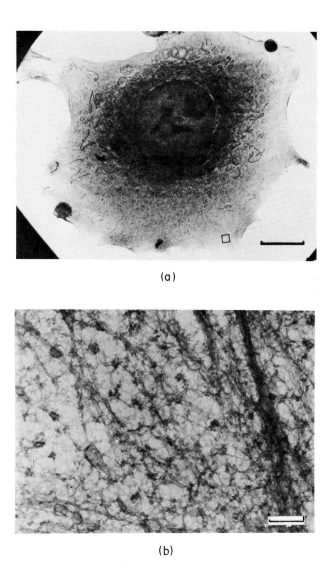

(a)

(b)

Fig. 5. Whole-mount, critical-point dried tissue-culture cell seen in the high-voltage electron microscope at 1 MeV. The bar in (a) indicates 10 μm. The small area in the box at the thin margin of the cell is shown at higher magnification in (b), where the bar now indicates 2000 Å. The high-magnification blowup reveals a complicated arrangement of formed structural elements, which are referred to as the "microtrabecular network" of the cytoplasm.[69] (Courtesy of Dr. Keith Porter.)

of the image intensity is greatly facilitated when the weak phase object approximation is valid. The limits of validity of the weak phase object approximation at 100 keV are not well known, but there are reasons to believe that serious discrepancies will occur for specimen thickness of 150 Å or more, particularly as regards the high-resolution details of the structure. At 500 keV or 1 MeV, one expects the weak phase object approximation to be valid for thicknesses of about twice the limit at 100 keV.

9.7.5. Aberration Correction

In closing, it is important to mention the possibility that the resolving power of the electron microscope might someday be significantly improved by the development of suitable optical elements which would be capable of reducing the spherical and chromatic aberration coefficients of the objective lens. It was shown by Scherzer[70] early in the development of electron optics that cylindrically symmetric magnetic lenses must always have positive aberration coefficients. Thus compound magnetic lenses containing elements of different "refractive index" cannot be used to correct these lens aberrations, as can be done with glass optics. However, Scherzer later enumerated the different ways in which aberration correction could be accomplished, including the use of electrostatic fields, time-dependent fields, and nonround optics.[71] Of these various alternatives, the most promising at the moment seems to be the use of multipole lens elements, typically involving a combination of magnetic and electrostatic elements. Koops and others[72] in Scherzer's laboratory have had great success in reducing the chromatic aberration by such an approach. However, practical implementation of an aberration corrector for high-resolution work remains to be accomplished.

Acknowledgments

I wish to thank the following individuals for graciously providing the prints used in the following figures: Dr. Nigel Unwin for Fig. 1a and Dr. Steven Hayward for Fig. 1b, Dr. Kenneth Taylor for Fig. 3, Mr. M. Ohtsuki and Dr. Albert Crewe for Fig. 4, and Dr. Keith Porter for Fig. 5.

[70] O. Scherzer, Z. Phys. **101**, 593 (1963).
[71] O. Scherzer, Optik **2**, 114 (1947).
[72] H. Koops, G. Kuck, and O. Scherzer, Optik **48**, 225 (1977).

10. VOLTAGE CLAMPING OF EXCITABLE MEMBRANES

By Francisco Bezanilla, Julio Vergara, and Robert E. Taylor

10.1. Introduction

"Animal electricity" was discussed by Galvani in his *Commentary* of 1791 (English translation by Green[1]) and, as pointed out by Hodgkin[2] in his Sherrington Lecture, Volta, in 1800, compared his battery to the stack of plates in the electric fish. Thus the observation that animals produce or respond to electrical changes is very old. Most of the early work on excitable cells, such as nerve and muscle,[3–5] was done by applying known currents and measuring the electrical or mechanical responses. Typically, no (or a very small) response occurs if the stimulus is below "threshold," and when the system "fires," one loses control. The excitability resides in a surface membrane in the form of channels which are embedded in a lipid bilayer with a rather large capacitance (about $1.0 \ \mu F/cm^2$). They are either open or closed, and the proportion of the time any one is open is potential dependent.

To see why control of the membrane voltage, or voltage clamping, is so important in the study of excitable membranes, consider the components of the current I_m through a membrane. There are two broad categories: current carried by ions and dielectric, or displacement, currents. For frequencies above a few kilocycles per second (kHz), the capacitance C_m of the membrane is almost independent of potential V_m,[6,7] so the capacitive current is $C_m \, dV_m/dt$. There is a small component of capacitive current, which is potential and time dependent, associated with the movement of

[1] R. M. Green, English translation of Luigi Galvani's "de Viribus Electricitatis in motu musculari commentarius" with introduction and other translations. Licht, Cambridge, Massachusetts, 1953.

[2] A. L. Hodgkin, "The Conduction of the Nervous Impulse." Thomas, Springfield, Illinois, 1964.

[3] B. Katz, "Electric Excitation of Nerve." Oxford Univ. Press, London and New York, 1939.

[4] A. L. Hodgkin and W. A. H. Rushton, *Proc. R. Soc. London, Ser. B* **133**, 444 (1946).

[5] R. E. Taylor, *Phys. Tech. Biol. Res.* **6**, 219 (1963).

[6] H. J. Curtis and K. S. Cole, *J. Gen. Physiol.* **21**, 757 (1938).

[7] K. S. Cole, "Membranes, Ions, and Impulses." Univ. of California Press, Berkeley, 1972.

METHODS OF EXPERIMENTAL PHYSICS, VOL. 20

charge within the membrane responsible for the initiation of the voltage-dependent ionic conductance changes which will be considered in Section 10.3.6.3. It is sometimes necessary to include the fact that the capacitance is lossy, probably because of the loss in the penetrating proteins (see Taylor et al.[7a]).

To a very large extent then, following a sudden change in V_m the membrane current will be purely ionic after the capacity transient. This transient will depend on the electronics employed in the feedback system used to clamp the voltage (Section 10.3.4).

The excitability properties of nerve and muscle membranes result from the fact that the membrane ionic conductances are voltage dependent. These conductances result from the flow of ions (most often sodium, potassium, and calcium) through imperfectly selective membrane channels, and the magnitude of the conductance depends upon the fraction of channels that are open (for reviews, see Ehrenstein and Lecar[8] and Taylor.[8a] For the squid axon membrane, the behavior is well described by the empirical equations of Hodgkin and Huxley,[9] and an excellent introduction can be found in Katz.[10] The important point for our purposes is that the current–voltage relations contain a region of negative resistance such that the system is stable under potential control.

One of the first attempts to determine the characteristics of a system of this kind using potential control was that of Bartlett[11] for an iron wire in contact with acid.[12-14] Bartlett was led to do this at the suggestion of K. S. Cole,[15] who later introduced this concept in the study of the squid giant axon with the use of electronic feedback.[7]

In order to proceed experimentally, it is necessary to establish conditions where a patch of membrane can be isolated over which the current and voltage are uniform (averaged over some smaller area); and much of the article will be concerned with this question. Cole[15] and Marmont[16] introduced techniques for isolating small regions of the squid giant axon membrane with the use of an internal current supplying electrode, and external guard system

[7a] R. E. Taylor, J. M. Fernández, and F. Bezanilla, *in* "The Biophysical Approach to Excitable Systems" (W. J. Adelman and L. Goldman, eds.). Plenum, New York, 1982.

[8] G. Ehrenstein and H. Lecar, *Ann. Rev. Biophys. Bioeng.* 1, 347 (1972).

[8a] R. E. Taylor, *Ann. Rev. Phys. Chem.* 25, 387 (1974).

[9] A. L. Hodgkin and A. F. Huxley, *J. Physiol.* (*London*) 117, 500 (1952).

[10] B. Katz, "Nerve, Muscle and Synapse." McGraw-Hill, New York, 1966.

[11] J. H. Bartlett, *Trans. Electrochem. Soc.* 87, 521 (1945).

[12] R. S. Lillie, *Biol. Rev. Cambridge Philos. Soc.* 11, 181 (1936).

[13] U. F. Franck and R. FitzHugh, *Z. Elektrochem.* 65, 156 (1961).

[14] R. Suzuki, *IEEE Trans. Bio-med. Eng.* 14, 114 (1967).

[15] K. S. Cole, *Arch. Sci. Physiol.* 3, 253 (1949).

[16] G. Marmont, *J. Cell. Comp. Physiol.* 34, 351 (1949).

and electronic feedback. This system was improved by Hodgkin *et al.*[17,18] by the addition of an internal voltage measuring electrode. They introduced the term "voltage clamp."

More recently, voltage clamp methods have been extended to other aspects of the membrane current. Some of the most notable recent extensions of the voltage clamp method are the measurements of channel noise, single-channel current jumps, and gating currents. Channel noise is the excess electrical noise created by the random opening and closing of the molecular ionic channels in the membrane. Under favorable circumstances the discrete current jumps caused by the opening and closing of single channels can be resolved.[18a] The gating currents are the displacement currents within the membrane which occur when charged groups of the channel macromolecule are rearranged during channel opening or closing.[18b]

In this paper we describe the principles of single-cell voltage clamping and discuss the basic theory and difficulties found in different preparations. We do not consider results of voltage clamp experiments. For reviews of voltage clamping in general, see Cole and Moore,[19] Moore and Cole,[20] Moore,[21] and Katz and Schwartz.[22] For reviews of results of voltage clamping, see Hodgkin,[2] Ehrenstein and Lecar,[8] Taylor,[8a] and Bezanilla and Vergara.[22a]

10.2. General Principles of Voltage Clamp

In an ideal system the current measured (without distortion) would be that which was flowing across a region of membrane, where the potential would instantaneously and accurately follow some time sequence (the command potential) determined by the experimenter.

The ideal system is never achieved in practice, but in many cases can be approximated quite closely. Some of the limitations of the real case can be illustrated by a simple example. Suppose we would like to voltage clamp a region of a cylindrical cell. In order to control the potential, it is necessary

[17] A. L. Hodgkin, A. F. Huxley, and B. Katz, *Arch. Sci. Physiol.* **3**, 129 (1949).

[18] A. L. Hodgkin, A. F. Huxley, and B. Katz, *J. Physiol. (London)* **116**, 424 (1952).

[18a] E. Neher and B. Sakmann, *Nature (London)* **260**, 779 (1976).

[18b] C. M. Armstrong and F. Bezanilla, *J. Gen. Physiol.* **63**, 533 (1974).

[19] K. S. Cole and J. W. Moore, *J. Gen. Physiol.* **44**, 123 (1960).

[20] J. W. Moore and K. S. Cole, *Phys. Tech. Biol. Res.* **6**, 263 (1963).

[21] J. W. Moore, *in* "Biophysics and Physiology of Excitable Membranes" (W. J. Adelman, ed.), p. 143. Van Nostrand-Reinhold, Princeton, New Jersey, 1971.

[22] G. Katz and T. L. Schwartz, *J. Membr. Biol.* **17**, 275 (1974).

[22a] F. Bezanilla and J. Vergara, *in* "Membrane Structure and Function" (E. E. Bittar, ed.), Vol. 2, p. 53. Wiley (Interscience), New York, 1980.

(a)

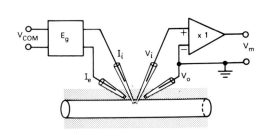

(b)

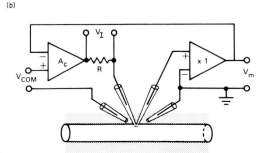

Fig. 1. Principle of controlling membrane voltage. (a) Schematic diagram of a system without feedback. (b) Schematic diagram of a voltage clamp system with negative feedback.

to measure it. We could introduce a microelectrode (Fig. 1a) and measure the membrane potential as the difference between the internal potential V_i and the external potential V_o, and supply current by another impaled micro-electrode (I_i) and collect it with an external electrode (I_e). If the voltage generator (E_g) had zero output impedance and there were no resistance in the current path through the electrodes, we might be able to record a membrane voltage V_m identical to the command V_{COM}. These conditions are seldom possible to achieve.

We can improve this arrangement by use of negative feedback (Fig. 1b), where the measured voltage V_m is compared to the command V_{COM}, and the difference is amplified by an amplifier A_c that supplies the current necessary to make $V_m - V_{COM} = 0$. The current I_m required to control the voltage can be measured as the voltage drop V_I across a resistance R. Clearly $I_m = V_I/R$.

For a cylindrical portion of a cell the arrangement of Fig. 1b suffers from a serious defect. Although the potential at the point of the impaled voltage measuring microelectrode may be well controlled, the measured current comes not only from the immediate neighborhood of this electrode, but also from other portions of the cell which may have nonuniform potential distributions. Under these conditions, the current is no longer being measured from a region of known and controlled potential. The errors could be more than just quantitative if the uncontrolled regions exhibit instabilities.[23] Spatial uniformity of current flow may be impossible to obtain for a membrane with nonuniform properties, but spatial uniformity of voltage can be approximated; when it is, it is referred to as a "*space clamp.*" Depending on the type and geometry of cells, the space clamp condition is obtained with a variety of methods. Axial wire is considered in Chapter 10.3; attempts with two and three microelectrodes in Chapter 10.4; patch isolation with pipettes in Chapter 10.5; and gap isolation is discussed in Chapter 10.6.

In the following discussion we assume perfect space clamp, and thus we picture the membrane subject to voltage clamp as a uniform patch of membrane. Later, we consider cases where the ideal is not met.

The basic circuit is pictured in Fig. 2. In this diagram we have represented one membrane patch as a box. The external electrode has a resistance of R_{eo} and the internal electrode a resistance R_{ei}. It is very common for the electrodes to be located at a certain distance from the surface of the membrane because there is connective tissue or adventitious cells surrounding the membrane under study. The electrical equivalent of these structures is a resistance in series with the membrane represented as R_s in Fig. 2. The resistance of the current electrode is represented by R_{ce}. A_c is an operational amplifier with open-loop gain A, and A_D is a differential amplifier with gain of 1. It is instructive to start with an idealized system where the input impedance of both amplifiers is infinite and the differential amplifier has a flat frequency response. Under these circumstances it is possible to derive the equation that relates V_M (the actual membrane voltage) to $-V_{COM}$ (the command voltage). The final result is

$$V_M = -\frac{-V_{COM} - I_m R_s - 2I_m(R_{ce} + R_s)/A}{1 + 2/A}.$$ (10.2.1)

This equation shows that the accuracy of the clamp not only depends on A but also includes membrane properties (unless R_{ce} and R_s are zero). In the ideal case, $A \to \infty$ for all frequencies, and

$$V_M = V_{COM} + I_m R_s.$$ (10.2.2)

Note that the actual membrane voltage is *never equal to the command voltage unless the membrane current or the series resistance is zero*. However, V_m,

[23] R. E. Taylor, J. W. Moore, and K. S. Cole, *Biophys. J.* **1**, 161 (1960).

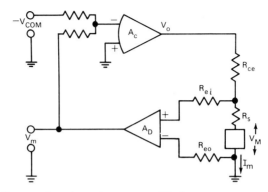

FIG. 2. Basic diagram of a voltage clamp system. The box represents a patch of membrane. V_M is the actual membrane potential, V_m is the measured membrane potential, I_m is the membrane current, and V_{COM} is the command voltage. R_{ei}, R_{eo} and R_{ce} are the resistances of external voltage, internal voltage, and current electrodes, respectively. R_s is the resistance in series with the membrane.

the measured voltage, will be equal to the command voltage because $V_m = V_M - I_m R_s$. The effect of series resistance can be serious in the determination of the current–voltage characteristics of a membrane because the voltage across the membrane is not controlled and becomes current dependent.[18] For this reason the voltage electrodes should be as close as possible to the membrane surface. However, in some cases the biological preparation contains a permanent barrier that effectively prevents positioning the electrodes right at the surface; in this case, some positive feedback can be introduced to compensate for the effect of series resistance as we shall see in the chapter on axial-wire voltage clamping (Section 10.3.5).

In modern operational amplifiers A is very large at dc and low frequencies, and Eq. (10.2.2) is a good approximation for that frequency range. However, at high frequencies A decreases, and Eq. (10.2.1) should be used. In practice this means that all fast changes in imposed V_{COM} will not be followed by the membrane as can be seen by inspection of Eq. (10.2.1) when A is decreased, and the clamp will perform far from ideal. In this situation the current-electrode resistance R_{ce} becomes very important.

Another very important source of error in practical voltage clamp originates in the capacitances of the voltage electrodes, which, in combination with their resistances, act as low-pass filters. This filtering effect is normally aggravated by the differential amplifier, which never has a flat frequency response as assumed in the example. Under these conditions, it is clear that the control amplifier will not receive the correct measurement of membrane voltage, but it will be distorted at high frequencies; this, in turn, will have the effect of producing a distorted error signal at the input of the control amplifier, which will be unable to clamp the membrane voltage at the commanded value. In some extreme situations the phase lag introduced by the measuring-electrodes–amplifier–membrane combination may be enough to render the whole system unstable. Some stability characteristics of axial-wire voltage clamping are discussed in Section 10.3.5.3. Attempts to decrease the response time usually increase the tendency of the clamp to oscillate, as does series-resistance compensation. We consider below some detailed stability analyses designed to produce a clamp system with series-resistance compensation and fast rise time without overshoots or oscillations in the membrane potential.

The above example illustrates the main features of a basic voltage clamp system which we can now summarize:

1. Measurement of membrane potential should be made with low-impedance electrodes and positioned as close as possible to the membrane.

2. The current should be measured from a region where the voltage is controlled and uniform.

3. Current electrodes should be of low impedance.

4. Amplifiers should introduce minimum alterations and phase shift in the frequency range of interest.

10.3. Axial-Wire Voltage Clamp

In the case of long cylindrical cells, the space clamp condition can be approximated with the use of a long wire inserted axially as described in Section 10.3.1. This technique is restricted to cells of diameters large enough to allow penetration of the axial wire without damage to the membrane properties.

The large diameter of the giant axon of the squid makes it almost an ideal preparation to study the electrical properties of excitable membranes. In fact, the first voltage clamp system was built to control the membrane potential of squid axons.[15,16] Hodgkin et al.[17,18] and Hodgkin and Huxley[9] made their classic description of the ionic currents using a voltage clamp system in the squid giant axon. In the squid axon it is possible to introduce relatively large electrodes from one cut end of the fiber. A longitudinal low-resistance current electrode can also be introduced from the ends (axial wire) and serves the dual purpose of passing current and attaining space clamp conditions.

The axial-wire voltage clamp has been also used in other preparations such as *Myxicola* axons,[24] crayfish axons,[25] and barnacle muscle fibers.[26,27]

10.3.1. Cable Theory of an Axon with Axial Wire in Voltage Clamp

The axial wire is usually made of a solid platinum wire coated with platinum black. As a metal its resistance is very low, but when it is in contact with an electrolytic solution like the axoplasm, its surface impedance can be large, depending on frequency and the amount and direction of current passing through the metal–solution interface. The minimum representation of an axial wire inside an axon must include the surface resistance of the wire; an equivalent circuit[23] of a segment of length Δx is presented in Fig. 3a, where we have assumed zero external resistance to simplify treatment.† Note that points A and B are at the same potential V_a; therefore the representation of Fig. 3b is still equivalent which, in turn, can be represented in equivalent form by Fig. 3c. This is the familiar representation of a cable in which the membrane element (the box in the figure) is in parallel with a series combination of a voltage generator V_a (the axial-wire voltage) and a resistance r_a (the surface resistance per unit length of the axial wire).‡ Now we can find the value of the space constant of the combination axon and axial wire. For this purpose we represent the membrane by a Thevenin equivalent of a battery ε_m in series with the membrane resistance r_m as pictured in Fig. 4a. The final step is to obtain the equivalent of Fig. 4a as pictured in Fig. 4b, where

$$r = r_m r_a/(r_m + r_a), \qquad \varepsilon = (V_a r_m + \varepsilon_m r_a)/(r_a + r_m);$$

r is the parallel combination of the membrane resistance and the wire surface resistance. A good axial wire will have r_a much smaller than r_m; therefore r will be practically equal to r_a; furthermore, from the above

[24] L. Binstock and L. Goldman, *J. Gen. Physiol.* **54**, 730 (1969).

[25] P. Shrager, *J. Gen. Physiol.* **64**, 666 (1974).

[26] S. Hagiwara, H. Hayashi, and K. Takahashi, *J. Physiol.* (*London*) **205**, 115 (1969).

[27] R. D. Keynes, E. Rojas, R. E. Taylor, and J. Vergara, *J. Physiol.* (*London*) **229**, 409 (1973).

---

† The case of finite external resistance is given in Taylor *et al.*[23]

‡ In a more general treatment the surface impedance per unit length of the axial wire, z_a, replaces r_a; it is defined as the Laplace transform of the voltage divided by the Laplace transform of the current per unit length of wire. The same equations can be used by simply replacing r_a by z_a and the time solution can be obtained by the use of tables or the complex inversion formula.

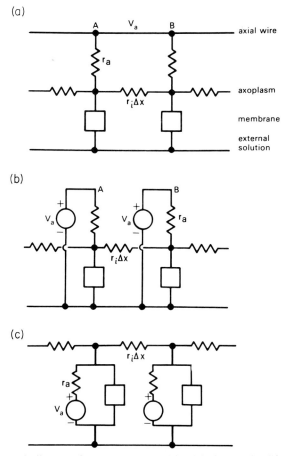

FIG. 3. Schematic diagram of an axon segment with axial wire: (a) simplified representation before reduction; (b) and (c) equivalent circuit after circuit reduction. For details see text.

equation it can be seen that ε will be very close to V_a. The space constant of the axon–axial wire combination will be

$$\lambda = \sqrt{r/r_i}. \tag{10.3.1}$$

The above equations show that as the axial wire is made of lower surface resistance, λ decreases. This is in direct contradiction with the general idea that the introduction of an axial wire in an axon makes its space constant larger. It is clear, however, from the derivation presented that the effect of the axial wire is exactly the opposite; it shortens the space constant, making each patch of membrane more independent of its neighbors but at the same time imposing a voltage V_a in each patch. This effect is desirable to isolate

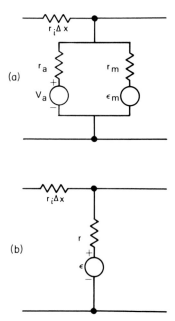

FIG. 4. Thevenin equivalent of a segment of an axon of length Δx containing an axial wire connected to a voltage source at potential V_a. (a) Equivalent circuit of a patch with axial wire. (b) Thevenin equivalent of a membrane patch with axial wire.

patches of membrane that do not have exactly the same properties as the rest of the membrane; in particular, it isolates end effects. This can best be illustrated solving the cable equation for a terminated cable using the circuit elements indicated in Fig. 4b.

The equation to be solved is

$$d^2V/dx^2 = (V - \varepsilon)/\lambda^2, \qquad (10.3.2)$$

where steady state and a constant value for r are assumed.† Normally, the axon is penetrated in both ends with electrodes and perfusion cannula. We can approximate the boundary conditions by the expression

$$V(x) = -\frac{1}{r_i}\frac{\partial V}{\partial x}R_g \quad \text{at} \quad x = 0 \quad \text{and} \quad x = l, \qquad (10.3.3)$$

where R_g is the short-circuit resistance at the ends and r_i is the internal resistance per unit length.

---

†Note that when the membrane resistance is negative, λ will be imaginary. However, the equations still hold and the solutions will contain trigonometric sines and cosines. For a discussion of this case see Cole[7] (p. 342).

The solution of the differential equation (10.3.2) with the above boundary condition (10.3.3) is

$$V = \varepsilon\left(1 - \frac{\sinh(x/\lambda)}{[1 + R_g/r_i\lambda]\sinh(l/\lambda)}\right.$$

$$\left. - \frac{\sinh[(l - x)/\lambda] + (R_g/r_i\lambda)\cosh[(l - x)/\lambda]}{[1 - (R_g/r_i\lambda)^2]\sinh(l/\lambda)}\right). \quad (10.3.4)$$

To simplify the situation, let us assume further that $R_g = 0$, which is equivalent to a short circuit at both ends. The membrane potential distribution is given by

$$V = \varepsilon\left(1 - \frac{\sinh(x/\lambda) + \sinh[(l - x)/\lambda]}{\sinh(l/\lambda)}\right). \quad (10.3.5)$$

As the membrane is voltage clamped, the membrane potential in the center of the fiber is equal to the command voltage V_{COM}

$$V_{COM} = V(l/2);$$

then

$$\varepsilon = V_{COM}\bigg/\left(1 - \frac{2\sinh(l/2\lambda)}{\sinh(l/\lambda)}\right),$$

and the voltage distribution in the voltage clamp will be

$$V(x) = V_{COM}\left(\frac{\sinh(l/\lambda) - \sinh(x/\lambda) - \sinh[(l - x)/\lambda]}{\sinh(l/\lambda) - 2\sinh(l/2\lambda)}\right). \quad (10.3.6)$$

Figure 5 shows the distribution of V/V_{COM} along the axon for different values of the space constant. It is clear that the smaller the space constant the more homogeneous is the potential along the axon, because end effects are circumscribed to smaller regions. Therefore the best space clamp is achieved when the surface resistance of the axial wire r_a is lowest. Very low-resistance axial wires can be prepared by platinizing the platinum wire,[20] but their resistance becomes very high when steady current is passed. This latter situation appears frequently when an axon is held at a potential different from its resting potential; under these conditions the space clamp may be far from ideal.

10.3.2. Giant Axon Preparation

As explained above, the axial-wire technique can only be used on large cells, and it has been successful with squid giant axon, crayfish axon, *Myxicola* axon, and barnacle muscle fibers. The experimental procedures are

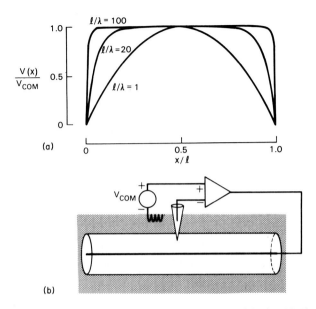

FIG. 5. Potential distribution inside an axon containing an axial wire. (a) The normalized steady-state potential distribution as calculated from Eq. (10.3.6) for three different values of l/λ, where l is the length of the fiber and λ is the space constant. The axon is voltage clamped, as indicated in (b), and the potential is controlled in the center.

similar for all these preparations. To familiarize the reader with this technique, we shall briefly describe the squid giant axon preparation.

10.3.2.1. Isolation of the Giant Axon. The giant axon from the stellate ganglion has been obtained from several species of squid: *Loligo forbesi, Loligo pealei, Loligo vulgaris, Loligo opalescens, Dosidicus gigas, Doritheutus plei,* and others. The axon diameter varies among different species and ranges between 200 nm and 1.3 mm. Normally, axons are dissected out of the squid mantle in running seawater, and they are later cleaned of other fibers and connective tissue under microscope dissection. For a more detailed description see, for example, Gilbert.[28] Cleaned segments from 2 to 8 cm long are tied at the ends with threads and can be kept for several hours at 4–8°C in seawater.

10.3.2.2. Experimental Chamber and Internal Perfusion. The details of the experimental chamber differ depending on whether the axon is held horizontally or vertically. Several designs have been used successfully.[18, 20, 29, 30]

[28] D. L. Gilbert, *in* "A Guide to Laboratory Use of the Squid, *Loligo pealei.*" Mar. Biol. Lab., Woods Hole, Massachusetts, 1974.

[29] W. K. Chandler and H. Meves, *J. Physiol. (London)* **180**, 788 (1965).

[30] C. M. Armstrong, F. Bezanilla, and E. Rojas, *J. Gen. Physiol.* **62**, 375 (1973).

The method used to internally perfuse the axon has also influenced the design of the chamber. Two general types of perfusion methods have been described: the roller technique originally described by Baker *et al.*,[31] in which the axoplasm is extruded by a roller and the axon later reinflated by the perfusion solution; the cannulation technique originally described by Oikawa *et al.*,[32] in which the axoplasm is sucked into a cannula as the cannula is advanced along the axon.

In the following, we briefly describe the Tasaki technique,[32] as modified by Fishman,[33] to illustrate one of the methods to perfuse and voltage clamp a squid axon. The axon segment is positioned in the chamber as illustrated in Fig. 6, which is based on the chamber designed by Armstrong.[30] A cut is made with a microscissors at point A, and a glass cannula (PC) is inserted into the axon and advanced with the aid of a micromanipulator. As the cannula is advanced, microscope observation is required to maintain the cannula centered in the axis of the axon. Normally, a small prism is used to visualize the position of the cannula in the vertical plane. Simultaneously with the advancing of the cannula, mouth suction is applied, and the axoplasm is sucked into the cannula lumen. At point B another cut is made, and the cannula is pushed outside the axon. The axoplasm inside the cannula is blown away, and the flow of the internal perfusion solution is started through the cannula. The composite electrode described in Section 10.3.2.3 is introduced partially inside the cannula, and then the cannula is pulled back as the electrode is pushed in. Normally, the air gap (region between B and C) is treated with a protease, such as papain or pronase, for about one and a half minutes by positioning the tip of the cannula at C and letting the internal perfusion solution containing the enzyme flow freely. Finally, the cannula is retrieved almost to point A as the electrode is pushed to the final position that requires the tip of the voltage pipette to be in the center of the chamber.

10.3.2.3. Internal Electrodes. Several types of internal electrodes have been used for voltage clamping. The classical combination electrode described by Hodgkin *et al.*[18] consists of two silver-chlorided silver wires twisted into a double spiral on the glass rod. One of the wires was used to measure potential (voltage electrode) and the other was used to pass current (axial wire). The main disadvantage of this electrode is that inside the unperfused axon it does not measure steady dc potential because the voltage electrode is not stable. One of the most commonly used combination

[31] P. F. Baker, A. L. Hodgkin, and T. I. Shaw, *J. Physiol. (London)* **164**, 330 (1962).

[32] T. Oikawa, C. S. Sypropoulos, I. Tasaki, and T. Teorell, *Acta Physiol. Scand.* **52**, 195 (1961).

[33] H. M. Fishman, *Biophys. J.* **10**, 799 (1970).

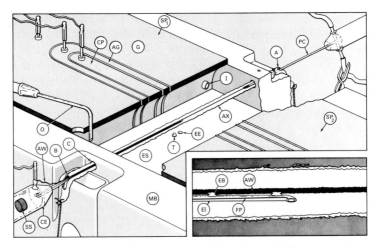

FIG. 6. Squid axon chamber. The main body of the chamber (MB) is made of Lucite. The two silver plates SP_1 and SP_2 are movable and can be approached to enclose the axon (AX). SP_1 is thermally connected to a Peltier cooler that is connected to a feedback loop to control the chamber temperature measured by thermistor T. The external solution (ES) is precooled by passing it inside the silver block before it enters the chamber through inlet I. The solution is sucked away by outlet O. The plates are made of several sections (G, guard; AG, auxiliary guard; CP, central or measuring plate) separated among them by thin sheets of Mylar. The faces of the plates in contact with the external solution are platinized (darker regions). The perfusion cannula (PC) enters in the axon at point A and the combination electrode (CE) at point B. Inside the combination electrode the silver–silver chloride pellet (SS) can be observed. The tip of the external electrode is at the bottom in the center of the chamber (EE). (The length of the perfusion cannula PC has been drawn shorter to fit the diagram. Its real length should span at least from point A to B, which is about 20 mm.) Inset: detail of the internal combination electrode inside the axon. AW is the axial wire, EI is the cannula of the potential-measuring electrode, and FP is a floating platinum wire. The glass cannula EI is glued to the platinized axial wire AW with small epoxy beads (EB).

electrodes is that described by Chandler and Meves,[29] which consists of a platinized platinum wire (axial wire) used to pass current and a cannula attached to it to measure the potential. The cannula is normally filled with 0.6 M KCl and acts as a salt bridge between the axon interior and a reversible hemicell of Ag–AgCl or calomel. To decrease the high-frequency impedance of the cannula, a floating platinum wire is positioned inside (Fig. 6). Sometimes the floating Pt wire is platinized.[34]

10.3.2.4. External Electrodes. The external voltage electrode is simply a salt bridge with one end positioned as close as possible to the axon and the other end in contact with a Ag–AgCl or calomel hemicell.

[34] H. M. Fishman, *IEEE Trans. Biomed. Eng.* **BME-20**, 380 (1973).

The external current electrode has been made in several different ways. The most common approach is to position two platinized surfaces made of silver or platinum on either side of the axon. Each surface is divided into three electrically isolated plates: a central electrode or measuring electrode and two lateral electrodes also called guards. These large metal electrodes both pass current and, at the same time, contribute to the space clamping. Normally, the guard electrodes are grounded and the central electrode is held at virtual ground and current is measured only with that electrode. See Fig. 6 for an example of a central plate and several guards used to verify homogeneity along the axon.

10.3.3. Measurement of Membrane Current

As mentioned earlier, one would like to collect all of the current flowing through a region of membrane over which the voltage is uniform; several different methods have evolved for doing this. We shall discuss the use of lateral guards, external differential electrodes, external pipettes, and gap isolation techniques (Chapter 10.6).

When an axial wire is used to supply current, the effects of voltage non-uniformities near the ends can be reduced with the use of lateral guards. In this method three external electrodes are employed and the current measured only from the central one, either by passing it through an external resistance and measuring the potential across it,[16,17] or by means of a current-to-voltage converter.[19] Various modifications of the precise form of the external electrodes have been used,[19,35] and the type in most common use today[36,37] consists of three pairs of electrodes forming the sides of a rectangular trough. In some systems the three chambers are separated by partitions, and some Vaseline/oil mixture is used for insulation. Except for considerations of noise (Section 10.3.4), it is not clear that this is an improvement over using no partitions. If the axon is uniform and the guards are long enough, the partitions are not needed; unless the insulation is quite perfect, any current flowing through the gap could produce voltage drops across the membrane which might aggravate the spatial nonuniformities. A system with thin plastic partitions used to record ionic and gating currents has been recently reported.[46]

Regardless of the arrangement for measuring current, there is the important experimental question of how one knows that the current distribution is indeed uniform. This has been investigated with the use of two small closely spaced external differential electrodes for measuring current density over

[35] I. Tasaki and S. Hagiwara, *J. Gen. Physiol.* **40**, 859 (1957).

[36] E. Rojas and G. Ehrenstein, *J. Cell. Comp. Physiol.* **66**, Suppl. 2, 71 (1965).

[37] E. Rojas, R. E. Taylor, I. Atwater, and F. Bezanilla, *J. Gen. Physiol.* **54**, 532 (1969).

small regions of the squid giant axon.[19,20,23,37] At first this would seem an ideal way to measure current, but there are problems of sensitivity, noise, and calibration. It is still true, however, that we know of no other way to check membrane current uniformity except with the use of multiple external longitudinal electrodes (Section 10.3.2.4 and Fig. 6). Narrow C-shaped external electrodes have been used for current measurement[29] and for checking uniformity in the case of radioactive tracer measurements where the use of guards was not possible.[37a]

We do not know of a theoretical analysis for the precise geometry employed with external differential electrodes. Chandler and Meves[29] present an approximate solution in connection with their method of calibration using the simplifying assumption that the two external electrodes see the current arising from a point within the axon, as if they were on two equipotential concentric spheres. We have derived that an electrode in a medium of resistivity R_e (Ω cm) at a radial distance r from the center of a cylinder of radius a and longitudinal position z has a potential due to a thin ring of current at $z = 0$ of

$$V(r, z) = \frac{i_0 R_e}{\pi} \int_0^\infty \frac{K_0(\omega r)}{\omega K_1(\omega a)} \cos(\omega z)\, d\omega, \qquad (10.3.7)$$

where K_0 and K_1 are modified Bessel functions. Numerical computations of Eq. (10.3.7) give results that are different from those obtained with the simplified model of Chandler and Meves.[29]

Davies[38] and Jaffe and Nuccitelli[39] have used a single vibrating electrode in place of the differential pair. It would seem from the analyses that have been done that the potential of a single external electrode close to the membrane would be a good approximation to the underlying membrane current.

Another way to measure membrane current is to apply an external pipette to the surface. If the internal resistance of the pipette is small compared to the leak around the tip, the method is successful (Chapter 10.5). One must either use a cell whose membrane is not covered by "extraneous coats" (i.e., single muscle fibers treated with collagenase or tissue-cultured cells) or bathe the region around the tip with sucrose solutions which can penetrate the connective tissue or Schwann cell layer. The use of suction can be helpful (Chapter 10.5). If the tip is small enough, it is possible to observe single channels.[18a] This is extremely important in that it not only allows direct determination of the conductance of single channels but also provides a powerful tool for studying channel mechanisms.

[37a] I. Atwater, F. Bezanilla, and E. Rojas, *J. Physiol. (London)* **201**, 657 (1969).

[38] P. W. Davies, *Fed. Proc., Fed. Am. Soc. Exp. Biol.* **25**, 332 (1966).

[39] L. F. Jaffe and R. Nuccitelli, *J. Cell Biol.* **63**, 614 (1974).

The last, and very little tried, method is the use of a pipette (Westerfield,[40] Llano and Bezanilla[41]) or metal electrode (Taylor and Bezanilla[41a]) which is connected to a current-to-voltage converter and draws off some of the current on its way to the main external electrode. The so-called gap isolation techniques are probably the most widely used methods of measuring current at this time and are extensively discussed in Chapter 10.6.

10.3.4. Electronics for the Voltage Clamp System

Many voltage clamp diagrams used for squid axons have been published. The system used by Hodgkin et al.[18] has a very short settling time, but it cannot control the membrane potential for long periods of time because the voltage electrode does not measure steady-state potentials (see Section 10.3.2.3). However, the system is perfectly appropriate to impose changes of potential across the axonal membrane. The major changes introduced by Moore and Cole[20] were to incorporate a microelectrode to measure the true membrane potential and the use of operational amplifiers. Since then, operational-amplifier performance has improved noticeably, and very fast clamps can now be built with commercially available units using the combination electrode described in Section 10.3.2.3.

Figure 7 is a schematic diagram of one possible voltage clamp system that has been used successfully in our laboratory. The membrane potential is measured by the differential combination A_1, A_2, and A_3. Special care must be exercised to minimize capacitative loading at the input to prevent filtering of high-frequency components. The membrane voltage is summed to the negative of the steady membrane potential desired (the holding potential HP) plus the pulses or waveforms to be imposed on the axon (the imposed waveform V_p) at the summing junction of operational amplifier A_4. The combination C_V and R_V act as a lead compensator and C_L and R_L as a stabilizing network. We have also (at the suggestion of R. Levis) found that the use of a lead compensator in the current feedback for series-resistance compensation (Section 10.3.5) provided by C_I is helpful. The output of the operational amplifier A_4 is connected directly to the axial wire when the voltage clamp is on or connected to an auxiliary feedback when the clamp is off. The voltage clamp can be turned on remotely by replacing the switch S_1 by two field-effect transistors connected as switches or using two bipolar transistors.[42] This electronic switch makes it possible to interrupt the free course of the action potential to study the ionic conductances during the

[40] M. Westerfield, personal communication.
[41] I. Llano and F. Bezanilla, *Proc. Natl. Acad. Sci. U.S.A.* **77**, 12 (1980).
[41a] R. E. Taylor and F. Bezanilla, unpublished observations.
[42] F. Bezanilla, E. Rojas, and R. E. Taylor, *J. Physiol. (London)* **211**, 729 (1970).

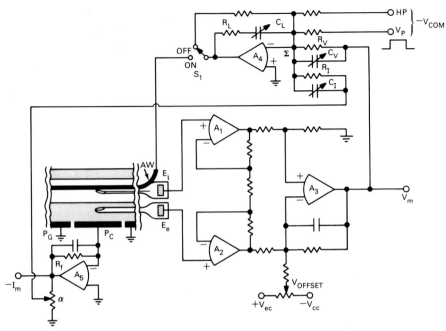

FIG. 7. Schematic diagram of a squid-axon voltage clamp. A_1 and A_2 are very high-input-impedance, low-input-capacitance, wide-bandwidth operational amplifiers wired in a voltage follower configuration to measure the potential of the internal electrode E_i and external electrode E_e. A_3 is an operational amplifier connected as a differential amplifier giving the difference between the outputs of A_1 and A_2 as V_m, the membrane voltage. V_{OFFSET} is used to compensate for electrode and junction potentials. Amplifier A_4 is the control amplifier, and its summing junction $\sum$ sums the negative of the desired potential (V_{COM}), which consists of a steady potential (HP) and a waveform (V_p) such as a pulse pattern, the membrane voltage (V_m), and a fraction of the negative of membrane current ($-\alpha I_m$) for series resistance compensation. Note that V_m is connected to $\sum$ through R_V in parallel, with C_V acting as a lead compensator. C_L and R_L are a stabilizing network. S_1 is a switch that connects the output to either the axial wire (AW): clamp in position ON, or to an auxiliary feedback loop: clamp in position OFF. Membrane current derived by the central plate (P_C) is connected to the negative input of operational amplifier A_5, whose output will be equal to the negative of the current derived by P_C times the feedback resistor R_f. The high gain of A_5 guarantees that P_C is held at the same ground potential as the guard plates P_G, which are connected directly to ground; however, at high frequencies the gain decreases and P_C may not be at ground potential Examples of commercially available amplifiers A_1 and A_2 are National Semiconductors LF356; of A_3, A_4 and A_5 are National Semiconductors LF357.

spike. Also, it can be used to protect the axon from electrical transients that could drive the membrane potential beyond safe limits. This is accomplished by measuring the membrane potential at all times and activating a bistable flip-flop whenever the membrane potential reaches unsafe limits. The output of the flip-flop is used to control the electronic switch to open the clamp loop.

The current amplifier A_5 is connected as a current-to-voltage converter with the input connected directly to the measuring plate (central). This makes the central plate potential equal to ground potential when the amplifier has a very large gain. However, it must be remembered that the open-loop gain of operational amplifiers decreases with frequency, and therefore at high frequencies the central plate will not be at ground potential, making a gradient of potential in the external solution along the axon (see Section 10.4.1). It is important then to select an amplifier with a large gain–bandwidth product to maintain space clamp over the entire frequency range of interest during voltage clamping.

Unless partitions are used, the resistance between the central plates and the guards is quite low because it is given by the external solution. This low-resistance pathway between the summing junction and ground drains a large current from the input voltage noise generator of the operational amplifier, producing significant noise at the output of the current amplifier. Frequently, this is the predominant source of noise in the voltage clamp system, and it can be minimized using current amplifiers with low input voltage noise.

Levis[43] has analyzed the noise performance for a clamp and concluded that the major source of noise is the voltage measuring electrodes if the resistance between the center chamber and the guard chambers is high. If this resistance is low (10 Ω, say), this may become the major source of noise. This problem may be avoided by not using guards[44,45] or by the use of insulating partitions.[46]

10.3.5. Series-Resistance Compensation

As mentioned above, the electrodes should sense the potential as close as possible to the membrane. In the squid axon, the Schwann cell layer constitutes a barrier that cannot be penetrated by the electrodes, and a significant resistance is included in series with the axolemma. The origins of this resistance are the narrow clefts between Schwann cells, which amounts to about $4 \ \Omega \ cm^2$,[18,47,56] but it varies considerably with the type of external solution and possibly with the state of the axon.

Hodgkin et al.[18] introduced in their clamp circuitry positive feedback that subtracted the voltage drop across the series resistance (compensated

[43] R. Levis, Doctoral Dissertation Department of Physiology, University of California, Los Angeles, 1981.

[44] E. Wanke, L. J. DeFelice, and F. Conti, *Pfluegers Arch.* **347**, 63 (1974).

[45] F. Conti, L. J. DeFelice, and E. Wanke, *J. Physiol (London)* **248**, 45 (1975).

[46] F. Bezanilla, R. E. Taylor, and J. M. Fernández, *J. Gen. Physiol.* **79**, 21 (1982).

[47] L. Binstock, W. J. Adelman, Jr., P. Senft, and H. Lecar, *J. Membr. Biol.* **21**, 25 (1975).

feedback). In Fig. 7 a similar arrangement has been incorporated. A voltage proportional to a fraction of the membrane current equal to $-\alpha I_m R_f$ is added to the measured V_m at the summing junction of amplifier A_4. If α is adjusted to make this voltage equal to the $I_m R_s$ voltage, the control amplifier A_4 will impose the command voltage plus holding potential on $V_m - I_m R_s$, which is the real membrane potential V_M [see Fig. 2 and Eq. (10.2.2)].

10.3.5.1. Effects of Series Resistance. If the electronics yielded a perfect clamp, i.e., the measured potential faithfully followed the command potential, there would still be errors introduced by uncompensated series resistance. These errors have been discussed by many people including Hodgkin et al.[18] (pp. 430, 435), Taylor et al.,[23] and Binstock et al.[47] We may distinguish three cases: (i) passive membranes which may be described by models containing elements which do not vary with voltage, current, or time; (ii) active membranes where the region of interest contains only current–voltage curves at any given time (isochronal curves), which are straight lines with positive slope; and (iii) systems in which the isochronal current–voltage curves in some region are either curved and varying with time or there is a transient or steady-state negative resistance region. In the first two cases, the data as recorded could be corrected with the use of a load line for a known series resistance. In case (iii), it may not be possible to make corrections. Not much can be added to the discussion of these points in Taylor et al.,[23] and the computer plots of current versus time for solutions of the equations of Hodgkin and Huxley in Binstock et al.[47] For more detailed studies of stability in the negative-resistance region as affected by series resistance, and the stability of the second patch in the model of Taylor, see Chandler et al.[48]

We may summarize here by saying that in studies of the squid giant axon in voltage clamp, the addition of series resistance produces changes in the shape of the current–time curve in response to an applied voltage step and changes in the peak inward and steady-state outward current–voltage curves. Load-line corrections, as predicted, correct for the linear portions of the peak inward and steady-state outward curves, but not for the maximum peak inward current or the shape of the peak inward current curves in the negative-resistance region. The time to peak of the transient inward current was corrected by the use of a load line over most of the curve. Taylor et al.[23] also demonstrated that, for their system, addition of negative-resistance compensation did correct, to a large degree, for the effects of removing the external reference electrode, which was, in fact, an addition of series resistance.

[48] W. K. Chandler, F. FitzHugh, and K. S. Cole, Biophys. J. 2, 105 (1962).

It is then advisable to perform experiments using series-resistance compensation in order to measure the real current–voltage characteristics of the membrane. The first step to compensate for series resistance is an accurate determination of its value; we address this problem in the next section

10.3.5.2. Determination of Series Resistance. The measurement of the resistance in series with the membrane and between the voltage measuring electrodes would be a fairly simple matter if (i) the membrane capacitances were loss free, (ii) the series elements were pure resistance, and (iii) the applied pulses and measurements were not distorted by electrode impedances and amplifier delays. In this case, the most accurate way would probably be to measure the impedance and extrapolate to infinite frequency. The most rapid would be to apply a rectangular pulse of current and extrapolate the measured voltage to zero time. It would also be possible to apply a voltage clamp pulse and fit the measured current. All of these methods have been used, with variable and confusing results.

10.3.5.2.1. LOSSY CAPACITANCE. Curtis and Cole,[6] using transverse currents with external electrodes, measured the impedance of the axon membrane and concluded that the capacitance was lossy and could be approximated over their frequency range by a constant-phase-angle impedance of the form $Z_{cc} = Z^* (j\omega\tau)^{-\alpha}$, where $j = \sqrt{-1}$, $\omega = 2\pi f$, and f is the frequency. They reported an average α of 0.85, representing a constant phase angle of $\phi = \alpha (90°) = 76°$. Using internal and external electrodes, Hodgkin et al.[18] concluded that their capacity transient for a voltage clamp pulse was roughly consistent with a phase angle of 80° (but see FitzHugh and Cole[49]). Taylor and Chandler,[50] using internal and external electrodes with a bridge, found that between 10 and 70 kHz the impedance closely fit the constant-phase-angle expression above, with comparable values for α.[51] Taylor[52] later showed that the same data could be fairly well approximated by a single relaxation-time Debye-type dielectric (and fit very well for a moderate distribution of relaxation times). FitzHugh and Cole[49] have computed voltage and current transients for the constant-phase-angle capacitance with parallel leakage conductance and series resistance. Their results could be compared with experimental results if such were available. A treatment of the transient response to an applied current of the form $I(t) = I_0(1 - e^{-t/\tau})$ can be found in Binstock et al.[47] for a capacitative element with a single-relaxation Debye-type dielectric with parallel leakage and series resistance.

[49] R. FitzHugh and K. S. Cole, Biophys. J. 13, 1125 (1973).
[50] R. E. Taylor and W. K. Chandler, Biophys. Soc. (Abstr.) TD1 (1962).
[51] N. Matsumoto, I. Inoue, and U. Kishimoto, Jpn. J. Physiol. 20, 516 (1970).
[52] R. E. Taylor, J. Cell. Comp. Physiol. 66, 21 (1965).

10.3.5.2.2. SERIES IMPEDANCE. The squid giant axon is surrounded by a Schwann cell layer which would not have a pure-resistance impedance. This is probably not serious in most cases because it approximates a pure resistance for frequencies below 100 kHz.[53]

10.3.5.2.3. FINITE RESPONSE TIME. Hodgkin et al.[18] reported the results of two experiments in which they determined the membrane capacitance and series resistance for the squid giant axon membrane using rectangular pulses of current. They considered that the amplifiers for measuring current and voltage had equal response times and that the effect would cancel. This is an important concept so we shall elaborate somewhat. Consider Fig. 7. The membrane potential is given by $V_M = I_m R_s + (1/C_m)_0 \int_0^t I_m \, dt'$. Denoting the Laplace transform with argument s by an overbar, for $I_m(0) = 0$ we find

$$\bar{V}_M(s) = \bar{I}_m(s)R_s + (1/C_m)\bar{I}_m(s)/s.$$

Say the electrodes plus amplifiers distort the measurement of V_M and I_m into V'_M and I'_m, where the transfer functions are given by $Y_v(s) = \bar{V}'_m(s)/\bar{V}_M(s)$ and $Y_i(s) = \bar{I}'_m(s)/\bar{I}_m(s)$. If $Y_v(s) = Y_i(s)$, they will indeed cancel, and we may use the equation we started with, but using the measured $V'_M(t)$ and $I'_m(t)$. In principle, for any applied current we may fit the measured curves and obtain values for C_m and R_s.

One way to reduce the effects of amplifier delays would be to slow down the time course of the applied current. Using a known finite-rise-time current pulse, Binstock et al.[47] derived the time course of the voltage for a capacity with parallel conductance and series resistance and presented results of measurements on squid axons and *Myxicola* central nerve cords. If the leak conductance is small, extrapolation of the linear portion of the voltage curve to zero time gives an intercept equal to $I_0 R_s = I_0 \tau/C_m$, where τ is the time constant of the rising phase of the current pulse.

A somewhat more elegant, but to our knowledge untried, scheme is presented by Cole and Lecar.[54] One may apply almost any shaped current pulse asymptotic to a constant and extract the series resistance in the following way: draw a straight line asymptotic to the late linear portion of the voltage record, pick a time t_0, integrate the difference between the straight line and the recorded voltage from $t = 0$ to $t = t_0$, and subtract this from the integral of the difference from $t = t_0$ to $t = \infty$; vary t_0 until the result is zero. The value of the voltage given by the straight line at $t = t_0$ is then equal to $I_0 R_s$, where I_0 is the value for a rectangular pulse giving the same voltage at long times.

[53] K. S. Cole, *Biophys. J.* **16**, 84 (1976).
[54] K. S. Cole and H. Lecar, *J. Membr. Biol.* **25**, 209 (1975).

For any of these approaches it is necessary to carefully consider the distortions produced by electrode impedances and amplifier delays for each particular case.

A very novel approach has been tried[55,56] in which voltage-sensitive dyes are employed. Application of a voltage clamp pulse which is rectangular as measured electrically results in a light response which, in the presence of a series resistance, will reflect the time course of the membrane current due to the $I_m R_s$ drop. For a nerve membrane where the sodium current component is large, the amount of negative-resistance compensation needed to make the response of the dye rectangular gives an estimate for the series resistance.

10.3.5.3. Voltage Clamp Stability and Series-Resistance Compensation.

A thorough analysis of the properties of a voltage clamp system as actually employed would be very complicated, and most voltage clamp systems in use today are constructed on the basis of intuition and cut and try.

Katz and Schwartz[22] have considered a very simplified circuit in which the resistance (R_s in Fig. 2) in series with the membrane and between the voltage measuring electrodes is ignored and only the time constants of the membrane and the control amplifier are considered. This gives a second-order system which is probably insufficient to say very much about an actual system. They analyze the effect of a feedback arrangement to reduce the effects of the access resistance R_{ce} between the output of the current-supplying amplifier and the preparation. This is important when using microelectrodes (Chapter 10.4) to supply current, and in some of the cases that they consider the series resistance R_s is small. For squid axon membrane clamping the access resistance is small, and R_s is a major problem.

Levis[43,57] has done many voltage clamp analyses and has concluded that an important consideration for stability and a fast clamp, using negative-resistance compensation (for R_s), is that the frequency response characteristics of the voltage and current measuring systems be comparable and combined before feeding back to the control amplifier through a compensating network.

Consider Fig. 2 with the additional feature that the current is measured as shown in Fig. 7 and a voltage $-\alpha I_m R_f$ proportional to this current is fed back to the control amplifier (Fig. 7). We have analyzed this system in some detail with the simplifying assumptions that the amplifiers are ideal operational amplifiers with zero output and infinite input impedance and neglecting the effects of the electrodes. We shall only make a few general comments here. Say that the control amplifier A_c (Fig 2) has a transfer function $Y_{opc}(s) =$

[55] L. B. Cohen and B. M. Salzberg, *Rev. Physiol., Biochem. Pharmacol.* **83**, 35 (1978).

[56] B. M. Salzberg, F. Bezanilla, and H. V. Davila, *Biophys. J.* **33**, 90a (1981).

[57] R. Levis, personal communication.

$v_{out}(s)/v_{in}(s)$ where $v(s) = \int_0^\infty V(t)e^{-st}\, dt$ is the Laplace transform of $V(t)$. Similarly, the transfer functions of amplifiers A_D (Fig. 2) and A_5 (Fig. 7) are Y_V and Y_I, respectively, and the admittance of the membrane is $Y_M(s)$. Also let $R'_s = R_s + R_{ce}$ (Fig. 2). It is important, even in simple analysis, to consider the effects of the resistance from the input of the current measuring amplifier A_5 to ground through the guard electrodes. Call this R_G, and let $R'_f = R_f R_G/(R_f + R_G)$.

With these definitions we can say that the transfer function for the relation between the membrane potential V_M and the command potential V_{COM} in Fig. 2 or Fig. 4 is a function of the complex variable s which has two zeros and four poles. The behavior of this system is completely determined by the values of the zeros and the poles (see, e.g., D'Azzo and Houpis[58]). We have found that there are always two real poles. The other two poles may be real or complex conjugates. The response to a step in V_{COM} will be a constant (the step introduces another pole at $s = 0$) plus the sum of four exponentials. For absolute stability the real part of any pole must be negative, so the response will be either the sum of a constant and four decreasing exponentials, or the sum of a constant plus two decreasing exponentials plus an exponentially decaying sinusoid.

There are several ways to approach the question of the behavior of the clamp system. One is to consider the frequency dependence of the transfer function and plot the magnitude of $v_M(j\omega)/v_{COM}(j\omega)$ versus $\omega = 2\pi f$, where f is frequency, on double-logarithmic paper. This is often referred to as a Bode diagram, and is particularly useful when one does not have an analytic expression for the system function. For stability the Nyquist approach is particularly useful when it is possible to readily measure the open-loop transfer function frequency characteristics [$H(j\omega)G(j\omega)$, see below]. For an analytical expression, stability can be determined with the use of the Routh criterion for the presence of positive real parts of the roots of the denominator. We have done this, but the most useful for us seems to be the so-called root locus procedure. (All of the above are well discussed in D'Azzo and Houpis.[58])

If we look at the distribution of the poles and zeros of the transfer function on the s plane and how they move when some parameter of the system is varied, we can observe the complete behavior of the system and obtain clues as to what one can do by adding more poles and zeros to improve it (i.e., with compensating networks). The major difficulty with this approach has always been the necessity of obtaining the complex roots of the denominator of the transfer function. For the simplified version we are considering there

[58] J. J. D'Azzo and C. H. Houpis. "Feedback Control System Analysis and Synthesis." McGraw-Hill, New York, 1966.

are four such roots. If the electrode impedances are included, as well as the compensation introduced by the elements in the feedback across and the input to the control amplifier (Fig. 7), there are seven roots to obtain. The availability of modern computers has changed this picture drastically, and such analyses are becoming feasible.

The term "stability" is used in more than one sense. Absolute stability refers to the situation in which there are no positive real parts of any pole and the system does not go to infinity with time. If the poles are all real and negative, the response is a sum of exponentials and there are no damped oscillations. This is a relative stability often referred to as critical damping. The stability of the system we are considering here depends very much on the bandwidths of the amplifiers employed, relative to the time constant of the membrane being clamped. For very wide-bandwidth amplifiers the system is unlikely to be stable for $R_s = 0$. For infinitely wide bandwidths with $R_s > 0$ the system may be stable, but no negative-resistance compensation is possible. On the other hand, it is quite possible to arrange things so that one can compensate for many times the series resistance. It is intuitively reasonable (and analytically true) that if the clamp system is slow, then only the low frequencies are involved and the presence of the capacitor in the membrane equivalent circuit will have very little effect; one could then compensate for $R_m + R_s$. If the time-constant of the amplifiers is $\tau_0 = 1/\omega_0$, one can compensate up to near $R_s + R_m/(1 + \omega_0^2 R_m^2 C^2)$.[59] For a usual case with the dc gain of the amplifiers equal to 10^5 and ω_0 equal to 2×10^7, the system is unstable for $R_s = 0$. If A is only 10^3 and $R_s = 5\,\Omega\,cm^2$, one can compensate for about twice R_s without instability.

One reason for the complexity of the analyses of clamp behavior is that the membrane for which one is attempting to control the potential is part of the feedback and not just a load as one finds in most textbooks. With the above definitions for our simplified system, it can be shown that the transfer function can be put into the standard form

$$v_M(s)/v_{COM}(s) = G(s)/[1 + G(s)H(s)]$$

if we let

$$G(s) = -\tfrac{1}{3}Y_{opc}(s)/\{1 + Y_M[R_s' + R_f'(1 - Y_I)]\}$$

and

$$H(s) = Y_V(1 + R_s Y_M) - Y_I \alpha R_f Y_M.$$

This is shown diagrammatically in Fig. 8. The open-loop transfer function is $H(s)G(s)$ and is the function that would be used for the Nyquist diagram approach to stability.

[59] R. E. Taylor and F. Bezanilla, *Program Abstr. Soc. Neurosci.*, p. 306 (1973).

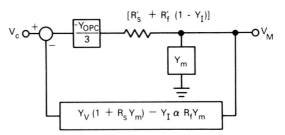

FIG. 8. Equivalent circuit for the transfer function $v_M(s)/v_{COM}(s)$ for the system shown in Fig. 2 with the addition of negative-resistance compensation as shown in Fig. 7. Here $V_c = v_{COM}$; Y_{OPC} is the transfer function of amplifier A_c (Fig. 2); Y_M is the admittance of the membrane with potential V_M; Y_V and Y_I are the total transfer functions of the voltage and current measuring arrangements: R'_s is the sum of the series resistance R_s and the access resistance R_{ce} (Fig. 2); αR_f is the amount of negative-resistance feedback (Fig. 7), where R_f is the feedback resistance of the current measuring amplifier A_5. $R'_f = R_f R_G/(R_f + R_G)$ where R_G is the resistance from the input of A5 to ground (not shown). The compensating networks R_V, C_V and R_I, C_I shown in Fig. 7 are not included here.

10.3.6. Pulse Generation and Data Acquisition

We now describe the basics of an electrophysiological setup used to study ionic and gating currents.

The setup contains (i) a unit to produce the pulse patterns to drive the membrane potential, (ii) voltage clamp electronics to control the membrane potential, (iii) a current measuring amplifier, and (iv) a data acquisition device to store currents obtained for the imposed membrane potentials. The different units of the setup can be easily assembled with conventional pulse generators, oscilloscope, and cameras, but digital computers are obtained at such reasonable prices today that the setup can best be assembled with a computer as the core block.

The description is then that of a particular computer-based system we have used successfully in our laboratory, although many similar systems have been used by many other investigators for many years.[60-62]

Basically, the computer is used to produce the pulse patterns used to drive the membrane potential and also to acquire and store the membrane currents. The computer, a Data General Nova 3 (Southboro, Ma) is programmed in a combination of FORTRAN IV and ASSEMBLER languages. ASSEMBLER is used to handle all the peripherals or input/output instructions and also in portions of the program when the FORTRAN execution times are excessively long.

[60] B. Hille, Ph.D. Thesis, Rockefeller University, New York, 1967, [University Microfilms (No. 68-9, 584), Ann Arbor, Michigan].

[61] C. M. Armstrong and F. Bezanilla, Ann. N.Y. Acad. Sci. 264, 265 (1975).

[62] W. Nonner, E. Rojas, and R. Stämpfli, Pfluegers Arch. 354 1 (1975).

10.3.6.1. Pulse Generation. The operator assembles the pulse pattern, and the program stores the sequence of amplitudes and durations in a bank of random-access memories. The random-access memories are communicated to a digital-to-analog (D/A) converter. Communication between the memories and the D/A converter takes place via optical isolators to prevent ground loops and, consequently, to decrease the pickup of digital noise in the analog side of the setup. The output of the D/A converter contains the pulse patterns, as decided by the investigator, and is applied as the command signal of the voltage clamp circuit. Most often the pattern is a series of rectangular pulses, but ramps[33,63] and other waveforms are sometimes employed.[64]

10.3.6.2. Data Acquisition. The membrane current is amplified and filtered conveniently to cover the dynamic range and speed of the sample and hold amplifier preceding the analog-to-digital (A/D) converter that digitizes the signal to be entered into the computer memory. Data from the A/D converter are transferred by way of optical isolators to the computer memory via the "data channel" or direct memory access (DMA); data transfer is not under direct program control, having the highest priority in the computer cycle operation. The stored current signal may or may not be processed before it is stored in hard or flexible magnetic disk.

An important design consideration in this setup is a clear separation between the analog and digital sections of the instruments. We have found that unless the computer is physically separated from the experimental table, significant digital noise is picked up by the analog circuitry, with adverse effect on the signal-to-noise ratio. In practice, the voltage clamp and current amplifier are set on or very near the experimental chamber. The D/A converter used to generate the voltage pulses is fed with its own power supply, and the same is done with the multiplexer, sample and hold, and A/D converter used to enter data into the computer. Both A/D and D/A converters are connected to the computer with multiple twisted pair cables. The analog signals, voltage, and current, are displayed in an oscilloscope different from the oscilloscope used to display data stored in the computer.

10.3.6.3. Instrumentation and Recording of Gating Currents. The use of a computer to generate the command pulses in a voltage clamp system is especially useful in the detection of gating currents. Gating currents are displacement currents produced by the movement of charge *inside* the membrane that are thought to be related to the opening and closing of the ionic channels.[18b, 65] (for a review, see Almers[66]). The currents are

[63] H. M. Fishman, *Nature* (*London*) **224**, 1116 (1969).
[64] Y. Palti and W. J. Adelman, Jr., *J. Membr. Biol.* **1**, 431 (1969).
[65] R. D. Keynes and E. Rojas, *J. Physiol.* (*London*) **239**, 393 (1974).
[66] W. Almers, *Rev. Physiol., Biochem. Pharmacol.* **82**, 96 (1978).

voltage and time dependent, and the method used to separate them from the larger capacitive currents is based on their nonlinear dependence on voltage. The first method ($\pm P$ procedure) used to detect them consisted of recording the membrane current for equal-magnitude but opposite-polarity pulses and adding the resultant currents. With the computer, it is an easy task to generate a sequence of positive (test) and negative (subtracting) pulses very well matched in amplitude and time course if the D/A converter has a resolution of 1 part in 4096 (12 bits). (The main problem is, however, to reduce the "glitches" to an acceptable minimum.) It was soon found that to measure gating currents more accurately, the subtracting pulse had to be started from a more hyperpolarized potential. Because strongly negative potentials can have deleterious effects on the membrane, the "$\frac{1}{4}P$" procedure was devised.[61,67] In this technique a test pulse of amplitude P is given, followed by four pulses of amplitude $\frac{1}{4}P$ riding on a very negative or very positive potential. Again, the computer can generate the sequence and amplitude of the pulses with a high degree of accuracy. Besides, the program can be made flexible enough to accommodate the "$\pm P$," "$\frac{1}{4}P$" procedures, or any other combination. When the $\pm P$ procedure is used, the currents are added; in the $\frac{1}{4}P$ procedure the currents from the four small pulses are added, and the result is subtracted from the current produced by the test pulse. With the computer it is quite simple to process the currents recorded, because under program control the current produced by each pulse can be stored and later added to or subtracted from the others. Furthermore, it is a normal procedure to signal-average the currents recorded to improve the signal-to-noise ratio; again, this procedure can be implemented with the computer with the same program that generates pulse amplitudes, durations, and sequences, and records the membrane currents.

The high gain required to record gating currents may produce saturation of the amplifiers. This problem has been solved by adding to the currents obtained, for both test and subtracting pulses, the current produced by a passive network which mimics the linear part of the membrane response.[67] If this transient generator is itself highly linear, the exact form produced is of no importance because it will be eliminated by the subtraction procedure.

10.3.6.4. Instrumentation for Studying Noise and Single Channels. The studies of membrane channel noise and single-channel current fluctuations have contributed precise values for the conductances of transmembrane ionic channels. As well as providing the most tangible proof of the existance of discrete molecular channels, the noise and channel-jump experiments provide unique insights into the kinetics of gating. Membrane noise analysis has by now become a large enterprise which has

[67] F. Bezanilla and C. M. Armstrong, *J. Gen. Physiol.* **70**, 549 (1977).

been reviewed extensively.[68]–[68e] Two of the main considerations in this important area of research are how to extract the noise spectrum generated by the activation of ionic channels from other extraneous sources of noise generated in a voltage-clamped preparation, and how to interpret the observed noise spectrum in terms of appropriate stochastic models of the gating process.

Single-channel currents observable with an isolated small (of the order of $1 \mu m^2$ in area) patch of membrane are the ultimate in resolution for membrane currents. The major considerations for this technique are how to fabricate electrodes which can be pressed up against the cell surface to make a seal and how to construct sufficiently low-noise virtual-ground current detectors for the picoampere-level currents involved. These techniques are presented in three excellent reviews by Neher and his collaborators.[68f]–[68h]

10.4. Voltage Clamp with Microelectrodes

10.4.1. Two Microelectrodes

In principle, it is possible to control the membrane potential using two intracellular electrodes, one to measure potential and the other to inject current. The technique, however, has two serious drawbacks: (i) it does not provide space clamp, and (ii) the high resistance of the microelectrodes make the system inherently slow.

The high resistance of the microelectrodes is in some cases not a serious problem. For example, the two-microelectrode technique has been used to voltage-clamp the end-plate potential in the frog neuromuscular junction.[69] Since in this preparation the end plate is usually activated by stimulation of the presynaptic terminal or by direct microiontophoresis of transmitter

[68] L. J. DeFelice, *Int. Rev. Neurobiol.* **20**, 169 (1977).

[68a] L. J. DeFelice, "Introduction to Membrane Noise." Plenum, New York, 1981.

[68b] V. E. Dionne, *in* "Techniques in Cellular Physiology" (P. F. Baker, ed.), in press. Elsevier, North-Holland, New York, 1982.

[68c] H. Lecar and F. Sachs, *in* "Excitable Cells in Tissue Culture" (P. G. Nelson and M. Lieberman, eds.), p. 137. Plenum, New York, 1981.

[68d] E. Neher and C. F. Stevens, *Ann. Rev. Biophys. Bioeng.* **6**, 345 (1977).

[68e] F. Conti and E. Wanke, *Q. Rev. Biophys.* **8**, 451 (1975).

[68f] E. Neher, B. Sakmann, and J. H. Steinbach, *Pfluegers Arch.* **375**, 219 (1978).

[68g] E. Neher, *in* "Techniques in Cellular Physiology" (P. F. Baker, ed.), in press. Elsevier North-Holland, New York, 1982.

[68h] O. P. Hamill, A. Marty, E. Neher, B. Sakmann, and F. J. Sigworth, *Pfluegers Arch.* **391**, 85 (1981).

[69] A. Takeuchi and N. Takeuchi, *J. Neurophysiol.* **22**, 395 (1959).

substance at constant potential, there is no need to charge the membrane capacity at high speed. In some other applications, when it is necessary to control the membrane potential in a small region but the recording of the current is not important, two-microelectrode voltage clamps have been used successfully (see, for example, Adrian et al.[70]).

However, the usual application of the voltage clamp to study the current–voltage characteristics of the membrane is a problem which requires special attention and varies with the type of cell considered because the potential distribution depends on the cable properties of the cell, which are partially related to its geometry.

To consider the problems in probably one of the worst cases, let us assume that we wish to voltage clamp a long cylindrical cell with two microelectrodes as was briefly discussed at the beginning of this article. The membrane current will be measured as the total current injected by the current microelectrodes. The potential at the tip of the voltage microelectrode will be controlled by the feedback circuitry, but the potential in other regions of the fiber will be given by the potential distribution in a long cylinder. In particular, in the steady state and assuming that the conductances are independent of voltage (not a very interesting case but useful for illustrative purposes), the voltage will be distributed according to $V_0 e^{-x/\lambda}$, where V_0 is the potential at the current injection point and λ is the space constant. It is easy to see that in the case of length long compared to λ, the current measured will not correspond to the membrane current in the controlled patch of membrane, but will be a mixture of the currents of different membrane patches all at different membrane potentials. For example, the length of a frog sartorius muscle fiber is about 20 times the space constant at rest, making it inappropriate for two-microelectrode voltage clamping. This situation is acceptable if the variation of the membrane potential with distance is negligible, a condition that can be approximated in two common cases: the short cylindrical cell and the spherical cell.

10.4.1.1. The Short Cylindrical Cell. It is intuitively clear that a short fiber will show less potential variation along its length the shorter it is in comparison to its length constant λ. Weidmann[71] has presented solutions of the distribution of potential in a short cable for the steady-state case. It can be easily shown that the potential distribution will be more homogeneous when the control of membrane potential is done in the center rather than at the end of the fiber.

To get an idea of the homogeneity of potential control, we show the prediction made by the solution of the cable equation for the case of a short

[70] R. H. Adrian, W. K. Chandler, and A. L. Hodgkin, J. Physiol. **204**, 207 (1969).
[71] S. Weidmann, J. Physiol. (London) **118**, 348 (1952).

cable of length $2l$ with both ends open circuited and with potential control at the center of the fiber.

The equation to be solved is (e.g., Taylor[5])

$$\lambda^2 \frac{\partial^2 V}{\partial x^2} - \tau \frac{\partial V}{\partial t} - V = 0, \qquad (10.4.1)$$

where x is the distance measured from the center to *either* end; V is the membrane voltage, t is time, λ is the space constant equal to $\sqrt{r_m/r_i}$; τ is the membrane time constant equal to $r_m C_m$; and r_i, r_m, and C_m are the internal resistance, membrane resistance, and membrane capacitance for unit length, respectively.

The boundary conditions at the ends require no current circulation:

$$i(x = l) = \frac{1}{r_i} \frac{\partial V}{\partial x}\bigg|_{x=l} = 0. \qquad (10.4.2)$$

At the center, the membrane potential is set equal to V_{COM}, and the total current injected I_0 is divided equally into the two segments of length l (total fiber length is $2l$). The solution of the above equations for V_{COM}, an arbitrary function of voltage $V_{COM} = f(t)$, was obtained by the use of the Laplace transform, the complex inversion formula, and the convolution theorem, giving

$$V(x, t) = \frac{\pi \lambda^2}{l^2 \tau} \sum_{n=1}^{\infty} (-1)^n (2n - 1) \cos \frac{(2n - 1)\pi(l - x)}{2l}$$

$$\times \int_0^t f(t - z) \exp\left[\frac{-z}{\tau} \left(\frac{\pi^2 (2n - 1)^2 \lambda^2}{4l^2} \right) \right] dz. \qquad (10.4.3)$$

Several driving functions can be analyzed with this equation. The most obvious case is the step function: $f = 0$ for $t < 0$ and $f = V_0$ for $t \geq 0$. In this case the distribution is given by (see also Waltman[72])

$$V(x, t) = V_0 \left(\frac{\cosh[(l - x)/\lambda]}{\cosh(l/\lambda)} \right.$$

$$\left. - 4\pi \sum_{n=1}^{\infty} \frac{(-1)^{n+1}(2n - 1) \exp\{-(t/\tau)[\pi^2(2n - 1)^2\lambda^2/4l^2 + 1]\}}{4l^2/\lambda^2 + \pi^2(2n - 1)^2} \times \cos[(2n - 1)\pi(l - x)/2l] \right).$$

$$(10.4.4)$$

Figure 9 shows an example of $V(x, t)$ when $l/\lambda = 0.44$. It is clear that the potential rises at the ends more slowly than at the center, and at long times

[72] B. Waltman, *Acta Physiol. Scand.* **66**, Suppl. 264 (1966).

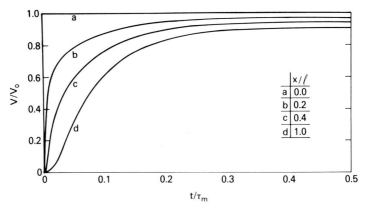

FIG. 9. Normalized membrane potential as a function of time in a voltage-clamped short fiber. At $x = 0$ (center of the fiber) a step function of voltage of magnitude V_0 is applied (curve a) and at distances x/l of 0.2 (curve b), 0.4 (curve c), and 1.0 (curve d); the potential is plotted as a function of time. Abscissa is normalized time t/τ, with τ the membrane time constant. Ordinate is normalized membrane potential defined as the membrane potential divided by the controlled potential V_0. λ is the space constant, and l/λ was taken as 0.44. Curves computed with Equation (10.4.4).

the potential at the ends is different from the potential at the center. The parameters used in Fig. 9 were obtained from experimental measurements performed in the lumbricalis muscle of digiti IV of the frog.[73] When a third microelectrode was inserted at the end of a voltage-clamped fiber, a very similar voltage waveform as observed in Fig. 9 was recorded.[74,74a]

It is interesting to compare the current predicted by the above model with the current expected for the case in which there is no decrement along the fiber. The total current I_0 injected by the current microelectrode is given by

$$I_0 = \frac{2V_0}{r_i}\left\{\frac{\tanh(l/\lambda)}{\lambda} + \frac{4\pi^2}{l}\sum_{n=1}^{\infty}\frac{(2n-1)^2}{4l^2/\lambda^2 + \pi^2(2n-1)^2}\right.$$
$$\left. \times \exp\left[-\frac{t}{\tau}\left(\frac{\pi^2(2n-1)^2\lambda^2}{4l^2}\right)\right]\right\}. \tag{10.4.5}$$

If one analyzes the ideal case in which $r_i = 0$ (perfect space clamp) the total current is given by

$$I_0 = 2V_0 l/r_m, \quad t > 0. \tag{10.4.6}$$

[73] F. Bezanilla, C. Caputo, and P. Horowicz, *Acta Cient. Venez.* **22**, Suppl. 2, 72 (1971).
[74] P. Heistracher and C. C. Hunt, *J. Physiol. (London)* **201**, 589 (1969).
[74a] F. Bezanilla, C. Caputo, and P. Horowicz, personal communication.

$I_0 \to \infty$ at $t = 0$, and $I_0 = 0$ at $t < 0$. It should be noted that Eq. (10.4.5) approaches (10.4.6) when $r_i \to 0$. The current measured in a typical two-microelectrode clamp is the total current I_0, and from the above discussion it is clear that it will not represent the properties of a single patch of the membrane under study, but instead the charging of the distributed membrane capacitance. The case approaches ideality when the space constant is very large compared to the length of the fiber. The error can be computed with Eqs. (10.4.5) and (10.4.6). If the length and diameter of the cell are known, it is possible to roughly estimate whether two-microelectrode clamping could be used. The value of λ is given as

$$\lambda = \sqrt{r_m/r_i} = \sqrt{R_m \pi a^2 / 2\pi a R_i} = \sqrt{R_m a / 2 R_i},$$

where a is the radius, R_m is the specific membrane resistance ($\Omega\,cm^2$); and R_i is the resistivity of the internal medium ($\Omega\,cm$). We can approximate λ to

$$\lambda \approx \sqrt{20a} \quad \text{(cm)}$$

for typical average values of R_m and R_i. For an appropriate space clamp with two microelectrodes, l/λ should be less than or equal to 1, and using the above approximation the length of the fiber should be

$$l \le \sqrt{20a}.$$

It should be noted, however, that the theoretical expressions presented above have been obtained with a simple core-conductor model which does not take into account current spread in three dimensions. Three-dimensional considerations are important near the site of current injection, and this may lead to error in the controlled potential at the tip of the voltage electrode when both electrodes are very close together.[75] It is also known that when the radius a of the fiber approaches λ, the three-dimensional equation should be used instead.[75] It is probably not appropriate to discuss that case in any detail because when $a \approx \lambda$ the space clamp will be a complete failure, as can be seen from the previous equations, unless $l \approx a$, in which case we are approaching the case of the spherical cell.

   10.4.1.2. The Spherical Cell. In the approximation of the core-conductor model made in the previous paragraph, the spherical cell would be considered an ideal case. We can test the validity of the assumption of potential uniformity by approaching the problem in the three-dimensional case. The solution to the problem of a spherical cell with two microelectrodes, one to inject current and the other to record potential, has been presented

[75] R. S. Eisenberg and E. A. Johnson, *Prog. Biophys. Mol. Biol.* **20**, 1 (1970).

in detail by Eisenberg and Engel,[76] Peskoff and Eisenberg,[77] and Peskoff and Ramirez.[78] Although their solution has not been produced for voltage clamp, we can assume that their voltage electrode is in our case the site of membrane potential control, and then ask for the potential in other places of the cell, without moving the current electrode and without changing the injected current, that will maintain the controlled potential in the original position. It should be noted that in the derivation by Eisenberg and Engel,[76] the current source was assumed ideal, that is, of infinite impedance. This is true in the case of "current clamping" because the electronics in that case are designed to have very high-impedance current electrode. When the objective is voltage clamping, the electrode resistance is made low purposely to be able to inject as much current as needed to control the membrane potential in the shortest possible time. However, a very low source resistance cannot be obtained because in practice the minimum resistance is given by the current microelectrode, which typically will vary between 5 and 100 MΩ. In any case, the problem discussed by Eisenberg and Engel[76] can be applied without modifications for the steady state because the impedance of the current source is not important provided the amount of current it supplies is known. In the transient solution, the situation is more involved, and it has not been treated in a form convenient for the discussion of the voltage clamp.

Let us consider the steady-state situation. Using Eqs. (1) and (2) of Eisenberg and Engel[76] we find that the potential distribution $V_m(\theta)$ is given by

$$V_m(\theta) = (I_0 R_m/4\pi b^2) f(\theta, b/\Lambda),$$

where θ is the angular separation between the current and voltage electrode; b is the radius; Λ is the space constant, equal to R_m/R_i with R_m membrane resistance per unit area (Ω cm^2) and R_i internal resistivity (Ω cm); and I_0 is the total injected current. If the potential is homogeneous in the cell surface, it is given by

$$V_0 = I_0 R_m/4\pi b^2,$$

whence it is clear that the factor $f(\theta, b/\Lambda)$ is defined as

$$f(\theta, b/\Lambda) = V_m(\theta)/V_0.$$

The situation sought in voltage-clamping a spherical cell is a minimum variation of potential with angle separation. It must be remembered that during voltage clamping the potential will be controlled at the voltage electrode; therefore, we must ask what is the deviation from a constant membrane potential with any angle between current and voltage electrode.

[76] R. S. Eisenberg and E. Engel, *J. Gen. Physiol.* **55**, 736 (1970).
[77] A. Peskoff and R. S. Eisenberg, *J. Math. Biol.* **2**, 277 (1975).
[78] A. Peskoff and D. M. Ramirez, *J. Math. Biol.* **2**, 301 (1975).

Eisenberg and Engel (Ref. 76, Table II) have published a table of $f(\theta, b/\Lambda)$ as a function of angle θ and the ratio b/Λ. It must be emphasized that in the vicinity of the current electrode there is a steep rise in membrane potential, but for most of the cell (for example, angles larger that $20°$, and $b/\Lambda < 0.01$), the variation in membrane potential will be less than 5%. This is not very difficult to achieve because a cell $100\,\mu m$ in diameter with a membrane resistance of $5\,k\Omega\,cm^2$ and an internal resistivity of $100\,\Omega\,cm$ will have $b/\Lambda = 0.0001$. This value could be reduced 100 times (for example, during excitation) and still result in a potential distribution with less than 5% variation for angles larger than $20°$. It should be noted, however, that a variation of 5% could produce local circuit currents near the threshold region for a very steep conductance versus voltage curve and, consequently, produce a total current that would be practically impossible to interpret in terms of the properties of an elementary membrane patch.

The above analysis assumes that the electrodes are inserted just under the membrane surface and that the external medium is isopotential. The first assumption is reasonable in the majority of the cases because it is not desirable to introduce the tip of the electrode too deep to prevent cell damage. In very small cells, however, this assumption may not be valid. The external isopotentiality has been demonstrated to be a very good approximation by Peskoff and Eisenberg,[77] who examined the difference it would make in their solution by including a finite conductivity of the external solution.

The discussions about space clamping in two-microelectrode arrangements have been limited to the case where the system is linear. In practice, the most interesting preparations will have conductances which are dependent on membrane potential; consequently, the above equations are not directly applicable. However, some insight can be obtained by changing the membrane parameters in the steady-state situation. The situation is even more difficult when regenerative phenomena are present because voltage-dependent negative conductances are involved and, in general, numerical solutions are required to analyze each particular case. Complications arise when the spherical cell has dendrites or axons, because current will be drained into cablelike structures, sometimes excitable, which could result in very large nonuniformities.

Having examined the problem of space clamp when voltage clamping with two microelectrodes, let us consider now the limitations imposed by the use of micropipettes. Microelectrodes are made by pulling glass tubing, leaving tips less than $1\,\mu m$ in diameter, which are then filled with a concentrated electrolyte solution. Normally, electrodes to record membrane potential are filled with $3\,M$ KCl. Electrodes to pass current are normally not filled with KCl because they cannot pass enough current for most applications; it has been found that electrodes filled with $2\,M$ K-Citrate exhibit lower resistance to current injection. Because of the small diameter near the tip, the resistance of these electrodes ranges between 5 and 500 MΩ depending on tip diameter, shank shape, and the electrolyte filling them. The high resistance of microelectrodes imposes serious limitations on the implementation of a voltage clamp system. Voltage recording must be done with very high-input-impedance amplifiers so that the electrode–amplifier combination does not drain current from the cell. The distributed capacitance of the wall of the microelectrode, the electrical shield, and the input capacitance of the amplifier, together with the high resistance of the pipette, constitute a low-pass filter that gives an erroneous measurement of the actual membrane potential when varying with time. The approach commonly used to solve this problem is to electronically compensate for the input capacitance using positive feedback through a capacitor (e.g., Bak[79]). The adjustment of the compensation is not a simple matter (for a discussion of methods, see Moore[21]), and as some of the capacitance is distributed, it cannot be compensated. Although the result is a faster response than the case without capacity compensation, the transfer function of the amplifier introduces phase lags that are important in the overall performance and stability of the voltage clamp feedback circuit. To decrease the need for negative capacity compensation, "driven shields" have been used which consist of connecting the shields at the input to the output of the amplifier. As the output is a low-impedance point with respect to ground, it shields effectively, and being at the same potential as the shielded conductor (the input), the capacity of the shield does not need to be charged or discharged when the potential of the microelectrode changes.

A very ingenious way to decrease the influence of the electrode capacitance has been devised by Eisenberg and Gage[80] based on the idea of bringing the interior of the cell to virtual ground.[81] In this arrangement a high-input-impedance operational amplifier is used, the microelectrode is connected directly to the inverting input, the output of the amplifier is connected to the bath. As the positive input of the amplifier is grounded, the negative input will be at "virtual" ground, and the output will be at minus the membrane potential. That is, measuring the potential in the bath with respect to ground gives the cell membrane potential. Meanwhile, all the effects of capacity to ground have been eliminated because the electrode is at ground potential, and only the capacitance across that portion of the wall of the microelectrode in the bathing solution remains, which can be reduced significantly, lowering the level of the external solutions. This configuration is recommended for measuring membrane potential, especially when used as part of the feedback loop in a voltage clamp system.

[79] A. F. Bak, *Electroencephalogr. Clin. Neurophysiol.* **10**, 745 (1958).
[80] R. S. Eisenberg and P. W. Gage, *J. Gen. Physiol.* **53**, 279 (1969).
[81] B. Frankenhauser, *J. Physiol.* (*London*) **135**, 550 (1957).

The current electrode also has a high resistance which imposes problems in the design of the voltage clamp. It has been found that to inject enough current to charge the membrane capacitance following a step change in voltage, a high-voltage amplifier is required because the high resistance of the pipette effectively limits the amount of current to be passed. Some investigators have used the internal amplifier of the oscilloscope,[82] and others have used operational amplifiers with large output-voltage range.[74]

However, even the use of high-voltage control amplifiers does not solve the problem of speed encountered when the membrane capacitance is charged through a large resistor. Katz and Schwartz[22] have proposed a compensating network for reducing the effect of the access resistance as discussed in Section 10.3.5.3. If the amplifier had infinite gain at all frequencies, it would be possible to impose a fast rise in potential across the membrane, but the limited bandwidth and slow rate of practical amplifiers limit even further the speed of the voltage clamp system.

The interaction ("cross talk") between the voltage and current electrodes due to capacitative coupling between them is important during fast transients, because it introduces undesired transients in the voltage clamp feedback circuit which may render it unstable. The coupling can be effectively reduced by positioning a shield between the electrodes and impaling the cell with a wide angle between them.

10.4.2. Three Microelectrodes

Adrian et al.[83] devised a technique to control the membrane potential of a muscle fiber and measure the current in the neighborhood of the controlled region. Their technique is based on the fact that membrane current is roughly proportional to the voltage gradient along the fiber. They impaled three microelectrodes near the natural end of a fiber, and, controlling the potential at the electrode closer to the end (control electrode) by passing current through the electrode farthest from the end, they measured the membrane current as the potential difference between the central electrode and the control electrode. Using the linear cable equation, they derived the exact expression for the current as a function of the potential difference between the two electrodes, and they found that if l/λ is less than 2, the error is less than 5%, if the current is measured just as the potential difference between the electrodes (l is the distance between the end of the fiber and the control electrode and between the two voltage electrodes, λ is the space constant). The method works very well to measure delayed rectification, but fails in the range of potentials needed to measure the sodium current. The reader is

[82] L. L. Costantin, J. Physiol. (London) 195, 119 (1968).
[83] R. H. Adrian, W. K. Chandler, and A. L. Hodgkin, J. Physiol. (London) 208, 607 (1970).

referred to the original paper for details of the sources of the errors of this technique.[83] Kootsey[84] has calculated the errors of this technique using numerical procedures for the case of negative conductance and has found that $l/\lambda = 2$ produces unacceptable errors. The error decreases to small but still noticeable values when $l/\lambda = 0.86$.

Recently, the technique of making the interior of the fiber virtual ground has been applied to the three-microelectrode voltage clamp.[84a]

10.5. Voltage Clamp of an Isolated Patch Using External Pipettes

The smaller the region in which the membrane current is measured, the closer one gets to the ideal situation of recording current from an isopotential region. In this chapter we review briefly some of the attempts that have been made to isolate a small patch of surface membrane using external pipettes to record the current through it.

10.5.1. External Patch Isolation

Strickholm[85,86] manufactured fire-polished pipettes and applied them to the surface of skeletal muscle fibers of the frog. He relied on the fact that the leakage conductance produced by imperfect contact with the membrane was much larger than the membrane conductance of the patch, and with this the patch of external surface membrane may be considered to be "clamped" at a constant potential. However, the lack of a feedback circuit to hold the patch voltage controlled when the patch conductance changed made this arrangement of limited value.

A significant improvement was made by Frank and Tauc.[87] These authors used two internal microelectrodes to voltage-clamp the membrane potential of a mollusk neuronal body and brought an external pipette near the external surface of the soma. The interior of this pipette was connected to the input of an operational amplifier connected in the current-to-voltage configuration. The other input of the amplifier was connected to the bath, making the output of the amplifier proportional to the current flowing through the patch

[84] J. M. Kootsey, *Fed. Proc. Fed. Am. Soc. Exp. Biol.* **34**, 1343 (1975).

[84a] P. C. Vaughan, J. G. Mclarnan, and D. D. F. Loo, *Can. J. Physiol. Pharmacol.* **58**, 999 (1980).

[85] A. Strickholm, *J. Gen. Physiol.* **44**, 1073 (1961).

[86] A. Strickholm, *J. Cell. Comp. Physiol.* **60**, 149 (1962).

[87] K. Frank and L. Tauc, *in* "Cellular Function of Membrane Transport" (S. Hoffman, (ed.), Prentice-Hall, Englewood Cliffs, New Jersey, 1963.

under the pipette tip. The comparison between total current and current through the patch revealed that the total current had contributions from the axonal membrane which was not under voltage control. The patch revealed current patterns that increased smoothly with membrane depolarization as was expected from a well-controlled membrane patch.

A further improvement was made by Neher and Lux[88] when they incorporated a feedback loop in the extracellular pipette using a partitioned pipette, one half to measure the pipette potential and the other to inject current. In their arrangement, the potential *inside* the pipette was maintained at bath potential by means of a feedback circuit. The current needed to control the potential of the pipette was measured as the current through the patch, the cell membrane potential still being controlled by an independent feedback circuit with two microelectrodes. Using feedback in the extracellular pipette effectively decreases the access resistance (pipette resistance) to the input of the amplifier, decreasing significantly the effect of the shunt resistance produced by the leakage pathway between the pipette wall and the cell surface.

The original approach of Strickholm using the extracellular pipette to control the membrane potential of the patch of membrane under the pipette was notably improved by Fishman.[89] He made a double-barreled concentric pipette, the internal to measure patch voltage and to inject current, and the external to introduce sucrose to improve the seal of the pipette against the membrane. As the preparation used was the squid giant axon, which is surrounded by a coat of Schwann cells and connective tissue, the sucrose was essential to decrease the leakage under the walls of the central pipette. Even with the sucrose flow the seal resistance was of the order of 2 MΩ, which is about ten times smaller than the membrane resistance under the pipette. The voltage clamp was established as a feedback circuit to control the potential inside the pipette at a predetermined voltage, and the current needed to maintain this potential was considered to be the current of the membrane patch under the pipette plus the leakage current due to the imperfect seal. This latter current could be subtracted off electronically because it was found that the leakage resistance was ohmic. The membrane currents recorded by Fishman resemble the current recorded with the conventional axial-wire technique, and he provided tests for isopotentiality inside the axon by inserting a voltage measuring pipette. This technique has been used to record membrane noise taking advantage of the fact that the currents are measured from a small patch of membrane.[90]

[88] E. Neher and H. D. Lux, *Pfluegers Arch.* **311**, 272 (1969).
[89] H. M. Fishman, *J. Membr. Biol.* **24**, 265 (1975).
[90] H. M. Fishman, D. J. M. Poussart, and L. E. Moore, *J. Membr. Biol.* **24**, 281 (1975).

The isolation of a patch of membrane with an external pipette has been perfected by Neher and Sakmann[40] and Neher et al.[68f] By using collagenase-treated denervated muscle fibers, the seal resistance was increased up to 50 MΩ, and they were able to record the opening and closing of single postsynaptic channels. In their arrangement an extremely low-noise operational amplifier was used as a current-to-voltage converter with its input connected to the pipette. Using 500 MΩ as a feedback resistor, the main source of noise in that configuration is the input voltage noise of the operational amplifier that appears as a current in the leakage resistance under the seal between the pipette and the cell surface. With 50-MΩ seal resistance the noise was low enough to observe a single acetylcholine-activated channel with a good signal-to-noise ratio.

Recently, Sigworth and Neher[91] and Horn and Patlak[91a] have achieved seal resistances above 1 G Ω using tissue culture cells. Sigworth and Neher have reported fluctuations of single sodium channels. Horn and Patlak have been able to excise a membrane patch attached to the pipette, enabling them to readily change the medium on the inner side of the membrane. These matters, and others, are discussed in recent reviews.[68g,68h]

10.5.2. Internal Access

Several groups of investigators[92–94] have used the external pipette to gain access to the interior of the cell (Fig. 10). Although these techniques are not strictly regarded as patch isolation, we have included them here because they utilize external pipettes (or perforated plastic partitions) that have to be pressed against the membrane to obtain a good seal.

Once the seal is in contact with the pipette, suction is applied to increase seal resistance and if the suction is increased even more, it is possible to break the cell membrane (or break it with a wire inside the pipette) and, consequently, have access to the internal medium (Fig. 10a). Under these conditions when the interior of the pipette is voltage clamped, the whole cell will be voltage clamped. The accuracy in the current recordings will depend on membrane parameters and geometry of the cells as discussed in the section on microelectrode clamping. This technique has the added advantage of making possible the exchange of the internal solution by known solutions loaded and circulated through the pipette shank.

[91] F. J. Sigworth and E. Neher, *Nature* (*London*) **287**, 447 (1980).
[91a] R. Horn and J. B. Patlak, *Proc. Natl. Acad. Sci. U.S.A.* **77**, 6930 (1980).
[92] P. G. Kostyuk and O. A. Krishtal, *J. Physiol.* (*London*) **270**, 545 (1977).
[93] K. Takahashi and M. Yoshii, *J. Physiol.* (*London*) **279**, 519 (1978).
[94] K. S. Lee, N. Akaike, and A. M. Brown, *J. Gen. Physiol.* **71**, 489 (1978).

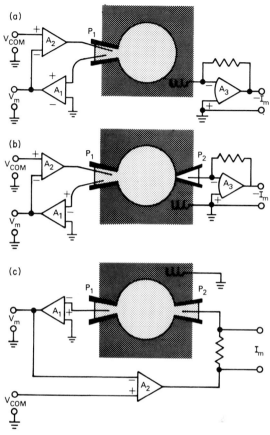

FIG. 10. Schematic diagrams of patch clamps with internal access. (a) Simple pipette arrangement. The interior of pipette P_1 is in direct contact with the cell contents. The voltage clamp circuit (A_1, A_2) controls the internal membrane potential and the total cell current is measured by the current to voltage converter A_3. (b) Voltage clamp as in (A) but current is only measured out of a small membrane patch with the smaller pipette P_2 that is against the *external* surface of the cell.[95] (c) Two pipettes have access to the interior of the cell. P_1 is used to measure the membrane potential with amplifier A_1 and then compared to the command voltage at amplifier A_2 that injects current through pipette P_2. Current I_M is measured as the total current injected into the cell.[96]

Kostyuk *et al.*[95] have carried this technique one step further (Fig. 10b). They have manufactured plastic pipettes that, when applied with suction against a molluscan neuronal soma, resulted in seal resistances of up to $10^9 \, \Omega$. They use one pipette to gain access to the cell interior by applying

[95] P. G. Kostyuk, O. A. Krishtal, and V. I. Pidoplichco, *Dokl. Akad. Nauk SSSR* **238**, 478 (1978).

enough suction to break the cell membrane and to voltage clamp the soma. Another pipette with smaller tip diameter is applied to the intact part of the cell soma; enough suction is applied to obtain a good seal, but not enough to rupture the membrane. This second pipette is connected to a current-to-voltage transducer to record the membrane current of the patch isolated under the pipette. With this technique they have been able to record current noise related to calcium channels.

Krishtal[96] has used the same type of pipettes to gain access to the cell interior with two pipettes in the same cell (Fig. 10c). One pipette is used to measure voltage and the other to pass current, approaching an almost ideal situation when the cells are small and have high membrane resistance. This technique makes possible the internal perfusion of the soma, as solution is passed from one pipette to the other removing the contents of the cell.

Conti and Neher[96a] have used an L-shaped pipette on the inside of the squid giant axon to isolate a small patch of membrane and record the fluctuations of single potassium channels.

Recently a new preparation has been described in which a squid axon had been cut open and the resulting membrane sheet positioned in a chamber separating two compartments. Normal-looking ionic and gating currents have been reported from this cut axon,[41] and fluctuations from a few sodium channels have been observed by applying patch pipettes to the internal surface membrane.[96b]

10.6. Voltage Clamp with Gap Isolation Techniques

Some of the most rewarding voltage clamp techniques can be gathered under the heading gap isolation techniques. In this case, cylindrical cells are mounted across partitions and patches of membrane are isolated by means of Vaseline, sucrose, Vaseline–sucrose, or Vaseline–air gaps. Voltage clamping of excitable cells using gap isolation techniques has provided sustained experimental evidence about electrophysiological mechanisms or preparations in which microelectrodes, pipette patch isolation, and axial-wire techniques cannot be used. The methods in general were initially developed for electrical recordings from the naturally isolated patch of excitable membrane in the node of Ranvier, but they have been modified to artificially define a small patch of membrane in unmyelinated nerve axons and muscle fibers.

---

[96] O. A. Krishtal, *Dokl. Akad. Nauk SSSR* **238**, 482 (1978).

[96a] F. Conti and E. Neher, *Nature (London)* **285**, 140 (1980).

[96b] I. Llano and F. Bezanilla, *Biophys. J.* **37**, 101a (1982).

10.6.1. Node of Ranvier

Myelinated nerve fibers propagate electrical impulses (action potentials) not as a continuous wave of depolarization as the unmyelinated axons do, but in a saltatory fashion in which action potentials are regenerated in small patches of bare membrane called nodes of Ranvier.[97] We will discuss first the limitations imposed by the characteristics of this preparation on the electrical techniques used to record resting potentials, action potentials, and later in voltage clamping. As said above, essentially all the patch isolation techniques were derived from those developed for the node of Ranvier, and thus the natural node will be described rather extensively. Voltage clamping of the node of Ranvier has become a standard technique in which low-noise current records, fast settling times, and accurate membrane potential control can be obtained without the use of extremely sophisticated electronics. Although studies of the node of Ranvier began in order to see if the insights developed by Hodgkin and Huxley were valid for the node, this preparation has become so popular in the past ten years that many of the recent developments in electrophysiology have been made on this preparation.[97]

10.6.1.1. General Characteristics. The techniques for dissection and mounting of single myelinated axons from the frog and toad have been discussed extensively by Stämpfli and Hille[97] and will not be covered here. The design of the chamber used to mount the nerve fibers and to make electrical recordings from the node of Ranvier depends on the technique employed, and its characteristics are described in connection with each of the available methods.

In order to orient the reader to the technical requirements of the node of Ranvier as an electrophysiological preparation, we have included in Table I the typical electrical parameters of a frog's myelinated fiber. From Table I we see that the node represents a very small patch of excitable membrane ($22\ \mu m^2$) with a very high resting resistance (40–80 MΩ).

10.6.1.2. Membrane Potential Measurement in a Node of Ranvier. We recall that the requirements of a good voltage clamp technique are that the potential measurement must be *accurate, fast*, sampled at an *isopotential patch of membrane*. We discuss now to what extent this has been achieved in myelinated nerve fibers.

A straightforward method for measuring transmembrane potential in a biological preparation is the use of micropipettes. However, as suggested by the work of Woodbury,[97a] they cannot be successfully used in the node of Ranvier.

---

[97] R. Stämpfli and B. Hille, *in* "Frog Neurobiology" (R. Llinas and W. Precht, eds.), Springer-Verlag, Berlin and New York, 1976.

[97a] J. W. Woodbury, *J. Cell. Comp. Physiol.* **39**, 323 (1952).

TABLE I.   Electrical Characteristics of a Frog's Myelinated Fibers

Fiber diameter	14 μm
Thickness of myelin	2 μm
Distance between nodes	2 mm
Area of nodal membrane (not measured directly)	22 μm^2
Resistance per unit length of axis cylinder	140 MΩ/cm
Specific resistance of axoplasm	110 Ωcm
Capacity per unit length of myelin sheath	10–16 pF/cm
Capacity per unit area of myelin sheath	2.5–5 nF/cm^2
Resistance times unit area of myelin sheath	0.1–0.16 MΩ cm^2
Specific resistance of myelin sheath	500–800 MΩ cm^2
Capacity of node of Ranvier	0.6–1.5 pF
Resistance of resting node	40–80 MΩ
Action potential amplitude	116 mV
Resting potential	-71 mV
Peak inward current density	20 mA/cm^2

An alternative method for measuring the potentials in this preparation was introduced by Huxley and Stämpfli.[98] Figure 11a shows a shematic drawing of the preparation and their electronic arrangement. The rationale of the Huxley–Stämpfli method is that of a simple potentiometric recording. The membrane potential V_m of node 0 in Fig. 11b can be measured without attenuation if the current source H supplies current through the partition AB until the meter G shows no deflection, indicating no current flow across the resistance R_{BC}. Since node -1 is not contributing to the membrane potential (because it is depolarized by KCl), then $V_{AB} = V_m$. If there is current flowing through the external resistor R_{BC}, the potential at C does not represent V_m, but is an attenuated membrane potential value. If there is no current flow, though, $V_C = V_B = V_D = V_m$. Huxley and Stämpfli[98] were able to measure accurately the resting potential and the peak value of the action potential in nodes of *Rana esculenta* by setting the current at H manually by trial and error to a value which blocked current flow. Their method has been called *static potentiometric* since the zero-current condition was only met at specific points of the electrical cycle.[97]

In the experiments of Huxley and Stämpfli the gap BC, measuring approximately 600 μm, was filled with paraffin oil and had a resistance of about 10 MΩ. The resistance of the whole loop DCBAD was between 60 and 90 MΩ; the voltage drop across the gap BC is only a small fraction of V_m, and an action potential will seem very small because of this voltage dividing effect (attenuation) unless the potentiometric method is used. In order to record action potentials, this attenuation has to be reduced dramatically, and two approaches have been followed in order to achieve

[98] A. F. Huxley and R. Stämpfli, *J. Physiol. (London)* **112**, 476 (1951).

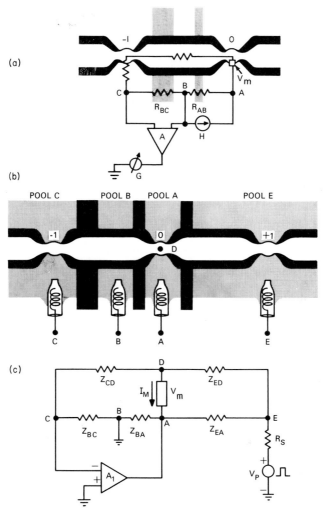

FIG. 11. Potentiometric methods of recording membrane potential. (a) Schematic diagram of the Huxley–Stämpfli[98] static potentiometric methods. Node 0 has a resting potential V_M and node -1 is depolarized by isotonic KCl. Shaded regions represent partitions. The BC partition is filled with paraffin oil and has a resistance of about 10 MΩ. The AB partition is a small trough with a resistance of about 20 KΩ. H is a current source, G is a galvanometer indicating the zero-current condition, and A is a differential amplifier measuring the voltage drop across R_{BC}. (b) Schematic diagram of Frankenhauser[81] dynamic potentiometric method. Node 0 has a resting potential V_M and nodes -1 and $+1$ are depolarized by isotonic KCl. Stippled regions represent petroleum jelly seals with resistance of about 5–10 MΩ. Important points are labeled and correspond to those in the circuit diagram shown in (c). (c) Equivalent circuit of Frankenhauser's method. The electrode impedance and balancing circuits are not considered; Z indicates generalized impedances as used in Section 10.A.1. They are drawn and described in the text as resistances for simplicity only. R_s is the output impedance of the stimulator (about 500 Ω). Amplifier A_1 is a high-input-impedance, wide-bandwidth amplifier. For complete discussion see text and Section 10.A.2.

this goal: (i) improved gap insulation and (ii) the dynamic potentiometric method.

10.6.1.2.1. GAP INSULATION. The resistance of the gap across which the potential is measured has to be made very high. In one method, Stämpfli[99] replaced the conductive solution in pool B of the Huxley–Stämpfli recording apparatus by deionized isotonic sucrose solution. In another method, Tasaki and Frank[100] eliminated the pool B and made a large air gap in the internodal region. With these methods the problem of attenuation has been largely reduced even though the use of an air gap, for example, requires very extensive drying of the preparation to obtain less than 10% attenuation. The disadvantages of gap insulation methods for voltage clamp techniques reside not only in the problem of the static or dc attenuation factors that violate our requirement of accuracy of potential measurement, but also in the dynamic attenuation factors introduced by stray capacitances. In the gap insulation techniques, the pool at which node -1 is located (pool C) is effectively insulated from the measuring pool (pool A), which is usually grounded. This means that pool C is a very high-impedance pool with respect to ground and that any measurement of potential made there will be affected by stray capacitances to ground, slowing down the recording system. Action potentials at room and even low temperatures are fast processes that can be seriously distorted if the frequency response of the recording apparatus is below 5 kHz. With high-impedance gaps of the order of 100 MΩ, only 2 pF are necessary to cut the frequency response below 1 kHz. In these cases it is necessary to introduce negative-capacity compensation techniques or driven-shield techniques that introduce instability in the voltage clamp circuits and/or complicate the electronic circuitry.

10.6.1.2.2. POTENTIOMETRIC METHODS. We have already discussed the Huxley–Stämpfli method as a static-potentiometric method in which the action potential time course cannot be continuously displayed because the amount of current supplied by H is manually changed. Their method, though, did not require a very high gap resistance R_{BC} since the current source H was adjusted until no current flowed through the partition BC. From the diagram shown in Fig. 11a it can be seen that if the potential drop across R_{BC} is differentially measured by an amplifier that in the case of Huxley and Stämpfli was used to inject current into pool B in a negative feedback configuration, then the potential drop across R_{BC} will be *dynamically* maintained at zero. This condition has been successfully used by Derksen[101] to record action potentials potentiometrically. The method still has the

[99] R. Stämpfli, *Experientia* **10**, 508 (1954).
[100] I. Tasaki and K. Frank, *Am. J. Physiol.* **182**, 572 (1955).
[101] H. E. Derksen, *Acta Physiol. Pharmacol. Neerl.* **13**, 373 (1965).

limitation that there is a relatively high impedance between the voltage measuring circuit and ground. This limits the bandwidth of the recording system and negative-capacitance compensation is required to avoid attenuation of fast signals such as the action potential. A related but new method of potential measurement in the node of Ranvier that has overcome many of these difficulties has been designed by Frankenhauser.[81]

The schematic diagram of the experimental arrangement of Frankenhauser's Vaseline gap technique is shown in Fig. 11b, and the equivalent circuit is shown in Fig. 11c.† It can be seen in Fig. 11b that there are three nodes (-1, 0, and $+1$) involved in the circuit instead of the two nodes used in the Huxley–Stämpfli method. The third node (in pool E, Fig. 11b) is used to pass current and stimulate the preparation. The circuit equations of Frankenhauser's method can be deduced from Fig. 11c.

It was shown by Frankenhauser[81] [see Section 10.A.1, Eq. (10.A.1)] that the potential in pool A is given by:

$$V_A = -V_m A R_{BC} / [R_{BC}(A + 1) + R_{CD}], \qquad (10.6.1)$$

where V_m is the membrane potential of the node, A is the gain of amplifier A_1, and R_{BC} and R_{CD} are the resistances between the respective points in the circuit diagram‡ (Fig. 11c). This result demonstrates the effectiveness of the potentiometric recording of V_m since from Eq. (10.6.1), when the open-loop gain of the amplifier A_1 becomes large, V_A approximates the membrane potential,

$$V_A = -V_m \qquad A \to \infty. \qquad (10.6.2)$$

In this situation it can be seen that V_D and V_C are both equal to zero. This is the same as saying that V_C is virtual ground, and since B is at ground potential, no current can flow through R_{BC}, in which case, no current flows through the resistance R_{CD}, verifying that the recording of V_m is potentiometric. In fact, the effective loop resistance defined as $(V_A - V_C)/I_{CD}$ is very high, given approximately by

$$R_{loop} = R_{BC}(A + 1). \qquad (10.6.3)$$

In the derivation of Eqs. (10.6.1)–(10.6.3), we did not consider any current drain at the input of the amplifier A_1, for which purpose a high-impedance amplifier has to be used. Field-effect-transistor input amplifiers are currently available with impedances of 10^{12}–10^{13} Ω; this effectively prevents any

---

† In this diagram and the following ones, electrode potentials and balancing circuits are not included. In the actual implementation of the Frankenhauser method, a balancing circuit is required to cancel electrode potentials. If this balance potential is misadjusted, V_C will be different from zero; consequently, V_D will also deviate from zero potential, but in a larger proportion than V_C (for a normal node $V_D \approx 20 V_C$, Hille[60].

‡ In Section 10.A.1 and in Figs. 11–13 the linear circuit elements are labeled as generalized impedances. For simplicity, we use the same resistance notation in the text.

significant current from leaking into the amplifier itself. (Frankenhauser[81] used a grid pentode that drained less than 10^{-11} A.) It is interesting to note that the point D inside the node is held at virtual ground by the amplifier A_1 and that the external solution at pool A is at $-V_m$, which is the inverse of the situation in conventional methods of recording membrane potential. In the three Vaseline gap techniques of Frankenhauser, node +1 (pool E) is used to stimulate node 0. Vaseline insulation is necessary to inject current into the fiber, preventing current leaks between pools E and A. It is clear, however, by inspection of Fig. 11c that in this case the membrane potential V_m would be shunted by the resistance given by the series combination $R_{ED} + R_{EA}$. This would produce current flow through the membrane, defeating the whole purpose of the potentiometric method unless the potential at pool E is forced to be close to zero, preventing current flow across the R_{ED} resistance. In the actual implementation of the Frankenhauser technique, a stimulator is connected between E and ground, providing a low-resistance pathway given by the output impedance of the unit. In this case current will circulate through R_{ED} only during the stimulus when V_E is made different from zero. Frankenhauser[81] pointed out this problem and used a 500-Ω-output impedance stimulator (R_s) to correct it. It is concluded, then, that the stimulator used in the Frankenhauser method should be a voltage rather than a current generator.

Probably the most important improvement of Frankenhauser's method of measuring the potential of the node of Ranvier over previous methods is the fact that pool C, a high-impedance pool, is electronically at virtual ground. This means that stray capacitances to ground would be insignificant alternative current pathways, thus improving the frequency response of the recording circuit without the need of negative-capacity compensation and other speeding-up techniques. Frankenhauser[81] reported a bandwidth for his apparatus of 30 kHz which seems satisfactory for most electrophysiological studies.

10.6.1.3. Voltage Clamp of the Node of Ranvier. Several arrangements to control the membrane potential in a single node of Ranvier have been reported.[102–108] Of these, the most commonly used today is that introduced by Dodge and Frankenhauser[105] with modifications described by Nonner[108]

---

[102] J. del Castillo, J. Y. Lettvin, W. S. McCulloch, and W. Pitts, *Nature* (*London*) **180**, 1290 (1957).

[103] B. Frankenhauser and A. Pearson, *Acta Physiol. Scand.* **42**, Suppl. 145, 45 (1957).

[104] I. Tasaki and A. F. Bak, *J. Neurophysiol.* **21**, 124 (1958).

[105] F. A. Dodge and B. Frankenhauser, *J. Physiol* (*London*) **143**, 76 (1958).

[106] F. A. Dodge and B. Frankenhauser, *J. Physiol.* (*London*) **148**, 188 (1959).

[107] C. Bergman and R. Stämpfli, *Helv. Physiol. Pharmacol. Acta* **24**, 247 (1966).

[108] W. Nonner, *Pfluegers Arch.* **309**, 176 (1969).

and Hille.[109] Del Castillo and Moore (see Moore and Cole[20]) cut the nerve fibers in pool E to be able to inject current more efficiently into the node 0. Nonner[108] included as standard in his methods the cut of the fiber at the internodes; Hille[109] also cut the fibers at the internodes, but used the Dodge–Frankenhauser electronic arrangement using two amplifiers for voltage clamping instead of one as described by Nonner.[108] Figure 12a shows Hille's[109] modification of the Dodge–Frankenhauser system. The circuit diagram of the Dodge–Frankenhauser voltage clamp is shown in Fig. 12b.

We have previously shown that if A (the voltage gain of amplifier A_1) is large, the following statements are approximately valid:

  (a) $V_C = V_D = 0$,
  (b) $V_A = -V_m$,
  (c)   no current flows in circuit ADCB; therefore, current injected through R_{ED} flows only through circuit EDA.

It is easy to see how Frankenhauser's potentiometric voltage measuring circuit can be used in a voltage clamp configuration if we connect a second amplifier A_2 in the circuit as shown in Figs. 12a and 12b. The following equation can be written when the second amplifier A_2 of gain A' is connected:

$$V_E = A'(V_+ - V_-) = A'(V_A + V_{com}).$$

From this equation it follows that $V_A = -V_{COM}$; since V_A is maintained at $-V_m$ by amplifier A_1,

$$V_m = V_{COM}.$$

The conclusion is that amplifier A_2 passes current through the membrane to keep $V_m = V_{COM}$. In order to calculate the value of the current I_m we can observe that point D is at zero potential, and since no current flows through DCB, then

$$V_E = I_m R_{ED}, \qquad I_m = V_E/R_{ED}.$$

In Section 10.A.1 a general derivation is presented with Laplace-transformed variables considering A and A' as functions of frequency. The above equations are only valid in the limiting case when A and $A' \to \infty$ for all frequencies.

One of the limitations of the Dodge–Frankenhauser voltage clamp is related to the attenuation factor. When we discussed Frankenhauser's potentiometric method for measuring potential from the node of Ranvier, we found that the loop resistance with amplifier gain A was $R_{BC}(A + 1)$

[109] B. Hille, *J. Gen. Physiol.* **58**, 599 (1971).

(a)

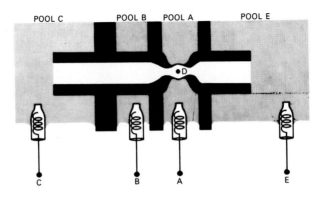

(b)

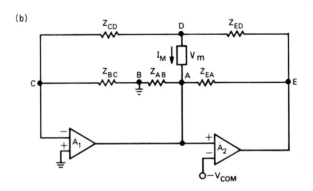

(c)

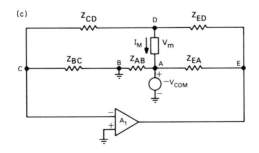

[Eq. (10.6.3)], which minimized the attenuation factor. A limitation in this method arises because pool B cannot be made wide enough to prevent the attenuation; a wide pool leads to a large phase lag in the potential recording system. This phase lag was acceptable for recording the membrane action potential but seriously limits the stability characteristics of the closed-loop configuration when the voltage clamp amplifier (A_2) is incorporated in the loop. A very simple rule defining the stability characteristics of a feedback control circuit is that the feedback should never become a positive feedback for gains larger than 1. Phase lags introduced by the loop network and the amplifier itself may transform the negative feedback into positive feedback if the signal frequency becomes higher than certain values. A phase lag of 90° between the input and output signals of a network is produced by each "pole" in a transfer function equation. A system with two poles, then, can produce a phase lag of 180°, and if such a network is in the feedback loop, the closed-loop system may become unstable for gains larger than 1. When the second amplifier (A_2) is added in order to voltage-clamp the preparation the situation becomes more critical. We can observe that the input signal to A_2 is V_A, which already has a phase lag introduced at least by the product $A(s)Z_{BC}(s)$. Amplifier A_2 will add at least another pole to the system, making necessary a careful analysis of the network closed-loop equations in order to

---

FIG. 12. Hille's modification of the Dodge–Frankenhauser[105] voltage clamp of the node of Ranvier and Nonner voltage clamp. (a) Schematic diagram: Myelinated fibers of about 16 μm internodal diameter and 1.5 mm internodal length are used. The nerve is mounted in an acrylic chamber in which the size of the pools is adjustable.[108] Four pools (shaded areas) are formed by laying three strings of Vaseline (black areas). The fiber is cut in pools E and C. Pools A and B contain Ringer's solution and pools E' and C contain isotonic KCl or CsF solution. The sizes of the gaps are 500 (BC), 200 (AB), and 200 μm (EA); the sizes of the pools are 250 (pool B) and 150 μm (pool A). Seal resistances are about 5 MΩ. The electrical connections are made through 1 M KCl agar bridges to calomel electrodes to which the electronic equipment is connected. V_{COM} includes the pulse and holding potential. (b) Equivalent circuit of the method: Amplifier A_1 has gain $A(s)$ and amplifier A_2 has gain $A'(s)$. Amplifiers A_1 and A_2 need to be phase-lag compensated. Membrane current (I_M) is measured as V_E/Z_{ED} and membrane potential as $-V_A$. To measure the membrane potential under current-clamp conditions, amplifier A_2 is disconnected, and E is connected to a pulse generator to stimulate the preparation as shown in Fig. 11c. For details see text and Section 10.A.1. (c) Equivalent circuit of Nonner's method: Same as (b) except that only one amplifier is used to voltage-clamp the preparation. The voltage generator used to impose the membrane potential (V_{COM}) had an output impedance of 200 Ω and the feedback amplifier A_1 had high input impedance (100 MΩ, 7 pF), low bias current (10^{-12} A), and low output impedance (100 Ω) in addition to wide bandwidth and high amplification (up to 86 dB). The amplifier was compensated based on measurements of the open-loop characteristics of the circuit using the Nyquist stability criterion.[58] The membrane current is measured indirectly (V_E) and the membrane potential is measured as the negative of the potential at pool A. In order to measure the voltage in a current-clamp configuration, the output of amplifier A_1 has to be connected to A (instead of E), disconnecting the stimulator V_{COM}. V_M is measured as $-V_A$ under those conditions.

evaluate its stability. Such a study has been done by Nonner,[108] who introduced a new voltage clamp system for the node preparation.

Nonner's method represents an improvement of the Dodge–Frankenhauser method mainly in two respects: (i) the attenuation artifact was eliminated even though the size of pool B was made very small, and (ii) by studying carefully the loop impedances and eliminating one of the amplifiers of the loop, the frequency response of the voltage clamp could be increased to at least 10 kHz, which allows one to observe the rapid ionic conductance transients occurring at room temperature. A circuit diagram of Nonner's technique[108] is shown in Fig. 12c, and a general derivation of the circuit equations is presented in Section 10.A.1. The air gap included in his technique helped to eliminate the attenuation artifact described by Dodge and Frankenhauser (discussed above), probably because the leak through the Schwann cell space disappeared with the drying of a certain portion of the internode. To record membrane potential under current clamp condition, Nonner[108] used directly Frankenhauser's potentiometric method, but to voltage-clamp the nodal membrane he switched the output of the potentiometric amplifier from pool A (current clamp) to pool E (voltage clamp). The voltage clamp method is also based on the fact that C becomes virtual ground when the gain A of the amplifier is large. In that condition the current through R_{CD} is zero, and point D is also virtual ground. The potential at pool A becomes $-V_m$. If a low-impedance voltage source is used to set the value of V_A to $-V_{COM}$, it follows from the equations discussed in Section 10.A.1 that $V_m = V_{COM}$. In order to measure the membrane current I_m we demonstrate also in Section 10.A.1 that $I_m = V_E/R_{ED}$. It is interesting to observe that the final voltage clamp condition is achieved with Nonner's technique without several of the stability complications present by the use of two amplifiers in the Dodge–Frankenhauser technique. The second amplifier (A_2) of the latter technique is redundant in the voltage clamp situation; A_2 can be replaced by connecting the output of amplifier A_1 to E and adding a low-impedance source to set the value of V_A as Nonner did. The equations can be compared with the Dodge–Frankenhauser loop equations to verify that the elimination of one of the amplifiers simplifies the equations for the loop and thus makes easier the design of a fast and stable voltage clamp. For an analysis of the stability characteristics, it is possible to set values for the terms included in the loop equations and study system performance using the stability criteria commonly used in feedback control systems (see, for example, D'Azzo and Houpis[58] and Nyquist[110]). Nonner[108] analyzed his voltage clamp system extensively and the reader is referred to his paper for an explicit discussion.

[110] H. Nyquist, *Bell Syst. Tech. J.* (1932).

Instrumental and thermal noise associated with Nonner's voltage clamp technique were analyzed by Conti *et al.*[111] when they used this technique to measure Na current fluctuations in the node of Ranvier. They found that the background thermal noise contributions from the partitions and passive membrane patch may be reduced by increasing Z_{ED}, Z_{BC}, and membrane impedance Z_m, and by decreasing Z_{CD}. Besides, the amplifier's voltage (e_n^2) and current (i_n^2) input noise are multiplied by factors (containing the impedances Z_{ED}, Z_{BC}, Z_{CD}, and Z_m) that are minimized by the same procedure.

10.6.2. Vaseline-Gap Techniques in Single Muscle Fibers

Frankenhauser *et al.*[112] demonstrated that the principles involved in potentiometric recording of membrane potential at the node of Ranvier can be applied to single muscle fibers. The error in the potential measurement in this latter preparation is about 1% at low frequencies, but can reach about 10% at frequencies of 50 kHz. Since a 50-kHz bandwidth is adequate for most purposes, their method has been attractive enough to encourage muscle investigators to use it for voltage clamp studies. Moore[113] applied for the first time the potentiometric method for triple-Vaseline-gap voltage clamp studies in single muscle fibers.

A significant improvement on the three-Vaseline-gap voltage clamp, and on muscle voltage clamps in general resulted from the work of Hille and Campbell.[114] These authors applied the Dodge–Frankenhauser[105] techniques to short segments of muscle fibers cut at both ends instead of one end as Moore[113] did. They also cut the fibers in a solution of isotonic CsF, which helped to keep the cut ends unswollen and left the muscle membrane in pool B depolarized but with high resistance. The diagram and schematic circuit of the Hille–Campbell voltage clamp technique is shown in Fig. 13. The circuit equations for that configuration are developed in Section 10.A.1

In order to have a fast and stable voltage clamp circuit, Hille and Campbell[114] employed the discrete amplifiers previously used by Nonner[108] that can be externally compensated. The compensation network for amplifiers A_1 and A_2 in the Hille–Campbell voltage clamp can be calculated theoretically from the open-loop equations and/or determined by trial and error until the open-loop Bode plot has a rolloff of less than 20 dB/decade.

[111] F. Conti, B. Hille, B. Neumcke, W. Nonner, and R. Stämpfli, *J. Physiol. (London)* **262**, 699 (1976).

[112] B. Frankenhauser, B. D. Lindley, and R. S. Smith, *J. Physiol. (London)* **183**, 152 (1966).

[113] L. E. Moore, *J. Gen. Physiol.* **60**, 1 (1972).

[114] B. Hille and D. T. Campbell, *J. Gen. Physiol.* **67**, 265 (1976).

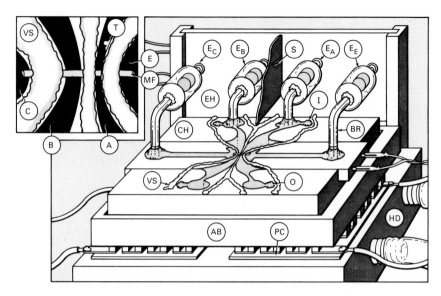

FIG. 13a. Voltage clamp of single muscle fibers with Vaseline-gap technique. Experimental chamber: The chamber (CH) is built of Lucite and rests on an aluminum block (AB) cooled by Peltier coolers (PC) connected to a feedback circuit to control the temperature measured by a thermistor (T) installed in pool A (see inset). HD is the heat dissipator for the Peltier coolers. The four electrodes (E_C, E_B, E_A, and E_E) are built with pellets of sintered Ag/AgCl in 1 M KCl connected to the chamber with agar bridges (BR) filled with 1 M KCl and containing a floating platinum wire to decrease the high-frequency impedance. The electrodes are mounted in an electrode holder (EH) that can be removed for storage of the electrodes. Pools are separated by Vaseline strings (VS), and solutions are changed by supplying solution at I and sucking it away at O. Inset: detail of a muscle fiber segment (MF) installed in the chamber. The pools are marked (A, B, C, E). VS are the Vaseline seals, and T is the thermistor.

The reason trial and error may be necessary is that the phase lags introduced by the preparation arise from a cable structure and not a lumped RC network.

One of the limitations of the Vaseline-gap voltage clamp (common to any voltage clamp system in which the length constant of the preparation is not modified by impalement with an axial wire) is that the membrane current is collected from a patch of membrane of finite length, while the potential control is restricted to a membrane ring at the AB partition. We shall discuss this problem extensively below in Section 10.6.4 and Section 10.A.2 as a general limitation of gap voltage clamps with special reference to a discussion by Hille and Campbell[114] of the muscle-fiber voltage clamp.

The Hille–Campbell technique for muscle-fibers voltage clamps has been used successfully to record sodium channel gating currents and excitation–contraction (EC) coupling charge movements.[115] It has further been

[115] J. Vergara and M. Cahalan, *Biophys. J.* **21**, 167a (1978).

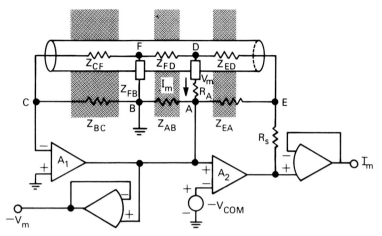

FIG. 13b. Voltage clamp of single muscle fibers with Vaseline-gap technique. Equivalent circuit: The diagram is similar to that of Fig. 12b (the Dodge–Frankenhauser voltage clamp). In this case there is an additional impedance Z_{FB} that represents the membrane patch in pool B. A resistance in series with the output of amplifier A_2 (R_s) is included to measure the total current I_E, but is selected of a small value (1 KΩ) and is not included in the derivation found in Section 10.A.1. Another resistance that has not been included in the derivations of Section 10.A.1 is a series resistance R_A with the membrane in pool A that should be compensated for by positive feedback.[114] It can be readily incorporated by replacing V_m by $V_m + I_m R_A$ in Eq. (10.A.14) of Section 10.A.1. A discussion of series resistance compensation is presented in Section 10.3.5.3.

used to study optical events related to the EC coupling in muscle fiber.[116,117] Vergara *et al.*[116] report a modification of the Dodge–Frankenhauser[105] chamber to include an optical fiber in pool A to illuminate the muscle fiber and verified that it can still be used for voltage clamp experiments. They also improved the Hille–Campbell technique in three aspects.

(1) The fibers were cut in relaxing solutions containing K-Aspartate instead of CsF or KF. Under these conditions the fibers are able to contract in the region of pool A (and only there) when they are depolarized. The movement associated with the depolarization is blocked by adding 2–4 mM EGTA in pool E and C and waiting about 20 min for diffusion.

(2) The fibers were depolarized for the first time, after dissection, in a high-K solution keeping the fibers stretched, thus preventing the over-shortening that occurs when they are depolarized without being held. Under those conditions the fiber contracts and relaxes as described by Hodgkin and Horowicz.[118] The sarcomere length of the fibers, mounted after

[116] J. Vergara, F. Bezanilla, and B. M. Salzberg, *J. Gen. Physiol.* **72**, 775 (1978).

[117] P. Palade, *Biophys. J.* **25**, 142a (1979).

[118] A. L. Hodgkin and P. Horowicz, *J. Physiol.* (*London*) **153**, 386 (1960).

this precaution is taken, is about $2.4 \pm 0.2 \,\mu m$[116] instead of $1.6 \pm 0.2 \,\mu m$ reported by Hille and Campbell.[114]

(3)   When the fibers were treated with TTX, normal-looking delayed K currents were described, whereas one of the major difficulties in the Hille–Campbell[114] report was that they were not able to record normal-looking K currents.

It should be pointed out finally that the fact that both ends of the fibers are cut means that both can be used to diffuse substances to the fiber interior and eventually replace the normal ionic content of the sarcoplasm. Undoubtedly, a good blockage of the K current can be obtained in this preparation by soaking the cut ends in a solution containing CsF.[114]

The Hille–Campbell technique and subsequent modifications of it constitute a reasonably accurate voltage clamp system for muscle fibers. We discuss below in Section 10.6.4 the limitations of gap isolation voltage clamps and give there error criteria for judgment of the longitudinal dispersion of the membrane potential of the fiber along the control pool. One conclusion to be drawn from that analysis is that if the gap is made small with respect to the radius of the fiber and the conductance of the membrane does not become too large, the membrane in the gap can be considered almost isopotential in the longitudinal direction.

10.6.3. Sucrose-Gap Methods

Stämpfli[99] introduced the use of deionized sucrose to increase the impedance of the recording gap. This technique was extended to voltage clamp by Julian et al.[119, 120] for nonmyelinated giant axons of the lobster and by Moore et al.[121] for the squid giant axon.

Two sucrose gaps and three pools (two lateral and one central) are defined for voltage-clamping a cylindrical cell. Julian et al.[119,120] called the lateral pools I pool (at which current was injected to the fiber) and V pool (at which voltage was recorded). The central pool was used for current recording and was kept at virtual ground by a current-to-voltage converter. The length of their sucrose gaps was about $600 \,\mu m$, and the central pool could be made as narrow as $50 \,\mu m$. The gap resistances were at least $25 \, M\Omega$, and under these conditions the potential measurements are accurate with less than 5% error. The sucrose-gap technique has the problem that there is a $20-60 \, mV$ hyperpolarization in the recorded potential[120] that has been suggested to arise from a liquid junction potential between sucrose and

[119] F. J. Julian, J. W. Moore, and D. E. Goldman, *J. Gen. Physiol.* **45**, 1217 (1962).

[120] F. J. Julian, J. W. Moore, and D. E. Goldman, *J. Gen. Physiol.* **45**, 1195 (1962).

[121] J. W. Moore, T. Narahashi, and W. Ulbricht, *J. Physiol. (London)* **172**, 163 (1964).

sea water.[122] Despite this problem, the currents recorded with the sucrose-gap technique in giant axon suggest that good potential control is achieved provided that the central gap length is not longer than the fiber diameter (see discussion below).

The sucrose-gap technique described by Julian et al.[119,120] has been used almost without modification in skeletal muscle fibers by Nakajima and Bastian.[123] These authors found, though, that the Julian–Moore–Goldman techniques could not be used directly with single muscle fibers of the frog but could be applied to Xenopus muscle fibers. Nakajima and Bastian[123] also found, as expected and discussed above, that an important bandwidth limitation (in the voltage clamp) is induced by stray capacitances to ground; their potential recording pool V is a high-impedance pool.

A modification of the double-sucrose-gap technique was made by Ildefonse and Rougier[124] to voltage-clamp single muscle fibers of the frog. Their system is not only a sucrose-gap technique but also uses Vaseline and sucrose.

Recently, Duval and Leoty[125] have used a double-sucrose-gap technique in mammalian muscle fibers in which they cut the ends of the fibers, thus improving the performance of the voltage clamp. Duval and Leoty[125] also measured the potential in the test gap lengthwise with a microelectrode, verifying that there was a partial lack of longitudinal control. They decided that gap lengths of 100 μm for fibers of 50–70 μm were safe values to prevent this lack of control. In the next section we discuss this problem extensively.

10.6.4. Errors Introduced by the Finite Length of the Gap

The measurement of current in the gap voltage clamp will only be exact if the length of the gap is made infinitesimally small. We consider in this section the errors introduced by making the gap of finite length.

This problem has been considered theoretically by Cole (Ref. 7, p. 418) and is analyzed in detail in Section 10.A.2. We show there that in most experimental situations the core-conductor approximation is valid.† We give formulas to calculate both the maximum error in the voltage at the EA

[122] M. P. Blaustein and D. E. Goldman, Biophys. J. 6, 453 (1966).

[123] S. Nakajima and J. Bastian, J. Gen. Physiol. 63, 235 (1974).

[124] M. Ildefonse and O. Rougier, J. Physiol. (London) 222, 373 (1972).

[125] A. Duval and C. Leoty, J. Physiol (London) 278, 403 (1978).

† Our three-dimensional model in Section 10.A.2 does not include a T-system network; consequently, the core-conductor cable model may not be a good description of a muscle fiber in a gap.

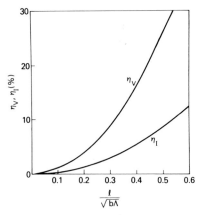

FIG. 14. Errors produced by finite gap length: Maximum error in membrane voltage η_V and in membrane current η_I produced by the length of the gap as computed by Eq. (10.A.28) and (10.A.29) derived from the core-conductor model in Section 10.A.2.

partition with respect to the controlled voltage and the error in the measured current with respect to the current across a controlled patch of membrane. These errors are plotted in Fig. 14 as a function of a single parameter $l/\sqrt{b\Lambda}$, in which l is the length of the gap at pool A, b is the fiber diameter, and Λ is the generalized space constant defined in Section 10.A.2. The experimentalist can estimate from Fig. 14 the maximum gap length that assures him voltage and current control, within a selected tolerance, when the fiber parameters b and Λ are known. Hille and Campbell[114] calculated the longitudinal variations of the potential inside the fiber assuming that the membrane current is constant and that the fiber in pool A can be considered as a semi-infinite cable. They derived

$$V(l) = I_M R_i l^2/A. \tag{10.6.4}$$

In the potentiometric method, the internal potential $V_D(z)$ can be defined as

$$V_D(z) = V(z) - V_0,$$

in which $V(z)$ is the membrane potential as a function of the distance from the AB partition and V_0 is the potential at the AB partition ($z = 0$). If we make this change of variable in Eq. (10.A.27) of Section 10.A.2, we obtain (see also Cole[7])

$$V_D(z) = V_0(\cosh(z\sqrt{2}/\sqrt{b\Lambda}) - 1). \tag{10.6.5}$$

A first-order approximation of the Taylor-series expansion of Eq. (10.6.5) evaluated at $z = l$ gives the Hille–Campbell result [Eq. (10.6.4)] provided that the membrane current $I_M = V_0/R_m$ is constant over the entire length of the gap. It is interesting to observe in Fig. 14, though, that only at $l/\sqrt{b\Lambda}$ smaller than 0.05 can the membrane current be assumed constant, but at larger and still realistic values of $l/\sqrt{b\Lambda}$ (for an active membrane for example) this approximation no longer holds and the exact formula should be used. Within the restrictions imposed by the first-order approximation, Hille and Campbell[114] showed that $V_D(l)$ can be roughly estimated from the potential at pool E (V_E) by the formula

$$V_D(l) = V_E l/(l + 2k),$$

in which k is the length of the fiber segment from $z = l$ to the cut end at pool E. An exact formula relating $V_D(l)$ and V_E can be derived from Eq. (10.6.5) and other equations for the core-conductor model included in Section 10.A.1. The result

$$V_D(l) = V_E \frac{\cosh(l\sqrt{2}/\sqrt{b\Lambda}) - 1}{\cosh(l\sqrt{2}/\sqrt{b\Lambda}) + (k\sqrt{2}/\sqrt{b\Lambda}) \sinh(l\sqrt{2}/\sqrt{b\Lambda}) - 1}$$

shows that, in the exact case, it is necessary to know the membrane parameters to estimate $V_D(l)$ as a function of V_E. The above equations may give the wrong impression that the potential $V_D(l)$ diminishes as k is made larger. This is not the case because the feedback amplifier (A_2) makes V_E larger if k increases, compensating for the change. The deviations of $V_D(l)$ are only functions of the error η_V in Fig. 14 and the imposed potential V_0.

The error criteria developed above can be used to roughly estimate the errors in voltage-clamping an active membrane. For example, when the K conductance increases upon depolarization Λ decreases, and if the current has reached a steady state (that is, when there is no K inactivation), a fairly good estimate of the errors in the records can be made from Fig. 14. In a non-steady-state situation Fig. 14 may still be used to obtain a rough estimate of the errors. For the case of negative conductance we have included in Section 10.A.2 Eqs. (10.A.31) and (10.A.32) to estimate η_V^a and η_I^a, respectively. It can be noticed that η_V^a and η_I^a in this case have opposite signs with respect to those calculated in the positive conductance case. Numerical computations with Eqs. (10.A.31) and (10.A.32) show that for values of $l/\sqrt{b\Lambda}$ in the range used in Fig. 14, η_V^a and η_I^a do not differ significantly, in absolute value, from η_V and η_I. These equations could be used in the steady state for cases in which Na inactivation is absent, but care should be exercised in using them when normal sodium conductance is present. For both positive and negative conductances in the non-steady-state situation, a better estimation of the errors can be obtained with numerical computations of the cable equations using the Hodgkin and Huxley equations. This has been done for the sucrose gap case by Moore *et al.*[126,127]

10.7. Concluding Remarks

The concept that all cells are bounded by a superficial layer with very special properties was derived from measurements of electrical and osmotic properties. These cell membranes are highly permeable to water and to lipid soluble substances, and they transport ions, sugars, and amino acids up and down electrochemical gradients. In this article we have been concerned with certain techniques which are used to study the electrical properties of some cell membranes which are excitable, i.e., they respond to electrical or chemical stimuli and produce electrical effects. In order to study these electrical properties it is necessary to measure or control the voltage across and the current through the membrane over an area in which they are relatively uniform. This has been done most successfully with the giant axon of the squid, the node of Ranvier in myelinated axons, and by the development of gap techniques and patch isolation with external electrodes. Some very good work has been done using microelectrodes in both cylindrical and spheroidal cells, but the high resistance of such electrodes and problems of uniformity make quantitative studies difficult.

Much of this paper would be relevant to attempts to control current (current clamp), but the voltage clamp is emphasized because it is the natural quantity to control in a system with large capacitance and voltage-dependent elements.

10.A. Appendix

10.A.1. Circuit Equations of Vaseline-Gap Voltage Clamp

The purpose of this section is to develop the circuit equations of the potentiometric method of Frankenhauser, and three commonly used voltage clamp systems: (i) the Dodge–Frankenhauser voltage clamp of a myelinated-fiber node of Ranvier, (ii) the Nonner voltage clamp of a node of Ranvier, and (iii) the Hille and Campbell voltage clamp of a skeletal muscle fiber. The equations developed here can be adapted to modifications of these methods and we think they will help in understanding the basic

---

[126] J. W. Moore, F. Ramon, and R. W. Joyner, *Biophys. J.* **15**, 11 (1975).
[127] J. W. Moore, F. Ramon, and R. W. Joyner, *Biophys. J.* **15**, 25 (1975).

principles involved in their application. Voltage and currents are Laplace-transform variables, and the circuit elements are defined as generalized impedances Z_{xy} that may behave as pure resistors or a combination of linear elements. The potential at a point W of the circuit is called v_W. The transformed membrane potential and current are called v_m and i_m, respectively. The diagrams shown in Figs. 11–13 are used in the derivation of the circuit equations. Neither potentials associated with the electrodes nor balancing potentials will be considered here for two reasons: (i) the resulting equations are simpler, and (ii) every electrode potential should be balanced electronically to reach the configuration analyzed in the circuit diagrams that corresponds to the ideal case.

10.A.1.1. Potentiometric Method of Frankenhauser.

The experimental arrangement and circuit diagrams are shown in Fig. 11c with $V_p = 0$. Using elementary circuit analysis and the relation

$$v_A = -Av_C,$$

we find

$$v_A = -v_m A Z_{BC}/(Z_{BC}(A + 1) + Z_{CD}) \tag{10.A.1}$$

and

$$i_m = -v_m(1 + K)/[Z_{BC}(A + 1) + Z_{CD}], \tag{10.A.2}$$

with

$$K = [(Z_{BC} + Z_{CD})(R_S + Z_{EA}) + AR_S Z_{BC}]/[R_S(Z_{EA} + Z_{ED}) + Z_{ED}Z_{EA}]. \tag{10.A.3}$$

In order to record the actual membrane potential without attenuation, i_m must be zero; this condition is only met when both $A \to \infty$ and $R_s \to 0$.

In practice when $A \to \infty$ we get, for R_S small,

$$i_m \to -v_m R_s/Z_{EA} Z_{ED}, \tag{10.A.4}$$

from which we can estimate how large R_s can be to keep i_m under a specified tolerance. For the node of Ranvier $R_s \approx i_m/v_m \times 10^{14} \, \Omega$. To drain less than 10^{-11} A from the node at $V_m \approx 100$ mV, $R_s = 10$ kΩ. In the case of the potentiometric method applied to muscle fibers, Z_{ED} is at least 100 times smaller than in myelinated nerve fibers and one can allow a maximum current drain from the membrane patch of about 10^{-10} A. This requires a resistance R_s of at most 1 kΩ.

10.A.1.2. Dodge–Frankenhauser Voltage Clamp of the Node of Ranvier. We consider the experimental arrangements and circuit diagrams shown in Figs. 12a and 12b, respectively. The final equations will contain the terms A and A', but only after the final derivation is made will the limits A and $A' \to \infty$ be taken. Ideally, A and A' are complex gains of the form $A(s) = A_0/(1 + s\tau)$, and the reader can use the equations derived here making that replacement to study the performance of his particular voltage clamp giving the appropriate form to every impedance element as well. In this case, we can analyze the circuit equations with the following equations for the amplifier configurations:

$$v_E = A'(v_A + v_{COM}) \tag{10.A.5}$$

$$v_A = -Av_C. \tag{10.A.6}$$

We obtain for v_m the following equation:

$$v_m = (v_{COM} A' - i_m Z_{ED})$$
$$\times [Z_{BC}(A + 1) + Z_{CD}]/[(Z_{BC} + Z_{CD} + Z_{ED}) + AA'Z_{BC}]. \tag{10.A.7}$$

For stability analysis of the type described earlier it is necessary to compute v_m/v_{COM} which can readily be obtained from Eq. (10.A.7) and to assume a linear membrane impedance to relate i_m and v_m. Thus i_m is given by

$$i_m = \{v_E[AA'Z_{BC} + (Z_{BC} + Z_{CD} + Z_{ED})]$$
$$- A'v_{COM}(Z_{BC} + Z_{CD} + Z_{ED})\}/AA'Z_{ED}Z_{BC}. \tag{10.A.8}$$

If we take the limits $A \to \infty$ and $A' \to \infty$ in Eqs. (10.A.7) and (10.A.8) we obtain respectively

$$v_m = v_{COM} \tag{10.A.9}$$

and

$$i_m = v_E/Z_{ED}. \tag{10.A.10}$$

Equations (10.A.9) and (10.A.10) have always been considered the fundamental equations of the Dodge–Frankenhauser voltage clamp, but it should be kept in mind that Eqs. (10.A.7) and (10.A.8) are the more general equations because they consider A and A' finite and frequency dependent.

10.A.1.3. Nonner's Voltage Clamp. We shall follow the same scheme as in part (a) but the circuit equations will be solved for the diagram of Fig. 12c. The gain of amplifier A_1 is connected in such a way that

$$v_E = -Av_C \tag{10.A.11}$$

We can calculate v_m:

$$v_m = v_{COM} - i_m Z_{ED}(Z_{BC} + Z_{CD})/[(A + 1)Z_{BC} + Z_{CD} + Z_{ED}]$$

(10.A.12)

and

$$i_m = v_E[1/Z_{ED} - (Z_{CD} + Z_{BC} + Z_{ED})/A Z_{BC} Z_{ED}]. \quad (10.A.13)$$

If we take the limit $A \to \infty$, we get Eqs. (10.A.9) and (10.A.10) derived for the Dodge–Frankenhauser voltage clamp. It can be observed though that in Nonner's clamp, the general expressions for V_m and I_m are simpler than in the Dodge and Frankenhauser method. Because a single amplifier is used in the voltage clamp circuit the product of AA' present in Eqs. (10.A.7) and (10.A.8) does not appear in Eqs. (10.A.12) and (10.A.13).

10.A.1.4. **Hille–Campbell Voltage Clamp of Single Muscle Fibers.** The difference between the Hille–Campbell and Dodge–Frankenhauser methods is that in the former there is a patch of membrane in pool B not existing in the latter. This is shown in Fig. 13b as the impedance Z_{FB}. This term slightly changes the equations derived in part (b), giving

$$v_m = (v_{COM} A' - i_m Z_{ED})\alpha/\beta, \quad (10.A.14)$$

with

$$\alpha = Z_{BC} Z_{FB}(A + 1) + Z_{FD}(Z_{BC} + Z_{CD}) + Z_{CD} Z_{FB}$$

and

$$\beta = Z_{ED}(Z_{FB} + Z_{BC} + Z_{CF}) + (Z_{FB} + Z_{FD})(Z_{CD} + Z_{BC}) + AA' Z_{BC} Z_{FB}.$$

If we take the limit $Z_{FB} \to \infty$ in α and β, Eq. (10.A.14) becomes identical to Eq. (10.A.7). The equation for i_m is

$$i_m = v_E \beta/(AA' Z_{ED} Z_{BC} Z_{FB}) - v_{COM}[\beta/A Z_{BC} Z_{AB} Z_{ED}) - A'/Z_{ED}]. \quad (10.A.15)$$

If we take the limits $A \to \infty$ and $A' \to \infty$ we get Eqs. (10.A.9) and (10.A.10), which are then common for the three voltage clamp techniques discussed in this section. It should be noted that in Eqs. (10.A.14) and (10.A.15) the gain A of amplifier A_1 appears always multiplied by the impedance term $Z_{FB} Z_{BC}$ (e.g., $AZ_{FB} Z_{BC}$) instead of Z_{BC} as is the case for the node of Ranvier. This difference is important in designing a compensating network for stabilization of the voltage clamp of muscle fibers.

10.A.2. Potential Distribution for a Fiber in a Gap Voltage Clamp

In this section we first consider models that describe the membrane potential distribution at the control pool (pool A) in nonmyelinated fibers voltage clamped using a Vaseline- (or sucrose-) gap technique. We then study the influence of the gap length and fiber characteristics (space constant and diameter) on the accuracy of the voltage control and current measurements when the Vaseline- (or sucrose-) gap voltage clamp is used.

The first model used is a three-dimensional cylinder extending from partition AB ($z = 0$) to partition EA ($z = l$) in the case of a Vaseline-gap arrangement.

The potential $V(r, z)$ anywhere in the fiber can be obtained by solving the Laplace equation in cylindrical coordinates:

$$\frac{\partial^2 V}{\partial r^2} + \frac{1}{r}\frac{\partial V}{\partial r} + \frac{\partial^2 V}{\partial z^2} = 0, \qquad (10.A.16)$$

where r is the radial coordinate from the axis of the cylinder and z is the longitudinal coordinate along the axis. There is no circular dependence because there is circular symmetry imposed by the following boundary conditions at the ends:

$$\partial V/\partial z = 0, \qquad z = 0, \qquad (10.A.17)$$

$$\partial V/\partial z = R_i I_1, \qquad z = l, \qquad (10.A.18)$$

where R_i is the internal resistivity (in Ω cm) and I_1 is the density of current (in A/cm^2) injected at pool E which has no radial or angular dependence.

The membrane boundary condition is given by

$$\frac{1}{R_i}\frac{\partial V}{\partial r} + \frac{V}{R_m} = 0, \qquad r = b, \qquad (10.A.19)$$

where we have assumed that the potential distribution has reached a steady state and that the external solution is isopotential at a potential zero.† R_m is the membrane resistance (in Ω cm^2) and b is the fiber radius.

---

† The condition of external potential equal to zero has been used to simplify the mathematical treatment and is directly applicable to the case of sucrose gap when using electrometric potential measurement. When using potentiometric potential measurement the *internal* potential is made virtual ground (as in the Dodge–Frankenhauser or Nonner methods; to use this treatment the internal potential $V_D(z, r)$ can be calculated as $V_D(z, r) = V(z, r) - V_0$.

We solved this boundary value problem applying the Hankel transform with the kernel (Özisik,[128] p. 135)

$$\sqrt{2}\,J_0(\beta_m r)/\{b[(R_i/R_m \beta_m)^2 + 1]^{1/2}J_0(\beta_m b)\},$$

where J_0 is the Bessel function of zeroth order and β_m are the roots of the equation

$$\beta J_1(\beta b) - \Lambda^{-1}J_0(\beta b) = 0,$$

where J_1 is the Bessel function of order one and Λ is the generalized space constant (see, e.g., Eisenberg and Johnson[75]) given by

$$\Lambda = R_m/R_i.$$

The resultant transformed ordinary differential equation

$$-\beta_m^2 \overline{V} + d^2\overline{V}/dz^2 = 0,$$

in which $\overline{V}$ indicates the Hankel transform of $V(r, z)$, was integrated, giving

$$\overline{V} = (\overline{I}_1 R_i/\beta_m)[\cosh(\beta_m z)/\sinh(\beta_m l)], \qquad (10.A.20)$$

in which $\overline{I}_1$ is the Hankel transform of I_1. The potential $V(r, z)$ can be obtained by inversion of Eq. (10.A.20) (Ref. 128, p. 135), giving

$$V(r, z) = 2I_1 R_i b \sum_{m=1}^{m=\infty} \frac{J_1(\beta_m b)J_0(\beta_m r)\cosh(\beta_m z)}{J_0^2(\beta_m b)[(b/\Lambda)^2 + b^2\beta_m^2]\sinh(\beta_m l)}. \qquad (10.A.21)$$

The membrane potential $V(b, z)$ is given by

$$V(b, z) = 2I_1 R_i \sum_{m=1}^{m=\infty} \frac{b/\Lambda}{\beta_m[(b/\Lambda)^2 + (b\beta_m)^2]}\frac{\cosh(\beta_m z)}{\sinh(\beta_m l)}. \qquad (10.A.22)$$

As a second model[7] we consider the simple core conductor (see, e.g., Taylor[5]),

$$d^2V/dz^2 = 2V/b\Lambda \qquad (10.A.23)$$

with the boundary conditions

$$dV/dz = 0, \quad z = 0; \quad dV/dz = I_1 R_i, \quad z = l. \qquad (10.A.24)$$

The solution is

$$V(z) = I_1 R_i \sqrt{\tfrac{1}{2}b\Lambda}\,\frac{\cosh(z\sqrt{2/b\Lambda})}{\sinh(l\sqrt{2/b\Lambda})}. \qquad (10.A.25)$$

[128] M. N. Özisik, "Boundary Value Problems of Heat Conduction." International Textbook Co., Scranton, Pennsylvania, 1968.

It can be demonstrated algebraically that when b/Λ is very small, Eq. (10.A.22) becomes Eq. (10.A.25). We have numerically computed and compared the results of Eqs. (10.A.22) and (10.A.25) and found that the core-conductor model [Eq. (10.A.22)] does not deviate more than 3% from the three-dimensional model [Eq. (10.A.25)], provided that two conditions are met simultaneously: (i) $b/\Lambda < 0.1$ and (ii) $l/b < 5$. These conditions are fulfilled in most experimental cases; therefore, we shall discuss the accuracy of the gap voltage clamp only considering the simpler case of the core-conductor model.

From Eq. (10.A.25) we can see that the membrane potential at $z = 0$, defined as V_0, is equal to

$$V_0 = I_1 R_i \sqrt{\tfrac{1}{2}b\Lambda}/\sinh(l\sqrt{2/b\Lambda}). \qquad (10.A.26)$$

The voltage clamp circuit is imposing V_0 at $z = 0$ and the potential as a function of the longitudinal parameter is then

$$V(z) = V_0 \cosh(z\sqrt{2/b\Lambda}). \qquad (10.A.27)$$

This equation allows us to quantify the deviation of the membrane potential from the controlled potential V_0, being maximal at $z = l$. We can estimate this maximal deviation of the membrane potential (given as percentage error, η_V) as a function of the gap length l, the fiber diameter b, and the space constant Λ. This is given by

$$\eta_V(\%) = [(V(l) - V_0)/V_0] \times 100;$$

then,

$$\eta_V = 100[\cosh(l\sqrt{2/b\Lambda}) - 1]; \qquad (10.A.28)$$

η_V has been plotted as a function of $l/\sqrt{b\Lambda}$ in Fig. 14. We can also estimate the deviation of the measured current from the current circulating at a controlled membrane patch ($z = 0$). At this point the membrane current density is given by

$$I_0 = V_0/R_m.$$

The measured membrane current density is experimentally defined as

$$I_m = I_1(\pi b^2)/2\pi b l = I_1 b/2l.$$

Now we can define the percentage error in the current (η_I) as

$$\eta_I = [(I_m - I_0)/I_0] \times 100,$$

and replacing in Eq. (10.A.26),

$$\eta_I = 100\,\frac{\sinh(l\sqrt{2/b\Lambda})}{l\sqrt{2/b\Lambda}}\,; \qquad (10.A.29)$$

η_I is plotted as a function of $l/\sqrt{b\Lambda}$ in Fig. 14.

The above discussion is valid for steady-state positive membrane conductance. In order to roughly estimate the error during the activation of a negative conductance[7] we may use the same derivation and replace, in Eq. (10.A.25), Λ by Λ_a, defined as

$$\Lambda_a = -R_m/R_i.$$

When this replacement is done we obtain

$$V(z) = I_1 R_i \sqrt{\tfrac{1}{2}b\Lambda_a}\,\frac{\cos(z\sqrt{2/b\Lambda_a})}{\sin(l\sqrt{2/b\Lambda_a})}\,. \qquad (10.A.30)$$

The errors η_V^a and η_I^a can be calculated in an equivalent way as done above:

$$\eta_V^a = 100[\cos(l\sqrt{2b\Lambda_a}) - 1], \qquad (10.A.31)$$

$$\eta_I^a = 100[\sin(l\sqrt{2b\Lambda_a})]/l\sqrt{2b\Lambda_a}. \qquad (10.A.32)$$

Acknowledgments

We thank Dr. Richard FitzHugh for checking the derivation of the expression for the differential electrodes. This work was supported by UPHS AM25201.

11. LIPID MODEL MEMBRANES

By G. Szabo and R. C. Waldbillig

11.1. Biological Membranes

The barrier between life and death is only two molecules thick. That is to say that the plasma membrane which envelops all living cells is an extremely thin film composed of an apposed pair of lipid molecules. Since the cell membrane serves as the boundary between the orderliness of life and the splendid chaos of the nonliving universe, it is not entirely surprising that the membrane system is functionally complex. In its essence the cell plasma membrane is a vectorial barrier which modulates the interaction of the cytoplasmic substance with the external world.

That individual cells are ordinarily small and that the plasma membrane is a vanishingly thin structure make the experimental analysis of membranes problematical. To overcome the cell size limitation, biologists have identified large-cell oddities of nature, such as the squid giant axon. To overcome limitations imposed by the complexity of plasma membranes, biophysicists have developed membrane model systems. These simplified model membranes represent a compromise between the separate demands of physics, chemistry, and biology. The main advantages of the artificial membrane are its physical and chemical simplicity and its low background conductance. The objective of this article is to provide a limited overview of current methods in model membrane science; it is meant to provide the reader with a noncritical introduction to the utilitarian aspects of lipid model membranes.

The notion that the cell plasma membrane is a modified lipid bilayer is so pervasive in biology that the origin of this concept is sometimes not fully appreciated. There is a strong historical link between interfacial science and membrane biology. As early as 1774 Benjamin Franklin demonstrated that a few drops of fat were sufficient to becalm a large area of a lake.[1]

---

[1] V. J. Bigelow, "The Complete Works of Benjamin Franklin," Vol. 5, p. 253. Putnam, New York, 1887.

METHODS OF EXPERIMENTAL PHYSICS, VOL. 20

Observation led to experimentation, and by the late 1800s physicists such as Pockels[2] and Rayleigh[3] had demonstrated that lipids form monomolecular films at an air/water interface. During the same period biologists came to recognize that living cells were bounded by a thin oily diffusion barrier.[4,5] By 1917 Langmuir[6] had completed an elegant series of physical studies describing the equation of state of a monolayer at the air/water interface. Langmuir's biologically esoteric studies of the interactions of lipids with water are now recognized as a critical link between interfacial physical chemistry and membrane biology.

A turning point in membrane biology came in 1925 in a laboratory of pediatrics where Gorter and Grendel[7] measured the surface area of monolayers made from lipid extracts of red blood cells. They found the area of a lipid monolayer to be about twice the surface area estimated for the intact cells. They correctly concluded that each cell was enveloped by a pair of monolayers. Working independently, Fricke[8] measured the impedance of red blood cells and found that the membrane behaved as a parallel-plate capacitor of bimolecular thickness. It was known from x-ray diffraction studies[9,10] of lipid crystals that the molecules were arranged in a lamellar orientation, and in 1934 Blodgett demonstrated[11,12] that lipid monolayers could be joined together to form bilayers having the lipid molecules arranged in a tail-to-tail configuration. Recent electron-microscopic evidence[13] has confirmed this structure. In 1935 Danielli and Davson[14] proposed a molecular model of cell membranes. Studies of oil/protoplasm interfaces led them to conclude that the cell membrane was organized as a pair of apposed lipid monolayers in close association with surface active protein. The concept of a modified lipid bilayer plasma membrane structure has remained largely intact, but it is now recognized that much of the membrane protein is embedded in the bilayer.

[2] A. Pockels, *Nature (London)* **43**, 437 (1891).
[3] Lord Rayleigh, *Philos. Mag.* [5] **48**, 321 (1899).
[4] H. Meyer, *Arch. Exp. Pathol. Pharmakol.* **42**, 109 (1899).
[5] E. Overton, *Vierteljahrsschr. Naturforsch. Ges. Zuerich* **44**, 88 (1899).
[6] I. Langmuir, *J. Am. Chem. Soc.* **39**, 1848 (1917).
[7] E. Gorter and F. Grendell, *J. Exp. Med.* **41**, 439 (1925).
[8] H. Fricke, *Phys. Rev.* **26**, 682 (1925).
[9] A. Muller, *J. Chem. Soc.* **123**, 2053 (1923).
[10] A. Muller and G. Shearer, *J. Chem. Soc.* **123**, 3156 (1923).
[11] K. B. Blodgett, *J. Amer. Chem. Soc.* **56**, 495 (1934).
[12] K. B. Blodgett, *J. Amer. Chem. Soc.* **57**, 1007 (1935).
[13] R. E. Pagano and N. L. Gershfeld, *J. Phys. Chem.* **76**, 1238 (1972).
[14] J. Danielli and H. Davson, *J. Cell. Comp. Physiol.* **5**, 495 (1935).

11.2. Lipid Monolayers

Since the historical origin of modern membrane chemistry and physics stems from studies of lipid monolayers, it is important to understand the fundamental characteristics of the air/water interface. The basic techniques for forming lipid monolayers are very well established[15] and need not be detailed here. The study of monolayer spread at the air/water interface rests upon the fact that the lipid molecules adsorb to an interface between polar (water) and nonpolar (air) media. In practice the lipid must be nonvolatile and virtually insoluble in water. Thus the polar head end of the lipid projects into the water phase, while its apolar fatty acyl tail projects into the air. The interaction of the lipid with the water phase alters the interfacial tension and interfacial dipole potential.[15] By analogy to three-dimensional states of matter the monomolecular lipid film can be thought to exert a surface pressure π determined by the difference in the interfacial tension in the absence γ_0 and presence γ_m of the lipid monolayer:

$$\pi = \gamma_0 - \gamma_m.$$

Thus compression of the monolayer with a moveable surface piston increases the surface pressure as the lipid packing density increases. The pressure/area curve of the monolayer is analogous to the ordinary pressure/volume curve with the monolayer exhibiting gaseous, liquid, and condensed states.[15–19] In an elementary way the monolayer pressure/area curve describes the energetics of the molecular kinetics in the two-dimensional surface film.

For purposes of illustration let us consider a representative example of recent studies involving the use of monolayers as highly simplified models of the cell membrane. Although the mixing properties of organic bulk solutions are fairly well understood, this is not the case for lipids in thin films. An effort has been made to better understand the interactions of lipids in cell membranes by systematically studying the mixing characteristics of amphipathic molecules in monolayers.[20]

An example of the utilitarian nature of monolayers can be found in the results of a series of studies of lipid interactions at very low film pressures

[15] G. L. Gains, "Insoluble Monolayers at Liquid-Gas Interfaces." Wiley (Interscience), New York, 1966.

[16] D. A. Cadenhead, *Biochim. Biophys. Acta* **455**, 169 (1970).

[17] D. G. Dervichian, *Prog. Chem. Fats Other Lipids* **2**, 193 (1954).

[18] D. O. Shah, *Adv. Lipid Res.* **8**, 347 (1970).

[19] D. O. Shah, *Prog. Surf. Sci.* **3**, 221 (1973).

[20] Y.-C. Lim and J. C. Berg, *J. Colloid Interface Sci.* **51**, 162 (1975).

(i.e., infinite dilution). It has been shown[21,22] that compression of a gaseous lipid monolayer (i.e. 10^5 Å^2/molecule) to a more condensed state changes the structure of the associated water. The thermodynamics of the monolayer liquid-expanded to liquid-condensed transition suggest that the tail region of the lipid is energetically similar to an ordinary bulk liquid hydrocarbon. Importantly, when two types of lipids coexist in the same monolayer, their mixing properties are generally consistent with the expectations of regular solution theory.[13] That is, the heat of vaporization and the molar volume of the lipid tail moiety are the primary determinants of the miscibility of the two lipids within the interfacial film. This is reflected in the fact that different lipids having the same class of π–area curve (i.e., liquid-condensed, liquid-expanded, etc.) tend to mix ideally, whereas dissimilar lipids exhibit positive deviations from ideality and immiscibility.[23] Membrane-specific lipids, such as lecithin and cholesterol, are immiscible in monolayers above their equilibrium spreading pressure; consequently, the two components probably coexist as separate phases in the mixed film. Low-pressure equilibrium studies have also revealed that lecithin monolayers formed at temperatures below the lipid liquid-crystalline phase transition temperature are metastable.[24] This suggests that the interpretations of the results of many earlier monolayer studies of lecithin and lecithin/cholesterol mixtures[16,19] must be reconsidered. Moreover, recent studies of the surface vapor pressure of gaseous monolayers[24] suggest that the internal energy of the monolayer system may be very different from that of lipid bilayers.[23] Caution must therefore be exercised in assuming that a lipid monolayer at the air/water interface is simply one-half of a lipid bilayer.

11.3. Spherical Lipid Bilayers

Cell plasma membranes are curved thin films which envelope the aqueous cytoplasm and serve to isolate it from the external aqueous environment. Spherical lipid vesicles are model membranes optimized to simulate the geometry of living system. The properties and preparation of spherical vesicles have been reviewed in detail by others,[25–27] so it is appropriate to limit the present discussion to several of the main points of concern.

[21] N. L. Gershfeld and R. E. Pagano, *J. Phys. Chem.* **76**, 1231 (1972).

[22] B. W. MacArthur and J. C. Berg, *J. Colloid Interface Sci.* **168**, 201 (1979).

[23] N. L. Gershfeld and K. Tajima, *J. Colloid Interface Sci.* **59**, 597 (1977).

[24] L. W. Horn and N. L. Gershfeld, *Biophys. J.* **18**, 301 (1977).

[25] A. D. Bangham, M. W. Hill, and N. G. A. Miller, *Methods Membr. Biol.* **1**, 1 (1974).

[26] T. E. Thompson, C. Huang, and B. J. Litman, *in* "The Cell Surface in Development (A. Moscona, ed.), p. 1. Wiley, New York, 1974.

[27] D. A. Tyrrell, T. D. Heath, C. M. Colley, and B. E. Ryman, *Biochim. Biophys. Acta* **457**, 259 (1976).

Plasma membrane lipids exhibit soaplike lamellar structures which can be regarded as smectic liquid crystals. For example, when thoroughly dry phospholipids are exposed to water, the lipid phase begins to swell because of the intercalation of water between the layers of lipids. If swelling continues, the lipid lamellar phase tends to lose long-range three-dimensional order, and the array degenerates into swirls of parallel bilayers. Mild agitation of these complexes leads to the formation of multilamellar spherical structures known as liposomes or "bangosomes." These concentric spheres of bilayers are the most elementary form of vesicular model membrane.

Multilamellar liposomes can be formed in a straightforward manner.[28,29] The usual procedure involves spreading a thin layer of lipid on the interior of a rotor evaporation flask by extraction of a volatile solvent such as chloroform. After evaporation of the lipid solvent a few glass beads and a small quantity of water are added, and the flask is agitated gently to disperse the lipid. The resultant multilamellar liposomes can be examined using electron-microscopic negative-stain techniques to estimate the size distribution of the closed spheres. The fact that the liposomes are both multilamellar and of nonuniform size makes it difficult to determine the total volume of the internal compartments. This in turn makes it difficult to evaluate the flux of compounds between the internal and external aqueous compartments. The situation is further complicated by the fact that the diffusion barrier presented by the concentric multibilayer structure is unlike that expected for a cell's single membrane.

Multilamellar liposomes can be converted into unilamellar vesicles by exposing the dispersion to ultrasonic radiation.[30] This treatment sufficiently disrupts the liposomal material to yield a vesicle population comprised primarily of single-walled spheres. The sonication is best carried out using a bath-type apparatus so as to maintain the lipid at a temperature above the liquid-crystalline transition and to prevent probe metal fragments from entering the preparation.[19,31] Prolonged sonication has the disadvantage of leading to chemical degradation of the lipid but the advantage of yielding a vesicle population of more uniform size.

Homogeneous populations of uniform-size single-walled spherical lipid vesicles can be prepared by passing a preparation of sonicated liposomes through a molecular sieve column such as Sephrose 4B.[26,32] The detailed properties of these unilamellar vesicles have been exhaustively studied by a

---

[28] A. D. Bangham and R. W. Horn, *J. Mol. Biol.* **8**, 660 (1964).

[29] A. D. Bangham, M. M. Standish, and J. C. Watkins, *J. Mol. Biol.* **13**, 238 (1965).

[30] L. Saunders, J. Perrin, and D. Gammack, *J. Pharm. Pharmacol.* **14**, 567 (1962).

[31] H. O. Hauser, *Biochem. Biophys. Res. Commun.* **45**, 1049 (1977).

[32] C. Huang, *Biochemistry* **8**, 344 (1969).

TABLE I.   The Physical Properties of Lipid Vesicles

Property	Value	Reference
Diameter	~ 20 nm	[a]
Diffusion coefficient	2×10^{-7} cm^2/sec	[a]
Sedimentation coefficient	2.6×10^{-13} sec	[a]
Viscosity	0.04 dl/g	[a]
Expansion coefficient	2×10^{-2} ml g^{-1} °C^{-1}	[b]
Internal volume	2×10^{-18} cm^3	[c]
Wall density	~ 1.02 g/cm^3	[a]
Wall internal microviscosity	$\sim 30 \times 10^{-2}$ poise	[d]
Outer:Inner lipid ratio	2.45:1	[c]
Lipid inversion half-time	6.5 h	[e]
Lipid inversion activation energy	20–30 kcal/mol	[e]
Lipid rotational diffusion coefficient	6×10^{-9}/sec	[d]
Lipid self-diffusion coefficient	2×10^{8} cm^2/sec	[d]
Wall Na$^+$ permeability coefficient	1.2×10^{-14} cm/sec	[c]

[a] T. E. Thompson, C. Huang, and B. J. Litman, *in* "The Cell Surface in Development" (A. Moscona, ed.), p. 1. Wiley, New York, 1974.

[b] D. L. Melchior and H. J. Morowitz, *Biochemistry* **11**, 4558 (1972).

[c] J. Brunner, P. Skrabal, and H. Hauser, *Biochim. Biophys. Acta* **455**, 322 (1976).

[d] A. G. Lee, *Prog. Biophys. Mol. Biol.* **29**, 3 (1975).

[e] R. D. Kornberg and H. M. McConnell, *Biochemistry* **10**, 1111 (1971).

wide variety of physical methods, and excellent reviews are available.[25-29,33,34] Of the general properties summarized in Table I it is evident that the fractionated lipid vesicles are most remarkable because of their very small diameter (approximately 20 nm). The extreme curvature of the lipid bilayer is physically interesting yet biologically unnatural when compared to the plasma membrane. Although vesicles prepared by prolonged sonication, fractionation, injection,[35] and detergent[36] techniques are subjected to different formation forces, they also have a minimum diameter near 20 nm. The reasons for this limiting diameter are not known. Unilamellar vesicles are particularly useful when a large area of membrane is required, as, for example, in spectroscopic studies or flux measurements.

There is good evidence that there are fewer lipid molecules in the inner monolayer of the microvesicle lipid bilayer.[26] Indeed, simple geometric considerations suggest that there are only about half as many lipid molecules with polar groups facing inward as there are facing outward. Nuclear magnetic resonance studies indicate that the inner head groups are packed

[33] A. G. Lee, *Prog. Biophys. Mol. Biol.* **29**, 3 (1975).

[34] A. G. Lee, *Biochem. Biophys. Acta* **472**, 285 (1977).

[35] S. Batzri and E. D. Korn, *J. Cell Biol.* **66**, 621 (1975).

[36] J. Brunner, P. Skrabal, and H. Hauser, *Biochim. Biophys. Acta* **455**, 322 (1976).

more closely that those in the external monolayer.[37,38] As an academic exercise one can assume the vesicle to be a unique variety of inverted soap bubble. Young's law[39] predicts that either there is a very substantial pressure gradient across the vesicle bilayer or that the apposing monolayers have different interfacial tensions (near zero). The possibility of a transbilayer pressure gradient is unlikely both because the bilayer is permeable to water and because the sign of radius of curvature is opposite in the two monolayers. We are left with the more realistic alternative that the lipid packing and therefore the interfacial tension against water are dissimilar in the opposed monolayers. This speculation is consistent with the fact that different kinds of lipids become asymmetrically distributed in microvesicles made from lipid mixtures.[38] Since the inner radius is a function of the bilayer thickness, one might expect that the length of the lipid tail would exert a considerable influence upon the interfacial tension and head-group packing of the inner layer. These matters remain to be resolved, but it is reasonable to exercise caution when comparing the physical properties of highly curved microvesicle models to those of less curved cell membranes and planar bilayer models.

11.4. Planar Lipid Bilayers

Cell plasma membranes typically have only a modest degree of curvature where the enclosed cytoplasmic substance is of the order of microns in diameter. The size of the intracellular compartment is sufficiently large to permit direct access by microelectrode techniques; consequently, it is a relatively straightforward matter to systematically study the transport of ions and other compounds across the plasma membrane barrier. Planar lipid bilayers represent a class of simplified model membranes which have been optimized to provide direct access to the aqueous compartments on opposite sides of the artificial membrane.

Physical studies of planar lipid films began in the year 1672 when Hook first described the formation of "black spots" in soap bubbles. A short time later Newton found that these low reflectance areas corresponded to very thin regions (9.5 nm) of the bubble wall.[40] By 1925 efforts were being

[37] H. Hauser, M. C. Phillips, B. Levin, and J. P. Williams, *Nature (London)* **261**, 390 (1976).

[38] C. Huang, J. Sipe, S. Chow, and R. B. Martin, *Proc. Natl. Acad. Sci. U. S. A.* **71**, 359 (1974).

[39] A. W. Adamson, "Physical Chemistry of Surfaces." Wiley, New York, 1976.

[40] P. Mueller, D. O. Rudin, H. Tien, and W. C. Wescott, *in* "Recent Progress in Surface Sciences" (J. F. Danielli, K. Pankhurst, and A. Riddiford, eds.), p. 379. Academic Press, New York, 1964.

made to measure the electrical properties of hydrated lipid films suspended on ring frames.[41] The early interest in using interfacial films as models of cell membranes peaked in the late 1930s when methods were developed to form hydrated lipid bilayers from monolayers[42,43] and from an oil/water interface.[44,45] For example, it was found[46] that films formed from phospholipids adsorbed at the oil/water interface had a specific geometrical capacitance of $1 \mu F/cm^2$. Although both the tension and capacitance of the model were similar to those of the cell membrane,[8,14] the utilitarian aspects of the experimental method went unappreciated. This neglect is somewhat surprising because these developments coincided with the emergence of the lipid bilayer concept of cell membrane molecular organization. In the 1960s Mueller–Rudin, and colleagues[47–49] rediscovered and refined the techniques of forming planar lipid bilayers in an electrolyte-filled chamber.

11.4.1. Lipid/Solvent Systems

The basic method of forming planar bilayers on a submerged aperture is elegant in its simplicity.[47–49] The chamber consists of two electrolyte-filled compartments separated by a hydrophobic partition which has a small (1 mm diameter) hole in it. An organic bulk mixture consisting of a surface-active lipid (lecithin, monoolein, etc.) in an appropriate solvent ($CHCl_3/CH_3OH$, alkane, etc.) is painted over the submerged aperture with the aid of a small brush[49] or Pasteur pipette.[50] The organic phase wets the hydrophobic partition and spans the hole to yield an oil/water interfacial film exhibiting vivid interference colors. As the lipid adsorbs to the interfaces, the film tension decreases, leading to the formation of low reflectance similar to those observed during the drainage of soap films.[51] That is, the Gibbs–Plateau border creates an interfacial pressure difference which leads to the drainage of the thick film until the bifacial film consists of two monolayers having their apolar ends apposed. As the black spots (i.e., low reflectance) spontaneously grow in area, the excess bulk material is

[41] W. Hardy, *J. Chem. Soc.* **127**, 1207 (1925).

[42] J. J. Bikerman, *Proc. R. Soc. London, Ser. A* **170**, 130 (1939).

[43] I. Langmuir and D. R. Waugh, *J. Gen. Physiol.* **21**, 745 (1938).

[44] J. F. Danielli, *J. Cell. Comp. Physiol.* **7**, 393 (1936).

[45] T. Teorell, *Nature (London)* **137**, 994 (1936).

[46] R. B. Dean, H. Curtis, and K. S. Cole, *Science (Washington, D.C.)* **91**, 50 (1940).

[47] P. Mueller, D. Rudin, H. Tien, and W. Wescott, *Circulation* **26** (II), 1167 (1962).

[48] P. Mueller, D. Rudin, H. Tien, and W. Wescott, *Nature (London)* **194**, 979 (1962).

[49] P. Mueller, D. Rudin, H. Tien, and W. C. Wescott, *J. Phys. Chem.* **67**, 534 (1963).

[50] G. Szabo, G. Eisenman, and S. Ciani, *J. Membr. Biol.* **1**, 346 (1969).

[51] J. Overbeek, *J. Phys. Chem.* **64**, 1178 (1960).

forced to the film perimeter to form a supporting collar or torus. This annulus of bulk organic material serves as a fluid reservoir which satisfies both the mechanical and interfacial boundary conditions required[52] for bilayer stability. Conceptually, planar black bimolecular lipid membranes (BLMs) of the Mueller–Rudin type are analogous to inverted soap films surrounded by water and supported by a bulk-phase organic fluid "frame." However, this interfacial model may be an oversimplification in that it is inadequate to fully explain the apolar composition of the bilayer.

11.4.2. Solvent-Filled Bilayers

The earliest Mueller–Rudin type bilayers were "stabilized" with linear organic solvents such as squalene ($C_{30}H_{50}$) or tetradecane.[49] The initial thought was that this solvent drained from the black film upon thinning. Subsequent studies have shown that a considerable volume of organic solvent remains within the bilayer membrane.[3,13,53–59] This circumstance is particularly unfavorable from the biological point of view because compounds such as n-alkanes are known to be membrane anesthetics.[60,61] A portion of the residual solvent is trapped in microlenses, but more problematical is the portion of solvent which chemically penetrates into the hydrophobic core of the bilayer to mix with the lipid acyl chains. For example, direct sampling of the black film has shown that as much as 50% of the volume of bilayers made from monoolein/decane solutions is solvent.[13] For many kinds of experiment, the presence of solvent is unimportant, but in some cases the anestheticlike properties of the solvent are clearly undesirable. Also, solvents impede reconstitution. Thus there has been a considerable effort made to rid the model membrane of solvents.

Lipid/alkane interaction in planar bilayers has been studied in some detail by taking full advantage of the fact that the bilayer, like the cell membrane, behaves as a parallel-plate capacitor.[46,62] By making exact measure-

[52] H. Ti Tien, *J. Phys. Chem.* **72**, 2723 (1968).

[53] D. Andrews, M. E. Manev, and D. Haydon, *Trans. Faraday Soc., Spec.* **1**, 46 (1970).

[54] R. Fettiplace, D. Andrews, and D. Haydon, *J. Membr. Biol.* **5**, 277 (1971).

[55] J. Requena, D. F. Billett, and D. Haydon, *Proc. R. Soc. London, Ser. A* **347**, 141 (1975).

[56] J. Requena and D. Haydon, *Proc. R. Soc. London, Ser. A* **347**, 161 (1975).

[57] J. Requena, D. E. Brooks, and D. Haydon, *J. Colloid Interface Sci.* **58**, 26 (1977).

[58] S. H. White, *Biochim. Biophys. Acta* **356**, 8 (1974).

[59] S. H. White, *Ann. N.Y. Acad. Sci.* **303**, 243 (1977).

[60] C. Difazio, R. E. Brown, C. G. Ball, C. Heckel, and S. Kennedy, *Anesthesiology* **36**, 57 (1972).

[61] D. Haydon, B. Hendry, S. Levinson, and J. Requena, *Biochim. Biophys. Acta* **470**, 17 (1977).

[62] H. Coster and R. Simons, *Biochim. Biophys. Acta* **203**, 17 (1970).

ments of the bilayer specific geometrical capacitance C_g and by assuming a value for the dielectric coefficient for the interior apolar region of the bilayer, it is a straightforward matter to calculate the dielectric thickness of the membrane.[59,63] It is clear from studies of this kind that the *thickness of the internal "oily" region of the bilayer is directly proportional to the length (i.e., volume) of the lipid fatty acyl chain.*[64,65] For example, bilayers made using an 18-carbon chain-length lipid (monoolein) in decane have a dielectric thickness of about 4.8 nm, whereas those made using the 14-carbon chain lipid (monomyristolein) are only 4.0 nm in thickness.[64] Interestingly, these values are in close agreement with those predicted for pairs of apposed, untilted, and fully extended lipid fatty acyl chains. This coincidence is the primary basis for the common presentation of schematic illustrations of the lipid bilayer as a tail-to-tail arrangement of "lollypop" entities having circular heads and sticklike tails. It will become apparent that such diagrammatic representations may be misleading.

It is now clear that the presence of residual organic solvent within the planar bilayer has a profound effect upon the internal structure of the membrane. Empirical studies[59] have also established that the *thickness of the Mueller–Rudin-type bilayer is inversely proportional to the length (i.e., molar volume) of the solvent n-alkane ($n \geq 10$).* Since C_g is essentially a measure of the bilayer volume, the observed decrease in dielectric thickness when hexadecane rather than decane is used as a solvent is attributable to the presence of fewer hexadecane molecules within the bilayer core.[59] Although the physicochemical basis for these effects remains somewhat obscure, they have nevertheless provided a practical means of minimizing the amount of undesirable solvent within the model membrane. Indeed, it is the inverse proportionality of solvent molecular size and bilayer solvent content that led to the recent reintroduction of the long alkene compound squalene as a solvent having a low affinity for the interior of the black membrane.[49,66] However, the contention that monoolein/squalene dispersions yield socalled solventless bilayers has been questioned.[67,68] It should also be noted that the use of lipid solvents stems more from tradition than from any rigorous physicochemical considerations. Indeed, it will be noted later that we have recently demonstrated that hydrocarbon solvents can be totally eliminated from the Mueller-type planar bimolecular lipid membrane.[68]

[63] F. Henn and T. E. Thompson, *Annu. Rev. Biochem.* **38**, 241 (1969).

[64] R. Benz, O. Frohlich, P. Lauger, and M. Montal, *Biochim. Biophys. Acta* **394**, 323 (1975).

[65] J. Taylor and D. Haydon, *Discuss. Faraday Soc.* **42**, 51 (1966).

[66] S. H. White, *Biophys. J.* **23**, 337 (1978).

[67] R. C. Waldbillig and G. Szabo, *Biophys. J.* **25**, 11a (1979).

[68] R. C. Waldbillig and G. Szabo, *Biochim. Biophys. Acta* **557**; 295 (1979).

11.4.3. Solvent-Depleted Bilayers

Montal and his collaborators[64,69] have reported that planar bimolecular lipid membranes can be formed by bringing together two monolayers over a submerged aperture. This procedure of forming bilayers from two lipid monolayers is basically an adaptation of the early Langmuir–Blodgett multilayer deposition technique[11] modified to allow apposing monolayers to join together over a submerged hole in a solid support.[42,69] The main apparatus consists of two monolayer troughs having a common wall of very thin (10 μm) Teflon which has a small-diameter (0.1 mm) hole in it. It must be noted that the formation of bilayers from monolayers is something of an art in that certain chemically and physically inexplicable maneuvers seem to facilitate membrane formation. For example, it is necessary to precoat the hole with a grease such as petroleum jelly.[64] Moreover, it is advantageous to raise the excess lipid surface films prior to the complete evaporation of the spreading solvent. Care must also be exercised to avoid bulges in the bilayer due to differences in the electrolyte levels of the two troughs. Since the bilayer is too small to be seen easily, the water levels must be adjusted to bring the film total capacitance to a minimum[64] to assure a flat membrane.

Another approach to ridding planar bilayers of residual organic solvent is to use squalene. It has been shown by independent methods[70] that squalene is insoluble in phospholipid bilayers (vesicles), but no evidence is available regarding the miscibility of squalene with monoglycerides. Nevertheless, it has been suggested by inference that squalene is virtually insoluble in monoglyceride bilayers.[66] This conclusion must be viewed cautiously in that the estimated area per molecule in the lipid bilayers made from squalene dispersions is less than that reported from rigorous studies of monoolein adsorption at the oil/water interface[53,55–57] and from studies of chemical samples of monoglyceride bilayers.[13] Since the lipid area per molecule in the bilayer is calculated from the specific capacitance using estimates for the dielectric coefficient and the dielectric volume per molecule,[56] a slight increase in the intramembrane volume due to residual squalene would lead to an underestimation of the lipid area per molecule. Although it would appear that planar bilayers made from monoglyceride/squalene dispersions may contain appreciably less residual solvent than planar bilayers made from monolayers, it would be inappropriate to regard bilayers made by either of these methods as solvent free.

[69] M. Montal and P. Mueller, *Proc. Natl. Acad. Sci. U. S. A.* **69** 3561 (1972).
[70] P. Lauger and B. Neumcke, *Membranes* **2**, 1 (1973).

11.4.4. Pure Lipid Bilayers

The most direct way of ridding the planar lipid bilayer of biologically obtrusive organic solvents is the straightforward elimination of hydrocarbons from the starting materials. This simple and pragmatic approach has been overlooked largely because of the historical relationship between model membranes and studies of the oil/water interface.[47, 52–54] The suggestion that solvents (e.g., n-alkanes) are unnecessary for stable planar bilayer formation becomes more plausible when it is recognized that lipids have many of the physicochemical attributes of oils.[71] We thus have recently made a departure from convention by demonstrating that a lipid can serve as a "solvent."[68]

Planar bilayers comprised solely of lipids are easily formed by a slight modification of the standard Mueller–Rudin method.[49,68] The modification involves the use of highly purified ($>99\%$) triglycerides as "solvents" for monoglycerides. Consequently, the membrane-forming bulk phase contains only lipids. A tool or spatula (i.e., Teflon, polyproplene) is used to transfer the binary monoglyceride/triglyceride lipid mixture to a Teflon aperture immersed in a conventional Mueller–Rudin-type bilayer chamber. The tool is used to brush the viscous lipid mixture over the aperture for bilayer formation. In general, *membranes formed from pure lipid (i.e., solvent-free) mixtures are large in area ($\simeq 1.25$ mm^2) and more stable than conventional planar bilayers made with lipid/alkane solutions.*

The dielectric properties of bilayers made from pure lipids are significantly different than lipid/alkane bilayers and from other so-called solvent-free bilayers. For example, bilayers made from monopalmitolein/triglyceride mixtures have a specific capacitance near $1 \mu F/cm^2$ as compared to a reported value[57] of $0.46 \mu F/cm^2$ for membranes made from monopalmitolein/decane solutions. Biological membranes such as nerve axons also have specific capacitance values near $1 \mu F/cm^2$.[72] Pure monoolein bilayers have a capacitance of $0.86 \mu F/cm^2$, whereas bilayers made from monoolein/squalene have a value of $0.79 \mu F/cm^2$, and membranes made from monolayers[64] have a value of $0.75 \mu F/cm^2$. If other factors are equal, then prevailing concepts of model membrane structure lead to the conclusion that the lower-capacitance bilayers are swollen owing to the presence of a residual volume of organic solvent in the core of the bilayer.

What are the relative proportions of monoglyceride and triglyceride in these pure lipid membranes? It is well known that bilayer capacitance and therefore dielectric thickness can be related to the length of the membrane lipid fatty acyl chain.[57] When this fact is used as a criterion for characterizing

---

[71] E. S. Lutton, *J. Am. Oil Chem. Soc.* **49**, 1 (1972).

[72] S. Takashima, *Biophys. J.* **22**, 115 (1978).

the chemical composition of the thin film, it is found that the bilayer thickness *depends* solely upon the length of the monoester component of the binary bulk mixture. Conversely, the thickness of the pure lipid bilayer is *independent* of the chain length (11–22 carbons/per chain) of the triglyceride, suggesting that the torus bulk phase disproportionates upon thinning to a yield a virtually pure monoglyceride bilayer phase. Additionally, the bilayer thickness is *independent* of the torus concentration of triester, suggesting that the triglyceride is insoluble in the bilayer phase. Thus this pragmatic solution to the bilayer solvent problem may provide a means of relating lipid structure to membrane function and may also provide a new avenue for developing a better understanding of the self-mixing characteristics of other membrane lipids.

11.5. Ion Transport Mechanisms

Lipid bilayers are excellent insulators. Indeed, the equivalent volume resistivity of the unmodified bilayer ($\simeq 10^{15}\ \Omega$ cm) is comparable to that of a pure hydrocarbon barrier. Cell membranes, in contrast, are relatively poor insulators because of the presence of nonspecific "leakage" pathways and specific pathways which mediate the transport of a variety of charged species, such as amino acids, Na^+, K^+, and Ca^{2+}. Since changes in membrane ionic permeability are known to trigger such diverse cellular phenomena as fertilization, nerve conduction, and muscle contraction,[73-75] it is apparent that the elucidation of the molecular mechanisms of ion transport is a matter of considerable interest. Studies of chemically modified model membranes have provided important clues about the physicochemical constraints which control the movements of molecules across the membrane barrier.

Considerations of membrane permeability for a large number of solutes through lipid bilayer membranes have revealed three broad categories of permeation mechanisms. The simplest of these, shown in Fig. 1a, consists of direct, unassisted translocation of the solute through the ultrathin bilayer membrane. Transport via this intrinsic mechanism is pratically negligible for hydrophilic molecules such as sugars, amino acids, or alkali-metal cations. This is because the nonpolar, hydrocarbonlike membrane interior presents a large energy barrier against the transfer of polar solute molecules from the high-dielectric-constant aqueous phases.[70,76] However, in the

[73] K. S. Cole, "Membranes, Ions and Impulses." Univ. of California Press, Berkeley, 1968.
[74] S. Hagiwara and L. A. Jaffe, *Annu. Rev. Biophys. Bioeng.* **8**, 385 (1979).
[75] B. Hille, *Biophys. J.* **22**, 283 (1978).
[76] O. S. Andersen and M. Fuchs, *Biophys. J.* **15**, 795 (1975).

(a) $\quad I' \xleftrightarrow{\text{Diffusion}} I_0 \underset{k_d}{\overset{k_a}{\rightleftharpoons}} I_0^{\bullet} \underset{k''}{\overset{k'}{\rightleftharpoons}} I_d^{\bullet} \underset{k_a}{\overset{k_d}{\rightleftharpoons}} I_d \xleftrightarrow{\text{Diffusion}} I''$

(b) $\quad I' \xleftrightarrow{\text{Diffusion}} I_0 \xrightleftharpoons{k_D}$ ...

$$IS_0^{\bullet} \underset{k''_{IS}}{\overset{k'_{IS}}{\rightleftharpoons}} IS_d^{\bullet} \xrightleftharpoons{k_R} \quad \quad k_R$$

$$S_0^{\bullet} \underset{k''_S}{\overset{k'_S}{\rightleftharpoons}} S_d^{\bullet}$$

$$\cdots \xrightleftharpoons{k_D} I_d \xleftrightarrow{\text{Diffusion}} I'$$

Solution	Membrane	Solution

FIG. 1. Kinetic schemes for (a) direct and (b) carrier-mediated transport of ions. k_a and k_d are the rates of adsorption and desorption, respectively, of the lipophilic ion, and k' and k'' are the translocation rates across the membrane interior. For carrier-mediated transport k_D and k_R are the rates of dissociation and association of the ion with the carrier. k'_{IS} and k''_{IS} are the translocation rates of the ion-carrier complex, and k'_S and k''_S are the translocation rates of the uncomplexed carrier.

particular circumstance of large and/or nonpolar solute molecules the energy barrier may be small enough to allow a significant movement of solute through the membrane. Thus, for example, tetraphenyl borate, dipicrylamine, or tetraphenylphosphonium have large membrane permeabilities[77-79] owing to their size and hydrophobicity.[80,81] It should be apparent that the cell membrane has little selective control over this intrinsic kind of solute movement, which is, however, of secondary importance because of the hydrophylic nature of most cytoplamic solutes.

The second transport mechanism, shown in Fig. 1b, utilizes a carrier molecule to reduce the energy barrier for the translocation of polar molecules. In the case of well-studied carriers such as valinomycin and nonactin, this is accomplished by the carrier S providing polar residues that replace the solvation shell of the solute.[82-84] The external surface of the ion–carrier complex IS is covered with hydrophobic alkyl residues, facilitating the translocation of the complex through the hydrocarbon interior of the bilayer membrane. Dissociation of the complex releases the transported solute at the opposite membrane surface, and the hydrophobic carrier

[77] O. H. LeBlanc, Biochim. Biophys. Acta 193, 450 (1969).

[78] E. A. Liberman and V. P. Topaly, Biofizika 14, 452 (1969).

[79] P. Mueller and D. O. Rudin, Curr. Top. Bioeng. 3, 157 (1969).

[80] O. S. Andersen, in "Transport Across Biological Membranes" (D. C. Tosteson, G. Giebisch, and H. H. Ussing, eds.), Vol. I, p. 369. Springer-Verlag, Berlin and New York, 1980.

[81] D. A. Haydon and S. B. Hladky, Q. Rev. Biophys. 5, 187 (1972).

[82] E. Grell, T. Funck, and F. Eggers, Membranes 3, 1 (1975).

[83] Y. A. Ovchinnikov, Eur. J. Biochem. 94, 321 (1979).

[84] D. W. Urry and M. Onishi, in "Spectroscopic Approaches to Bimolecular Conformations" (D. W. Urry, ed.), p. 263. Am. Med. Assoc., Chicago, Illinois, 1970.

is free to return for the transport of another solute molecule. Note that the carrier as well as the solute molecule may be electrically charged. However, for energetic reasons, the complex must be either univalent or neutral. Carrier-mediated transport has two notable features. First, it permits selective translocation, which is primarily governed by the formation and dissociation rates of the complex. Second, it allows a coupling between the flow of the transported solute and the electrical potential across the membrane whenever the carrier and/or the complex has an electrical charge.

The third class of transport mechanisms, shown in Fig. 2, is comprised of a continuous polar pathway or channel that spans the membrane. Implicit in the notion of a channel is the requirement that a large number of ions move through this pathway before it undergoes structural changes. The formation and dislocation of such conductive pathways as well as voltage-dependent gating have been detected in bilayer membranes treated with a variety of compounds such as gramicidin A, excitability-inducing material (EIM), and alamethicin.[80]

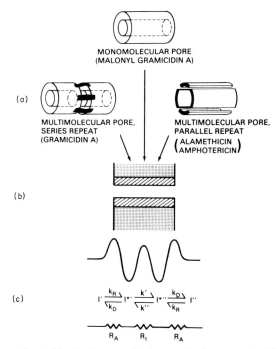

FIG. 2. Structural and kinetic diagrams for ion-conductive pores in bilayer membranes. (a) Different molecular arrangements result in (b) an aqueous pore lined with polar residues (hatched lines) that span the membrane (dotted structure). (c) Energy barriers that an ion encounters during its transit through the pore and the corresponding kinetic transport scheme.

Both carrier and channel mechanisms can be studied in lipid bilayers simply by adding the appropriate material to one or both aqueous phases in contact with the bilayer membrane.

It is generally accepted that ion transport through cell and organelle membranes may be mediated by carriers as well as channels.[85] Circumstantial evidence, based on transport kinetics, indicates that the transport of sugars and amino acids, as well as ion transport which derives its energy directly from chemical reactions (e.g., active transport) may be accomplished by a carrierlike mechanism.[85-87] Direct evidence exists, in contrast, for the presence of ion-conductive channels in most electrically active membranes such as muscle and nerve.[75] Thus, for example, quantal jumps of the electrical current, corresponding to the opening of ion channels by acetylcholine and other agonists, have been observed in the postsynaptic membranes of muscle and nerve.[88] With the development of these techniques the elucidation of the molecular basis of ionic selectivity, conductance, and pore opening and closing has become a major goal of current research in membrane biophysics.[75]

11.6. Techniques for the Measurement of Ion Transport

Several types of relaxation measurements have been used to determine the kinetic parameters of ion transport through bilayer membranes. In each case, a current electrode and a voltage electrode are placed on each side of the bilayer. The current electrode is usually connected to an impedance converter in order to facilitate measurement of the current through the high-impedance lipid bilayer.

The simplest "voltage clamp" method involves holding the membrane potential at some steady value V (usually at the equilibrium potential) and then increasing it to a new steady value $V + \Delta V$.[89] The membrane relaxation current is usually described well by one or several single exponentials. Thus the relaxation current is characterized by the initial currents and the time constants of relaxation. A special case of voltage clamp, where the current response is quite different, occurs when the membrane contains a single channel (or a few channels). This is discussed further in Section 11.9.1.

A second "charge pulse" method has been developed for the study of

[85] S. B. Hladky, *Curr. Top. Membr. Transp.* **12**, 53 (1979).

[86] J. Silverman, *Biochim. Biophys. Acta* **472**, 302 (1976).

[87] S. J. Singer, *J. Supramol. Struct.* **6**, 313 (1977).

[88] E. Neher and C. F. Stevens, *Annu. Rev. Biophys. Bioeng.* **6**, 345 (1977).

[89] P. Lauger, *Science (Washington, D.C.)* **178**, 24 (1972).

fast transport kinetics.[90] A small amount of charge is deposited rapidly (<1 μsec) on the membrane. The time course of the membrane potential is then recorded under zero-current (e.g., open-circuit) conditions. The potential decays as the membrane capacitance is discharged by ionic currents across the membrane. In principle, this technique permits the resolution of very rapid transients. In practice, the interpretation of the results is made difficult by the fact that both the bilayer capacitance and the rate constants of most transport systems vary as a function of the membrane potential, which, of course, also varies during the relaxation event. For small potential differences these difficulties are minimal.

There are two other methods which have a superficial similarity in that both utilize noise. However, these methods are fundamentally different in concept. In the "conductance fluctuation" method, a constant voltage is placed across the bilayer, and the fluctuations of the resultant current, as well as its mean value, are recorded. For gated channels, these current fluctuations arise from the random opening and closing of individual channels. For gramicidin channels, the current fluctuations arise from the reversible dimerization of gramicidin monomers. This is discussed further in Section 11.9.2.

The "transfer function" method involves the measurement of the membrane current induced by fluctuations of the membrane potential. The applied membrane potential, for example, a pseudorandom Gaussian noise filtered to decrease the magnitude of the high-frequency components (e.g., pink noise), and the resulting membrane current are both digitally recorded. The magnitude and the phase of the system transfer function (admittance) are calculated from the ratio of the Fourier-transformed current output and voltage input signals.

11.7. Direct Transport of Hyhrophobic Ions

It has been noted that lipid bilayer membranes are made electrically conductive by the presence of certain organic ions. Dipicrylamine, tetraphenylborate, tetraphenylphosphonium, and many organic dyes are typical and well-documented examples of these.[91–93] A large ionic size and a delocalized electrical charge, as well as the presence of hydrophobic residues, appear to be the necessary attributes for enhanced intrinsic membrane

[90] S. W. Feldberg and H. Nakadomari, *J. Membr. Biol.* **31**, 81 (1977).
[91] A. Pickar and R. Benz, *J. Membr. Biol.* **44**, 353 (1978).
[92] G. Szabo, *Nature (London)* **252**, 47 (1974).
[93] A. S. Waggoner, *Annu. Rev. Biophys. Bioeng.* **8**, 47 (1979).

permeability. All of these factors are expected to decrease the energy required to transfer the ion from the aqueous phase to the nonpolar membrane phase having a low dielectric constant.

Voltage clamp, charge pulse, and transfer function measurements on lipid bilayer membranes exposed to lipophilic ions indicate that the kinetic scheme shown in Fig. 1a accurately describes the salient features of direct transport of ions. In this scheme lipophilic ions (I) are adsorbed at opposite sides (I_0^* and I_d^*) of the bilayer. They move through the membrane interior with voltage-dependent rate constants k' and k''. Relaxation of the membrane current arises whenever the rate of translocation of the adsorbed ion across the membrane interior is more rapid than the rate of its adsorption and/or diffusion through the aqueous boundary layers next to the membrane surface. This results in a local depletion of the surface reservoir of adsorbed ions available for transport.

The quantitative details of direct transport of hydrophobic ions have been reviewed recently.[85] It has been shown in that review and elsewhere that, under the simplest circumstances, the system is simply characterized by the voltage-dependent rate constants k' and k'' and by the coefficient of adsorption β for the lipophilic ion at the membrane surface.

11.8. Carrier-Mediated Ion Transport

The overall mechanisms of carrier-mediated transport are well understood and have been reviewed extensively.[81,83,85,89,94,95] Figure 1 shows the kinetic scheme that has been used to describe quite accurately both the steady-state and relaxation behavior of carrier-mediated transport. The hydrophobic ion–carrier complex IS is formed at the membrane surfaces by the reaction:

$$I + S \; \underset{k_D}{\overset{k_R}{\rightleftharpoons}} \; IS,$$

where k_R is the rate constant of formation and k_D that of dissociation for the ion–carrier complex. The free carrier moves across the membrane interior with rate constants k_S' and k_S''. It should be noted that the translocation rates k_{IS}' and k_{IS}'' of the ion–carrier complex are analogous to those for hydrophobic ions.

The host carrier S, its guest substrate I, and the ion–carrier complex IS may all be charged. For example, neutral carriers of cations (valinomycin,

[94] S. Hladky, *J. Membr. Biol.* **46**, 213 (1979).
[95] S. McLaughlin and M. Eisenberg, *Annu. Rev. Biophys. Bioeng.* **4**, 335 (1975).

macrotetralid actins; positively charged complex[89,96]), negatively charged carriers of divalent cations (monensin, lasalocid; neutral or positive complex[97]) have all been observed in lipid bilayer membranes. Certain neutral carriers, such as N,N'-diheptyl-N,N',5,5'-tetramethyl-3,7-dioxanonan diamide may act in several distinguishable modes by complexing both cations and anions.[98]

The heterogeneous reaction permits selective transport of substrates. For example, the neutral cation carrier valinomycin transports potassium ions 3×10^4 times more effectively than sodium ions.[99] This is because the equilibrium constant for the formation of the K^+–Val complex is 3×10^4 times larger than that of the Na^+–Val complex so that, everything else being equal, K^+ produces 3×10^4-fold more complexes than Na^+.

Selective transport of cations has been extensively characterized for the neutral cation carriers valimomycin, macrotetralid actins, and enniatins.[50,89,96,99,100] Over a wide range of experimental conditions, the heterogeneous reaction is at equilibrium so that the number of complexes at the membrane surface N_{IS} is fixed by the aqueous cation activities a_I and the membrane surface concentration of carrier N_S.

$$N_{IS} = K_{IS} N_S a_I, \qquad K_{IS} = k_R/k_D.$$

In this so-called equilibrium domain[50] the selectivity and conductance of the bilayer obey simple relationships. For example, let us consider the membrane selectivity as measured by the zero-current potentials for bilayers doped symmetrically† with a neutral carrier and exposed to aqueous solutions of two cations I^+ and J^+. The steady-state electrical current density is given for each ion by

$$J_I = -zqL(k'_{IS}N'_{IS} - k''_{IS}N''_{IS}) = -zqLK_{IS}N_S(k'_{IS}a'_I - k''_{IS}a''_I),$$

$$J_J = -zqL(k'_{JS}N'_{JS} - k''_{JS}N''_{JS}) = -zqLK_{JS}N_S(k'_{JS}a'_J - k''_{JS}a''_J),$$

where L is Avogadro's number. The translocation rate of the charged complex is expected to be influenced by the electrical potential V across the thin bilayer membrane. Theoretical and experimental considerations[70,76,81]

[96] S. Krasne and G. Eisenman, *J. Membr. Biol.* **30**, 1 (1976).

[97] B. C. Pressman and N. T. deGuzman, *Ann. N. Y. Acad. Sci.* **264**, 373 (1975).

[98] R. Margalit and G. Eisenman, *Proc. Am. Pept. Symp., 6th, 1979*, p. 665 (1980).

[99] B. W. Urban and S. B. Hladky, *Biochim. Biophys. Acta* **554**, 410 (1979).

[100] R. Benz, *J. Membr. Biol.* **43**, 367 (1978).

[101] S. Ciani, *J. Membr. Biol.* **30**, 45 (1976).

† This is done by dissolving the carrier in the lipid solution or the aqueous solutions at equal concentrations at opposite sides (′ and ″) of the membranes. More rigorous treatments of carrier-mediated transport have been presented by a number of authors.[85,101]

show that the potential dependence of the translocation rates is approximated by the following expression:

$$k' = ke^{-zqV/2kT}, \qquad k'' = ke^{zqV/2kT}.$$

Setting $J_I + J_J = 0$, the zero-current membrane potential V_0 is obtained:

$$V_0 = \frac{kT}{q} \ln \frac{a_I' + \beta_{IJ}a_J'}{a_I'' + \beta_{IJ}a_J''}, \qquad \beta_{IJ} = \frac{K_{JS}k_J}{K_{IS}k_I}.$$

The selectivity parameter β_{IJ} is formally equivalent to the permeability ratio P_I/P_J of the Goldman–Hodgkin–Katz equation[99,102] frequently used to describe ion selectivity in biological membranes.

Selectivity of ion transport is also evident in the membrane conductances. The specific membrane conductance is obtained directly from the current densities J:

$$G_I = \frac{J_I}{V} = 2zqLN_Sk_IK_{IS}a_I \frac{\sinh(zqV/2kT)}{V}.$$

The zero-potential conductance is

$$G_I^0 = \lim_{V \to 0} G_I^0 = (Z^2q^2/kT)LN_Sk_IK_{IS}a_I.$$

Thus at the same carrier and ion concentrations

$$G_J^0/G_I^0 = k_JK_{JS}/k_IK_{IS} = \beta_{IJ}.$$

That is, the permeability and conductance ratios should be equal. This relationship, as well as the expected proportionality between G_0, salt concentration a_i, and carrier concentration N_S, holds over a range of several orders of magnitudes for valinomycin,[103] the macrotetralid actins,[99] and the enniatins.[100]

Structural studies[82,90,104] have shown that the ion is buried in the interior of the neutral carrier, its hydration shell being replaced by residues from ester, ether, or peptide linkages. As the exterior of the charged complex is hydrophobic, the entire complex is similar to a lipophilic ion. The size and shape of the complex should be relatively independent of the type of ion that is in the interior of the complex. That is, complexes of different cations should be "isosteric."[50] This implies that translocation rates for complexes

[102] A. L. Hodgkin and B. Katz, *J. Physiol.* (*London*) **116**, 473 (1949).

[103] G. Stark and R. Benz, *J. Membr. Biol.* **5**, 133 (1971).

[104] M. Pinkerton, K. Steinrauf, and P. Dawkins, *Biochem. Biophys. Res. Commun.* **35**, 512 (1969).

bearing different ions should be similar, so that $k_I/k_J \simeq 1$. The selectivity coefficient β_{IJ} then becomes

$$\beta_{IJ} = K_{JS}/K_{IS}.$$

That is, transport selectivity arises mainly from the specificity of the ion–carrier reactions at the surface of the membrane. In light of this, it is not surprising that, in the equilibrium domain, the selectivity coefficients can be related directly to the equilibrium constants K_{IS}^* for the extraction of ions by the carrier into bulk organic phases[99] (BS denotes bulk solvent)

$$I_{aq} + S_{BS} \xrightleftharpoons{K_{IS}^*} IS_{BS}$$

by the following relationship:

$$\beta_{IJ} = K_{JS}^*/K_{IS}^*.$$

As these relationships indicate, the selectivity of carrier-mediated transport may be related to simple chemical equilibria and, therefore, to differences in free energy of ion–water and ion–carrier interactions.[50] Although there are no convincing *ab initio* calculations for these free-energy differences, the success of simple electrostatic considerations in predicting ion selectivity[105] indicates that it provides a quantitative understanding from first principles at the molecular level.

For all of the well-characterized carriers (valinomycin and analogues, macrotetralid actins, enniatins) there are experimental conditions under which the membrane current exhibits conspicuous time dependences.[70,81,100] Under these circumstances the heterogeneous reaction and/or the return of the free carrier are driven out of equilibrium. The relaxation properties of the membrane currents have been studied by voltage clamp,[106-108] charge pulse,[90,109,110] and small-signal admittance[111,112] techniques. In all cases the results could be analyzed within the framework of the carrier model, and by extending the studies over a range of ion concentrations and membrane potentials, the rate constants for each step in the transport process could be determined at least approximately.[106,107] A comprehensive theoretical treatment for current transients in carrier-mediated transport has recently been presented.[94]

[105] G. Eisenman, S. Krasne, and S. Ciani, *Ann. N. Y. Acad. Sci.* **264**, 34 (1975).

[106] S. B. Hladky, *Biochim. Biophys. Acta* **375**, 327 (1975).

[107] R. Laprade, S. Ciani, G. Eisenman, and G. Szabo, *Membranes* **3**, 127 (1975).

[108] G. Stark, B. Ketterer, R. Benz, and P. Lauger, *Biophys. J.* **11**, 981 (1971).

[109] R. Benz and P. Lauger, *J. Membr. Biol.* **27**, 171 (1976).

[110] S. W. Feldberg and G. Kissel, *J. Membr. Biol.* **20**, 269 (1975).

[111] E. Neher, J. Sandblom, and G. Eisenman, *J. Membr. Biol.* **40**, 97 (1978).

[112] M. P. Sheetz and S. I. Chan, *Biochemistry*, **24**, 4573 (1972).

TABLE II. Typical Rate Constants for K^+ Transport by the Neutral
Carriers Valinomycin, Nonactin, and Enniatin B

	k_{IS} (sec^{-1})	k_S (sec^{-1})	k_R (M^{-1} sec^{-1})	k_D (sec^{-1})
Valinomycin[a]	1.1×10^6	1.1×10^5	2.7×10^5	2.1×10^5
Nonactin[b]	2.3×10^4	3.7×10^4	1.9×10^5	4.6×10^4
Enniatin[a]	3.1×10^4	3.6×10^4	1.7×10^5	2.6×10^5

[a] Data for monolinolein/decane bilayers from R. Benz, *J. Membr. Biol.* **43**, 367 (1978).

[b] Data for monolein/decane bilayers from S. B. Hladky, *Curr. Top. Membr. Transport* **12**, 53 (1979).

Table II shows typical values of the rate constants for potassium transport by valinomycin, nonactin, and enniatin B. The rates of ion–carrier association and dissociation, of the order of 10^4 per second, are remarkably rapid despite the large solvation energies, of the order of 200 kJ/mol, which would imply large activation energies and slow reaction rates. Evidently, the ionic solvation sphere is exchanged gradually between water and the carrier. This sequential replacement of ion hydration water by the carrier has been documented for the macrotetralid actins and valinomycin (see Grell et al.[82]).

For carriers, simple reasoning based on Fick's law indicates that tanslocation rates of the complex (k_{IS}) and of the carrier (k_S) should both be increased by either decreasing the thickness or increasing the fluidity of the membrane interior. Charged species moving through the thin (nanometers) membrane will encounter a significant image force due to interactions between the ion and aqueous dipoles. The thinner the membrane, the more these forces reduce the electrostatic barrier of the membrane interior and therefore the more rapid the ionic translocation rates become.[85] In model membranes it is possible to alter the thickness of the membrane interior either by varying the length of lipid hydrocarbon tails or by utilizing alkane solvents that remain trapped there.[68] A decrease in membrane thickness is generally observed to increase ion transport by carriers.[85] Moreover, kinetic measurements with valinomycin or macrotetralid actins confirm that both k_{IS} and k_S are inversely related to membrane thickness. For example, Lapointe et al.[113] find with trinactin and 1 M KCl that k_{IS} decreases from 4.5×10^5 to 2.8×10^3 sec^{-1} and k_S decreases from 1.8×10^5 to 8.6×10^4 sec^{-1} when the membrane thickness is increased from 2.3 to 4.8 nm by including decane in the monoolein membrane. The relatively large change of k_{IS} is consistent with the results of image-force calculations.[85]

[113] J. Y. Lapointe, R. Laprade, and F. Grenier, *Fed. Proc., Fed. Am. Soc. Exp. Biol.* **39**, 1991 (1980).

Under the simplest circumstances, ion–carrier association k_R and dissociation k_D rates should not depend on compositional changes in the membrane interior since the complex is formed in the polar region. This is also found to be the case for the macrotetralid actins. Thus, in the experiments of Lapointe *et al.* k_R and k_D appear to be independent of membrane thickness. The situation is more complex for valinomycin. Both k_R and k_D also vary with membrane thickness, indicating complexities in the detailed mechanisms of valinomycin-mediated transport.[85]

Considerable evidence, recently reviewed,[85] indicates that charged polar head groups create an electrostatic potential as well as an accumulation of counterions at the membrane surface. The surface potential ψ_0 and the surface concentration of counterions C_0 are related to the surface density σ of the charged polar head groups by the well-known results of electrical double-layer theory:

$$\psi_0 = (2RT/zF)\sinh^{-1}[\sigma\sqrt{\pi/2RTD\sum C}],$$

$$C_0 = C\exp(-zF\psi_0/RT)$$

where C is the bulk ion concentration and $\sum C$ is the ionic strength.[114] The presence of membrane lipids with charged polar head groups may, therefore, influence transport activity by virtue of an altered concentration of the permeant ion at the membrane surface.

The observed effects of surface charges on the kinetic and steady-state properties of carrier-mediated transport are in agreement with the predictions of double-layer theory for an increased surface activity of the permeant ion. The effects of the charged lipid head group are often large. For example, in decimolar monovalent salt solutions the nonactin-mediated potassium conductance is larger by a factor of 100 in negatively charged phosphatidyl serine bilayers compared with the amphoteric phosphatidyl choline membranes.[114,115] Indirect effects, for example titration or screening of the surface charges, are also important, as indicated by the strong dependence of ψ_0 on surface charge density and ionic strength.[114]

11.9. Molecular Channels

It is now well established that excitation in nerve and muscle occurs by means of the opening and closing of molecular channels that are embedded in the cell membrane. The main line of evidence for this is the observation of discrete conductance changes for several types of biological channels.

[114] S. McLaughlin, *Curr. Top. Membr. Transp.* **9**, 71 (1977).
[115] G. Szabo (1977). *Ann. N.Y. Acad. Sci.*, **303**, 266 (1977).

The attempt to understand the properties of these channels—notably their voltage dependence and ionic selectivity—is the rationale behind the considerable effort to study channels in lipid bilayers. The advantages of such studies include the small background conductance of the bilayer, the possibility of doping the bilayer with a small concentration of channels in order to observe single channels more easily, and the possibility of utilizing channels of known molecular structure.

11.9.1. Single Channels in Bilayers

The measurement of ionic currents through single pores of molecular dimensions constitutes the most direct method for the characterization of conductive channels.[81,88,116-120] These measurements are difficult, however, not only because single-channel currents are small (picoamperes), but also because of the large membrane capacitance, typically of the order of nanofarads for a membrane of 0.01 cm² area. Since sensitivity is limited by the electrical current noise originating in the input amplifier and filtered through the membrane capacitance, small membrane areas ($\simeq 50 \, \mu m$) which reduce input capacitance, as well as low-noise amplifiers which reduce the input voltage noise, are helpful for the measurement of single-channel currents, particularly for the resolution of high-frequency components.

One goal of single-channel experiments in bilayers is the understanding of molecular configurations that provide pathways for ions through channels that are open. Ion-conductive pores may arise from several types of molecular arrangements, as shown in the upper part of Fig. 2. The simpler of these, exemplified by malonyl-bis-desformylgramicidin (MG) is a cylindrical structure formed by a single molecule.[121] Polar residues, in MG formed by the β-helical polypeptide backbone, line the inner wall to form the conductive pore, while the hydrophobic residues are distributed on the outer surface to insure compatibility with the nonpolar environment of the membrane interior. Once inserted into a membrane, a stable monomolecular pore should provide a long-lasting channel for ion conductance since it can be removed only by desorption, an expectedly slow process. Figure 3 (upper

[116] R. C. Bean, W. C. Shepherd, H. Chan, and J. T. Eichner, *J. Gen. Physiol.* **53**, 741 (1969).

[117] G. Boheim and H. A. Kolb, *J. Membr. Biol.* **38**, 99 (1978).

[118] Kh. M. Kasumov, M. P. Borisova, L. N. Emirshkin, V. M. Potselvyev, A. Ya. Silberstein, and V. A. Vainshtein, *Biochim. Biophys. Acta* **551**, 229 (1979).

[119] H. A. Kolb and P. Lauger, *J. Membr. Biol.* **37**, 321 (1977).

[120] R. Latorre, O. Alvarez, G. Ehrenstein, M. Espinoza, and J. Reyes, *J. Membr. Biol.* **25**, 163 (1975).

[121] D. W. Urry, A. Spinsi, M. A. Khaled, M. M. Long, and L. Masotti, *Int. J. Quantum Chem., Quantum Biol. Symp.* **6**, 282 (1979).

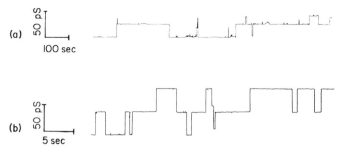

FIG. 3. Typical time histories of single-channel conductances observed in the presence of nanomolar quantities of (a) malonyl-bis-desformyl gramacidin A (MG) and (b) gramicidin A. The bilayers were formed from a dispersion of monoolein in hexadecane 1 M KCl, 100 mV, 23°C. Note the large difference between the time scales.

trace) shows that the pore formed by MG conforms to this expectation. It forms conductive channels of almost indefinite lifetime.[122]

Discrete conductance jumps, such as those shown in Fig. 3, also occur with channels that are voltage dependent. In these cases, the single-channel records provide not only the magnitude of the single-channel conductance, but also the number of conductance levels and the voltage dependence of the opening and closing rates. This is discussed in more detail in Ehrenstein and Lecar.[122a] Single-channel experiments have demonstrated, for example, that EIM in oxidized cholesterol is a simple two-state channel, whereas alamethicin forms individual channels with approximately six conductance sublevels.

Several molecular arrangements have been suggested for the structure of conductive pores formed from subunits.[121] Two extreme cases of these, a series and a parallel repeat of subunits, are shown in the upper part of Fig. 2. The series construct arises when doughnutlike monomers, too small to span the membrane individually, can stack to create a pore of the appropriate length. According to the structure proposed by Urry[123] and confirmed by subsequent experimental data,[124] gramicidin A forms conductive pores of this type by N-terminal apposition and linkage, via six hydrogen bonds, of two monomers. Although the monomers have a polar channel formed by a helical arrangement of the pentadecapeptide backbone, they are not long enough (~ 1.3 nm) to span the bilayer (~ 3.0 nm). It should be noted that the dimer is stabilized by relatively weak, noncovalent

[122] E. Bamberg, H. J. Appell, H. Alpes, E. Gross, J. L. Morell, J. F. Harbaugh, K. Janko, and P. Lauger, *Fed. Proc., Fed. Am. Soc. Exp. Biol.* **37**, 2633 (1978).

[122a] G. Ehrenstein and H. Lecar, *Q. Rev. Biophys.* **10**, 1 (1977).

[123] D. W. Urry, *Ann. N.Y. Acad. Sci.* **195**, 108 (1972).

[124] H. J. Appell, E. Bamberg, H. Alpes, and P. Lauger, *J. Membr. Biol.* **31**, 171 (1977).

bonds. Thus it is expected to dissociate readily, giving rise to relatively short-lived channels. This is confirmed qualitatively by the single-channel records of Fig. 3, which shows lifetimes for gramicidin-A channels (lower trace), much shorter than those of MG (upper trace), a covalently linker pore. An accurate measure of channel stability may be obtained by calculating the mean lifetime τ for a large population of channels. Under the conditions of Fig. 3, $\tau > 10$ min for MG, while $\tau = 3.3$ sec for gramidicin A, in agreement with the qualitative results.

Structural considerations indicate that some pore-forming ionophores, for example alamethicin, nystatin, and amphotericin B, are likely to assume a rodlike conformation having polar and nonpolar faces.[125,126] It has been suggested that subunits of this type form a pore by a parallel association of several monomers, arranged like the staves of a barrel in such a way that the polar face of the monomers line the pore and the nonpolar faces contact the hydrocarbon membrane interior.[127] The properties of alamethicin channels are in agreement with the parallel-repeat hypothesis. Conductance measurements suggest, for example, that pore size increases stepwise, in agreement with the notion that monomers are recruited into the parallel pore one by one.[125] The steep dependence of the number of alamethicin pores on the membrane potential further suggests that monomers must span the membrane to form a pore, again in agreement with a parallel-subunit arrangement.[127]

It is the polar interior of a pore that provides an ion channel. This essential feature is depicted in the center of Fig. 2, where the channel is taken to be a water-filled cylinder having an internal lining of polar residues. Electrostatic arguments predict that an ion in the middle of the pore should encounter an energy barrier as a result of its proximity to the apolar membrane core.[128] Energy barriers should also exist for ion movement into and out of the pore as a result of partial ion dehydration and/or diffusional processes.[129] In contrast, some polar groups in the pore may tend to bind ions. The trans-membrane free-energy profile shown in the lower part of Fig. 2 takes into account these general considerations. Following standard procedures of chemical kinetics, the movement of ions through the pore has been described by a series of voltage-dependent first-order processes, each corresponding to an energy barrier.[130] Thus, in the lower part of Fig. 2 rate constants k'

[125] M. Eisenberg, M. E. Kleinberg, and J. H. Shaper, *Ann. N. Y. Acad. Sci.* **303**, 281 (1977).

[126] A. Finkelstein and R. Holz, *Membranes* **2**, (1973).

[127] G. Bauman and P. Mueller, *J. Supramol. Struct.* **2**, 538 (1974).

[128] D. G. Levitt, *Biophys. J.* **22**, 209 (1978).

[129] O. S. Andersen and J. Procopio, *Acta Physiol. Scand. Supp.* **481**, 27 (1979).

[130] B. W. Urban, S. B. Hladky, and D. A. Haydon, *Fed. Proc., Fed. Am. Soc. Exp. Biol.* **37**, 2628 (1978).

and k'' represent ion translocation across the central barrier, while k_R and k_D represent ion entry and exit rates, respectively. An alternative representation, also shown in Fig. 2, takes into account the rate process by resistances of access for the channel (R_A) and translocation through the channel (R_I). It is comforting that this simple three-barrier, two-binding-site model predicts the salient features of channel electrical properties. In particular, the model predicts concentration-dependent current–voltage characteristics, saturation of channel conductance at high salt concentration, blocking of channel conductance by traces of highly permeant ions, and a flux ratio exponent larger than one. All of these have been observed for the gramicidin-A channel.[130] It has been pointed out, however, that a more complex kinetic scheme is necessary for detailed quantitative fitting of most experimental data.[111]

Beyond the previous general arguments, the relationships between pore structure and single-channel properties are not well understood, although much progress is being made in this direction.[122] It is important to note, however, that channel properties are influenced not only by the chemical composition of the ionophore but also by that of the lipid bilayer membrane.[124,126] It is therefore essential to consider the channel together with its membrane lipid environment.

11.9.2. Multichannel Bilayers

Direct electrical observation of single molecular channels is possible only in carefully controlled circumstances. Usually there are many conductive channels in the membrane, and it is the overall membrane conductance that is experimentally accessible. Since the number of channels and/or the conductance of each channel varies with time, the macroscopic membrane conductance will also be time variant. A standard voltage clamp experiment will indicate the way in which this conductance changes with time and also its steady-state values. This enables one to determine the voltage dependence of conductance and the voltage dependence of the rate of transitions between conducting states. This approach is discussed in more detail in Part 10 of this book, "Voltage Clamping of Excitable Membranes."

Standard voltage clamp experiments will not, however, allow the determination of such microscopic properties as channel conductance. In order to obtain this type of information from multichannel bilayers, conductance fluctuation measurements have been employed. The noise fluctuations can be considered to be the sum of many single-channel discrete conductance jumps. The mean square, autocorrelation, and spectral densities of these conductance fluctuations have been used, together with the mean value of the membrane conductance, to characterize microscopic membrane conductive properties.[131-133] It is possible to use noise measurements to determine microscopic channel properties provided certain characteristics of the system (e.g., the number of conductive states or whether or not the pore breaks down into subunits) are known. It must be emphasized, however, that it is not always possible to infer the conductive properties of single channels from multichannel experiments. Thus single-channel measurements are essential for a complete characterization of the conductive properties.

Gramicidin A forms a pore by the association of two monomers according to the reaction[121]

$$\text{monomer + monomer} \quad \underset{k_B}{\overset{k_F}{\rightleftarrows}} \quad \text{conductive dimer,}$$

where k_F and k_B are the rate constants for formation and breakdown, respectively. It has been repeatedly shown that the pore's conductive properties may be deduced from multichannel conductance measurements.[131] The autocorrelation $C(\tau)$ of current fluctuations δJ is related to the single-channel parameters by the following equations (for a derivation, see Conti and Wanke[131]):

$$C(\tau) = \overline{(\delta J)^2} \exp(-\tau/\tau_c),$$

with

$$1/\tau_c = k_B + 4\sqrt{k_F k_B \lambda/L\Lambda} \tag{11.9.1}$$

and

$$\overline{(\delta J)^2} = \Lambda V \bar{J}(1 + 4\sqrt{k_F \lambda/k_B L\Lambda})^{-1}$$

where λ is the membrane specific conductance ($\Omega\,\text{m}^{-1}\,\text{cm}^{-2}$), V is the membrane potential, Λ is the single-channel conductance, and L is Avogadro's number. Equivalently, the spectral density of the current fluctuations is given by

$$S_I(f) = 4\tau_c\overline{(\delta J)^2}/[1 + (2\pi\tau_c f)^2].$$

For a given aqueous ionic composition and membrane potential, a gramicidin-A pore is characterized by its conductance and by the rate constant of its formation and breakdown. These channel parameters can be determined by autocorrelation or spectral analysis of the macroscopic fluctuations in the membrane current. Macroscopic and microscopic determinations yield equivalent results.[132,133] In some instances, however,

[131] F. Conti and E. Wanke, *Q. Rev. Biophys.* **8**, 451.
[132] H. A. Kolb and E. Bamberg, *Biochim. Biophys. Acta* **464**, 127 (1977).
[133] H. P. Zingsheim and E. Neher, *Biophys. Chem.* **2**, 197 (1974).

the macroscopic parameters do not agree with those measured directly from single channels. For example, in bilayers made from long-chain ($n = 18$) monoglycerides in $1\,M$ CsCl, pore conductance from autocorrelation analysis is only 39% of the directly measured value.[132] Interaction between the pores is believed to be the reason for this discrepancy. Clearly the assumption of noninteracting pores must be verified if macroscopic fluctuation analysis is to yield reliable channel parameters.

It is often observed that the number of conducting pores in a membrane is altered by the membrane potential.[4] For some pore formers, for example, alamethicin and hemocyamin, this voltage dependence is extremely steep, a change of a few millivolts in the membrane potential being able to double the average number of open channels in the membrane. The phenomenon is extremely important in its own right since it provides a realistic model system for the electrical gating of ion channels in excitable cell membranes.[120,134] Conductance fluctuation measurements have been used for the characterization of channel open–close kinetics for a number of systems.[131]

When a simple system, such as gramicidin A, is voltage clamped, an exponential time course of the membrane current is both expected and found.[135] The time constant of the relaxation is given by Eq. (11.9.1) so that k_F and k_B can be calculated by measuring τ_cs for a range of different membrane conductances. Kinetic parameters measured by this relaxation method are not significantly different from those measured by fluctuation analysis. For example, Kolb and Barnberg[132] have obtained $k_R = 1.3 \times 10^4$ and $0.62 \times 10^4\,\mathrm{cm^2\,mol^{-1}\,sec^{-1}}$, $k_D = 1.9$ and $1.6\,\mathrm{sec^{-1}}$, respectively, using relaxation and fluctuation analysis. The equivalence of relaxation and fluctuation measurements is not unexpected; it follows from the fluctuation–dissipation theorem.[131] It has been noted, however, that fluctuation measurements are more reliable since the system is not perturbed. Furthermore, for gramicidin the relatively weak dependence of rate constants on membrane potential and the possibility of time-dependent single-channel conductances may complicate measurements and interpretation in the relaxation method.

11.10. Relationship between Channel Structure and Function

Resolution of the relationships between pore molecular structure and ion permeability is essential for a molecular understanding of membrane function. Initial efforts in this direction have concentrated on establishing

[134] G. Ehrenstein, R. Blumenthal, R. Latorre, and H. Lecar, *J. Gen. Physiol.* **63**, 707 (1974).
[135] E. Bamberg and P. Lauger, *J. Membr. Biol.* **11**, 177 (1973).

the salient features of ion permeability for channels formed by molecules of known chemical composition.[81,121,122] Certain pores, such as those formed by the polypeptide alamethicin, appear to have an effective channel diameter which is much larger than the size of a hydrated cation. Pores of this type are characterized by a conductances nearly independent of the membrane potential† and proportional to aqueous salt concentration. Such channels show practically no selectivity for small ions.[125] It is likely that considerations of electrodiffusion through an aqueous hole would suffice to take into account the permeability characteristics of large pores. Specific interactions of the permeant ion with the pore wall should become increasingly important as the size of the pore approaches that of the ion. Other effects should also become important in the narrow pore. For example, an ion may not be able to jump over another one also present in the pore, resulting in a single-file ion movement.[136] There are at least four lines of evidence for pore size effects in the gramicidin channel. First, anions, multivalent cations, and large monovalent cations such as tetramethylammonium cannot move through the pore to any appreciable extent.[81] Second, the single-channel conductance is voltage dependent and is not always proportional to salt concentration; this would be expected for narrow but not for wide pores. Third, trace amounts of permeant ions such as Tl^+ or Ag^+ can reduce or "block" the movement of the alkali-metal cations through the pore.[111,137] Fourth, the flux ratio exponent n, defined by the relationship

$$\vec{J}/\overleftarrow{J} = (C_1/C_2)\exp(nqV/kT)$$

is larger than unity under certain circumstances. Here $\vec{J}$ and $\overleftarrow{J}$ are the isotopically measured unidirectional fluxes of ion I having concentrations C_1 and C_2 across the membrane. A flux ratio exponent larger than unity has been interpreted to mean multi-ion interactions, not expected for wide pores.[136]

Ion selectivity of conductive pores may be assessed both by single-channel conductance ratios and by zero-current potential measurements much in the same way as is done for ion carriers. The selectivity behavior of gramicidin A, the only narrow pore for which extensive data is available, is rather complex. Not only do the conductance and permeability ratios differ in magnitude but also both of these selectivity parameters depend on the

---

[136] B. Hille and W. Schwarz, *J. Gen. Physiol.* **72**, 409 (1978).
[137] D. McBride and G. Szabo, *Biophys. J.* **21**, 25a (1978).

---

† It should be emphasized that the macroscopic membrane conductance may nevertheless be strongly voltage dependent, as is the case for alamethicin.[117] This is because the number of open channels (but not the single-channel conductance) depends strongly on the membrane potential.

ionic concentrations as well as the membrane potential.[111,130] Similar complexities are observed also for biological channels and in this regard gramicidin A is a good model. Unfortunately neither the gramicidin-A pore nor any other well-documented pores show selectivities comparable to those of biological membranes.

There have been impressive recent advances in our understanding of both the electrophysiology and the biochemistry of membranes. The synthetic lipid bilayer has contributed to this progress by providing clearly defined model systems for specific membrane functions. Thus, for example, the detection in planar bilayers of ionic currents through pores of molecular dimensions has paved the way for the observation of corresponding events in muscle and nerve membranes. It is likely that bilayer model systems will continue to provide both theoretical and experimental guidance for a more detailed understanding of the highly complex processes of cell and organelle membranes. Thus, for example, the presently obscure subject of peptide conformation in the membrane phase is presently being elucidated in lipid bilayer membranes.[121]

The bilayer should also continue to provide insights into the molecular mechanisms of transport. It is already possible, for example, to use powerful biophysical techniques, such as NMR, to infer the exact location of ions in model pores having a variety of molecular structures. These techniques should eventually be applied to the pores of electrically excitable cell membranes.

Using recent biochemical techniques, it has been possible to isolate and to purify the macromolecules responsible for specific transport functions in cell membranes. The lipid bilayer should prove to be the substrate necessary for the reconstitution of these isolated transport functions in a purified and functional form. In the more immediate future, specifically designed peptides in lipid bilayers should provide several model systems, well understood at the molecular level, for such physiological membrane functions as chemical and electrical gating of ion channels.

AUTHOR INDEX

Numbers in parentheses are reference numbers and indicate that an author's work is referred to although the name is not cited in the text.

A

Abragam, A., 2, 78
Adams, D. H., 132
Adams, R. N., 107
Adamson, A. W., 519
Adar, F., 152, 154(136)
Adelman, Jr., W. J., 436, 464(47), 465(47), 466(47), 471
Adrian, R. H., 474, 481, 482(83)
Aggerbeck, L., 351, 373, 375(43), 378(43), 380(43, 105)
Ahmed, F. R., 244
Akaike, N., 484
Alberi, J. L., 357, 358
Albrecht, A. C., 128, 129, 130
Alden, R. A., 265
Aleksanyan, V. I., 366
Alfano, R. R., 186
Alkins, J. R., 154
Allemand, R., 357
Allerhand, A., 18, 19, 33, 34
Almers, W., 471
Alpert, S. S., 329
Alpes, H., 537, 539(122, 124), 542(122)
Alvarez, O., 536, 541(120)
Ambesi, D., 227
Ambesi-Impiombato, F. S., 226, 227
Amos, L. A., 293, 411
Anatassakis, E., 149
Anderegg, J. W., 340, 343(3), 362(3), 369(3), 370, 371
Andersen, O. S., 525, 526, 527(80), 531(76), 538
Anderson, R. G. W., 197

Anderson, W. A., 8
Andrew, E. R., 31
Andrew, J. R., 149
Andrews, D., 521, 523, 524(53, 54)
Antanaitis, C., 154
Appell, H. J., 537, 539(122, 124), 542 (122)
Applebury, M. L., 186
Aragon, S. R., 322
Armstrong, C. M., 447, 456, 457, 470, 471(18b), 472
Arndt, U. W., 248, 250(23), 251, 252 (25)
Arnott, S., 276, 287
Arthur, J. W., 131
Asakura, S., 323, 329(47)
Asano, S., 323
Asch, R., 314
Asher, I. M., 149
Asher, S. A., 152, 153
Atkinson, D., 352
Atkinson, G. H., 156
Aton, B., 130, 155, 158, 186
Atwater, I., 459, 460
Auchet, J. C., 198, 218, 222(5)
Aue, W. P., 39
Auston, D. H., 167, 168(11), 192(11), 193(11), 194(11)
Avivi, A., 226, 227
Avouris, P., 182, 185(31)
Axelrod, D., 198, 205(14), 209(14, 15, 16, 17), 210(16), 211(15, 16), 212, 215, 216(14, 17), 217, 218, 227
Azhderian, E., 45(54), 46
Azzi, A., 199

B

Babcock, G., 154
Bachman, P., 39, 40(44), 41
Bachovchin, W. W., 42, 43
Bacon, G. E., 340, 341(13), 352(13), 353(13), 357(13), 359(13), 375(13)
Bagyinka, C., 156
Bahr, G. F., 396, 425
Bahr, W., 198, 209(10), 217, 221(10)
Bak, A. F., 480, 492
Baker, P. F., 457
Baker, T. S., 430
Balaram, P., 36
Baldwin, J. P., 379
Ball, C. G., 521
Bamberg, E., 141, 537, 539(122, 124), 540, 541, 542(122)
Bandekar, J., 147
Bangham, A. D., 516, 517, 518(25, 28, 29)
Banks, G., 329
Bansil, R., 135(c), 136, 145
Barak, L. S., 216, 218, 221, 223(70), 227
Barber, J., 182, 183(29)
Barenholtz, Y., 202, 207(28)
Barisas, B. G., 224
Baro, R., 351, 359(32)
Barrett, A. N., 290, 291(67)
Barrett, T. W., 146
Barrington Leigh, J., 290, 291(67)
Barron, L. D., 154
Bartholdi, E., 39
Bartholdi, M., 224
Bartholomew, R. M., 122
Bartlett, J. H., 446
Baskin, R. J., 142, 317
Bastian, J., 501
Bates, S. R., 354
Batterman, B. W., 360
Batzri, S., 518
Baudy, P., 379
Bauer, R. S., 88
Bauman, G., 538
Baumann, W. J., 20, 21, 22
Baumeister, W., 425
Bazett-Jones, D. P., 426
Bazhulin, P. A., 169
Bean, R. C., 536
Beaudry, P., 379
Becker, B., 158

Becker, E. D., 15, 16, 26, 27
Beeman, W. W., 340, 343, 348, 351(25), 359(31), 362, 369(3), 386
Beer, M., 438
Behr, J. P., 22
Behringer, J., 129
Benedek, G. B., 301, 316, 318(38), 319, 333, 334, 335(83)
Bennett, V., 225
Bennett, Jr., W. R., 163
Benoit, H., 373
Ben-Sira, I., 333
Benz, R., 522, 523(64), 524(64), 529, 531, 532, 533, 534
Berg, H. C., 327
Berg, J. C., 515, 516
Bergé, P., 330
Berger, J., 382
Berger, M. J., 422
Bergman, C., 492
Berlin, R. D., 199
Berliner, L. J., 54, 79, 88, 93(1), 94, 97(1), 110
Bernal, J. D., 244
Berne, B. J., 300, 308, 328, 329(60), 333(60)
Bernstein, H. J., 132, 136, 138, 139, 141(52), 142(52), 151
Bezanilla, F., 446, 447, 456, 457(30), 459, 460, 461, 463, 467, 469, 470, 471(18b), 472, 476, 486, 499, 500(116)
Bianco, A. R., 222
Bielka, H., 370
Bigelow, V. J., 513
Bikerman, J. J., 520, 523(42)
Billett, D. F., 521, 523(55)
Binstock, L., 452, 463, 464, 465, 466
Bittner, J. W., 438
Bjarnason, J. B., 149
Black, J. J., 160
Blackwell, J., 145
Blagrove, R. J., 314
Blaise, J. K., 221
Blake, G. C. F., 260
Blandau, R. J., 330, 336(66)
Blaurock, A. E., 221
Blaustein, M. P., 501
Blazej, D. C., 143, 145(76)
Bleaney, B., 78
Bloch, F., 1, 7, 9(1), 73

Blodgett, K. B., 514, 523
Bloembergen, N., 16, 163, 190(3)
Blomberg, F., 43
Bloomer, A. C., 293
Bloomfield, V. A., 300
Blout, E. R., 202
Blow, D. M., 229, 234, 235, 255, 261, 263, 285, 286
Blumenthal, R., 216, 541
Blundell, T. L., 229, 231, 258, 264, 269(1)
Boeyens, J. C. A., 79
Bogomolni, R. A., 158
Boheim, G., 536, 542(117)
Bolton, J. R., 54
Bond, P. J., 143, 144
Bonner, R. F., 331, 332, 333(78a)
Bonse, V., 352
Boon, J. P., 329
Borders, Jr., C. L., 253
Borejdo, J., 224
Borg, D. C., 54
Borisova, M. P., 536
Borkowski, C. J., 350
Boseley, P. G., 379
Bothner-By, A. A., 36
Bottger, M., 370
Bourdel, J., 357
Bowen, P. D., 331, 332
Bowman, R. L., 331, 332
Bradbury, E. M., 379
Bradley, D. J., 167
Brahms, J., 145
Bram, S., 366, 379
Brand, L., 203, 204
Brdiczka, D., 141
Breit, G., 81
Breslow, E., 36
Bricogne, G., 293
Brooks, D. E., 521, 523(57), 524(57)
Broomfield, C. A., 63
Brown, A. M., 484
Brown, D. A., 327
Brown, E. B., 144, 150
Brown, G., 247
Brown, K. G., 150
Brown, M. S., 197
Brown, R. E., 521
Brown, T. R., 52
Brûlet, P., 109, 111, 112, 115, 116, 119(74), 122(64, 74)

Brumberger, H., 340, 342
Brunner, H., 152
Brunner, J., 518
Budinger, T. F., 428
Bueche, A. M., 367
Buerger, M. J., 239, 242, 248
Bunemann, H., 366
Bunow, M. R., 137, 138, 139, 140, 141, 142(50), 156(46)
Buras, B., 354
Burke, C. A., 102
Burke, J. M., 152, 154(138)
Bürkli, A., 205, 207(48), 208(48), 218, 221(48)
Burns, A. R., 157, 158(184)
Busch, G. E., 179, 186
Busslinger, M., 205, 207(48), 208(48), 218, 221(48)
Butler, P. J. G., 293
Butler-Browne, G., 379
Bye, J. M., 221, 222(69), 223(69)

C

Cadenhead, D. A., 515, 516(16)
Cael, J. J., 145
Cahalan, M., 498
Caille, J. P., 149
Calhoun, Jr., R. R., 351
Callender, R. H., 130, 132, 155, 156(171), 157, 158, 186
Caloin, M., 323
Camejo, G., 387
Camerini-Otero, R. D., 317
Campbell, D. T., 497, 498, 499(114), 500, 502, 503, 504, 507
Campbell-Smith, P. J., 287
Campion, A., 156, 157, 158(176, 179), 159
Candau, S., 335
Cantor, C. R., 224
Capaldi, R. A., 219
Caputo, C., 476
Cardoso, M. F., 307, 313(22)
Carew, E. B., 142
Carey, P. R., 136, 150, 155
Carlson, B. C., 422
Carlson, F. D., 300, 317, 323, 331, 335, 336
Carpenter, D. K., 373, 386

Carpentier, J.-L., 226
Carr, H. Y., 26, 29
Carriere, R. G., 136, 155
Carrington, A., 71, 80
Carter, Jr., C. W., 265
Caspar, D. L. D., 288, 355, 358(49)
Castaing, R., 440
Chabay, I., 160
Chabre, M., 379
Champion, P. M., 151, 152, 153(129), 154(129)
Champness, J. N., 293
Chan, H., 536
Chan, S. I., 533
Chance, B., 158, 184, 185(35)
Chandler, W. K., 456, 458, 460, 464, 465, 474, 481, 482(83)
Chang, K. W., 222
Chang, R. C. C., 152, 153
Changeux, J. P., 198, 218, 222(5)
Chapman, D., 32, 198, 205, 207(47), 218, 221(47)
Chauvin, C., 371
Chen, F. C., 322
Chen, L. A., 204
Chen, M. C., 143, 147, 150
Chen, S. H., 300, 314, 322, 323, 325, 327, 328, 329, 330(57)
Cherry, R. J., 204, 205, 207(45, 47, 48), 208(48), 218, 221(47, 48), 225
Childers, R. F., 33, 34(32)
Chimosky, J. E., 332
Chinsky, L., 145
Chipman, D. R., 360
Chiu, W., 435
Chou, C. H., 145
Chow, S., 519
Chrzeszczyk, A., 322
Chu, B., 300, 322
Church, G. M., 267
Chylack, Jr., L. T., 336
Ciani, S., 520, 531, 532(50), 533
Cjeka, Z., 384, 385(151)
Claessen, H. H., 127
Clark, W. R., 225
Clarke, J., 308
Cochran, W., 229, 276
Cogan, U., 202, 207(26)
Cogdell, R. J., 183, 184(34)

Cohen, C., 293, 294(72), 295(72)
Cohen, H., 342, 363, 366
Cohen, L. B., 467
Cohen, R. B., 316, 319(32)
Cohen, R. J., 318(38), 319
Cohen, S., 226
Cohn, M., 54, 104, 109
Cole, K. S., 445, 446, 447, 449, 451(15), 452(23), 454, 455(20), 456(20), 459(19, 20, 23), 461, 464, 465, 466, 493, 501, 502, 509(7), 511(7), 520, 521(46), 525
Cole, M. D., 438
Colley, C. M., 516, 518(27)
Collins, D. W., 151
Collins, R. J., 164
Colman, P. M., 261
Cone, R. A., 198, 204, 207(46), 217, 218, 225
Connelly, J. A., 351
Connick, R. E., 43
Conrad, V. H., 387
Conti, F., 463, 473, 486, 497, 540, 541(131)
Convert, P., 357
Copeland, B. R., 119
Cornell, D. W., 85, 88(36)
Cosslett, V. E., 426
Costantin, L. L., 481
Coster, H., 521
Cotton, J.-P., 357, 373
Coulombe, L., 148
Cowley, J. M., 400
Craig, T., 328, 330
Cramer, F., 367, 370(92)
Cramer, S., 160
Crawfoot, D. C., 244
Crepeau, R. H., 411
Crespi, H. L., 384, 385(151)
Crewe, A. V., 420, 436, 437, 438, 444
Crichton, R. R., 373, 376, 377(112), 378(112), 379, 380, 381
Crick, F. H. C., 255, 276
Cross, D. A., 306
Cross, P. C., 128
Crouch, B., 132
Crouch, R., 158
Crowther, R. A., 263, 411, 415, 418
Crunelle-Cras, M., 157
Cser, L., 355
Cuatrecasas, P., 218, 222, 223(80)

Cummins, H. Z., 300, 312(1), 315(1), 323
Curatolo, W., 141
Curtis, H. J., 445, 465, 520, 521(46)

D

Dadok, J., 36
Dale, R. E., 204
Dalton, L. R., 97
Damaschun, G., 351, 359, 361, 370, 382, 387
Damaschun, H., 387
Daniell, G. J., 415
Danielli, J. F., 514, 520
Danielmeyer, H. G., 169
Daune, M., 323
David, G., 330
Davies, P. W., 460
Davila, H. V., 463(56), 467
Davson, H., 514, 520(14)
Dawkins, P., 532
Dawson, M. J., 51, 52(64)
Day, E. P., 322
D'Azzo, J. J., 468, 495(58), 496
Dean, R. B., 520, 521(46)
Debye, P., 341, 367
Decius, J. C., 128
DeFelice, L. J., 463, 473
Degenkolb, E. O., 179, 182, 185(31)
DeGrip, W. J., 149
deGuzman, N. T., 531
DeKruijff, B., 156
de Laat, S., 206, 217
Delbruck, M., 220
del Castillo, J., 492
Delhaye, M., 161
Dellepiane, G., 155
De Marco, J. J., 360
DeMaria, A. J., 164
Demel, R. A., 156
DeMeyts, P., 222
den Hollander, J. A., 52
Derksen, H. E., 490
DeRosier, D. J., 411, 418
Dervichian, D. G., 515
Desmeules, P. J., 148
DeToma, R. P., 203
Deutsch, M., 361

Devaux, P., 108, 109, 112, 113
De Wolf, B., 379
de Wolff, P. M., 349
Dhamelincourt, P., 161
Diamond, R., 264, 265, 291
Dianoux, A. C., 202, 207(24)
Dickerson, R. E., 252, 253, 254, 255(31), 269
Dietrich, I., 419
Difazio, C., 521
Dinerstein, R. J., 102, 103
Dionne, V. E., 473
Doddrell, D., 18, 19(15), 34
Dodge, F. A., 492, 493, 495, 496, 497, 499, 504, 506, 507, 508
Doerman, F. W., 81
Doskocilova, D., 32
Doukas, A. G., 130, 132, 155, 158
Dousmanis, G. C., 80, 82(27)
Downes, N. W., 377
Downing, K., 439
Dragsten, P. R., 225
Dubochet, J., 426
DuBois, M., 330
Duguay, M. A., 177
Duquesne, M., 145
Durchschlag, H., 387
Durig, J. R., 128
Dutta, P. K., 136, 160
Dutton, P. L., 183, 184, 185(35)
Duval, A., 501
Dwek, R. A., 2, 54, 110
Dykes, G., 411

E

East, E. J., 147
Easter, J. H., 203
Ebrey, T. G., 158
Edelman, G. M., 217
Edelstein, S. I., 411
Edgell, W. F., 131
Edidin, M., 198, 199, 207(20), 209, 217, 218, 225
Eggers, F., 526, 532(82), 534(82)
Ehrenberg, W., 351
Ehrenstein, G., 446, 447, 459, 536, 537, 541

Eicher, H., 387
Eichner, J. T., 536
Eisenberg, D., 430
Eisenberg, H., 342, 363, 366
Eisenberg, M., 530, 538, 542(125)
Eisenberg, R. S., 477, 478, 479, 480, 509
Eisenberger, P., 153
Eisenman, G., 520, 531, 532(50), 533, 539(111), 542(111), 543(111)
Eldridge, C. A., 217, 219, 227
El-Sayed, M. A., 156, 157, 158(176, 179, 183, 184), 159(179)
Elson, E. L., 198, 205(14), 209(14, 15, 16, 17), 210(16), 211(15, 16), 212(15), 213, 214(52, 53, 54), 216, 217, 218, 219, 221, 223(70), 224(53), 226, 227
Emirshkin, L. N., 536
Engel, A., 426
Engel, E., 478, 479
Engelman, D. M., 344, 373, 375, 377, 378(112, 118), 379(112), 381(112, 118), 382, 384, 385, 390
Englander, S. W., 377
Engstrom, S. K., 145
Entine, G., 198, 217
Epp, O., 261
Erfurth, S. C., 143, 144
Erickson, H. P., 293, 407, 411
Ernst, R. R., 8, 39, 40(44), 41
Espinoza, M., 536, 541(120)
Esser, A. F., 122
Evtuhov, V., 168
Ewing, J. J., 169

F

Fahey, P. F., 215, 216, 227
Fairhurst, S. A., 154
Fanconi, B., 146
Farnoux, B., 357
Farquharson, S., 154, 156(147)
Farrar, T. C., 15, 16, 26, 27
Federov, B. A., 366
Fehlhammer, H., 261
Feigin, L. A., 383, 386
Feke, G. T., 333
Feld, M. S., 163, 190(2)
Feldberg, S. W., 529, 532(90), 533
Felton, R. H., 150, 152, 153

Fermi, E., 81
Fernández, J. M., 446, 459(46), 463
Fernandez, S. M., 199
Fettiplace, R., 521, 524(54)
Fichtner, P., 382
Finch, J. T., 411
Finkelstein, A., 156, 538, 539(126)
Fishbach, F. A., 371
Fisher, J., 357, 358
Fisher, K. A., 412(23), 413, 429(23)
Fishman, H. M., 457, 458, 471, 483
Fitchen, D. B., 151
FitzHugh, R., 446, 464, 465
Fleischer, S., 220
Flynn, G. W., 315
Folger, R. C., 331
Folkhard, W., 387
Foord, R., 314
Ford, N. C., 300, 314, 316, 330
Forrest, G., 144
Fournet, G., 340, 352, 363(2), 367(2), 368(2), 388(2)
Fowler, V., 225
Fox, F., 419
Fox, C. F., 64
Franck, U. F., 446
Frank, J., 404, 405(11), 429, 430
Frank, K., 482, 490
Frankenhauser, B., 480, 489, 491, 492, 493, 495, 496, 497, 499, 504, 505, 506, 507, 508
Franklin, R. E., 289
Franklin, R. M., 317
Franks, A., 351
Fraser, A. B., 331, 336
Freed, J. H., 102
Freedman, T. B., 135(e), 136
Freeman, S. K., 128, 132(10), 134(10)
Freer, S. T., 249, 252, 265
Frey, T. G., 411
Fricke, H., 514, 520(8)
Friedman, J. M., 133, 152, 154(136)
Fringeli, U. P., 425
Frohlich, O., 522, 523(64), 524(64)
Frushour, B. G., 146, 147, 148
Frye, L. D., 198, 217
Fuchs, M., 525, 531(76)
Fuchs, P., 202
Fuess, H., 381, 386
Fujime, S., 323, 329(47), 331

Funck, T., 526, 532(82), 534(82)
Funfschilling, J., 134

G

Gaber, B. P., 129, 138, 139, 141, 142(51), 147, 148(18), 150(18), 154(18), 155(18)
Gabler, R., 316
Gaddum-Rose, P., 330, 336(66)
Gadian, D. G., 51, 52(64)
Gaffney, B. J., 54, 93, 100
Gage, P. W., 480
Gains, G. L., 515
Gallwitz, U., 291
Gammack, D., 517
Gardner, K. H., 145
Garrett, R., 387
Gaskin, F., 317
Geissler, E., 335, 336
Gelman, R. A., 336
Gershfeld, N. L., 514, 516, 521(13), 523(13)
Gethner, J. S., 315, 317
Giaconnetti, G., 82
Gibby, M. G., 37
Giddon, D. B., 331
Giege, R., 143, 380
Gilbert, D. L., 456
Gilbert, P., 414
Gill, D., 130, 155, 156(170), 158
Gilson, T. R., 160
Giordmaine, J. A., 190
Gitler, C., 202, 207(24, 25), 218, 221(25)
Gladkikh, I. A., 371
Glaeser, R. M., 397, 400(7), 412(23), 413, 426, 428, 429(23), 431, 433, 435
Glatter, O., 361, 380
Glushko, V., 34
Golan, D. E., 225
Goldfarb, W., 430
Goldman, D. E., 500, 501
Goldman, L., 452
Goldman, R., 293
Goldstein, H., 78
Goldstein, J. L., 197
Gonzalez-Rodriguez, J., 205, 207(47), 218, 221(47)
Goodisman, J., 342

Goodman, J. W., 233, 234, 235(6), 262, 264, 397, 398(5)
Goodwin, D. C., 145
Gorczyca, L. E., 149
Gorden, P., 226
Gorter, E., 514
Gould, R. W., 354
Grano, D. A., 412(23), 413, 429(23)
Grant, D. M., 35
Green, B. H., 186
Green, D. W., 245
Green, R. M., 445
Greenaway, A. H., 439
Greenhalgh, D. A., 160
Grell, E., 48, 526, 532(82), 534
Grendell, F., 514
Grenier, F., 534
Greve, K. S., 179
Greytak, T., 301
Griffin, R. G., 38
Griffith, O. H., 54, 85, 88, 102, 219
Gross, E., 537, 539(122), 542(122)
Grunberg-Manago, M., 379
Grund, S., 431
Guillot, J.-G., 145
Guinier, A., 247, 338, 340, 352, 362, 363(2), 365(1), 367(2), 368(2), 388(2)
Gull, S. F., 415
Gunsalus, I. C., 151, 152, 153(129), 154(129)
Gupta, R. K., 44
Gutowsky, H. S., 24

H

Haas, J., 373, 376, 377(112), 378(112), 379, 380, 381
Haeberlen, U., 29
Hafeman, D. G., 114, 122
Hagiwara, S., 452, 459, 525
Hahn, E. L., 36, 37, 39
Hahn, M., 425
Hainfeld, J. F., 438, 439
Hall, C. E., 391
Hallett, F. R., 328, 329, 330
Hamill, O. P., 473, 484(68h)
Hamilton, M. G., 135(b), 136
Hamlin, R., 252
Hammes, G. G., 218, 221, 223(70)

Hansen, W. W., 1, 9(1)
Hanssen, K. J., 439
Harada, I., 147, 148, 150
Harbaugh, J. F., 537, 539(122), 542(122)
Harburn, G., 282
Hardy, W., 520
Harney, R. C., 142
Harris, R. K., 35
Harris, W. C., 128
Harrison, P. M., 371
Harrison, S. C., 248, 371, 381(100, 101)
Hart, M., 352
Hartman, K. A., 144, 145
Hartmann, S. R., 36, 37
Harvey, A. B., 160
Harvey, R. C., 142
Hasan, F., 49, 50(60)
Haskell, R. C., 331
Hauser, H. O., 517, 518, 519
Hayashi, H., 452
Haydon, D. A., 521, 522, 523(53, 55, 56, 57), 524(53, 54, 57), 526, 530(81), 531(81), 533(81), 536(81), 538, 539(130), 542(81), 543(130)
Hayward, S. B., 412, 413, 428, 429, 444
Haywood, B. C. G., 358
Heard, H. G., 168
Heath, T. D., 516, 518(27)
Hecht, A. M., 335, 336
Heckel, C., 521
Heide, H. G., 419, 431
Heidenreich, R. D., 391
Heistracher, P., 476, 481(74)
Hekman, M., 227
Hellwarth, R. W., 163
Helmreich, E. J. M., 227
Henderson, R., 221, 260, 412, 413, 428, 429, 432
Hendrichs, R. W., 350, 351, 360
Hendrickson, W. A., 268
Hendry, B., 521
Henis, Y. I., 226, 227
Henkart, P., 216
Henn, F., 522
Henry, L., 440
Henry, N., 122
Herbeck, R., 145
Herbert, T. J., 323
Hermann, P., 379

Hermans, Jr., J., 257, 264
Herriott, D. R., 163
Herriott, J. R., 261, 269, 270(53), 271, 273, 274(54)
Herskovitz, T., 154
Herz, J., 335
Herzberg, G., 128, 154(2)
Hess, B., 158
Hestenes, M. R., 268
Hetherington, III, W., 160
Heumann, H., 384, 385(151)
Heyde, M. E., 155, 156(170), 158
Heynau, H., 164
Hild, G., 335
Hill, I. R., 138
Hill, M. W., 516, 518(25)
Hill, W. E., 370
Hille, B., 470, 487, 488(97), 491, 493, 495, 497, 498, 499(114), 500, 502, 503, 504, 507, 525, 528(75), 542
Hilton, B. D., 46
Hirakawa, A. Y., 129, 143, 145, 151
Hjelm, B. P., 379
Hladky, S. B., 526, 528, 530, 531, 532, 533, 534, 535(85), 536(81), 538, 539(130), 542(81), 543(130)
Ho, J. T., 322
Hoard, J. L., 153
Hobbs, L. W., 426
Hocker, L. O., 334, 335(83)
Hodgkin, A. L., 445, 446, 447, 450(18), 451, 456(18), 457, 459(17), 461, 463, 464, 465, 466, 474, 481, 482(83), 499, 504, 532
Holasek, A., 380
Holbrook, S. R., 267
Holloway, Jr., G. A., 332
Holm, C. H., 24
Holm, R. H., 154
Holmes, K. C., 229, 234, 285, 286, 288, 289, 290, 291, 292(63)
Holz, M., 314, 322, 327, 328(57), 329(43), 330(57)
Holz, R., 156, 538, 539(126)
Honig, B., 130, 155, 156(171), 157(171)
Hoppe, W., 383, 384, 385(151)
Horio, T., 153, 154(142)
Horn, L. W., 516
Horn, R., 484
Horne, R. W., 517, 518(28)

Horowicz, P., 476, 499
Horváth, L. I., 156
Hossfeld, F., 361, 387
Hou, Y., 216
Hoult, D. I., 41
Houpis, C. H., 468, 495(58), 496
Hruby, V. J., 149
Hsieh, C.-L., 157, 158(183, 184)
Hsu, S. L., 140, 146, 147
Huang, C., 516, 517, 518, 519
Hubbell, W. L., 63, 86(12), 91(12), 101
Huber, L. M., 29
Hudson, A., 107
Hudson, B. S., 160
Humphries, G. M. K., 111, 112(64), 114, 115(64, 70), 117, 118, 120, 122(64, 71, 75)
Hunkapiller, M. W., 42
Hunt, C. C., 476, 481(74)
Hurd, R. E., 45(54), 46
Hutley, M. C., 130
Huxley, A. F., 446, 447, 450(18), 451, 456(18), 457(18), 459(17), 461(18), 463(18), 464(18), 465(18), 466(18), 488, 489, 490, 491, 504
Huxley, H. E., 411
Hvidt, A., 377
Hyde, J. S., 97
Hyman, A. S., 342, 364, 374(83)
Hynes, R. O., 221, 222(67, 69), 223(69)

I

Ibel, K., 357, 358, 376, 377, 378(117), 379, 381, 384, 385(151), 386(117)
Iizuka, T., 153
Ikegami, A., 204
Ildefonse, M., 501
Inagaki, F., 129
Inbar, M., 202, 207(37, 38), 208, 218, 221(37, 38)
Ingram, V. M., 245
Inoue, I., 465
Ippen, E. P., 158, 171, 172, 178, 191
Isaacson, M. S., 394, 420, 437, 438, 439, 441
Ishimoto, C., 336
Ishimura, Y., 153
Ishiwata, S., 336

J

Jack, A., 371, 381(101)
Jacobs, S., 222
Jacobson, K., 198, 202, 207(14, 27), 209, 215, 216, 217
Jacrot, B., 340, 359(10), 371, 377(10), 378, 379, 380
Jaffe, L. A., 525
Jaffe, L. F., 460
Jakeman, E., 307, 308(21), 313, 314
James, R. W., 234, 240, 269(7), 340, 341(11)
James, T. L., 2, 12, 28
Janko, K., 537, 539(122), 542(122)
Jap, B. K., 397, 400(7)
Javan, A., 163
Javin, T. M., 226
Jensen, L. H., 229, 247(3), 265, 269, 270(53), 271, 273, 274(54)
Johnson, B. B., 129, 130, 131(12, 25), 150(12), 151(25)
Johnson, D. E., 420, 440, 441
Johnson, E. A., 477, 509
Johnson, L. N., 229, 231, 258, 260, 264, 269(1)
Johnson, M., 217
Johnston, P. D., 46
Jones, F. W., 361
Jones, R. P., 174, 175(22), 179
Jost, P. C., 54, 102, 219
Jouannet, P., 330
Jovin, T. M., 224
Joyner, R. W., 504
Jubb, J. S., 412(24), 413
Julian, F. J., 500, 501

K

Kaesberg, P., 340, 343(3), 362(3), 369(3)
Kahn, C. R., 222
Kaiser, B., 387
Kammerer, O. F., 355, 358(49)
Karhan, J., 39
Karplus, M., 155
Kasai, M., 198, 218, 222(5)
Kasha, M., 164
Kasumov, K. M., 536

Katz, B., 445, 446, 447, 450(18), 451(17, 18), 456(18), 457(18), 459(17), 461(18), 463(18), 464(18), 465(18), 466(18), 467, 481, 532
Katz, J. J., 384, 385(151)
Kaufman, H. W., 142
Kaufmann, K. J., 182, 183, 184, 185(31, 35)
Kawato, S., 204
Kayushina, R. L., 386
Keana, J. F. W., 102, 103
Keiser, H. R., 332
Keller, R., 382
Kendrew, J. C., 252
Kennedy, S., 521
Keonig, J. L., 143
Kerker, M., 302, 319
Ketterer, B., 533
Keynes, R. D., 452, 471
Khaled, M. A., 536, 537(121), 540(121), 542(121), 543(121)
Kiefer, W., 132, 133, 151(36)
Kilponen, R. G., 130, 155, 156(170), 158(170)
Kim, J. J., 370
Kim, S. H., 45, 267, 370
Kincaid, B. M., 153
King, R. W., 155
Kinosita, K., 204
Kirchauser, T., 387
Kirschner, K., 387
Kirste, R. G., 373, 374(107)
Kiser, J. K., 144
Kishimoto, U., 465
Kissel, G., 533
Kitagawa, T., 153, 154(141, 142)
Kito, Y., 186
Kitzes, A., 379
Klauminzer, G. K., 160, 161(200)
Kleinberg, M. E., 538, 542(125)
Klesse, R., 357
Kley, G., 351
Klotz, I. M., 155
Klug, A. L., 276, 288, 293, 407, 411, 415, 418
Knapek, E., 419
Kneale, G. G., 379
Knowles, P. F., 2
Knzazev, I. N., 169

Kobayashi, T., 176, 177, 179, 180(28), 182, 185(31), 187, 188, 189, 192
Koch, M. H. J., 373, 376, 377, 378(112), 379, 380, 381
Koenig, J. L., 145, 146, 148, 149
Kolb, H. A., 536, 540, 541, 542(117)
Komoroski, R., 18, 19(15)
Konnert, J. H., 268
Koops, H., 444
Kootsey, J. M., 482
Kopf, D., 438
Kopka, M. L., 253
Kopp, F., 425
Kopp, M. K., 350
Koppel, D. E., 198, 207(14), 208(14, 15, 16, 17), 210(16), 211(15, 16), 212(15), 216, 217, 218, 224, 225, 226, 227, 317, 318
Korn, E. D., 518
Kornberg, R. D., 109, 112, 113, 198, 518
Kostner, G., 380
Kostorz, G., 357
Kostyuk, P. G., 484, 485
Kotlarchyk, M., 323
Koyama, Y., 141
Kozlov, Z. A., 371
Krakow, W., 436
Kranold, R., 361
Krasne, S., 531, 533
Kratky, O., 340, 347, 351, 352, 360, 362, 367, 370, 371, 380, 383, 387, 388
Kraut, J., 265
Kretsinger, R. H., 248
Krigbaum, W. R., 387
Krimm, S., 140, 146, 147
Krishtal, O. A., 484, 485, 486
Kruger, G. J., 79
Kruuv, J., 225
Kuck, G., 444
Kugler, F. R., 387
Kuhlmann, K. F., 35
Kuhlmann, W. R., 350
Kumar, K., 155
Kumosinski, T. F., 340, 351
Kunz, C., 168
Kunze, Jr., R. K., 322
Kuo, I., 428
Kuriakose, T. J., 131
Kurtz, Jr., D. M., 155
Kyogoku, Y., 141, 144, 153, 154(141, 142)

L

Lagerkvist, V., 387
Laggner, P., 380
Lai, C. C., 314
Lajzerowicz-Bonneteau, J., 79
Lake, J. A., 361, 411
Lamola, A. A., 186
Landau, L. D., 302
Langer, J. A., 382, 384, 385
Langley, K. H., 330
Langmore, J. P., 394, 436, 438
Langmuir, I., 514, 520, 523
Lapointe, J. Y., 534, 535
Lappe, D. L., 332
Laprade, R., 533, 534
Lardner, T. J., 198, 209(11), 217
Larsson, K., 137, 141
Latimer, P., 306
Latorre, R., 536, 541
Lattman, E. E., 261, 263
Lauger, P., 522, 523, 524(64), 525(70), 528, 530(89), 531(70, 89), 533, 536, 537, 539(122, 124), 541, 542(122)
Lauterbur, P. C., 41
Lauterjung, K. H., 350
Laycock, L. C., 160
Lebech, B., 354
Leberman, R., 290, 291(67)
LeBlanc, O. H., 526
Lecar, H., 446, 447, 463, 464(47), 465(47), 466, 473, 537, 541
Leciejewicz, J., 354
Lee, A. G., 518
Lee, K. S., 484
Lee, W. I., 323, 330, 331, 336(66)
Lehmann, W., 141
Lehn, J. M., 22
Leigh, J. S., 104, 183
Lenard, J., 112
Lentz, B. R., 202
Leopold, H., 352
Leoty, C., 501
LePault, J., 379
Lepock, J. R., 225
Letokhov, V. S., 163, 190(2)
Lettvin, J. Y., 492
Levi, A., 226
Levin, B., 519

Levin, I. W., 132, 137, 138, 139, 140, 141, 142(50), 156(46), 161
Levine, L., 118
Levinson, S., 521
Levis, R., 463, 467
Levitt, D. G., 538
Lewis, A., 151, 157, 158
Lewis, G. N., 164
Lewis, J. T., 122
Li, T.-Y., 155
Liberman, E. A., 526
Libertini, L. J., 88, 102
Licht, A., 371, 387(97)
Liebmann, P. A., 198, 217
Lifshitz, E. M., 302
Likhtenstein, G. I., 54
Lillie, R. S., 446
Lim, T. K., 300
Lim, Y.-C., 515
Lin, S. C., 323
Lin, S. D., 425
Lin, W.-J., 102
Linden, C., 64
Lindley, B. D., 497
Lindsay, R. M., 135(d), 136
Lippert, J. L., 135(d), 136, 137, 138, 141, 147, 148, 149
Lipson, H., 229
Litman, B. J., 202, 516, 517(26), 518
Llano, I., 461, 486
Lockwood, D. J., 131
Loehr, J. S., 135(e), 136, 155
Loehr, T. M., 135(e), 136, 154, 155
London, R. E., 22
Long, D. A., 161
Long, M. M., 536, 537(121), 540(121), 542(121), 543(121)
Long, T. V., 154
Longuet-Higgins, H. C., 80
Lonsdale K., 340, 349
Lontie, R., 382
Loo, D. D. F., 482
Lord, R. C., 143, 144, 146, 147, 148(101), 150
Love, W. E., 257, 261, 263
Lovenberg, W., 154
Lovesey, S. W., 340
Lowe, I. J., 8
Lozier, R. H., 158
Luban, M., 361

Luckhurst, G. K., 107
Lunacek, J. H., 301
Lurix, P., 131
Lutton, E. S., 524
Lutz, M., 154
Lux, H. D., 483
Luz, Z., 43
Luzzati, V., 351, 359, 366, 369, 373, 374,
    375(43), 376(95), 378(43), 380(43, 95,
    105, 111), 387, 390
Lynn, K. R., 155

M

MacArthur, B. W., 516
McBride, D., 542
McCallney, R. C., 94, 97
McClung, F. J., 163
Macco, F., 141
McConnell, D. G., 220
McConnell, H. M., 24, 54, 63, 64, 79,
    80(25), 84, 85, 86(12), 88, 91(12), 93,
    94, 97, 98, 100, 101, 108, 109, 111,
    112, 113, 114, 115(64, 70, 72), 116,
    119, 120, 121, 122, 198, 225, 518
McCulloch, W. S., 492
McDonald, G. G., 28
McDonald, J. R., 160
McDonald-Ordzie, P. E., 145
McFarland, J. T., 155
MacGillavry, D. H., 349
McLachlan, A. D., 71
Mclarnan, J. G., 482
McLaughlin, S., 530, 535
McPherson, A., 246, 370
McQueen, J. E., 257, 264
Magde, D., 213, 214(52, 53), 224(53)
Maiman, T. H., 163
Mair, G. A., 260
Maisano, J., 135(*a*), 136, 142
Makowski, L., 268
Malnig, R., 387
Mandelkow, E., 290, 291, 293, 294, 295
Manev, M. E., 521, 523(53), 524(53)
Mansfield, P., 29
Mansy, S., 145
Marcus, M. A., 157, 158
Margalit, R., 531
Margoliash, E., 253

Marguerie, G., 379, 381(132)
Marmont, G., 446, 451(16), 459(16)
Marsh, D., 2, 103, 104
Marsh, J., 328, 330(61)
Marshall, W., 340
Martin, R. B., 519
Marty, A., 473, 484(68h)
Masetti, G., 155
Masotti, L., 536, 537(121), 540(121),
    542(121), 543(121)
Masson, F., 366
Mataga, N., 192
Mateu, L., 351, 369, 373, 374, 375(43),
    376(95), 378(43), 380(43, 95, 105), 387
Mathies, R., 132
Mathis, A., 366
Matricardi, V. R., 431
Matsumoto, N., 465
Matsuura, H., 147, 148, 150
Matthews, B. W., 246
Mattice, W. L., 373, 386
Mattick, A. T., 177
Matwiyoff, N. A., 22
Maxfield, F. R., 149, 150(111), 226
Maxwell, D. E., 39
May, R., 384, 385(151)
Mayer, A., 357, 358(55), 382, 387
Mehring, M., 29, 30(26), 38(26)
Meiboom, S., 43
Meiklejohn, G., 149
Melchior, D. L., 518
Memory, J. D., 2
Mendelsohn, R., 135(a), 136, 138, 139,
    141(45, 52), 142, 147, 150, 151
Merlin, J. C., 157
Metzger, A., 218
Metzger, H., 121
Meves, H., 456, 458, 460
Meyer, H., 514, 541(4)
Michel-Villaz, M., 379
Mikke, K., 354
Milanovich, F. P., 142
Mildner, D. F. R., 355
Mildvan, A. S., 44
Milledge, H. J., 349
Miller, N. G. A., 516, 518(25)
Miller, T. A., 107
Misell, D. L., 439
Mitschele, C. J., 191

Miyazawa, T., 129, 147, 150
Mocker, H. W., 164
Moffatt, J. K., 260
Monger, T. G., 186
Montal, M., 522, 523, 524(64)
Moodie, A. F., 400
Moon, R. B., 51
Moore, J. W., 447, 449, 452(23), 455(20), 456(20), 459(19, 20, 23), 461, 464(23), 480, 493, 500, 501, 504
Moore, L. E., 483, 497
Moore, P. B., 229, 373, 375, 377, 378(112), 379(112), 381(112, 118), 382, 384, 385, 411
Moore, W. H., 146, 147
Moras, D., 380
Morell, J. L., 537, 539(122), 542(122)
Moretz, R. C., 431
Morowitz, H. J., 518
Morrisett, J. D., 63
Mortenson, L. E., 154
Moshe, Y., 387
Moss, T. H., 154
Mueller, J. J., 351, 359, 361, 370, 387
Mueller, K. W., 380
Mueller, P., 519, 520, 521(49), 522(49), 523, 524, 526, 538
Muller, A., 514
Munch, J. P., 335
Murari, R., 20, 21, 22
Murayama, M., 323, 329(47)
Mustacich, R. V., 330

N

Nafie, L. A., 130, 131(25), 151
Nagai, K., 154
Nagakura, S., 176, 177, 187
Nagayama, K., 39, 40(44), 41
Nakadomari, H., 529, 532(90), 533(90)
Nakajima, S., 501
Nakamura, J. K., 131
Nakanishi, K., 132, 158
Nakashima, N., 192
Narahashi, T., 500
Neeland, J. K., 168
Neher, E., 447, 460(18a), 473, 483, 484, 486, 528, 533, 536(88), 539(111), 540, 542(111), 543(111)

Nestor, J. R., 136, 151, 160, 161(200)
Netzel, T. L., 176, 177(23a), 183
Neufeld, E. J., 226
Neumcke, B., 497, 523, 525(70), 531(70), 533
Newman, J., 317
Newton, S. A., 330
Nezlin, R. S., 371
Nibler, J. W., 160
Nicholson, G. L., 47
Nickel, B., 329, 330
Nicolaieff, A., 359, 362(62)
Niederberger, W., 48, 49
Nielsen, C., 252
Nielsen, N. C., 220
Nielsen, S. O., 377
Ninio, J., 387
Nishida, T., 202, 207(26)
Nishimura, S., 145
Nishimura, Y., 143, 145, 151
Nockolds, C. E., 248
Nogami, N., 147, 150
Noggle, J. H., 34
Nonner, W., 470, 492, 493, 495, 496, 497, 504, 506, 507, 508
Norberg, R. E., 8
Nordio, P. L., 82
Norrish, R. G. W., 164, 166
North, A. C. T., 260
Nossal, R., 300, 325, 328, 329, 331, 332, 333(60, 78a), 336
Novak, J. R., 164
Nuccitelli, R., 460
Nunes, A. C., 358
Nyquist, H., 496

O

Ogievetskaya, M. M., 371
Ohtani, H., 177, 187
Ohtsuki, M., 437, 444
Oikawa, T., 457
Oldfield, E., 32, 33, 34(32)
Olins, D. E., 145
Oliver, C. J., 313, 314
Olson, G. L., 179
Onishi, M., 526
Orci, L., 226
Orly, J., 222

Osawa, T., 202
Osborne, H. B., 379
Oseroff, A. R., 132, 157
Osgood, R. W., 332
Ostanevich, Y. M., 371
Osterberg, R., 387
Osterholt, D., 157, 158
Osterhout, D. J., 135(d), 136
Ostroff, E. D., 29
Otnes, K., 354
Ottensmeyer, F. P., 426
Ovchinnikov, Y. A., 526, 530(83)
Overbeek, J., 520
Overton, E., 514
Owicki, J. C., 119
Ozaki, Y., 153, 154(141, 142)
Özisik, M. N., 509

P

Packard, M., 1, 9(1)
Packer, A. J., 130, 150(22), 151(22)
Padlan, E. A., 243, 257
Pagano, R. E., 514, 516, 521(13), 523(13)
Painter, P. C., 143, 149
Palade, P., 499
Pallotta, D., 145
Palmer, R. A., 254, 255(31), 269(31)
Palti, Y., 471
Papahadjopoulos, D., 215
Parce, J. W., 114, 121, 122
Pardon, J. F., 379
Parfait, R., 373, 376, 377(112), 378(112), 379, 380, 381
Parish, G. R., 205, 207(48), 208(48), 218, 221(48)
Parker, G. W., 2
Parker, N. W., 438
Parma, R., 332
Parola, A., 202
Parrish, W., 346, 349
Parson, W. W., 183, 184(34)
Parsons, D. F., 431
Pastan, I., 217, 218, 221, 222, 223, 226
Patlak, J. B., 484
Patterson, A. L., 240, 241, 242, 243, 244, 249, 252, 254, 261, 262, 263
Peacocke, A. R., 314
Pearson, A., 492

Pecora, R., 300, 301, 322
Perez, J. P., 440
Perkins, J. P., 222
Perrin, F., 164, 201, 202, 221(23), 224(23)
Perrin, J., 517
Person, W. B., 150
Perutz, M. F., 154, 245
Peskoff, A., 478, 479
Pessen, H., 340, 351
Peters, J., 198, 209(10), 217, 221(10)
Peters, K., 186
Peters, R., 198, 209, 217, 221(10)
Petersen, H. O., 379
Petersen, R. L., 155
Peticolas, W. L., 129, 130, 131(12, 25), 137, 138, 139, 141, 142(51), 143, 144, 145, 146, 147, 150(12), 151
Petit, V. A., 198, 217
Petrash, G. G., 169
Petsko, G. A., 257
Pézolet, M., 143, 145, 148, 149, 151
Phelps, D. J., 136
Phillips, D. C., 260
Phillips, M. C., 519
Pickar, A., 529
Piddington, R. W., 330
Pidoplichico, V. I., 485
Pigeon-Gosselin, M., 148, 149
Pike, E. R., 300, 307, 313, 314
Pilz, I., 340, 360, 362, 367, 370, 371, 382, 387, 388
Pines, A., 37
Pinkerton, M., 532
Pirenne, M. H., 367
Pitts, W., 492
Placzek, G., 128, 151(5)
Pockels, A., 514
Podleski, T. R., 218
Pollard, T., 293
Polnaszek, C. F., 49, 50(60)
Poo, M., 198, 217
Porod, G., 365, 368, 369, 374
Porte, M., 158
Porter, G., 164, 166, 182, 183
Porter, K. R., 442, 443
Poste, G., 198, 209(13), 217
Potselvyev, V. M., 536
Pound, R. V., 1, 16
Poussart, D. J. M., 483
Powers, L., 38

Prescott, B., 135(*b*), 136, 145
Pressley, R. J., 192, 193
Pressman, B. C., 531
Procopio, J., 538
Puchwein, G., 387
Purcell, E. M., 1, 16, 26, 29
Purschel, H. V., 382, 387
Pusey, P. N., 317

Q

Quigley, G. J., 370

R

Racey, T. J., 329
Radda, G. K., 218
Radeka, V., 357
Ramirez, D. M., 478
Ramon, F., 504
Ramsey, N. F., 81
Rand, R. P., 141
Rattle, H. W. E., 2
Ravdin, P., 218
Rayleigh, L., 514, 521(3)
Razi-Naqvi, K., 205, 207(47), 218, 221(47)
Redfield, A. G., 46
Reich, J. G., 382
Reichert, K. E., 244
Reid, B. R., 45, 46
Reidler, J., 217, 219, 227
Remy, J. M., 157
Rentzepis, P. M., 174, 175(22), 176,
    177(23a), 179, 182, 183, 184, 185(31,
    35), 186, 190, 191
Requena, J., 521, 523(55, 56, 57), 524(57)
Reuben, J., 54
Rey, P., 114, 115(72)
Reyes, J., 536, 541(120)
Rich, A., 45, 143, 370
Richards, B. M., 379
Richards, F. M., 257
Richards, J. H., 42, 51
Richards, P. M., 107
Rimai, L., 130, 154, 155, 156(170), 158
Riste, T., 354
Riva, C. E., 333
Robbin, Y. A., 386
Robbins, P. W., 142, 202

Roberts, B. W., 346
Roberts, J. D., 42, 43
Robertson, R. E., 84
Robson, R. R., 146
Rockley, M. G., 182, 183, 184(34)
Rogers, L., 357
Rohrlich, F., 422
Rojas, E., 452, 456, 457(30), 459, 460, 461,
    470, 471
Rose, A., 427
Rose, H., 394
Rosenbaum, J., 293
Ross, D., 330
Ross, P. A., 347
Rossmann, M. G., 235, 250, 261, 263
Roth, J., 222
Roth, S., 204
Rothman, J. E., 112
Rothschild, K. J., 149
Roudant, E., 357
Rougier, O., 501
Rousseau, D. L., 133, 152, 154(136)
Rozantzev, E. G., 82, 85
Rubassay, D., 371
Rubenstein, J. L. R., 119
Ruchpaul, K., 387
Rudin, D. O., 519, 520, 521(49), 522(49),
    524, 526
Rudy, B., 202, 207(25), 218, 221(25)
Rushton, W. A. H., 445
Rust, H. P., 426
Ruterjans, H., 43
Rutishauser, U., 217
Ryman, B. E., 516, 518(27)
Rymo, L., 387
Ryzhkov, V. M., 85

S

Sachs, F., 473
Sachs, L., 202, 207(37, 38), 208(37, 38),
    218, 221(37, 38)
Sackmann, E., 109, 112
Saffman, P. G., 220
Saito, M. I., 153
Sakato, N., 319
Sakmann, B., 447, 460(18a), 473, 484
Sakura, J. D., 141
Salares, V. R., 136

Saleh, B. E., 307, 313(22)
Salmeen, I., 154
Saludgian, P., 366
Salzberg, B. M., 463(56), 467, 499, 500(116)
Sandblom, J., 533, 539(111), 542(111), 543(111)
Sardet, C., 351, 373, 375(43), 378(43), 379, 380(43, 111, 112)
Sarma, V. R., 260
Sattelle, D. B., 330
Sauer, K., 152, 153
Saunders, L., 517
Saxena, A. M., 355, 358
Saxman, A. C., 191
Saxon, W. O., 429
Scandella, C. J., 108, 109
Scanu, A. M., 351, 373, 375(43), 378(43), 380(43, 105)
Schaefer, D. W., 308, 317, 329
Schäfer, F. P., 169
Schelten, J., 351, 357, 358, 361, 387
Scheraga, H. A., 146, 148(100), 149, 150(100), 155
Scherm, R., 387
Scherzer, O., 444
Scheule, R. K., 155
Schiff, L. I., 399
Schiffer, M., 261
Schimmer, B., 350
Schindler, D. E., 384
Schindler, M., 225, 226
Schirmer, R. E., 34
Schlect, P., 357, 358(55), 387(55)
Schlessinger, J., 198, 207(14), 209(14, 15, 16, 17), 210(16), 211(15, 16), 212(15), 216, 217, 218, 219, 221, 222, 223, 226, 227
Schlimme, E., 367, 370(92)
Schmatz, W., 357, 358, 387
Schmid, M., 261
Schmidlin, E., 131
Schmidt, P. J., 360, 361
Schmidt-Ullrich, R., 142, 156(68)
Schmitz, K. S., 323
Schneider, B., 32
Schneider, G., 205, 207(48), 208(48), 218, 221(48)
Schneider, H., 136, 155
Schneider, R., 387

Schoenborn, B. P., 340, 355, 357, 358, 375(9), 377, 378(118), 381(118), 385, 390
Schramm, M., 222
Schultz, R. I., 135(d), 136
Schurr, J. M., 300, 323
Schuster, I., 387
Schuster, T. M., 152, 153
Schwager, P., 261
Schwaiger, S., 387
Schwartz, M. A., 113, 121
Schwartz, S. E., 131
Schwartz, T. L., 447, 467, 481
Schwarz, W., 542
Searle, G. F. W., 182, 183(29)
Seelig, A., 48, 49
Seelig, J., 48, 49, 54
Sela, M., 371, 387(97)
Selig, H., 127
Selser, J. C., 317, 322(34)
Seltzer, S. M., 422
Senft, P., 463, 464(47), 465(47), 466(47)
Serdyuk, I. N., 378, 380
Seredynski, J., 425
Setser, D. A., 164
Sevely, J., 440
Shah, D. O., 515, 516(19), 517(19)
Shamir, J., 127
Shank, C. V., 158, 171, 172, 178, 191
Shaper, J. H., 538, 542(125)
Shapiro, S. L., 190
Shaw, T. I., 457
Shchedrin, B. M., 386
Shearer, G., 514
Sheats, J. R., 113, 114
Shechter, Y., 218, 222, 223, 226
Sheetz, M. P., 225, 226, 533
Shelnutt, J. A., 133
Shepherd, I. W., 128, 134(8)
Shepherd, W. C., 536
Shichida, Y., 177, 187
Shimada, H., 153
Shimanouchi, T., 147, 148, 150
Shimshick, E. J., 63, 64, 94, 97
Shinitzky, M., 202, 206, 207(24, 26, 28, 37, 38), 208, 218, 221(37, 38)
Shipley, G. G., 141
Shore, B., 142
Shrager, P., 452
Shriver, D. F., 155

Shulman, R. G., 52
Siamzawa, M. N., 147, 150
Siegel, D. P., 325
Sieker, L. C., 265, 269, 270(53), 271, 273, 274(54)
Sigworth, F. J., 473, 484
Silberstein, A. Y., 536
Silman, O., 155
Silverman, J., 528
Simon, I., 387
Simon, S. R., 154
Simons, R., 521
Singer, S. J., 47, 528
Sipe, J., 519
Sistemich, K., 350
Sjöberg, B., 388
Skala, Z., 351
Skrabal, P., 518
Slayter, H. S., 411
Slichter, C. P., 17
Small, D. M., 141
Small, E. W., 146
Smallcombe, S. H., 42
Smith, B. A., 119, 122, 225, 325, 326
Smith, D. S., 218
Smith, I. C. P., 49, 50(60), 102, 104
Smith, L. M., 122
Smith, R. S., 497
Sneden, D., 370
Snitzer, E., 168
Snyder, R. G., 137, 140
Soler, J., 386
Sombret, B., 157
Sosfenov, N. I., 383
Sparks, C. J., 354
Spaulding, L. D., 152, 153
Spencer, R., 361
Sperling, L., 366
Spiker, Jr., R. C., 138, 139, 141(48, 49)
Spinsi, A., 536, 537(121), 540(121), 542(121), 543(121)
Spiro, T. G., 129, 130, 131, 132, 136, 146(18), 147, 148(18), 150, 151, 152, 154, 155(18), 160, 161(200)
Spoonhower, J., 158
Springer, T., 358
Stabinger, H., 360
Stämpfli, R., 470, 487, 488, 489, 490, 491, 492, 497, 500
Standish, M. M., 517, 518(29)

Stanley, H. E., 135(c), 136, 142, 145, 149
Stark, G., 532, 533
Stein, J., 332
Stein, P., 129, 131(14)
Steinbach, J. H., 473, 484(68f)
Steinrauf, K., 532
Stenbuck, P., 225
Stenn, K., 425
Stepanov, A. P., 85
Stern, M. D., 332
Stevens, C. F., 473, 528, 536(88)
Stiefel, E., 268
Stock, G., 328, 329(59), 330(59)
Stockton, G. W., 49, 50
Stöckel, P., 382, 384, 385(151), 387
Stoeckenius, W., 157, 158, 221
Stoeffler, G., 387
Stoicheff, B. P., 128
Stout, G. H., 229, 247(3)
Strandberg, B. E., 252
Strathder, J., 79, 80(25)
Strekas, T. C., 130, 131, 132, 150(22), 151, 152
Strell, I., 384, 385(151)
Strickholm, A., 482, 483
Stryer, L., 132, 199, 224
Stubbs, G. J., 288, 289, 291, 292, 293
Stubbs, G. W., 202
Stuhrmann, H. B., 345, 369, 372, 373, 374, 376, 377, 378(104, 112, 117), 379, 380, 381, 386
Suau, P., 379
Suddath, F. L., 370
Sufrà, S., 155
Sugawara, Y., 148
Sugeta, H., 147, 149, 150
Summer, J. B., 244
Sun, S. T., 322
Sunder, S., 138, 139, 141(52), 142(53), 151
Sundstrom, V., 186
Sussman, J. L., 267
Sutcliffe, L. H., 154
Suzuki, K., 177
Suzuki, R., 446
Swain, S., 307, 313(20)
Swift, T. J., 43
Swinney, H. L., 300, 312(1), 315(1)
Sygusch, J., 254, 255(32)
Sypropoulos, C. S., 457

Szabo, A., 154
Szabo, G., 520, 522, 524(68), 529, 531(50), 532(50), 533, 534(68), 535, 542
Szalontai, B., 156

T

Tajima, K., 516
Takahashi, K., 452, 484
Takashima, S., 524
Takematsu, T., 147, 150
Takeuchi, A., 473
Takeuchi, N., 473
Tanaka, T., 333, 334, 335(83), 336
Tanford, C., 301, 302(13), 305,(13), 306(13), 316, 363
Tang, J., 128, 129, 130(6)
Tang, S.-P., 154
Tao, T., 224
Tardieu, A., 351, 366, 369, 373, 374, 375, 376(95), 378, 380
Tartaglia, P., 322, 329(43)
Tasaki, I., 457, 459, 490, 492
Tasumi, M., 129
Tatchell, K., 379
Tauc, L., 482
Taylor, C. A., 282
Taylor, J., 522
Taylor, J. S., 104
Taylor, K. A., 431, 432, 444
Taylor, R. E., 445, 446, 447, 449, 452, 459, 460(37), 461, 463, 464, 465, 469, 475, 509
Teitelbaum, H., 377
Teorell, T., 457, 520
Teraoka, J., 153, 154(141)
Terner, J., 156, 157, 158(176, 179, 183, 184), 159(179)
Tews, K. H., 198, 209(10), 217, 221(10)
Thach, R. E., 426
Thach, S. S., 426
Thamann, T. J., 155
Thierry, J.-C., 380
Thomas, D. D., 93, 97, 98
Thomas, G., 433
Thomas, Jr., G. J., 135(b), 136, 143, 144, 145
Thomas, H. P., 351, 387

Thomas, J., 293, 294(72), 295(72)
Thompson, J. D., 370
Thompson, J. E., 225
Thompson, T. E., 202, 516, 517(26), 518, 522
Thompson, W. S., 142
Tien, H., 519, 520, 521, 522(49), 524(47, 49, 52)
Timasheff, S. N., 340, 342, 351, 363
Tinker, D. O., 322
Tobin, M. C., 128, 132(9), 134(9)
Toda, S., 141
Tolles, W. M., 160
Topaly, V. P., 526
Topp, M. R., 174, 175
Torrey, H. C., 1, 94
Toyoshima, S., 202
Tramontano, D., 226, 227
Trauble, H., 10, 112, 113
Tredwell, C. J., 182, 183(29)
Tsernoglou, D., 257
Tsuboi, M., 129, 143, 145, 151
Tu, A. T., 142, 149
Tullock, A. P., 49, 50(60)
Turpin, P. Y., 145
Twardowski, J., 148
Tyminski, D., 148
Tyrrell, D. A., 516, 518(27)

U

Ugurbil, K., 52
Ulbricht, W., 500
Unwin, P. N. T., 411, 412, 413, 428, 429, 432, 444
Urban, B. W., 531, 532(99), 533(99), 538, 539(130), 543(130)
Urry, D. W., 526, 536, 537, 540(121), 542(121), 543(121)
Utlaut, M., 438, 441
Uzgiris, E. E., 325

V

Vainshtein, B. K., 275, 280, 281, 283, 383
Vainshtein, V. A., 536
Valdré, U., 426

Vallee, B. L., 154, 155
Vand, V., 268
Van de Hulst, H. C., 302, 319, 320(15),
    321(15), 323(15)
Vanderkooi, G., 219
van der Saag, P. T., 206, 217
Van Holde, K. E., 379
Van Wart, H. E., 146, 148(100), 149,
    154(100), 155
Varnum, J., 253
Vastel, D., 379
Vaughan, P. A., 364, 374(83)
Vaughan, P. C., 482
Vaughan, R. W., 29
Vaz, W. L. C., 224
Veatch, W., 225
Veldkamp, W. B., 314
Verdugo, P., 330, 331, 336(66)
Vergara, J., 447, 452, 498, 499, 500(116)
Verma, S. P., 141, 142, 156
Vernon, W., 252
Vickery, L. E., 152, 153
Vogel, H., 387
von de Haar, F., 367, 370(92), 387
von der Linde, D., 167
von Schulthess, G. K., 318, 319
von Sengbush, P., 290, 291(67)
Vorobev, V. I., 366
Voss, R. F., 308

W

Wade, R. H., 404, 405(11)
Waggoner, A. S., 54, 529
Wagner, G. C., 151, 152, 153(129),
    154(129)
Wahl, P., 198, 218, 222(5)
Wald, G., 185
Waldbillig, R. C., 522, 524(68), 534(68)
Walker, T. E., 22
Wall, J. S., 436, 438
Wallach, D. F. H., 141, 142, 156, 225
Wallert, F., 157
Walter, G., 361
Waltman, B., 475
Wanke, E., 463, 473, 540, 541(131)
Ward, K. B., 261
Ware, B. R., 300, 315, 325, 330, 331

Ware, D., 29
Warren, S., 288, 292(63)
Warshel, A., 129, 150(15), 152(15), 153(15),
    154(15), 155, 157(15)
Wasserman, E., 118
Watenpaugh, K. D., 265, 269, 270, 271,
    273, 274
Watkins, J. C., 517, 518(29)
Watson, H. C., 386
Watson, J. D., 287
Watters, K. L., 155
Waugh, D. R., 520
Waugh, J. S., 29, 31, 37
Wawra, H., 351, 360
Webb, M. B., 340, 343(3), 362(3), 369(3)
Webb, W. W., 198, 207(14), 209(14, 15, 16,
    17), 210(16), 211(15, 16), 212(15), 213,
    214(53, 54), 215, 216, 217, 218, 221,
    223(70), 224(53), 226, 227
Weber, G., 201, 202, 204, 207(26)
Weber, R., 141
Weber, S., 202, 207(24)
Wecht, K. W., 190
Weidmann, S., 474
Weinberg, D. L., 359
Weiner, R. S., 325
Weinstein, E., 385
Weinzierl, J. E., 253, 254, 255(31), 269(31),
    370
Weiss, B., 351, 375(43), 378(43), 380(43)
Weiss, K., 426
Welberry, T. R., 282
Wescott, W. C., 519, 520, 521(49), 522(49),
    524(47, 49)
Westerfield, M., 461
Westhead, E. W., 316
Weston, J. A., 221, 222(68)
Wey, C.-L., 225
Weyl, R., 419
Whitaker, D. R., 42
White, S. H., 521, 522, 523(66)
Whytock, S., 412(24), 413
Wiedekamm, E., 141
Wiegandt, H., 217, 219
Wiggins, J. W., 438
Wildermuth, G., 141
Wilhelm, B., 323
Wilkie, D. R., 51, 52(64)
Williams, D. F., 134

Williams, R. J. P., 154, 519
Willingham, M. C., 218, 222, 223, 226
Willis, B. T. M., 248, 252
Wilson, D. M., 22
Wilson, Jr., E. B., 128
Windsor, M. W., 164, 182, 183, 184(34)
Winick, H., 247
Wippler, C., 373
Wishner, B. C., 261
Witters, R., 382
Witz, J., 347, 351, 359, 362(62), 371
Wobschall, D., 202, 207(27)
Woessner, D. E., 24
Wojcieszyn, J., 216
Wolf, D. E., 216, 225
Wolosewick, J. J., 442, 443(69)
Wonocott, A. J., 248, 250(23), 251
Wood, E., 314
Woodbury, J. W., 487
Woodruff, W. H., 154, 156
Woods, G., 323
Woodward, C. K., 46
Wooley, J. C., 379
Worcester, D. L., 358, 379
Worthington, C. R., 221
Worthmann, W., 383
Wright, K., 64
Wu, E.-S., 215
Wu, G., 198, 209(13), 217
Wun, K. L., 335
Wunderlich, R. W., 331
Wüthrich, K., 2, 39, 40(44), 41
Wyatt, P. J., 306
Wyckoff, H. W., 268

X

Xuong, N. H., 249, 252

Y

Yager, P., 138, 139, 141, 142(51)
Yahara, I., 217
Yamada, K. M., 217, 218, 221, 222(66, 68), 223
Yamada, S. S., 217, 223
Yamanaka, T., 153, 154(141)
Yangos, J. P., 314
Yannas, I. V., 135(c), 136, 145
Yarden, Y., 226
Yeager, M., 379
Yeh, Y., 142, 317, 322(34)
Yellin, N., 137, 139(42), 141
Yesaka, H., 177
Yoshii, M., 484
Yoshino, S., 331
Yoshizawa, T., 177, 186, 187
Young, C. G., 168
Young, N. M., 136, 155
Yu, N.-T., 128, 146, 147, 148(99), 149(99), 150, 152, 153, 154(11), 155(11)
Yu, T.-J., 143, 145, 147

Z

Zaccai, G., 358, 380
Zagyansky, Y., 198, 209, 217
Zanchi, G., 440
Zavoisky, E., 66
Zeitler, E., 396, 418
Zemilin, F., 426
Zerbi, G., 155
Zidovetski, R., 226
Zillig, W., 384, 385(151)
Zimm, B. H., 305, 306(17)
Zingsheim, H. P., 540
Zinke, M., 387
Zipper, P., 366, 387

SUBJECT INDEX

A

Anisotropic motion, 24, 25, 98–102
  time-dependent hyperfine interaction, 98
Antibodies, 114, 115
Antigens, spin labels as, 110, 111
ATPase, mitochondrial, 245
Autocorrelation, 310–313
Axial wire (in voltage clamp), 452–455

B

Black lipid membranes, see Lipid model
  membranes
Bloch equations, 6, 7, 72–75, 108
Blood flow, 331–333
Bragg scattering vector, 303, 326
Bragg's law, 230–232
Brownian motion, 308–310, 316

C

Cable theory, 452–455
Carriers, ion transporting, 530–535
CARS, 159–161
Channels, 447, 472, 473, 484, 497, 536–539
  noise in, 447, 497
  single-channel recording, 484
  single channels in bilayers, 536–539
  single channels in excitable membranes,
    447, 472, 473
Characteristic function parameters, 368,
  369
Chemical shift, 23, 24, 39
Collimation of x rays, 347–349
Collision broadening, EPR, 104–109
Conformations, 137–150
  lipid, 137–142
  nucleic acid, 143–145
  polysaccharide, 145
  protein, 146–150

Contrast

Contrast, in small-angle scattering, 372–
  375
Contrast variation, 378–381
Correlation time, 13–16
Correlators, 314

D

Debye–Waller factor, 240
Depolarization ratio, 127, 130
Detectors, optical, 194, 195
Dichroism, 204, 205, 208
Difference Fourier synthesis, 259, 260
Diffractometer, four-circle, 251, 252
Diffusion coefficient, lateral, 217, 316
Double resonance, 32–38

E

Echelons, 174–176
Einstein–Debye theory (light scattering),
  363
Electron density maps, 256–258
Electron microscopy, 391–444
  aberrations, 392, 395, 398, 405, 444
  absorption coefficient, 396
  back projection, 413, 414
  dark-field methods, 433–438
  high voltage, 441–444
  hollow cone problem, 416–418
  hydration, 430–433
  image restoration, 404, 405
  radiation damage, 418–430
  resolution, 392–395
  shot noise, 427–430
  strong phase object approximation, 399
  three-dimensional reconstruction, 409–
    418
  weak phase angle approximation, 397–
    399
Energy transfer, between antenna pigments, 182, 183

Envelope function, 403, 404
Enzymes, 41–44
EPR spectrometer, 57
Ewald sphere, 233, 234, 402
Extended scattering, 380, 381

F

Fiber diffraction, 275–288
Flash photolysis, 166, 167
Fluorescence, 133, 134, 197–227
  anisotropy, 200
  correlation spectroscopy (FCS), 208,
    209, 212–216
  photobleaching recovery (FPR), 208–216
  polarization, 200–208
Fourier transforms, 409–413

G

Gap isolation (voltage clamp), 486–504
Gating currents, 447, 471, 472
Gels, 333–336
g factor, electron, 83, 84

H

Hamiltonian, spin electron, 78, 84–90
Haptens, spin labeled, 110–122
Harmonic generation, picosecond spectros-
  copy, 191, 192
Hemoglobin, 257, 387
Hemoproteins, 150–154
Holography, 438, 439
Hydrogen exchange, 376–378
Hydrophobic ions, transport of, 529, 530
Hyperfine splitting, nuclear, 59

I

Immunoassay, laser light scattering, 319
Infrared vibration spectrum, 127, 128
Intensity fluctuation spectroscopy, 306–
  310
Ion transport mechanisms, 525–541
Isomorphous replacement, 252–255
Isotopic labeling, 42, 43
Isotropic motion, effect on EPR spectra,
  93–98

K

Kerr effect, optical, 194

L

Larmor equation, 5, 6
Lasers, 164, 170, 171, 174–176, 315
  devices, 315
  mode locked, 164, 170, 171
  Q switched, 164
  single shot, 174–176
Laser scattering, 299–336
Lateral diffusion of lipid spin labels, 108,
  109
  spin label haptens, 112, 113
Length distributions in scattering, 367
Lipid model membranes, 513–543
  Mueller–Rudin type, 520–525
  planar bilayers, 519–525
  solvent containing, 521, 522
  solvent depleted, 523
  solvent free, 524, 525
  spherical bilayers, 516–519
Lipid monolayers, 515, 516
Lipid/solvent systems, 520, 521
Lipoproteins, 387
Lysozyme, 386

M

Macromolecular aggregates, scattering
  from, 381–385
Magnetic dipole–dipole interactions, 17–19
Magnetogyric ratio, 3, 67
Membranes, 47–49, 112–122, 137–142,
    197–228, 445–543
  current measurement, 459–461
  fluidity, 142, 156
  model, 513–543
  NMR in, 47–49
  Raman scattering in, 137–142
  spin labeling in, 112–122
  viscosity, 202, 204, 216–222
Metabolism, 49–52
Microelectrode (voltage clamp), 473–482
Microtubules, (structure), 293–296
Mobility, electrophoretic, 324–326
Model building programs, 264

Molecular motion, 197–200, 208–216
  macroscopic, 198, 199, 208–216
  microscopic, 197–199
  rotational, 199, 200
Molecular replacement, 260–263
Monochromatization
  neutrons, 353–355
  x rays, 346, 347
Motility, 326–330
Motional averaging, EPR, 94
Multiphoton absorption, 190
Multishot emission spectroscopy, 177–179
Muscle fiber (voltage clamp), 497–499
Myelinated nerve (voltage clamp), 492, 497
Myoglobin, 386

N

Neutrons, thermal, 338
Node of Ranvier, 487, 492–497
Noise, channel, 472, 473, 497
Nonlinear phenomena in picosecond spec-
  troscopy, 190–194
Nuclear hyperfine splitting, *see* Hyperfine
  splitting, nuclear
Nuclear hyperfine structure in EPR, 77–79
Nuclear magnetic moment, 2, 5
Nuclear magnetic resonance, 1–52, 198
  continuous wave, 7–9
  free-induction decay, 7–9
  Fourier transformed, 8, 10–12
  magic-angle spinning, 30–32
  multiple-pulse methods, 28–30
Nuclear Overhauser effect, 33, 34
Nuclear relaxation, enhancement by spin
  labels, 109, 110
Nucleic acids, 44, 45

O

Order parameter, 48, 49
Orientation effect on EPR spectra (single
  crystal), 85–90
Overhauser effect, *see* Nuclear Overhauser
  effect

P

Paramagnetic probes, 43, 44
Paramagnetic relaxation, 68–72

Patch voltage clamp, 482–484
Patterson function, 240–244
Patterson inversion, 367
Perfusion, internal, 456, 457
Phase contrast, 401–403
Phase determination, in XRDA, 252–255
Phosphorescence, 224, 225
Phosphorous NMR, 50–52
Photosynthesis, 181–185
  antenna pigments, 181
  reaction centers, 181
Polarizability tensor, 126, 130, 131
Polarization parameter, 200
Polyenes, 156
Porod characteristic function, 368
Porphyrins, 150–154
Position-sensitive x-ray detectors, 350
Potentiometric recording, 490–492
Powder spectra, EPR, 90–93
Precession camera, 248–251
Protein crystallography, 244–246
Protein Data Bank, 268
Proteins, fibrillous, 333–336

Q

Quadrupolar interactions, 22, 23

R

Radius of gyration, 364, 365, 378–380
Raman spectroscopy, 123–161
  anti-Stokes spectrum, 124, 125
  resonance, 125, 128–130
  Stokes spectrum, 124, 125
Rayleigh scattering, 126
Reciprocal lattice, 237–238
Refinement of structures from XRDA,
  263–268
Restricted motions, 25
R factor in XRDA, 264
Rubredoxin, 268–275

S

Saturation transfer, 33–36
Scanning transmission electron microscope
  (STEM), 436–438, 440, 441

Scattering length densities, neutron and x-ray, 375
Scattering, low angle, 337–390
Scattering signal, 341–343
Series resistance (voltage clamp), 463–465
Slit smearing, 345
Small-angle compromise, 344, 345
Solution scattering, 337–390
Space clamp, 449–452, 453–456
Space groups, 239
Spallation sources (for neutrons), 355
Spectrum analyzers, 314, 315
Spin angular momentum, 2
Spin decoupling, 33
Spin exchange, 102, 104–109
Spin-label spectroscopy, 53, 198
Spin–lattice relaxation, 13, 21
Spin–spin coupling, 14, 20–22, 39, 102–104
Stability (voltage clamp), 467–470
Stimulated Raman scattering, 193, 194
Streak camera, 179, 180
Structure factor, 238–241

Sucrose gap (voltage clamp), 500–504, 508–511
Synchrotron radiation, 339

T

T_1, 7, 9, 12, 13, 16, 19, 25, 26
T_2, 7, 9, 13, 14
TEMPO, 55
Tobacco mosaic virus (structure), 288–293
tRNA, 387
Two-dimensional detection, absorption spectra, 180, 181

V

Vibrational frequency, 124, 125
Vision, 185–189
Voltage clamp, 445–512

X

X-ray diffraction analysis, 229–236